# Support Systems For Buildings

# Support Systems For Buildings

Jack R. Lewis, Ph.D.
Architect

Professor Emeritus
California Polytechnic State University
San Luis Obispo, California

Prentice-Hall, Englewood Cliffs, New Jersey 07632

*Library of Congress Cataloging in Publication Data*

LEWIS, JACK R., (date)
Support systems for buildings.

Bibliography: p. 426
Includes index.
1. Buildings—Mechanical equipment. 2. Buildings—
Electric equipment. I. Title.
TH6010.L48 1985 696 85-9353
ISBN 0-13-877184-7

Editorial/production supervision
and interior design: *Theresa A. Soler*
Cover design: *Photo Plus Art*
Manufacturing buyer: *Rhett Conklin*

*Cover photograph courtesy of Marc P. Anderson.*





Printed in the United States of America
10 9 8 7 6 5 4 3 2 1

ISBN 0-13-877184-7 025

Prentice-Hall International (UK) Limited, *London*
Prentice-Hall of Australia Pty. Limited, *Sydney*
Prentice-Hall Canada, Inc., *Toronto*
Prentice-Hall Hispanoamericana, S.A., *Mexico*
Prentice-Hall of India Private Limited, *New Delhi*
Prentice-Hall of Japan, Inc., *Tokyo*
Prentice-Hall of Southeast Asia Pte. Ltd., *Singapore*
Editora Prentice-Hall do Brasil, Ltda., *Rio de Janeiro*
Whitehall Books Limited, *Wellington, New Zealand*

# Contents

# Preface

The construction of buildings by human beings has ranged in complexity from an early barrier of stones across the mouth of a cave to our present high-rise structures. With or without formal planning, support systems to make these buildings habitable have been included in one way or another. Cave dwellers considered the proximity of water supply, warmth from the sun or protection from extreme heat, light from cave entrances or through accidental shafts, and ventilation provided by multiple entrances or fissures in the surrounding rock.

Later, as groups of people drifted together to form tribes, pueblos, villages, and towns, the need for support systems for potable water, sewage disposal, heat, and light became more acute. In addition to individual family needs, the needs of the community had to be considered so that the waste of one family did not contaminate the supply of a neighbor and the distribution of water or the use of natural lighting was not interfered with unduly. With the coming of more fully enclosed structures came the candle, then the primitive lamp, then the oil-burning light, and finally the electric light. All these processes and advancements obviously took hundreds of years, and during that time experimenters, known and unknown, steadily tried out new ideas or developed theories that provide the basis for our present expertise in designing, installing, and using the various systems that supplement the actual building.

The author of this book does not pretend or envision that it could possibly cover every aspect of every support system that might be required or desirable in a building. Buildings differ in their construction, their use, and their support-system requirements. What is necessary in a high-rise office or apartment building is rarely required in a warehouse or manufacturing plant. Rather, this book is intended to outline, in a somewhat abbreviated form, the *principles* of electrical equipment, heating, ventilation, and plumbing so that with a general knowledge of the design and installation of support systems, those who are not specialists in these fields may be better able to understand the requirements for space, function, and control.

**ACKNOWLEDGMENTS**

The authors and the publishers wish to thank all of the following organizations, which have contributed material or allowed the publication of copyrighted illustrations for the production of this book. Without their help and permission, such a task would have been impossible to accomplish.

American Society of Heating, Refrigeration and Air-Conditioning Engineers, Inc. (ASHRAE), 1791 Tullie Circle NE, Atlanta, GA 30329

"Automatic" Sprinkler Corporation of America, P.O. Box 180, Broadview Heights, OH 44147

Bally Case and Cooler, Inc., P.O. Box 98, Bally, PA 19503

California Energy Commission, Sacramento, CA 95814

California Polytecnic State University Library, San Luis Obispo, CA 93407

Carrier Corporation, P.O. Box 4808, Syracuse, NY 13221

Construction Specifications Institute, *The Construction Specifier*, 601 Madison Street, Alexandria, VA 22314

ELJER Plumbingware, Wallace Murray Corporation, 3 Gateway Center, Pittsburgh, PA 15222

GTE Lighting Products, Sylvania Lighting Center, Danvers, MA 01923

International Association of Plumbing and Mechanical Officials, "Uniform Plumbing Code," 5032 Alhambra Avenue, Los Angeles, CA 90032

Kelvinator Appliance Corporation, 930 Ft. Duquesne Boulevard, P.O. Box 716, Pittsburgh, PA 15230

Lennox Industries, Inc., P.O. Box 809000, Dallas, TX 75380

National Fire Protection Association, Quincy, MA 02269, *National Electrical Code*®, copyright 1983. The title *National Electrical Code* and NEC are registered trademarks of the NFPA.

New Zealand Government Tourist Office, 10960 Wilshire Boulevard, Los Angeles, CA 90024

Pacific Gas and Electric Company, 894 Monterey Street, San Luis Obispo, CA 93401

Prentice-Hall Inc., *Architectural Draftsman's Reference Handbook*, by Jack R. Lewis, Englewood Cliffs, NJ 07632

U.S. Gypsum Company, 101 Wacker Drive, Chicago, IL 60606

United States National Park Service, Cortez, CO

Westinghouse Electric Company, Advanced Systems Technology, 777 Penn Center Boulevard, Pittsburgh, PA 15235

In addition, I wish to thank my son, David P. Lewis, for review and comment on the manuscript and my daughter, Holly K. Ramsey, who has again typed one of my books.

*Jack R. Lewis*

# 1

# Energy and Environment

## 1.1 INTRODUCTION

Mankind has been using energy of one kind or another for thousands of years, but not until fairly recently has major attention been given to the production, transportation, and consumption of energy. The use of energy to provide a more comfortable living has also created some harmful environmental conditions that are only now being considered. The high depletion of some forms of energy, harmful by-products resulting from energy use, and increases of population all contribute to the probability that the world must make some changes for the future.

The present rapid depletion of energy sources and increase in pollution resulting from energy use have become major fields for engineering research. Some of the more traditional uses of energy provide the most pollution, and even though there are still billions of pounds, gallons, or British thermal units (Btu) of energy left, some conservation for the future must be considered now. New technology must be developed to improve the efficiency of all types of power sources, and certainly air and water contamination by fumes, smoke, and waste of all types must be drastically curtailed if not eliminated entirely. Energy needs and the environment itself are so vast that it is almost impossible to think of any comfort-producing process that is not now affected, or that cannot be improved by changing future energy requirements.

## 1.2 ENERGY USAGE

Waterpower, supplied by streams and rivers, was used to turn waterwheels as early as the first century B.C., and as the need grew, waterwheels were used to turn grinding wheels and thus became a major source of power. Today a great deal of electricity is produced with the help of hydroelectric turbines driven by the same type of waterpower. One of the major disadvantages of the use of waterpower is the requirement for adequate flow, with the result that any generating station location is fixed and stationary, depending on that flow.

With the invention of the steam engine, the use of fossil fuels became a reality and coal and oil replaced the waterwheel to some extent. As more power machines were developed, gasoline and diesel engines became a major source of energy conversion for moving vehicles, and added to the pollution already created by coal-burning steam boilers. As the demand for products of all kinds has increased, so have the demands on energy resources. "Industrialize" became the rallying cry, until today every country in the world wants to produce more of something; and increased industrialization requires more and more energy (see Fig. 1-1).

Even as the demands for more energy increase, the types of energy in common use change also. In addition to waterpower, a major early source of energy was wood. Wood was usually available in large quantities in the forests surrounding cities and was, and still is, a renewable energy source, given enough time. Animals and human beings were also a source of power for pulling plows, carts, and carrying loads and also reproduced themselves. With the advent of the Industrial Revolution, other heat sources (energy) came into prominence and have continued evolving to this day with the introduction of nuclear energy. A breakdown of current energy supply and demand in the United States, shown in Fig. 1-2, points up some of the problems.

A major conflict exists in today's cultures between the use and the conservation of energy. On one hand, greater comfort in everyday life requires the channeling of more and more energy into lighting, air-conditioning, and

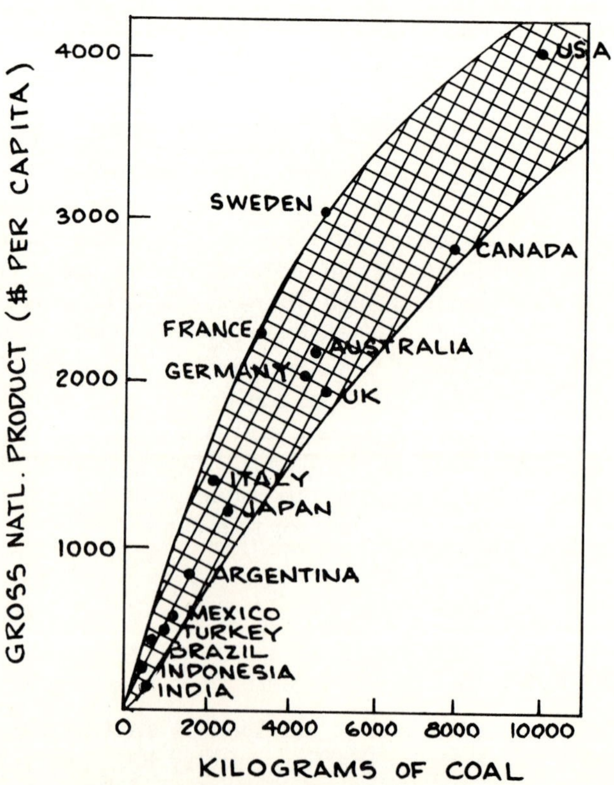


**Figure 1-1** World energy requirements.

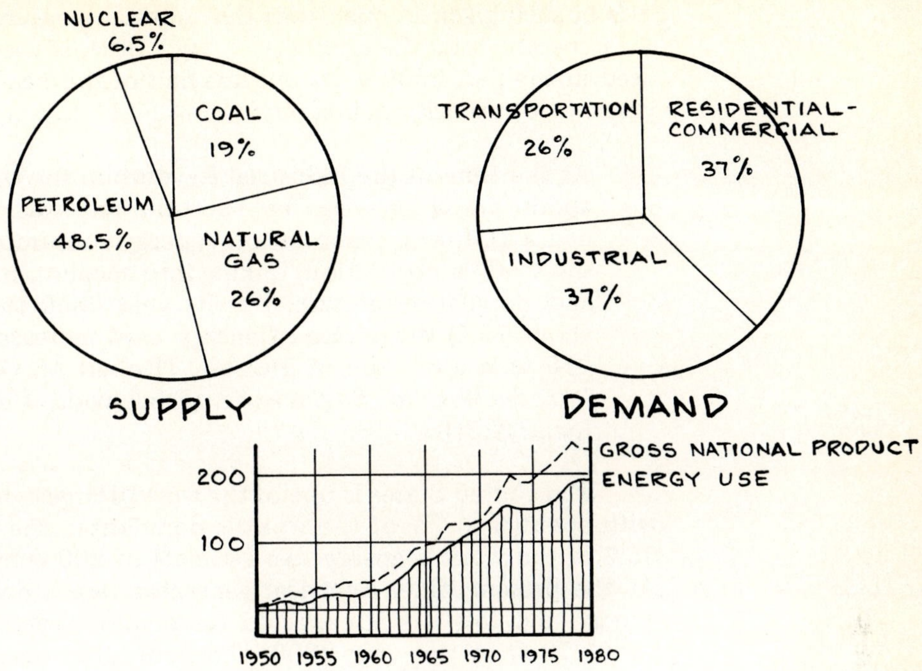

Figure 1-2

related systems. At the same time, more energy is being used to construct and demolish buildings, provide transportation to and from those buildings, and manufacture the materials used for construction. The use of energy creates contamination. Smoke is generated by the combustion of various fuels. Fly ash is created by the burning of coal. Changes during combustion also release into the air various chemical compounds that are detrimental to health. Waste, in its numerous forms, results from energy use and may appear as sewage, solid "dry" waste, or other contamination.

Research has shown that the energy used in the operation of a building is usually more than the energy used to construct the building. This has brought about several attempts to legislate requirements for a decrease in heat loss and a resulting decrease in the amount of fuel required. In fact, this has already happened in at least 8 states, with about 15 other states developing regulations. In addition to these state regulations, the American Society of Heating, Refrigerating and Air-Conditioning Engineers (ASHRAE) has made the most concerted effort toward a national code on energy standards in its Standard 90-75, developed in 1975. This recommendation is not a binding code but was designed to serve as a model for states and local authorities.

Most reports of energy use rely on the allowable input of energy per building area per unit per year, or Btu per square foot of floor area per year. It is estimated that active energy conservation could save up to 30% in the operation of existing buildings and perhaps 60% in properly designed new construction. Over a 10-year period, savings to consumers at present fuel prices could amount to $25 billion or more.

## 1.3 ENERGY CONSUMPTION

One of the most commonly used measurements for energy is the *British thermal unit* (Btu). One Btu is the amount of energy required to raise one pound of water one degree Fahrenheit. When overall energy use is considered, it is in such gigantic quantities that normal measurements are not adequate, so a special unit, the quad (Q; = $10^{15}$ Btu), is used. About 15 Q has been used in the past 2000 years but one-half of this has been used in the last 100 years. One authority points out:

> At the time of the Industrial Revolution the world was consuming only about $1/100$ Q per year: by 1960 the rate was $1/10$ Q per year. Thus, we had a 10-fold increase in total energy consumption while only doubling the world's population. Taking into account increasing industrialization and population growth, by the year 2050 the world will have spent about 75 Q if the rate of energy used increases at a 3% annual rate or 275 Q at a 5% rate of increase. [Robert M. Glorioso and Francis Hill, *Introduction to Engineering* (Englewood Cliffs, N.J.: Prentice-Hall, Inc., 1975)]

The United States is one of the world's largest users of energy. Currently, with about 6 to 7% of the world's population, the United States uses about 35% of the world's energy. An estimate of 200 Q by year 2050 is predicted. At the present time, electrical generators are a comparatively low user of energy, but this use is expected to increase from about 30 Q in 1982 to about 70 Q in the year 2000. Transportation use, mostly petroleum, will also increase to about 35 Q, but industrial and household use is expected to rise less rapidly. In 1974 nuclear reactors accounted for only 1% of available energy, but breeder reactors are expected to provide nearly 100% in the near future. Although it is now being used more extensively in some applications, solar power has not yet proven to be a major energy source for industrial use.

## 1.4 TREND OF ENERGY SOURCES

There are two major sources of energy: renewable (income) and nonrenewable (capital). Renewable sources are theoretically available over long periods of time at a definite rate of availability. Wood and solar energy are two examples of renewable energy sources, for entirely different reasons. Wood has been a fuel for centuries, is reproduced slowly but steadily if not overused, and is readily portable. On the other hand, solar energy has been available even longer, has a relatively steady rate of production, and is a possible energy source over most parts of the world. Solar energy does have a few disadvantages. For example, in very northern or southern areas the exposure is minimal. Further, the development of equipment that converts solar energy into usable form is in its infancy, and not all equipment now on the market is either energy-efficient or aesthetically acceptable.

With the steady increase in the world's population, there is a corresponding increase in the amount of land used for urban development and a decreasing amount of area left for wood production. In addition, the increased demand for building lumber decreases the amount of forest left for fuel production. Hardwood, which has the greatest Btu value per cord, is slow-growing and is used almost exclusively for veneer or finish lumber. Almost all hardwood fuel is from small woodlots or pastures and is so high priced that its use is not practical. While the availability and use of wood is decreasing, the use of solar energy is increasing, especially in the southern and southwestern United States. It is true that solar energy has been successfully harnessed in the far-northern states of Maine, Michigan, and Washington but more on an experimental and adjunctive basis than for total energy needs.

All types of buildings have solar collectors, of many varieties and shapes, and are in everyday use in some more southern locations.

Of the nonrenewable (capital) energy sources, petroleum is currently the leader, with coal close behind (see Table 1-1). Much of the petroleum used in the United States is imported, even though the United States has a considerable amount of available petroleum being produced and a large reserve yet undiscovered but assumed recoverable. According to a 1972 report, approximately 9000 Q of coal in the United States was known recoverable and another 38,000 Q was assumed recoverable. Crude oil and natural gas were each considered to have about 300 Q known recoverable but only about 5000 Q total assumed recoverable. Geothermal energy, including geysers and hot-water underground sources, currently has a known reserve of about 2000 Q and almost boundless submarginal resources. Some of the latter energy is being used to generate electricity in California, New Zealand, and several other locations.

**TABLE 1-1**
Energy Resources

| | *Reserves* | | | | | | *Production %* | | |
|---|---|---|---|---|---|---|---|---|---|
| | *Natural gas* | | *Coal* | | *Oil* | | | | |
| *Area* | *Cu. ft. × $10^{12}$* | *%* | *Million tons* | *%* | *Barrels × $10^9$* | *%* | *Gas* | *Coal* | *Oil* |
| North America | 279 | 10.8 | 257 | 25.4 | 32 | 4.9 | 44.3 | 18.8 | 15.6 |
| Soviet Union | 900 | 34.9 | 480 | 48.0 | 67 | 10.2 | 23.0 | 19.3 | 18.2 |
| Middle East | 740 | 28.7 | | | 365 | 55.8 | 2.7 | | 33.0 |
| Latin America | 144 | 5.6 | 3 | 0.3 | 58 | 8.8 | 3.2 | 0.4 | 8.3 |
| China | | | 143 | 14.2 | 20 | 3.1 | | 17.9 | 3.3 |
| Rest of world | 515 | 19.9 | | 12.1 | 112 | 17.2 | 26.8 | 43.6 | 21.6 |

The U.S. Energy Research Administration (ERA) has reported that with no new programs for energy conservation, energy imports will increase about fivefold from 1975 to 2000 (assuming that imports will continue to be available), to the equivalent of 30 million barrels per day. Another report indicates that with energy conservation the United States could have a surplus by the year 2000. Whatever situation finally develops, the Institute of Gas Technology developed a chart in 1975 showing that coal had a known and potential reserve of nearly 2000 billion tons; oil had a similar reserve of 400 billion barrels; and gas had a reserve of about 1600 trillion cubic feet. If projections of fuel use at a 3.1 to 4.3% increase per year are correct, the world supply of oil should last about 30 to 35 years, gas supplies about 38 years, and coal supplies around 110 years.

## 1.5 CONSERVATION

Do we have an energy crisis or at least an energy shortage? Everything would point to it. Until the Arab oil embargo in 1973, we used all types of available energy as though it would last forever, but scientists, manufacturers, and eventually most of the public "got the word." Our energy supplies, worldwide, will not last forever and conservation, by a number of different methods, is now the watchword. Conservation of energy offers the only hope.

The two principal sources of energy waste are (1) a major amount from commercial and industrial manufacturing processes, and (2) a lesser, but not negligible, amount from residential as well as other areas. In the first instance, many older buildings were poorly designed to conserve energy. Show windows in retail stores had to be large to display merchandise properly and overilluminated to attract the attention of the public. General interior lighting, adequate by itself, was, and still often is, augmented with spotlights, signs, local displays, and case lighting. Groceries and other markets display frozen food in long, open refrigerated cases with no thought to conserve the refrigeration energy expended. Manufacturing uses an excessive amount of fuel to supply its power plants, and many power plants operate with old and inefficient equipment simply because it has been the custom or because local fuel is "cheap." Here, too, the use of electricity for superfluous lighting and heat loss generated by processing machines are simply allowed to continue because few companies have had an incentive to apply conservation. In high-rise office buildings, entire floors are often brightly lighted at night when one custodian is cleaning one office, and factors of safety are exceeded by the use of cumulative small amounts of heating, power, or air-conditioning simply to assure adequate capacity.

Housing for people is a major contender for the questionable prize for the inefficient use of energy. Individual single-family homes that are "tract built" on a speculation basis are located to provide the most units on the least amount of land, with little or no thought or orientation for advantageous use of the sun or air movement. Multiple units (apartments and condominiums) are often bunched together with only one thought in mind—how much construction money can be saved by using common walls and identical plans. Most housing is wood-framed and rather poorly constructed, with ill-fitting doors and windows, minimum insulation or none at all, identical location of heating outlets regardless of building orientation, and substitution of cheaper, less efficient materials rather than those which are "first grade." Even in custom-built homes, most of these same problems may be found.

Owners and developers still follow a traditional path in the economics of construction. On a per capita basis, the United States wastes more construction energy than does any other nation. For the majority of real estate developers, the various tax and interest laws, both federal and local, mitigate against their self-interest in constructing anything except to minimum standards and with sales as immediate as possible. Concern with the ultimate owner's long-term operating expenses are not deemed part of a developer's problems. An additional original cost for more insulation than is absolutely required or a more efficient heating, ventilation, and air-conditioning (HVAC) system that will pay for itself in perhaps three years may offer no financial inducement whatever. Federal law also allows tax deductions for additional fuel costs in the year that they occur but only in very small increments over a long-term capital depreciation of the building. Federal law also allows accelerated depreciation during the early years of a building's life to the extent that the developer may shield his or her projected income from any tax whatever.

Nor are architects and engineers exempt from blame for poorly designed buildings. The multistory glass-faced office building which expresses profound disregard for energy conservation is to be found in most major cities. Poorly designed glass walls allow additional heat buildup inside the building, in addition to the heat load already provided by excessive lighting. Is the simple solution to add more air-conditioning to handle the heat, which in turn requires more space and more operating energy? Sprawling buildings of one or two stories may be the answer to more economical construction because they can be built of wood and require less construction equipment as they are closer to ground level, whereas multistory designs will usually have a smaller exposed roof area for sun-heat absorption and will shorten runs of pipe, ducts, and electrical conduit. Color, also long neglected in regard to energy (heat/cold), needs to be studied for both interior and exterior use. Light colors reflect unwanted heat and can increase illumination by as much as 30% in some spaces, thereby reducing the energy required for lighting. In the past, a building was "designed" by an architect and the nearly completed drawings given to electrical and mechanical engineers for them to fit their work to the proposed structure. That day is probably gone forever. Today the "design" of a building must be a team effort with as much consideration being given to electricity, elevators, HVAC, and other energy-related aspects as to the function and aesthetic qualities of a proposal.

## 1.6 ENERGY RECOVERY

Heating, ventilation, and air-conditioning consumes more than 20% of the energy used in the United States and accounts for about 60% of energy used in buildings. Electrical use consumes another 10% of building energy in general and as much as 50% in some office buildings. The remaining 30% is used up by poor insulation, building cracks, inefficient operation, improper orientation, some on elevators and escalators, and a bit on plumbing pumps. While some energy surveys have indicated that a savings of up to 30% may be possible, a great deal of energy savings comes from minute adjustments. Proper insulation and proper equipment are two of the initial factors that should be considered in the building design. Recovery of otherwise wasted heat energy may be a major consideration in the design of HVAC systems and their operation.

Over-illumination is one of the greatest wastes of energy and produces a considerable amount of heat, which in turn requires more energy for cooling. Two major methods may be used to reduce this waste: (1) by circulating cooling water through lighting fixture reflectors, or (2) by exhausting room air through lighting fixtures and into a plenum. Water-cooling systems are more effective than air-cooling systems because the latter simply adds more load to the air-conditioning system. One system of water cooling uses unchilled water in the lighting fixtures as well as in a series of tubes, which form a louver system at the windows. During the summer the heat is absorbed and removed by using an evaporative cooler. In winter the absorbed heat supplements the building heating system. Summertime operation captures about 90% of solar heat and 70% of lighting heat. In winter the system provides about 75% of heat required for the building. According to design engineers, this waste-heat recovery system can cut fuel bills by 60% and original costs by about 10%.

Exhaust-heat recovery reclaims interior heat from people, lights, solar gain, and other sources. Several types of heat exchanges are in common use: rotary wheel exchanges, air-to-air exchanges, water-cooled coils, and heat-pipe banks. The rotary wheel frame, packed with aluminum or stainless steel shavings, absorbs heat from the exhaust air duct and rotates it into the supply air duct, where the heat is transferred to the cooler air. Air-to-air exchanges transfer heat by conduction through the walls of cellular passages between exhaust and intake ducts. Water-cooled systems employ finned tubing in both exhaust and intake ducts, with water in the tubing transfering the heat as the water is pumped. The heat-pipe bank consists of several hundred small tubes, sealed at each end and protruding into both adjacent exhaust and supply ducts. A wick inside each tube allows a charge of refrigerant to absorb heat in the exhaust duct and release it into the supply duct. These systems have an efficiency of about 60 to 70%. Another waste-heat method makes use of a "double-bundle" condenser in which one coil acts in the normal manner in the air-conditioning system while a second coil either wastes part of the heat when not needed or retains heat and reuses it when heat is needed.

Many other potential sources for reclaiming energy loss are possible and will be developed in the future. These include rejected heat from refrigeration units, water heaters, kitchen equipment, boilers, incinerators, motors—in fact, nearly every energy-using device may be a prospect.

## 1.7 THERMAL INSULATION

Insulation, in its various forms as used in construction, is probably the most effective and most generally applicable means of conserving energy. Although construction of non-insulated buildings still continues, there are signs that this variety of energy waste is diminishing. Most common types of insulation are made for walls, ceilings, and floors and are probably in that order of importance to the builder or owner. However, other forms of insulation may also be used to reduce energy use in buildings. Pipes, ducts, and various equipment, when properly insulated, may reduce heat loss. Insulation in the form of caulking and weather stripping to close cracks and crevices is another major type.

Walls, ceilings, and floors treated with insulation retard the passage of heat flow through the building envelope and promote thermal comfort. Usually, this insulation is of batt or blanket mineral wool enclosed in a paper jacket and fitted snuggly between framing members. Masonry walls may be insulated with pellets or foamed plastic in the cavity between walls of brick construction or in open cells of concrete masonry units. Rigid board installed at foundation edges, on furring strips for concrete walls, and beneath built-up roofs is also extensively used. Such installation may provide a savings of up to 20% or more, depending on the type and thickness of insulation. Minimum $R$-values of recommended insulation are shown in Table 1-2.

**TABLE 1-2**
Thermal Properties of Insulating Materials

| *Material* | *Density (lb/ft³)* | *Resistance, R* |
|---|---|---|
| ¼-in. asbestos-cement board | 120 | 0.06 |
| ½-in. gypsum board | 50 | 0.45 |
| ½-in. Douglas fir plywood | 34 | 0.62 |
| ½-in. cane acoustic tile | | 1.25 |
| 1-in. loose sawdust fill | 8.0–15.0 | 2.22 |
| ½-in. mineral acoustic tile | 23 | 2.38 |
| 1-in. expanded perlite | 1.0 | 2.78 |
| 1-in. cellular glass | 8.5 | 2.86 |
| 1-in. mineral fiberboard | 15 | 3.45 |
| 3-in. mineral fiber batts | 0.3–2.0 | 11.0 |

In addition to supplying a more dense barrier against heat conductance in the building frame, insulation also provides a reduction in drafts due to small openings, eliminates the probability of condensation of air moisture on cold surfaces, balances radiated heat more readily, and improves acoustic loss by transmission through walls and ceilings. Reflective coverings of aluminum foil on some forms of batt insulation can also reduce radiated heat transmission by 60% or more, as the foil used is less absorptive of radiant energy. Preformed pipe coverings, duct wrappings, boiler and furnace insulation, vapor barriers, and waterproofing of concrete and masonry beneath-grade walls should not be overlooked, as they also provide barriers against heat loss or heat gain. In many instances it has been demonstrated that a little more money spent on insulation of a structure has not only provided a more pleasant and healthful living atmosphere but has paid for itself in reduction of fuel costs within a few years of installation.

## 1.8 POLLUTION

Pollution of the air, water, and land is progressing at an astounding rate and has fostered a great number of federal, state, and local statutes and ordinances to limit such pollution. On days when smoke, smog, fumes, and other irritants bother the public, there is a loud outcry, but on sunny, warm, pleasant days all is forgiven, and usually forgotten. Smoke belching from a manufacturing plant or the smog created by traffic bothers people momentarily, but their elimination costs money and little is done.

One of the most common forms of air pollution is that caused by smoke. Smoke from industrial plants includes a wide variety of chemicals, dust, and other contaminants. Oxides of sulfur and nitrogen, hydrocarbons, and carbon monoxide may combine with airborne moisture to form sulfuric acids, nitrogen dioxide, fluorine, or ozone, all detrimental to human and plant health. In 1968 approximately 33.5 million tons of sulfur oxide was emitted into the air in the United States alone, and it is estimated that about 60 million tons will be emitted by the year 2000, primarily by burning fossil fuels in electrical generating plants (76.5%). At the same time (1968) mobile transportation accounted for 41.0% of nitrogen oxide, while utility and industrial combustion accounted for 42.3%, about 20 million tons total, and expected to double by the year 2000.

Water sources are also being polluted at an increasing rate, by the dumping of waste and by the use of water in manufacturing. At the moment both of these are major offenders.

Electrical generating plants have a thermal efficiency of 35 to 40%; in other words, 60 to 65% of the heat generated is nonproductive and wasted. A small portion of this heat is discharged into the air, but the major part is released into water. Cooling ponds and water towers account for about 1 to 2% of evaporation loss, but an estimate by the Federal Power Commission indicates that the 4.75 million acre-feet per year that is used will be expected to increase to 10.64 million acre-feet per year by 1990. Chemicals added to cooling water (chlorine, sulfuric acid, calcium sulfate, chromates) have an adverse effect on fish and human beings. Up to a point, increases in the heat of wastewater tend to increase the size and type of fish and lower forms of aquatic life, but the addition of other gas or mineral pollution may have adverse effects in aquatic life destined for human consumption.

Pollution of the land itself is continuing at a similar rate. With more and more manufacturing, there is an increase in both solid and liquid waste in the manufacturing process. Increasing production also means more waste by the user, and the quantity of trash of all types is a major problem. Even the dry waste material removed from smoke and used chemicals results in millions of tons of waste per year. Underground burial in several forms and disposal at sea are not the answer, as these methods simply perpetuate the problem by changing the form. Increased use of chemical fertilizers, chemical sprays, disinfectants, and other water-soluble materials for agriculture may increase or protect crops but are hazardous when they eventually appear in water systems.

## 1.9 ENVIRONMENTAL CONTROL

Environmental information, suggestions, sworn records, and some misinformation fill the pages of newspapers, magazines, books, and even the *Congressional Record*. The United States has a National Council on Environmental Quality, an Environmental Protection Agency, and numerous congressional standing committees. In 1965, the Federal Water Quality Act set some standards for interstate streams and coastal waters, and in 1970 the National Air Quality Standards Act was passed. The Federal Radiation Council, the Federal Power Commission, the Atomic Energy Commission, the Clean Air Act of 1967 and its 1970 Amendments, and a wide variety of other "controls" have been recorded. Similar committees and statutes exist in most other countries of the world.

Each of these controls has created some constraint on the haphazard use of energy and reduction of waste. Objectives, however, may not always be as completely achieved as desired, mostly because of the cost factors involved. The possible ultimate costs have not even been completely estimated and probably never will be, as they affect the production of energy, its consumption, the recovery of waste or its proper disposal, and most of all, the cost of educating the public to reduce extravagance in energy use. Nongrowth, both in population and energy-use expansion, has been proposed, but this is a difficult concept as a solution, and could lead to ultimate chaos.

As a percentage of total energy consumption, the energy used in building construction and operation is relatively small. The greatest part of construction energy goes into the production of building materials (70%), with small amounts used as direct fuel costs (15%), into administering the project (10%), and into the transportation of materials and equipment (5%). The operation of a building during its first 8 to 10 years is estimated to cost approximately the same as the cost of construction. Then costs gradually increase as more energy leaks develop, until demolition at the end of the building's fortieth year nearly equals the cost of construction. Suggestions to conserve some of this energy include regulations to limit the number of Btu per square foot per year, to limit the maximum allowable input of energy into a building depending on its use, to limit the maximum thermal transfer of energy units, and to encourage better design of HVAC equipment so as to reduce lost energy or recover it. As energy becomes more costly and less available, the design professions must become more aware of energy use and waste, and design more efficient buildings. The American Institute of Architects (AIA) has estimated that a 30% savings in energy use could approximately equal the domestic energy output by the year 2000.

## 1.10 SUMMARY

Energy and the environment are now so closely related that no effort must be spared to conserve energy and reduce contamination of the atmosphere. Governmental regulations alone will not solve the problem. Since the energy shortages in the 1970s, more of the public is actively aware that something must be done *now*. Support systems for buildings must be more efficiently designed, manufactured, and operated, and dependence on current energy sources must be drastically reduced and new sources developed. Not everyone engaged in the design and construction of buildings is entirely aware (or sometimes even conscious at all) of the use or waste of energy. Many are of the opinion that "George will do it," relying on a design consultant or engineer to design an efficient system. Needless to say, much more active participation is needed on the part of every person involved in the design and construction of a building and its energy systems.

# 2

# Measurement

## 2.1 INTRODUCTION

Before considering the primary uses of energy in support systems for buildings, the reader should be aware that certain units of measure, conversions, formulas, and terms must be used in designing, or even discussing, these systems and their application. A thorough course in physics may be adequate in many cases, but such instruction is often related to the theory of forces and not directed to easy use in designing systems. It is also true that some of the support systems use a combination of measurements or engineering principles. A selected group of measurement units is discussed in this chapter and is included to reinforce other information and possibly awaken some forgotten knowledge.

## 2.2 MEASUREMENT UNITS

Two major types of measurement units are in common use. *Fundamental units* define certain physical qualities, whereas *derived units* are defined in terms of fundamental units. In the United States, the old "English" system defines a length of measurement in inches and feet and is a fundamental unit. The volume of a box, however, is measured by multiplying length by width by depth and is a derived unit. To make such simple calculations more difficult, two *measurement systems* are in current use in the United States. In its early organization as a nation, the United States "borrowed" many of the *English* measurement units, such as the inch, the foot, and the pound. On the other hand, France, and now most of the world, use a *metric* system, based on the meter, the gram, and the liter. In 1971, the U.S. Department of Commerce recommended that the United States change over to the metric system, which is a decimal system (based on the function of tens), and some progress has been made in the use of metric units in weights, road distances, and some other measurements. One of the principal reasons for this change has been the influx of such imports as automobiles, electrical equipment, raw materials measured by weight or volume, and the tools and equipment necessary to maintain and repair these imports, almost all of which are manufactured to the metric system of measurements.

## 2.3 CONVERSION

Although most unit values may readily be recognized, it is not always easy to compare units of the English system with those of the metric system. In some cases approximations are adequate, but in the design of support systems for buildings it is often necessary to use equivalent values that are precise. In these cases, you must either be able to convert from one system to another yourself or have on hand a conversion table. One of the problems inherent in converting units from one system to another is whether to multiply or divide. The table of conversion factors given in the Appendix to this book indicates multiplication, but some conversions may require more arithmetic if not all factors are known or if the figures with which you are working are indicated in either larger or smaller units than those shown in the conversion table.

**EXAMPLE**

Assume that it is desired to convert a known distance of 1.40 kilometers (km) to miles (mi).

$$1 \text{ km} = 3281.0 \text{ ft} \qquad 1 \text{ mi} = 5280.0 \text{ ft}$$

$$1.40 \text{ km} \times 3281.0 \text{ ft} = \frac{4593.4 \text{ ft}}{5280.0 \text{ ft}} = 0.869 \text{ mi}$$

or

$$1.40 \text{ km} \times 0.6214 \text{ (factor)} = 0.869 \text{ mi}$$

Areas given in square inches, square feet, or square yards may similarly be converted to square centimeters or square meters since each side of an equation may be squared without changing the equality.

**EXAMPLE**

Find the number of square centimeters (cm) included in an area 4 inches × 6 inches.

$$4 \times 6 = 24 \text{ in.}^2 \qquad 1 \text{ in.} = 2.54 \text{ cm}$$

$$\frac{24 \text{ in.}^2}{1} \times \frac{6.45 \text{ cm}^2}{1} = 154.8 \text{ cm}^2$$

Liquids are measured in metric liters or English gallons. According to the conversion table, one gallon (gal) equals 3.785 liters. Similarly, one gallon of water weighs 8.3453 English pounds.

**EXAMPLE**

If a water tank holds 500 liters of water, what is the weight of the water in English tons?

$$3.785 \text{ liters} \times 500 = 1892.5 \text{ gal}$$

$$1892.5 \text{ gal} \times 8.3453 \text{ lb} = 15{,}793.48 \text{ lb}$$

$$\frac{15{,}793.48 \text{ lb}}{2000} = 7.896 \text{ tons}$$

If the weight of a liquid or gas is known for a standard quantity such as a cubic foot, the pressure may be calculated for a given height of that material as follows:

$$\text{pressure} = \frac{\text{weight}}{\text{cubic foot}} \times \text{height}$$

**EXAMPLE**

A storage tank 50 ft high and 30 ft in diameter holds water to the 40-ft level. What is the pressure intensity at the bottom of the tank? The diameter of the tank is unimportant for establishing unit pressure. If total pressure is required, the unit pressure must be multiplied by the area of the tank bottom.

water weighs 62.4 lb per cubic foot

$$\text{pressure} = 62.4 \text{ lb} \times 40 \text{ ft} = 2496.0 \text{ lb}$$

Similarly, the volume of space may be determined, when the weight and quantity of a material are known, by the simple formula

$$\text{quantity in pounds} \times \frac{\text{cubic feet}}{\text{weight per unit}} = \text{volume}$$

**EXAMPLE**

Given 1000 gal of water, find the space it will occupy in cubic feet.

1 gal of water weighs 8.345 lb

1 cu. ft. of water weighs 62.4 lb

$$1000 \text{ gal} = 8345 \text{ lb} \times \frac{\text{cubic feet}}{62.4 \text{ lb}} = 133.73 \text{ cu. ft.}$$

And perhaps last but not of least importance is the conversion of temperature (see Fig. 2-1). In addition to the English scale of Fahrenheit (°F) and the metric Celsius (°C) scales, which are both related to zero there are the Rankin (°R) and Kelvin (°K) scales, which express temperature in terms of absolute zero. Both the Fahrenheit and Celsius scales use reference points related to the freezing point of water, with the Fahrenheit scale starting at 32° and the Celsius scale starting at 0°, and with the boiling point at 212° on the Fahrenheit scale and at 100° on the Celsius scale. The Rankin scale uses the Fahrenheit scale plus 460° for absolute zero, while the Kelvin method uses the Celsius scale plus 273° for absolute zero.

To convert Fahrenheit degrees to Celsius degrees, use the formula

$$\frac{\text{Celsius}}{100} = \frac{\text{Fahrenheit minus 32}}{180}$$

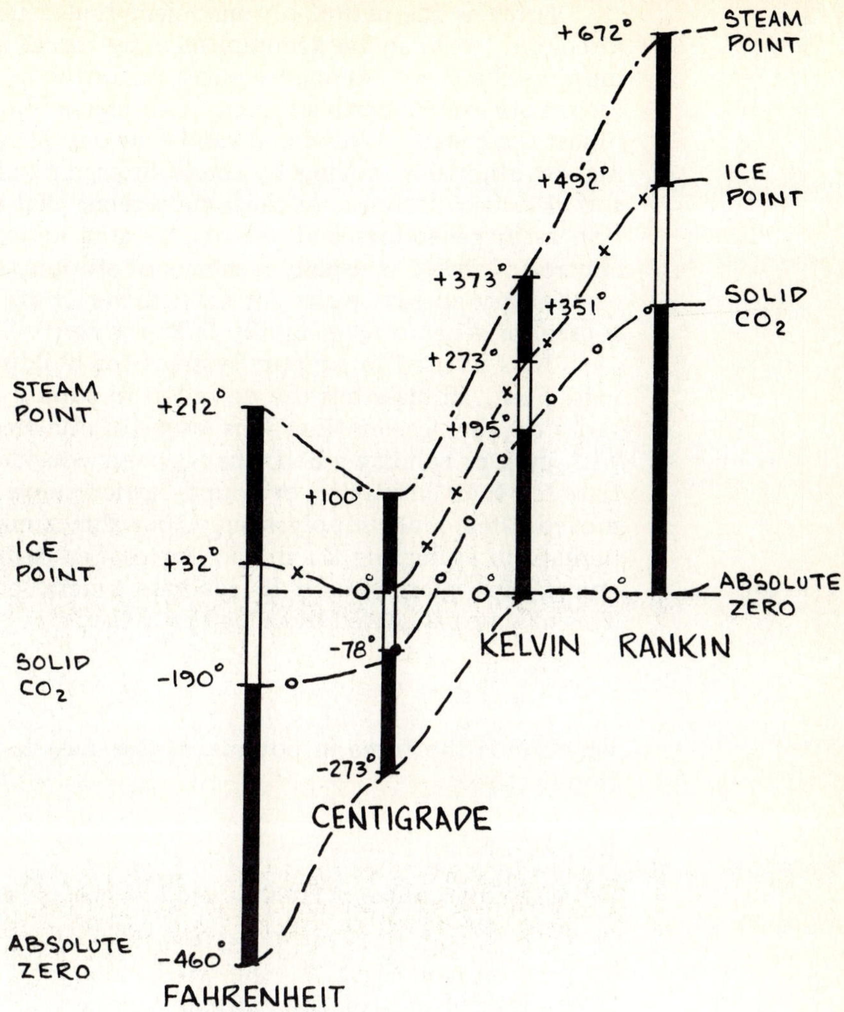


**Figure 2-1** Temperature conversion chart.

**EXAMPLE**

Convert 75° Fahrenheit to Celsius, Rankin, and Kelvin.

$$\frac{°C}{100} = \frac{75° - 32}{180} = \frac{43}{180} = 23.8°C$$

$$°R = 75°F + 460 = 535$$

$$°K = 23.8°C + 273 = 296.8$$

## 2.4 FORCE, MASS, AND WORK

All support systems for buildings depend on, or are a part of, energy transmission in one form or another. Energy may be stored as latent energy where no movement is taking place, but the instant that energy is activated, it causes the transfer of energy to another object or system. Potential energy is the energy of position or possibility, whereas kinetic energy is the energy of action or motion.

*Force* is the action on an object that tends to move it or change its direction. This can be accomplished by direct contact as is the case with most mechanical systems, by electrical attraction as with magnetic force, or by gravity of the earth's forces. With mechanical systems, movement of an object is generally desired and can be accomplished by pushing, pulling, turning, or otherwise moving by direct linkage or physical connection of some sort. Electrical force may cause movement that depends on magnetic attraction as in generators and motors, or stop action as used in most electrical controls. Gravity is usually a means of stopping action but may be an initial or auxiliary energy source for a pile driver or drop hammer and certainly is a source for generating energy by falling water.

*Mass* as used in support systems for buildings has two principal meanings. First, all mass has the potential to supply energy, but may not necessarily require acceleration. This is true of one meaning of mass in connection with bulk or solidity when what is being considered are walls or other materials for their insulating or sound-barrier value. If such masses were to be moved, they could supply energy, but this is not usually considered, except perhaps in evaluating the impact force of an earthquake. In more normal circumstances, mass is used to generate kinetic energy by some acceleration. This may be calculated by using the formula;

$$F = M \times a$$

where $F$ is the force in pounds, $M$ the mass in pounds, and $a$ the acceleration in ft/sec$^2$.

**EXAMPLE**

Assume a mass of 1000 lb and an acceleration of 10 ft/sec$^2$.

$$F = 1000 \times 10 = 100{,}000 \text{ lb}$$

If the gravity pull of the earth is considered, all objects in free-fall are attracted at the rate of 32 ft/sec$^2$, and the formula $W = M \times g$ may be used, where $W$ is the weight of the object in pounds, $M$ the mass of the object, and $g$ the acceleration due to gravity.

*Work* is the resultant when *force* combines with *motion*, and is expressed as distance times force—normally as foot-pounds. To be properly termed *work*, the force and the movement of the mass must be in the same direction. If the force acts at right angles to movement, the work is zero, or if the force is at an angle with the movement, only a portion of the work is used.

## 2.5 ELECTRICAL MEASUREMENT

Electricity plays such an important part in the lives of most of the people of the world that it is simply accepted without thought. Even those who think about electricity rarely know much about measurement of that energy source. In addition to the concept of length, mass, force, and temperature, the introduction of volts, amperes, watts, and related definitions comes with electricity. The function of the latter measurements is explained in more detail later, but some discussion of measurement is included here.

One of the most common formulas used in electrical measurement is that used to calculate *watts* (W), *volts* (V), or *amperes* (A). In a direct-current system, the formulas are simple.

$$\text{watts} = \text{volts} \times \text{amperes} \qquad \text{amperes} = \frac{\text{watts}}{\text{volts}} \qquad \text{volts} = \frac{\text{watts}}{\text{amperes}}$$

**EXAMPLE**

An electrical wall heater rated at 1200 watts operates on 120 volts.

$$1200\ \text{W} = 120\ \text{V} \times 10\ \text{Amp.} \qquad 10\ \text{Amp.} = \frac{1200\ \text{W}}{120\ \text{V}} \qquad 120\ \text{V} = \frac{1200\ \text{W}}{10\ \text{A.}}$$

This formula is also used for single-phase alternating current, but if a phase difference occurs, the formula changes to include the power factor:

$$\text{watts} = \text{volts} \times \text{amperes} \times \text{power factor}$$

The *power factor* is the ratio of the true power to the apparent power, and low power factors may cause excessive current (amperes) to be carried through motor and wiring systems. The ideal power factor is 1.0.

$$\text{power factor} = \frac{\text{true power}}{\text{apparent power}} = \frac{\text{watts}}{\text{volts} \times \text{amperes}}$$

**EXAMPLE**

Assume a power factor of 0.75 at 120 volts and current at 15 amperes. What is the true power?

$$\text{true power (watts)} = \text{volts} \times \text{amperes} \times \text{power factor}$$

$$120 \times 15 \times 0.75 = 1350\ \text{watts}$$

*Ohm's law* deals with resistance (ohms) as well as current (amperes) and voltage (volts). In this law, current is directly proportional to the applied voltage and inversely proportional to the current resistance, and may be calculated by the formula

$$I\ \text{(amperes)} = \frac{E\ \text{(volts)}}{R\ \text{(ohms)}}$$

**EXAMPLE**

Assume a simple circuit of 120 volts with one resistance unit of 1200 ohms ($\Omega$).

$$I = \frac{E}{R} = \frac{120\ \text{volts}}{1200\ \text{ohms}} = 0.1\ \text{A}$$

This formula is for a simple circuit with one resistance unit, but often multiple resistance units (lamps, motors, heaters) may be included in the circuit. The method of wiring the circuit (series, parallel, or series-parallel) changes the formula. In a *series* circuit, the incoming current is introduced at one terminal of a unit, passes through the unit, and exits from the other terminal. Each unit (lamp, motor, heater) provides resistance to the passage. Three simple formulas take care of total current ($I_{\text{total}}$), total resistance ($R_{\text{total}}$), and total voltage ($E_{\text{total}}$).

$$I_{\text{Total}} = I_1 = I_2 = I_3$$
$$R_{\text{Total}} = R_1 + R_2 + R_3$$
$$E_{\text{Total}} = E_1 + E_2 + E_3$$

**EXAMPLE**

Using the resistance values shown in Fig. 2-2 and 120 volts, these formulas are solved as follows:

$$I_T = \frac{E_T}{R_T} = \frac{120}{75} = 1.6 \text{ amps.}$$
$$R_T = 20 + 25 + 30 = 75 \text{ ohms}$$
$$E_T = I \times R$$
$$E_1 = 1.6 \times 20 = 32$$
$$E_2 = 1.6 \times 25 = 40$$
$$E_3 = 1.6 \times 30 = 48$$
$$E_T = 32 + 40 + 48 = 120 \text{ volts}$$

A *parallel* wiring system has several paths for the current and is governed by the following formulas:

$$I_T = I_1 + I_2 + I_3$$
$$\frac{1}{R_T} = \frac{1}{R_1} + \frac{1}{R_2} + \frac{1}{R_3}$$
$$E_T = E_1 = E_2 = E_3$$

**EXAMPLE**

Again using the resistances shown in Fig. 2-2 and 120 volts, the three values are:

$$\frac{1}{R_T} = \frac{1}{20} + \frac{1}{25} + \frac{1}{30} = 0.05 + 0.04 + 0.03$$

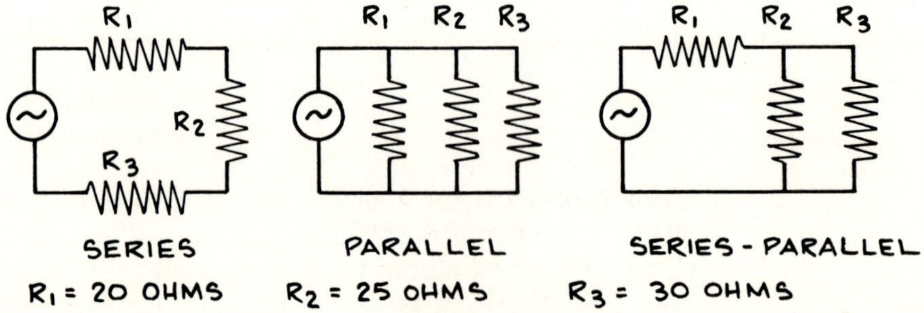


Figure 2-2

$$\frac{1}{R_T} = 0.123 \qquad R = 8.13 \text{ ohms}$$

$$I_T = \frac{E_T}{R_T} = \frac{120}{8.13} \qquad I_T = 14.76 \text{ amps.}$$

The voltage drop is 120 V across the resistor.

Calculations for *series-parallel* circuits become a little more complicated. First, replace the total circuit by a series circuit alone, so that the resistance formula reads

$$\frac{1}{R_e} = \frac{1}{R_2} + \frac{1}{R_3} = \frac{1}{25} + \frac{1}{30} = \frac{4}{100} + \frac{3.33}{100} = \frac{7.55}{100}$$

$$R_e = 13.64 \text{ ohms}$$

Next replace the parallel portion and solve as a series circuit. Watch the $R_e$ value!

$$R_T = R_1 + R_e = 20 + 13.6 = 33.6 \text{ ohms}$$

Apply Ohm's law and solve for current:

$$I_T = \frac{E_T}{R_T} = \frac{120 \text{ volts}}{33.6 \text{ ohms}} = 3.57 \text{ amperes}$$

Next solve for the *voltage drop* across each resistor, as resistance and current are now known:

$$E_1 = I_1 \times R_1 = 3.57 \text{ amperes} \times 20 \text{ ohms} = 71.4 \text{ volts}$$

$$E_e = I_e \times R_e = 3.57 \text{ amperes} \times 13.64 \text{ ohms} = 48.7 \text{ volts}$$

By adding 71.4 V and 48.7 V (= 120.1 V), proof of voltage is obtained. Now solve for current through each leg of the circuit, as follows:

$$E_e = E_2 = E_3 = 48.7 \text{ volts} \qquad I_2 = \frac{E_2}{R_2} = \frac{48.7}{25} = 1.95 \text{ amperes}$$

$$I_3 = \frac{E_3}{R_3} = \frac{48.7}{30} = 1.62 \text{ amperes}$$

Therefore,

$$I_2 + I_3 = 3.57 \text{ A} \qquad \text{which checks with } I_T \text{ above}$$

Where long runs in a circuit are necessary, some consideration of and calculation for the voltage drop from the energy source to the energy use should be made. Resistance and reactance of wires and cables are given in tables issued by manufacturers of these items, and the voltage drop should be designed for 3% or less. This may be calculated as indicated in the following example.

**EXAMPLE**

Use the formula

$$Z = \sqrt{R^2 + X^2}$$

where $Z$ = total impedance (ohms)
$R$ = total resistance of conductor (ohms)
$X$ = reactance of conductor (ohms)
Assume that $R$ = 5 ohms, and $X$ = 12 ohms in a 120-volt system.

$$Z = \sqrt{5^2 + 12^2} = \sqrt{25 + 144} = 13.0 \text{ ohms}$$

The voltage drop is then found by the formula $V = IZ$, where $V$ is the voltage drop (volts), $I$ the current in the conductor (amperes), and $Z$ the total impedance (ohms).

$$I = \frac{V}{Z} = \frac{120}{13} = 9.23 \text{ amperes}$$

$$V = 9.23 \times 5.0 = 46.15 \text{ volts}$$

To find the percentage of drop:

$$\frac{I \times Z}{V} = 9.23 \times 13.0 = 119.99$$

Therefore,

$$\text{percentage of drop} = \frac{119.99}{120.0} = 0.999\%$$

A shorter method may be used for small buildings or for short uncomplicated runs.

$$V \text{ (volts)} = \frac{I \text{ (amperes)} \times L \text{ (length, feet)} \times E \text{ (resistance, ohms} \times 1000)}{1000 (\text{constant } R \text{ per foot})}$$

**EXAMPLE**

A feeder 90 ft long has a resistance of 0.205 ohms per foot for No. 3 copper wire, and a value of 12 amperes for $I$.

$$V = \frac{12 \times 90 \times 0.205}{1000} = \frac{221.4}{1000} = 0.22 \text{ volts drop}$$

$$Z = 0.205 \times 90 = 18.45$$

$$\% \text{ of drop} = \frac{I \times Z}{V} = 12 \times 18.45 = \frac{221.4}{120} = 1.85\%$$

Calculating *lighting levels* requires a long formula, but many of the factors to be included are provided by luminaire manufacturers or by tables. The required footcandles (fc) must be determined from the work done or the space occupied. The following formula will give the average maintained illumination level:

$$\text{fc} = \frac{\text{LL/L} \times \text{CU} \times \text{LLD} \times \text{LDD}}{\text{A/L}}$$

where A/L = area per luminaire ($ft^2$)
LDD = luminaire dirt depreciation factor
LL/L = lamp lumens per luminaire
CU = coefficient of utilization
LLD = lamp lumen depreciation factor

**EXAMPLE**

Assume that a space is desired to have 100 fc and that

$$A/L = 20 \qquad CU = 70\% \qquad LLD = 85\%$$
$$LDD = 0.75 \qquad LL/L = 4500$$

Then

$$fc = \frac{4500 \times 0.70 \times 0.85 \times 0.75}{20}$$
$$= \frac{2008.125}{20} = 100.41$$

If the answer had been less than the desired 100 fc, the factor LL/L would have had to be increased by adding more lamps, selecting a different luminaire, or installing more fixtures.

Electrical *heating* requirements are derived in the same manner as those for other types of heating or for air-conditioning. That is, each room is calculated for heat loss (or gain), depending on exposure to the exterior; number of openings; type of construction of walls, ceilings, and floors; and related items. These factors and their application are considered in Chapter 23. However, the National Electrical Manufacturers Association (NEMA) has derived a formula for estimating the annual cost of electrical energy used for heating as follows:

$$\text{cost per kilowatt-hour} = \frac{\text{gross heat loss (kW)} \times \text{degree-days} \times C}{\text{design temperature difference}}$$

The gross heat loss is determined by figuring the loss as noted above for each room and thus for the entire building. The degree-days are obtained from data compiled by the U.S. Weather Service which are based on the daily difference between mean temperature and 56°F, when the mean temperature is less than 65°F. Factor $C$ is a correction obtainable from the local utility and is determined by the type of installation involved; it is usually about 17.0 average. The gross kilowatts may be derived using the formula

$$\text{watts} = \frac{\text{Btu/hr}}{3.413} \times \text{Btu/hr} = \text{watts} \times 3.413$$

**EXAMPLE**

Assume a calculated Btu/hr (British thermal unit per hour) value of 40,000; 6310 degree-days for Chicago, Illinois; a correction factor of 17.0; and a cost of $0.054 per kilowatt-hour (kWh); and a degree difference of 70°F between outdoor and indoor temperatures.

$$W = \frac{\text{Btu/hr}}{3.413} = \frac{40{,}000}{3.413} \times \frac{11{,}719.89\ \text{W}}{1000} = \frac{292.99}{24} = 12.20$$

$$\text{kWh} = \frac{12.2 \times 6310 \times 17.0}{70} = \frac{1{,}309{,}543.10}{70} = 18{,}707.76\ \text{kWh}$$

$$\text{cost} = 18{,}707.76 \times \$0.054 = \$1010.22 \text{ per year}$$

## 2.6 FLUID MECHANICS

Two "fluids"—liquids and gases—are generally considered in connection with support systems for buildings. Either may be at rest or in motion. The most common fluids are water, oil, air, and a variety of other gases. We first need to define several words used in fluid mechanics. *Viscosity* is the property of a fluid to withstand shear or resistance to movement within itself. *Density* is defined as mass per unit of volume and is expressed in pounds mass per cubic foot. The *specific weight* of a fluid is the weight per unit expressed in pounds per cubic foot. *Pressure* is a measure of force of a fluid against a given area, exerted equally in all directions, and is expressed as pounds per square foot or as pounds per square inch (psi). We also need to know the following symbols.

| | |
|---|---|
| Viscosity | lowercase Greek letter $\mu$ (mu) |
| Density | lowercase Greek letter $\rho$ (rho) |
| Weight | lowercase Greek letter $\gamma$ (gamma) |

Although the field of fluid mechanics is quite extensive, only a few problems arise in connection with support systems for buildings. The one that arises most often concerns pressure, or unit pressure, against tank walls.

**EXAMPLE**

If a vertical tank is 10 ft in diameter and contains gasoline to a height of 20 ft, what is the total pressure on the bottom, and what is the pressure per square foot? The weight of gasoline is 42.0 pounds per cubic foot.

Areas of tank bottom: $\pi r^2 = 3.1415 \times 5 = 78.54$ sq. ft.

Volume of tank: $A \times \text{height} = 78.54 \times 20 = 1570.8$ cu. ft.

Total pressure: $1570.8 \times 42.0 = 65{,}973.6$ ft-lb

Pressure per sq. ft.: $\dfrac{65{,}973.6}{78.54} = 840$ ft-lb

The pressure against the tank walls will vary from zero at the top level of a liquid to maximum at the bottom level. If it is desired to find the pressure against the tank walls at any vertical point, the following formula is used:

$$P = \gamma \times \text{height}$$

**EXAMPLE**

Assume the same tank filled with gasoline. What is the pressure intensity in pounds per square inch at a point 6 ft from the top?

$$P = \gamma \times h = 42.0 \times 14.0 = \frac{588}{144} = 4.083 \text{ psi}$$

To determine the strength of pipes, tanks, and similar cylindrical containers, it is often useful or necessary to determine the stress exerted against the walls when the pressure of the contained fluid is known. The internal force must be equalized by the combined and opposite reaction of the container walls to keep the container in equilibrium. By a series of calculations of the various forces at walls, thickness of materials, internal pressure, and length of exposed wall, we arrive at the formula

$$S = \frac{pd}{2t}$$

where $S$ = stress in container walls (psi)
$p$ = internal pressure (psi)
$d$ = cylinder inside diameter (inches)
$t$ = wall thickness (inches)

**EXAMPLE**

If the same tank of gasoline is used, find the stress when the pressure is 65,973.6 lb/ft and the wall thickness is ¼ in.

$$P = \frac{65{,}973.6}{144} = 444.26 \text{ psi}$$

$$d = 10 \times 12 = 120 \text{ in.}$$

$$S = \frac{444.26 \times 120}{2 \times 0.25} = \frac{53{,}311.2}{0.50} = 106{,}622.4 \text{ psi}$$

When a fluid is to be pumped from one location to another, the size or horsepower rating of the required pump must be known. The horsepower value is equal to 550 ft-lb/sec and the formula to find horsepower required is

$$\text{hp} = \frac{\text{force} \times \text{distance}}{550 \times \text{time}} \quad \text{or} \quad \text{hp} = \frac{Q\gamma H}{550}$$

where $Q$ is the rate of flow in $\text{ft}^3$/sec, $\gamma$ the specific weight of the fluid in $\text{lb/ft}^3$, and $H$ the energy head in feet. The quantity of fluid per cubic foot must be known or determined.

**EXAMPLE**

What is the required horsepower to pump water at the rate of 30 gal/min with an external head of 150 ft? Water has a specific weight of 62.4 $\text{lb/ft}^3$ and 7.48 $\text{gal/ft}^3$.

$$\text{Convert } Q \text{ to ft}^3\text{/sec} = \frac{30}{60} \times \frac{1 \text{ ft}}{7.48} \times \frac{1 \text{ min}}{60 \text{ sec}}$$

$$= 0.5 \times 7.48 \times 0.0166$$

$$= 0.062 \text{ ft}^3\text{/sec}$$

$$\text{hp} = \frac{Q\gamma H}{550} = \frac{0.062 \times 62.4/\text{ft}^3 \times 150}{550} = 1.055 \text{ hp}$$

Friction in the pipe and fittings would probably indicate that at least 1¼ hp be used.

## 2.7 MECHANICAL MEASUREMENT

Units of work may be expressed in terms of distance and force, or in foot-pounds. Work combines motion or movement with a given amount of energy and is equal to force (lb) times distance in the direction of the force (ft). Resistance to force is caused by friction and depends on the weight of the body being moved and the degree of roughness, or smoothness, of the touching surfaces. The coefficient of a particular surface varies from 0 to 1.0, as noted in Table 2-1, with no friction indicated as zero.

**TABLE 2-1**
Coefficient of Friction

| | |
|---|---|
| Aluminum on steel | 0.57 |
| Brass on steel | 0.44 |
| Copper on steel | 0.36 |
| Glass on steel | 0.40 |
| Metal on concrete | 0.30-0.70 |
| Metal on wood | 0.25-0.36 |
| Wood on wood | 0.30-0.50 |

When finding the force necessary to overcome friction across a surface, there are two fundamental situations to consider: movement of a weight along a horizontal surface, and movement of a weight along an inclined surface (see Fig. 2-3).

$$f = \gamma N \quad \text{or} \quad W = \gamma(W \cos \phi) \times \text{distance}$$

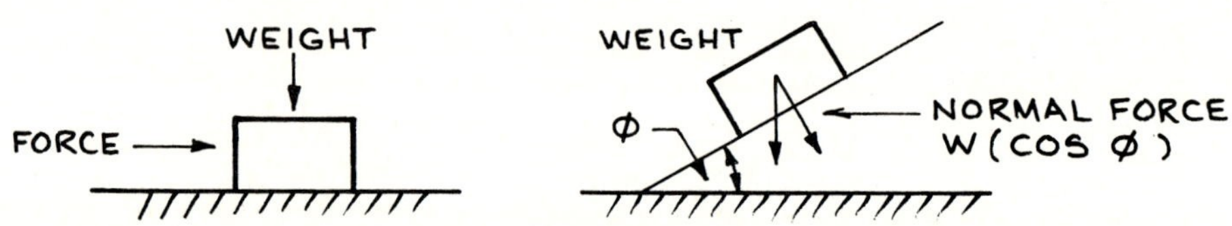


Figure 2-3

**EXAMPLE**

If the weight in each case is 100 lb and the angle for the inclined plane is $\theta = 30^\circ$, what is the force required to overcome a coefficient of friction of 0.40 (wood on wood) and a distance of 25 ft?

$$f = 0.40 \times 100 = 40{,}000 \text{ lb} \qquad W = 0.40 \times 86.6 \times 25$$

$$W = 0.40 \times 100 \times 25 = 1000 \text{ lb} \qquad = 866 \text{ lb}$$

More often, one needs to know the amount of power required to perform some work. This adds the factor of time needed to accomplish the task to the $f = \gamma N$ formula as follows:

$$\text{power } (P) = \frac{W}{t}$$

**EXAMPLE**

If it is desired to move a 100-lb weight a distance of 25 ft in 10 sec, what is the horsepower required?

$$f = \gamma N = 0.40 \times 100 \times 40.0 \text{ lb}$$

$$P = \frac{W}{t} = \frac{100 \times 25}{10} = 250 \text{ ft-lb/sec}$$

Since 1 hp = 33,000 ft-lb/min, 250 ft-lb/sec = 2.2 hp.

In addition to the two situations noted, there is the matter of weight or power in relation to a lever arm. For example, one may need to calculate the required strength or capacity of a beam (structural shape, pipe, duct) supported at each end or secured at one end so as to function as a cantilever. There are many possible loading conditions, with uniform loads distributed equally or with concentrated loads localized. The most simple is the uniform load supported at each end of the beam span. To find the reaction or necessary support to maintain the load, the formula is

$$R \text{ (reaction upward)} = \frac{W \text{ (weight)} \times L \text{ (length)}}{2}$$

with the maximum shear as

$$M = \frac{WL^2}{8}$$

at the center of the span. If the "beam" is simply supported at only one end and has a uniform load throughout its length, $R = W \times L$ and the maximum moment is $M = WL^2/2$ at the fixed end. These formulas and related information are readily available in a number of books concerning structures and may be easily applied to suspended pipes and other equipment.

## 2.8 HEATING AND COOLING

The transfer of heat is normally accomplished by conduction, convection, or radiation. *Conductance* is the process of heat transfer from a body of higher temperature to one of lower temperature where direct contact is made between the two bodies. *Convection* transfers energy by mixing molecules of a substance near heat with those farther from the heat. *Radiation* is a process by which heat is transferred from one body to another without physical contact and in all directions from the heat source. A building receives radiant heat from the sun.

Conductance is common in heat transfer through walls and ceilings as well as other "solid" surfaces and the rate of heat gain or loss may be calculated by the formula

$$Q = \frac{A(t_1 - t_2)}{\Delta X}$$

where $Q$ is the rate of exchange, $A$ the area, $t_1$ the temperature on one side, $t_2$ the temperature on the opposite side, and $\Delta X$ the thickness of the wall. To determine the amount of heat transfer, the quality $k$ must be introduced to find the Btu/hr value when $k$ equals

$$\frac{\text{Btu/hr-ft}^2\text{-}^\circ\text{F}}{\text{inch thickness}}$$

$$Q = \frac{k}{\Delta X} A(t_1 - t_2)$$

Convection and radiation are more complicated to calculate. Motion in a fluid due to convection is created by the difference in density and may be natural or free convection, or may be forced convection when coolers, heaters, or pumps are used. The temperature of the film surrounding a material or fluid, its viscosity, turbulence, and a number of other factors enter into the calculation for transfer. Somewhat similarly, radiation is affected by the angle of reflection, temperature, surface color, materials between surfaces, and absorption remittance. Calculations for the latter two are more scientific in nature than is generally necessary for use in buildings and are covered completely in the *ASHRAE Handbook*, 1981 Fundamentals.

Heating and cooling calculations, including the values of many types of materials used in the construction of walls, ceilings, and floors, and a more complete use of loss evaluation are included in Chapter 23 and in the Appendix to this book.

## 2.9 ACOUSTICS AND SOUND

Sound performs in some very predictable ways in free air and often in much the same way as light. Several basic formulas are used to determine fundamental values.

The *decibel* (dB) is equal to 10 times the logarithm of the ratio of a sound of different intensities.

$$\text{IL} = 10 \log \frac{I}{I_0}$$

where IL = intensity level
$I$ = *intensity* watts/cm
$I_0$ = reference intensity 10 watt/cm

Another characteristic is the *intensity* of sound, which is reduced by one-fourth each time the distance is doubled. Since the sound spreads in a spherical manner, the following formula is used:

$$I = \frac{P}{4\pi r^2} \quad \text{or} \quad \frac{P}{930 \times 4\pi r^2} \qquad (930 \text{ cm}^2/\text{ft}^2)$$

Most acoustic control materials have a predetermined absorbent coefficient value, a fractional number, indicating the incident energy that is absorbed by the surface. Total absorption of an area is calculated by the formula $A = S \times a$, where $A$ is the total absorption in sabins, $S$ the surface area in square feet, and $a$ the coefficient of the material or assembly.

**EXAMPLE**

Assume a room 20 ft × 30 ft (600 sq. ft. of ceiling area), a ceiling height of 10 ft, with acoustic mineral tile installed over gypsum drywall with a coefficient of 0.09.

$$A = S \times a = 6000 \times 0.09 = 540 \text{ sabins}$$

Another major calculation used in designing acoustic requirements is the *reverberation time* of sound, which varies inversely with the room absorption and directly with the room volume. The total absorption may be calculated but more often is taken from a graph or table showing room volume and sabins (S). See Chapter 29 for recommended values of total absorption. The formula is

$$T = \frac{0.05V}{a}$$

where $T$ is the reverberation time in seconds, $V$ the volume of the room in cubic feet, and $a$ the absorption in sabins.

**EXAMPLE**

Assume again a room 20 foot × 30 foot with a ceiling height of 10 feet and an absorption of 500 sabins is known.

$$20 \times 30 \times 10 = 6000 \text{ cubic feet}$$

$$T = \frac{0.05 \times 6000}{500} = \frac{300}{500} = 0.6 \text{ seconds}$$

## 2.10 SUMMARY

This chapter touches on only a few of the many measurements and formulas used to determine amounts of material for various support systems for buildings. For a great many designers, this information is adequate for preliminary or even final decisions. However, in most cases, complete calculations produced by competent engineers in the several fields should be used. More extensive and full details are also available in a great number of specialized books.

# 3

# Site Planning

## 3.1 INTRODUCTION

Every building must have a site. This is a rather obvious statement; however, almost every site is different, and site planning may have a considerable influence on the location, size, orientation, and general construction of the building. The proper use of on-site resources may also have a definite effect on the energy consumed in the building and the comfort of the occupants. Site planning is rather firmly regulated by local zoning and occupation codes and ordinances as well as health, fire, and safety requirements. In addition, adjacent construction or existing buildings, trees and other vegetation, slope of property, and wind direction also influence energy conservation, or lack of it.

Residential property is often purchased primarily for the view, or perhaps because it seemed "reasonable" in price. Although these sites may be satisfactory, or even outstanding, from the owner's point of view, it may be difficult to build on them and to utilize properly their natural resources. This type of property may present more problems, or more opportunities, than many other kinds of property. Commercial buildings are generally constructed right to the property lines in city locations, which practically eliminates the use of any on-site natural factors. Where commercial, industrial, schools, or other construction is built on property that will not occupy all the land, opportunities for proper development become available. More and more commercial construction is being built on larger tracts, and this requires consideration of the on-site conditions.

## 3.2 SITE PLANS

Location of property in the United States is based on a land survey called the *rectangular* or *government system*, originated in 1785. In this system the country is divided into a number of sections, as shown in Fig. 3-1, each section having east-west "base lines" and north-south "meridans." Supplemental

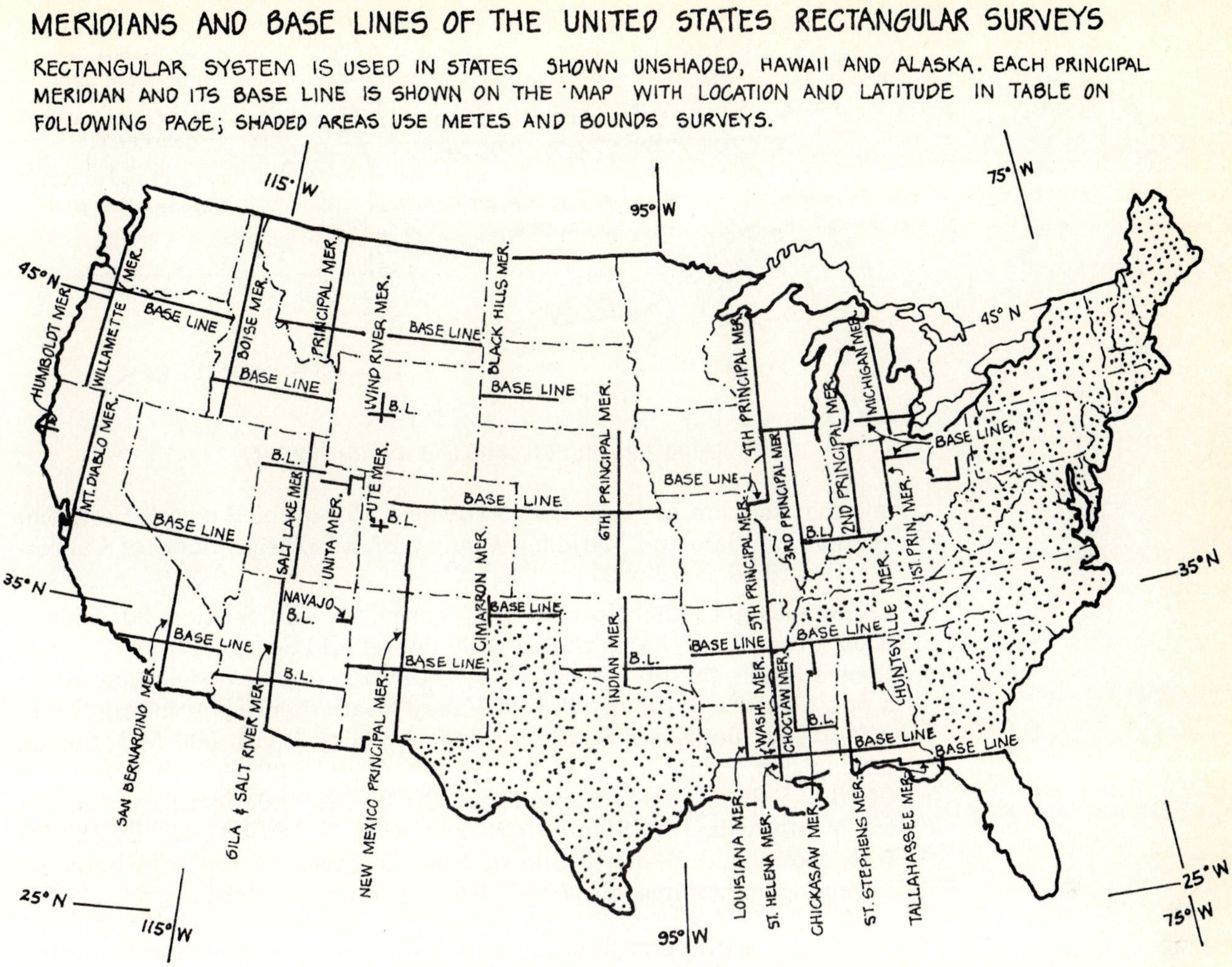


**Figure 3-1** *(From the Architectural Draftsman's Reference Handbook.)*

*guide* meridans and *correctional* lines locate *townships* 24 miles square, which in turn are subdivided into 36 sections each 1 mile square containing 640 acres. Sections in turn may be subdivided by federal law to 40-acre quarter-quarters. Corners of sections and quarters are generally marked with a concrete or pipe marker and the number of the licensed surveyor or U.S. Geodetic Survey. From these markers any portion of land may be accurately located.

Land parcels may be described by two principal methods: (1) metes and bounds surveys, or (2) reference to the rectangular system. Early descriptions were not actually surveyed but were based on the occupied land. A typical *metes and bounds* description might start at a known physical point, such as "the northeast intersection of Valley Road and Stenner Creek," and then proceed with "then along Valley Road 5000 paces to a white oak tree, then southwest to a pile of stones topped with a 6-in. log, then southeast to a fence post with a railroad spike, then northeast to Stenner Creek at three willow trees, and along the creek to the starting point." This obviously allows for considerable error, as the trees may eventually disappear, the stones may be scattered, or the creek may change course, but there are still many land parcels described in exactly this manner.

In contrast, a "legal" description based on the rectangular system may read as follows (see Fig. 3-2):

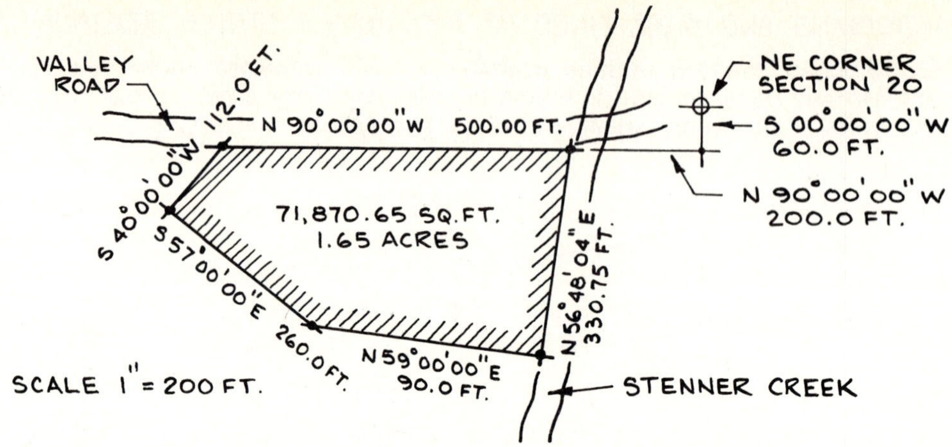


**Figure 3-2** Plot of metes and bounds property.

All that portion of Section 20, Township 5 South, Range 3 East, San Bernardino Base and Meridian, County of San Diego, State of California, described as follows:

Commencing at the Northeasterly corner of said Section 20; thence S 00° 00′ 00″ W, along the Easterly line of said Section 20, 60.00 feet; thence N 90° 00′ 00″ W, 200 feet to the true point of beginning, being a point on the Southerly line of Valley Road; thence continuing N 90° 00′ 00″ W, along said Southerly line of Valley Road, 500 feet; thence leaving said Valley Road, S 40° 00′ 00″ W, 112.00 feet; thence S 57° 00′ 00″ E 260.00 feet; thence N 59° 00′ 00″ E 90.00 feet to a point on the Westerly line of Stenner Creek; thence N 56° 48′ 04″ E, 330.75 feet, along said Westerly line of Stenner Creek to the true point of beginning, containing 71,870.65 square feet or 1.65 acres.

To carry this property description still further, most cities have adopted a system that locates property by lot, block, and subdivision. The original city plan is divided into a number of blocks. Each block is again divided into lots. A description of a typical lot might be ". . . Lot 10, Block A, original city." If subdivisions are eventually added to the original city, each such addition will be given a name, and a lot in one may be described as ". . . Lot 6, Block B, Smith's Addition." Boundaries of lots are indicated in feet and decimals of a foot and by direction in the same way as any other property. Property descriptions and maps are usually retained in the office of the proper county recorder or county assessor in the following format:

*Lot-Block Description:*

Lot 7, Block 45, Ramsey Subdivision in the City of San Diego, County of San Diego, State of California, according to Map thereof number 7501, filed in the office of the County Recorder of said San Diego County, March 1, 1982.

## 3.3 LAND CONTOURS

Very few parcels of land are dead flat. To indicate irregularities in the land, a series of contours are drawn on the plot plan. Each contour land represents a series of constant elevations on the lot (see Fig. 3-3). The vertical

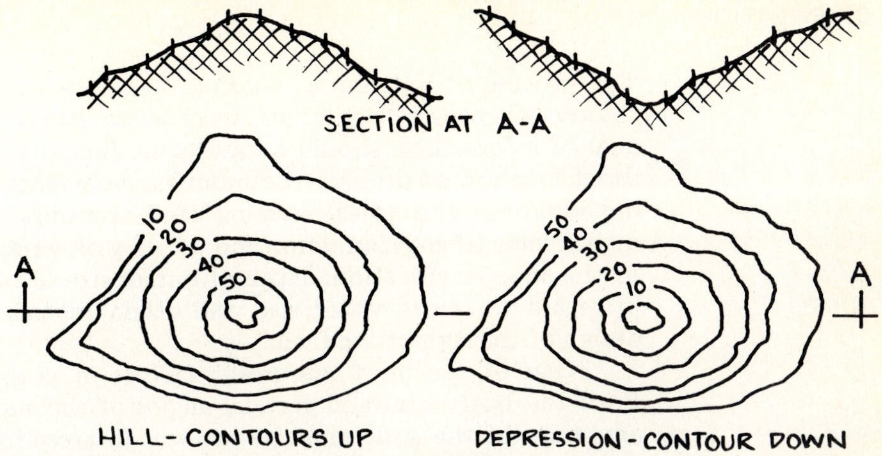


Figure 3-3

distance between adjacent contour lines is called the *contour interval*, and may represent 1 foot, 5 feet, or any other requirement. Contour lines of the *natural grade* (N.G.) of the earth are usually given as elevations above sea level but may be taken from any other convenient datum point. Similar lines for *finish grades* (F.G.) after the site work is completed are usually shown as dashed lines, with solid lines reserved for natural grades.

Both natural and finished grades are important in connection with storm drainage, sewers, and other underground work, as well as in planning for building location, pitch of driveways, and diversion of surface water. Where the building site is irregular, a certain amount of cut and fill may be necessary to provide an adequate building site or to accommodate underground piping. Both natural grades and finish grades are shown in Fig. 3-4. Note that the lines are closer together in some places. The closer together the lines, the steeper the land surface, and by the same token the farther apart the lines, the flatter the surface.

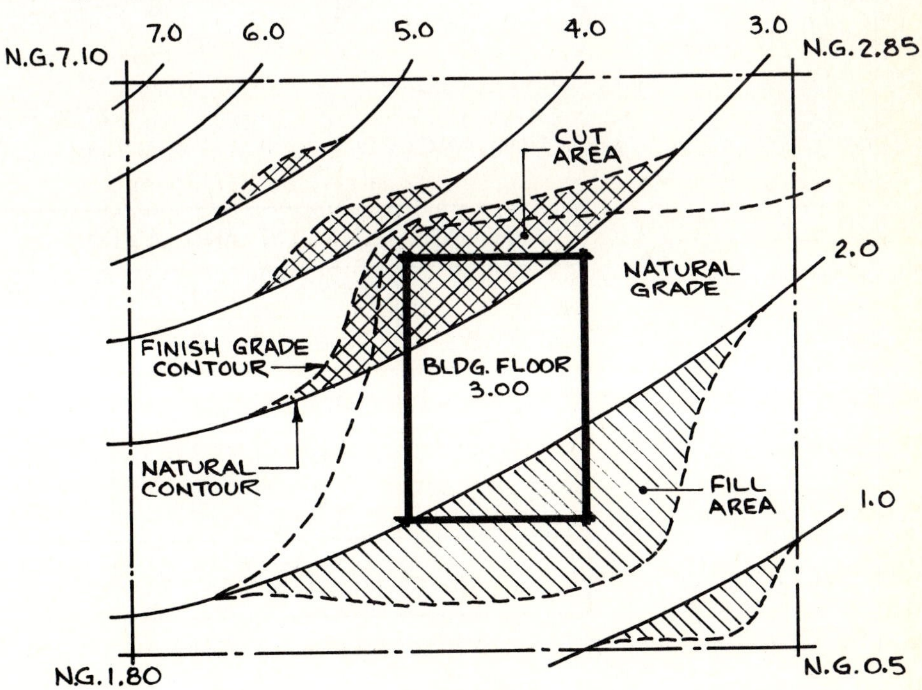


Figure 3-4

## 3.4 ORIENTATION

The process of locating a building on a site to take full advantage of the resources is called *orientation*. In order to orient a building properly, a great deal of information should be available. Included in this information are the exact location of property boundaries, easements, location and size of existing buildings and trees, existing land contours, property setback requirements, view (if any), direction and velocity of wind, and latitude and longitude of the site. The location, depth of burial, size of pipe, and other information in regard to sewer, water, gas, and other utilities are also necessary for the design of the supporting facilities.

One of the most important factors in proper orientation is the *sun*. Both the horizontal and vertical angles of the sun in relation to the site at any time of the year must be known. In areas in the northern hemisphere such as the United States, the vertical angle of the sun is greater in summer than in winter and greater in southern latitudes than in northern latitudes. The horizontal angle is also greater in summer than in winter, so the daylight hours are longer in summer. In addition to the sun angle, the roof overhang, the location and size of windows, landscape, and adjacent buildings must also be considered. The location and size of the windows is probably most important, as a comfortable sun exposure in the mornings is usually desirable, but a more direct and warmer exposure during the afternoon and early evening may require some shading. Proper roof overhand design will allow the sun to enter the buliding in the winter when the heat is more desirable but will shade the windows in summer to exclude undesirable heat. The

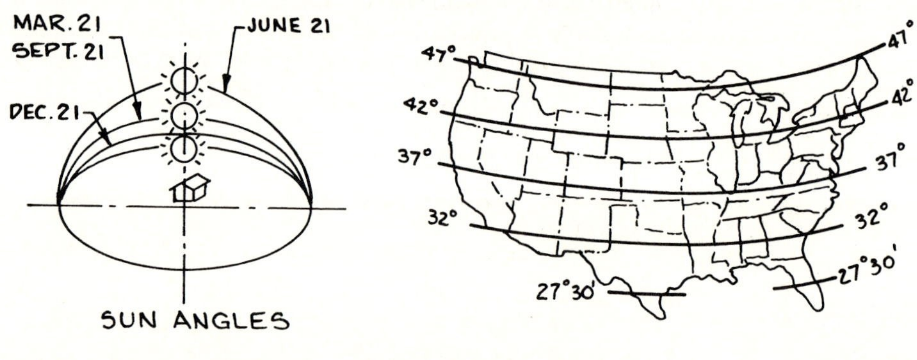


| SOLAR ALTITUDE AND AZIMUTH ANGLES | | | | |
|---|---|---|---|---|
| | MAR. 21 | JUNE 21 | SEPT. 21 | DEC. 21 |
| 27° - 32° | 60 | 83 | 60 | 37 |
| 32° - 37° | 56 | 77 | 56 | 31 |
| 37° - 42° | 50 | 73 | 50 | 27 |
| 42° - 47° | 45 | 69 | 45 | 22 |
| VALUES ARE FOR 180° ALTITUDE AND NOON SUN TIME. | | | | |

Figure 3-5

matter of turning a building or "flopping" it to reverse the exposure may solve part of the problem, but consideration must also be given to roof overhangs, trellis work, shades or louvers, and even trees and planting. The *sun angle* at various times of the year is obtainable from many sources and needs to be combined with the latitude and longitude of the site for proper shading calculations (see Fig. 3-5).

Although consideration of the sun's angles and its effects on the heating and cooling requirements of a residence are important, it is not usually possible to resite a commercial structure to take advantage of the sun. Air-conditioning is often simply overloaded or sun-facing walls are made solid, extensively shaded, or glazed with mirror glass to reduce or reflect the sun's light.

## 3.5 TEMPERATURE

The exterior temperature and its relationship to the interior temperature is also important. The actual shape of a building should receive some consideration, as a rectangular floor plan will generally expose less wall surface to the outside temperature than a rambling shape with the same floor area. The material used for construction is also very important, as a less dense material will not retain or retard heat as will a dense material. Extensive records have been kept of the average design temperatures for most of the areas in the United States (Appendix Table A10) as well as for a great many other locations in the world. These tables should be used to determine the load estimates for heating or air-conditioning. The design load for a building is the capacity of the equipment to provide and maintain a comfortable indoor condition. For example, if a building is located in Des Moines, Iowa, the design outside summer temperature is 94°F and the outside winter temperature −10°F (see Table A10, page 427). The design inside temperature for a residence should be in the range 68 to 72°F, so calculations would require heat in winter and cooling in summer to achieve the desired temperature indoors.

## 3.6 WIND

Wind is the natural movement of air and include *prevailing winds*, which are rather constant in both velocity and direction, and storm winds, which are intermittent and often seasonal. Wind velocity is considered as "light to gentle" from 1 to 10 mph up to "storm" at 50 mph. Winds that are frequent or prevailing and more than 10 mph should be deflected, but a velocity of less than 10 mph may be conveniently used for ventilation. Building height and bulk create barriers against the wind so that the wind is compressed as it passes between, around, or over a building (see Fig. 3-6).

Cold wind moving along the side of a building convects heat away. This also means that interior heat is pulled out of the building together with the wind. This may be an advantage in summer as a help to cooling, but in winter is undesirable. Location of the garage or other nonheated spaces on the windward side may eliminate some of this problem, as well as placement of other buildings, trees, or other obstructions that may block or deflect the wind.

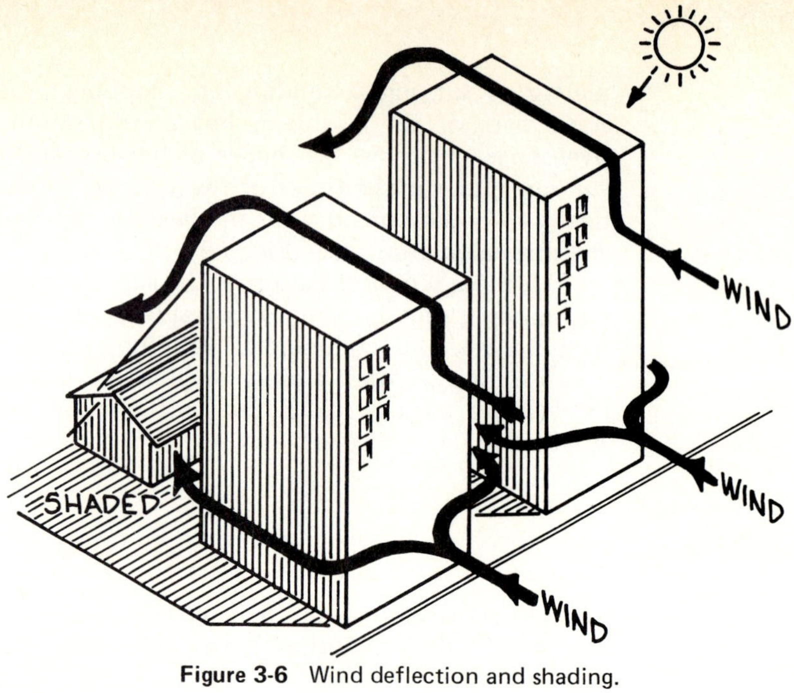


**Figure 3-6** Wind deflection and shading.

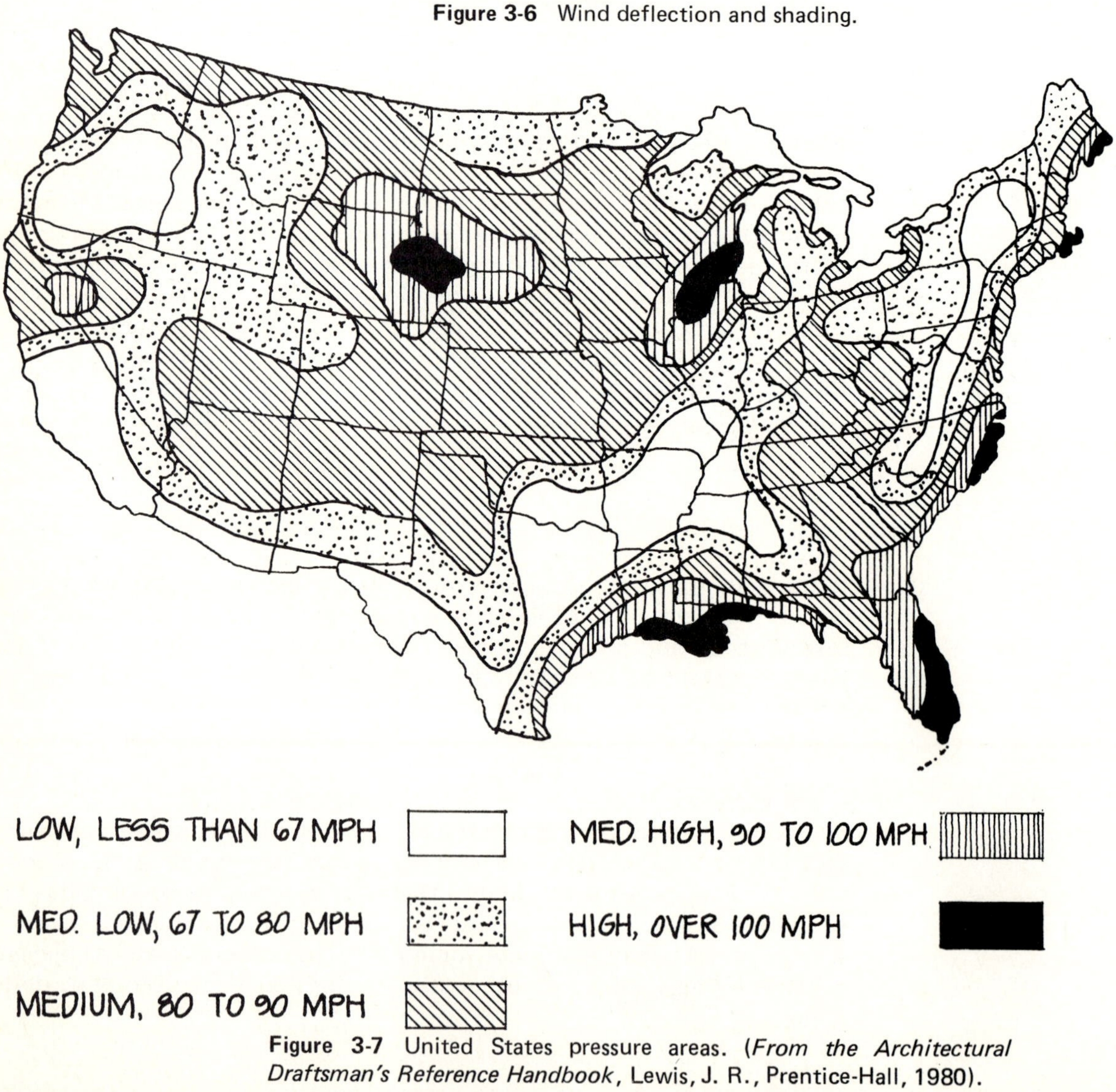


**Figure 3-7** United States pressure areas. (*From the Architectural Draftsman's Reference Handbook*, Lewis, J. R., Prentice-Hall, 1980).

Wind may also be used as a natural ventilator, either by a *gravity* system or by *cross-ventilation*. With gravity ventilation, the principle of warm air rising and cooler air falling or replacing the warm air is used. An opening near the ceiling will allow hot air to escape while cooler air replaces it near the floor. Cross-ventilation also makes use of this gravity principle but in addition employs openings on opposite sides of the room or building to allow a flow of air from one opening to another. The same effect may be achieved through proper location of doors and windows. Regulation of natural ventilation is accomplished with movable grilles, louvers, barrier walls, and plantings.

Figure 3-7 shows a map of the wind-pressure areas in the United States.

## 3.7 TERRAIN

The shape of the land surface determines the use of the land as well as the climate to a great extent. Climatic conditions in areas of uniformly flat land without mountains, forests, or bodies of water tend to be about the same throughout the area. In locations that are hilly or include trees, water, or other construction, the conditions of temperature, wind, sun exposure, and other factors may vary from one valley or hilltop to the next. The *shape* of the land may also dictate the type of construction; oriented almost any way within local ordinances on flat land, but required to be split-level, multistory, or on "stilts" on hilly property. In the case of hilly land, the view may be such a determining factor that proper orientation of rooms to take advantage of natural resources may be practically impossible.

The two composite sites shown in Fig. 3-8 demonstrate the possibilities of climate changes due to hills and valleys in the terrain. Note the compression of the wind around hills and wind expansion in the valleys. In addition, a certain amount of noise is direct or blanketed by the hills. Although these items undoubtedly affect the climate and comfort conditions at various sites, they also present certain problems for storm drainage, sewers, and water installation. If it can be assumed that the main sewer and storm drains are located in the valleys and that the valleys slope toward a disposal plant, there probably is no problem. However, if storm drains and sewers are located in a street partway up a slope, some buildings downhill from the sewer will

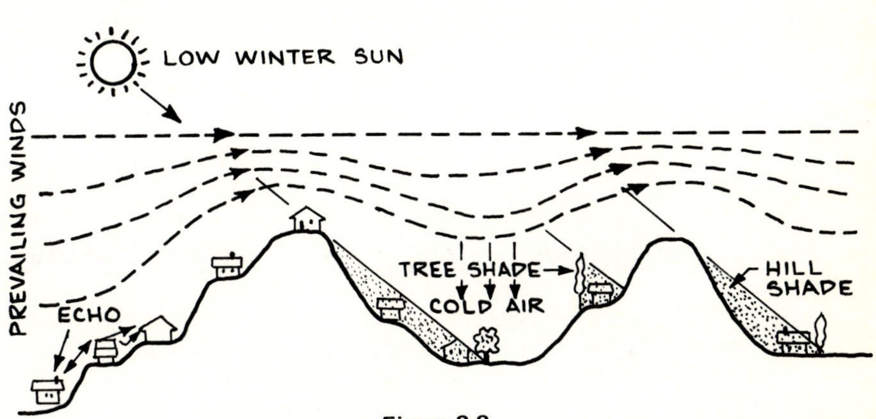


Figure 3-8

require sewage ejectors. Water pressure may also have to be increased to supply uphill branches, or water tanks may have to be situated on hilltops.

## 3.8 PRECIPITATION

Together with other site factors, precipitation (rain, hail, snow, fog) in the immediate site area must be considered (see Fig. 3-9). Rain is probably of more concern than other types of precipitation, as it has immediate runoff probability, but rain usually occurs in reasonably warm weather. Hail or snow occurs in subfreezing weather, thus presenting two important conditions: lower temperature during the falling of hail or snow, and the resulting water when the hail or snow melts. The first of these affects heating requirements, whereas the second affects drainage. Fog is of lesser importance, but persistent fog may require humidification consideration in air-conditioning design and may affect the amount of sun available as well as wind force or direction.

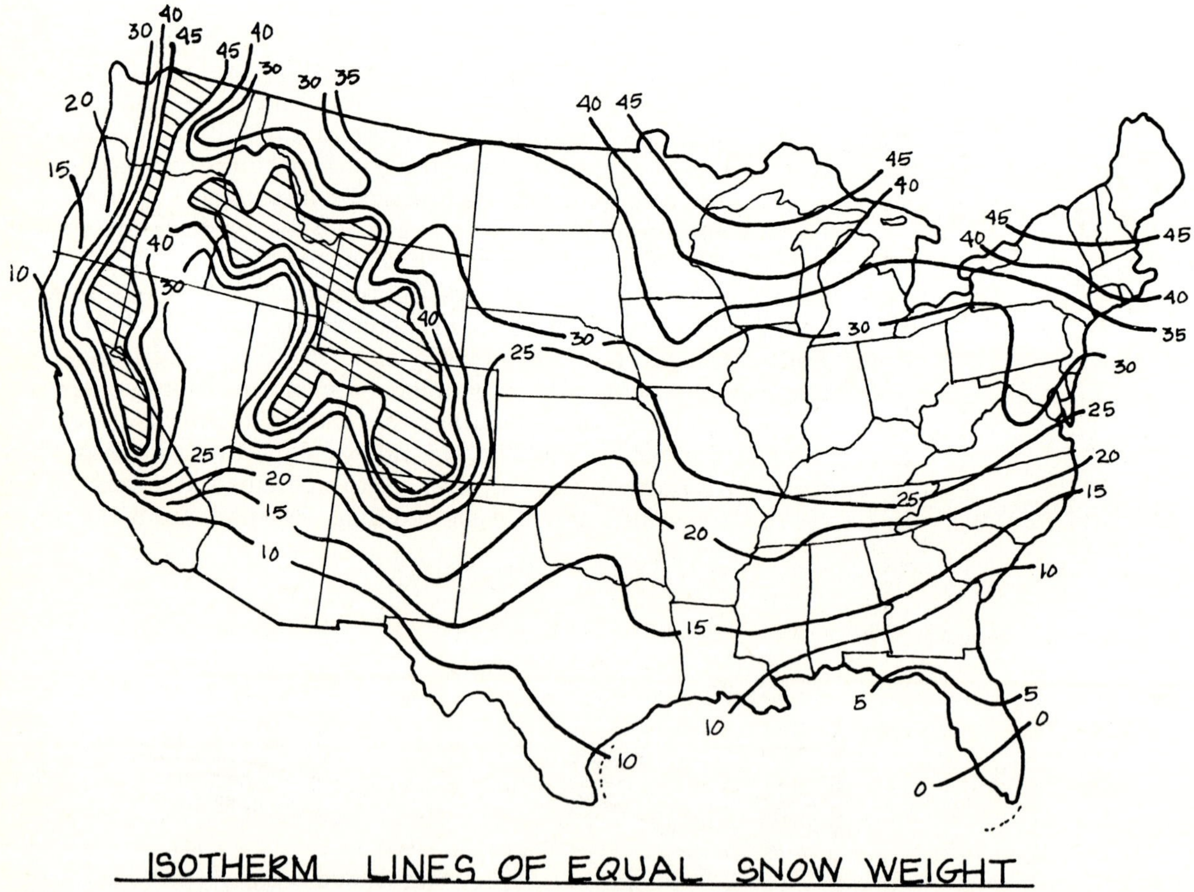


**Figure 3-9** (a) U.S. snow load; (b) U.S. rainfall. (*From the Architectural Draftsman's Reference Handbook.*)

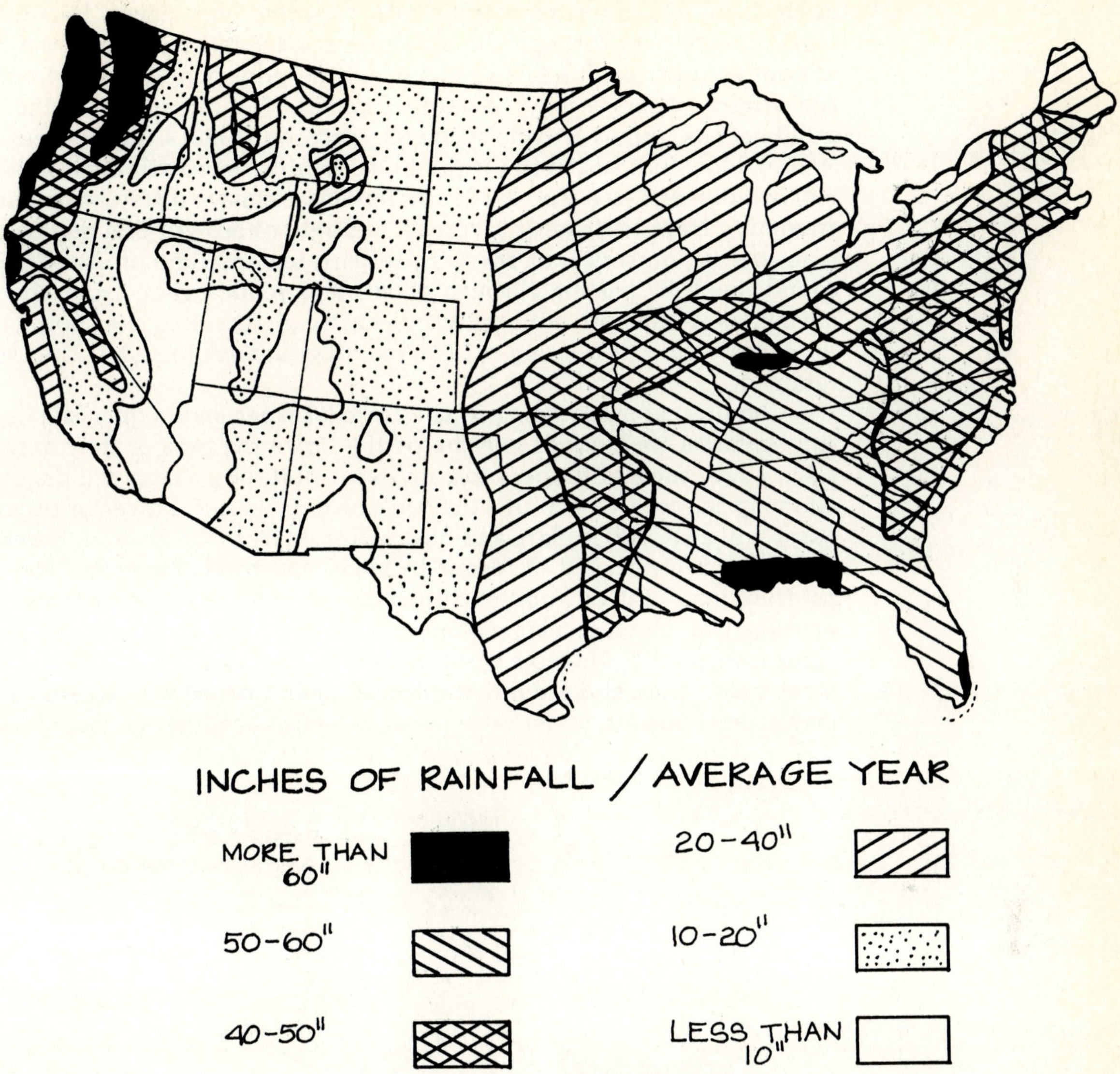


**Fig. 3.9** (continued)

## 3.9 LANDSCAPE

Trees and other vegetation take years to grow to any reasonable size, yet in a great many cases every bit of vegetation is removed before construction starts, especially in "tract" or speculative housing. Of course it is true that construction of mass housing is cheaper if there are no surface obstructions, such as trees, to have to retain and work around. It is also true that "landscaping" of these projects may often consist of a few 1-gallon-size trees (3/4 in. in diameter by 48-in. high) and a bit of half-grown lawn. The retaining of existing trees should be a major consideration from both the standpoint of appearance and the effect that they may have on the comfort of building occupants and on heating and cooling requirements.

There are two basic types of trees: deciduous, which shed their leaves each fall and grow new ones each spring, and evergreens, which retain their leaves all year long. In the first group are such species as oaks, maple, sycamore, poplar, and most fruit trees. The second group includes pines, fir, larch, hemlock, and similar trees with "needles" for leaves. Deciduous trees retain their leaves through the spring, summer, and fall, when they may act as protection from the hot sun, but expose only bare limbs and branches in the winter, when it may be more desirable for the sun to shine into the building. Evergreens provide shade all year around and so may be less desirable as they block out sun in the winter; but their close growth makes them excellent windbreaks (see Fig. 3-10). Different trees have different characteristics stemming from their height, spread, and rate of growth that can be obtained from books, nurseries, or the U.S. Department of Agriculture. Trees absorb sound, cool air passing through them, and reduce glare from the earth or paving, as well as providing privacy. Existing trees on a site should certainly be considered in orienting a new building and protected if they are to be retained.

Additional landscaping in the form of vines may reduce the heat of the sun against a wall. Vines may be of the flowering type or they may be vegetable vines, such as beans, peas, or sweet potatoes. These will die in the fall, allowing the wall to capture the sun's heat in winter. Lawns or other ground cover will also reduce glare and heat reflected from the ground. Several things should be remembered in regard to landscape work, however. Too much of anything may not be good and this applies here. A forest of trees will often provide too much shade, making a building cold and gloomy. Too many vines may tend to destroy wood or plaster surfaces with their creepers, and most trees, vines, and ground cover need maintenance with regard to dropped leaves, dead flowers, undesirable roots, as well as needing occasional trimming.

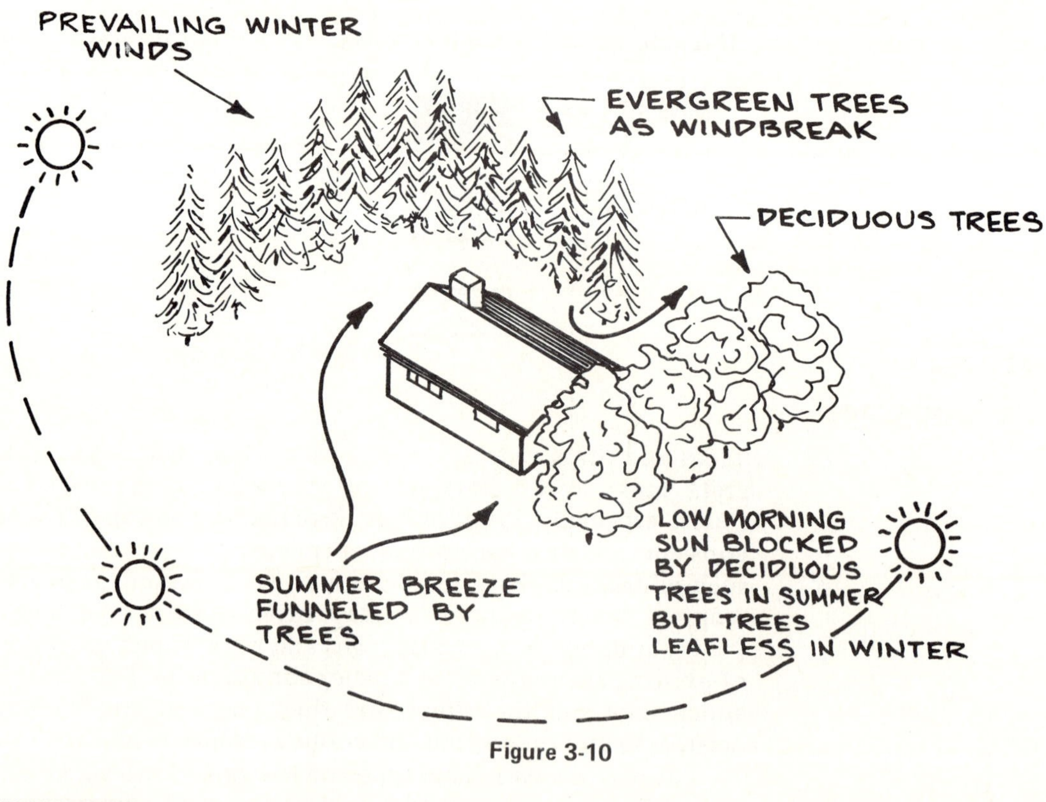


Figure 3-10

### 3.10 ADJACENT BUILDINGS

Comfort in a building is also influenced by adjacent buildings. Tall or bulky buildings disrupt the flow of wind so that it may be blocked completely, or an undesirable turbulence on the leeward side may be created. A street or passageway may funnel winds against a building with greater velocity than normal or a large flat building may create a suction area on the downward side. Any of these actions may change the heating or cooling load for the building.

Shading of one building by another is also a common problem, as shown in Fig. 3-6. A building of two or more stories on the sunny south side of another may easily cast an enveloping shadow over the lower building, especially in winter when the sun angle is low. A tall building, even though quite a distance away, may cast a long enough shadow to create a reduction of sun heat on a lower building. Shading may not always be detrimental. It is entirely possible to use shade from another building to provide a sun barrier to reduce the cooling load, although this may also increase the heating load.

## 3.11 SITING

Siting a building on flat land usually presents few problems for support systems unless the land is so flat that excessive excavation is required to obtain proper gravity pitch for sewer and storm drain lines. If this does occur, the municipal sewer may have to be installed in sections, with retaining basins and lift pumps at intervals to eliminate excessive depth of pipe burial. On sloping property, perhaps the most satisfactory solution is cut and fill, if this does not cause problems with surface drainage both above the site and below it. When this is done on steep hillsides, a problem may arise where street sewers or water lines have too much pitch or slope, but this is a matter for the municipality, not the building designer, to cure. Another minor problem for buildings built on "stilts" on sloping property is the amount and anchorage of pipe from ground level to the usable floor. The downhill height of some of this construction, totally unfinished and unoccupied, may be several stories and so might be treated as additional floors, but friction loss and static pressure will be greater than in more normal installations. Shading from the hill above or greater wind exposure may also be undesirable in these situations.

## 3.12 SUMMARY

The physical condition of a site, the vegetation or lack of it, the direction of sun movement, wind direction and velocity, and a number of other factors should determine the orientation of a building. Unfortunately, most of these are often *not* considered, simply because a building site has definite boundaries and it is impossible to build with desired sun and wind exposure. On sloping land the streets and underground municipal utilities are already installed and not subject to major changes, and "remodeling" of rolling land into flat building sites substantially changes the storm drainage, wind direction, existing vegetation, and other natural conditions.

Requirements, or advantages, of orientation in regard to support systems for buildings are generally overlooked in the design of the building. This is an unfortunate oversight, as a minor change, or a major one, may provide better opportunities for use of the sun or wind, may provide a change in heat loss or gain, and may actually save money on installation of service systems from municipal supplies. Residential considerations and commercial requirements are usually quite different, in that residential construction allows more freedom for advantageous orientation than does commercial construction, which usually covers the entire property, whether adequately oriented to use the sun and the wind or not.

While residential planning probably requires the greatest amount of design ingenuity to provide the most usable space in the least constructed space, the commercial building probably presents the greater challenge to designers of support systems. Even a multistory residence has few electrical, plumbing, or air-conditioning problems compared to those of a multistory office building or a sprawled industrial plant. In addition, some changes of siting are usually possible in residential planning, but little tolerance is allowed for commercial work. It becomes obvious that every advantage or opportunity should be thoroughly investigated, and used if possible, in properly orienting a building to take advantage of the natural resources of the site.

# 4

# Building Design

## 4.1 INTRODUCTION

The design of a building is a very flexible thing and probably should remain so, as people do not agree with each other on what design really is, nor do they agree in most cases on what good design is and what it is not. For many years there has been a saying that "form follows function," and this may very easily be applied to buildings as "design follows function." In other words, the design of a building is basically dictated by the function or use of the building. The arrangement of interior spaces, their relative size, requirements for windows and doors, and the support systems (electrical, plumbing, heating, air-conditioning, and vertical transportation) all force conditions on the building designer which must be adequately solved if the building is to function properly.

This book is not a design book from the standpoint of arrangement of doors, windows, rooflines, or other exterior features that concern the aesthetic value of a building. In this chapter these items will be considered in regard to their effect or relationship to the support systems depends largely on the arrangement of bathrooms, kitchens, and other areas that require plumbing work. Convenience of electrical outlets and adequate lighting are influenced by wall openings, room sizes, and power requirements. Heating and air-conditioning design is based on calculations of heat loss or heat gain of building materials, size and location of openings, and on the effects of sun and wind as they relate to interior comfort for building occupants. For some items to be considered, appearance may have a greater priority than utility, but the building designer will have to make decisions of this kind in any situation.

## 4.2 MATERIALS OF CONSTRUCTION

Density of a material is one of the major criteria that influence building construction in relation to interior comfort. The more dense a material, the more slowly that material will transfer heat, or cold, due to temperature changes. A material of high density will also retain heat longer than less dense materials. Various combinations of lighter-density materials may also be used to provide better resistance to the transmission of heat. In addition to density, the texture and the color of a material's surface have an effect on retention or rejction of heat.

Earth is one of the oldest dense materials that has ever been used. As a sun-baked brick or adobe unit, earth is compact and has been a major building material for many early cultures and is still used in some areas. Unfortunately, sun-baked earth is not readily available and has some undesirable characteristics that limit its general use. Concrete—a mixture of cement, sand, gravel, and water—is a dense material, readily available in most areas, easily

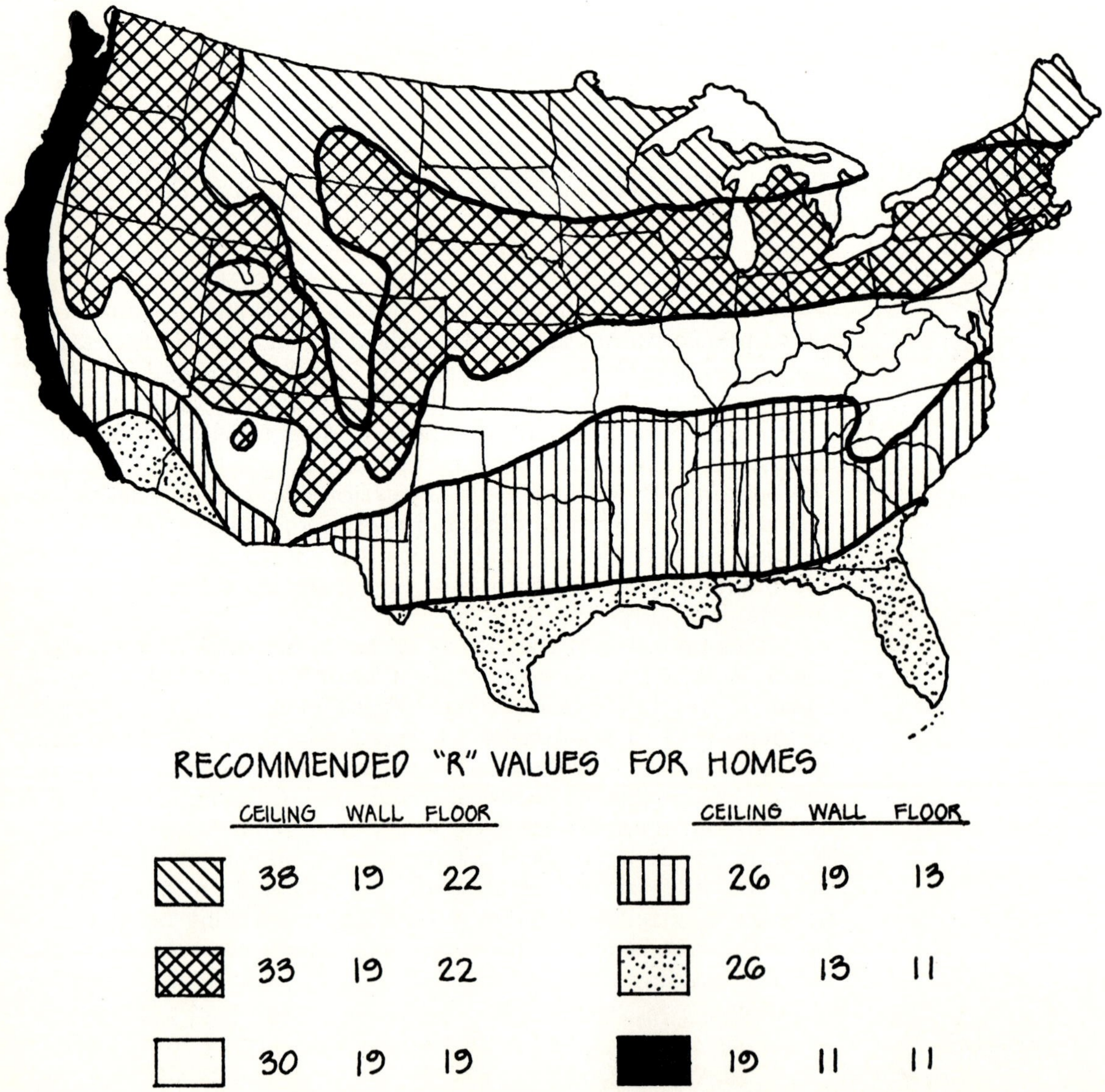


**Figure 4-1** (*From the Architectural Draftsman's Reference Handbook.*)

formed, and quite permanent. This material is used primarily for foundation work and in the construction of commercial buildings. When formed into block units, the blocks have most of the qualities of cast-in-place concrete but may be installed without formwork. Heat-fired bricks are another form of reasonably dense material used for constructing building walls and are available in a great number of shapes and finishes.

A more common but less dense material is wood. Wood construction is more often used in residential work and usually is limited to buildings three stories or less in height. If wood-framed buildings had solid walls, the density would be rather high, but this is not the case. The wood studs are normally spaced at 16 in. on center and the space between is covered on each side with plaster, gypsum drywall, or wood siding, none of which has high density. A similar situation exists in the construction of wood floors and ceilings covered with plaster or gypsum drywall. The air spaces generated by this type of construction are valuable in heat calculations but are not adequate in quantity to do very much good.

Insulation may be the answer in almost every type of construction. Insulation used in foundation and slab-on-grade installations helps prevent loss of heat to the outside. Walls, especially wood-framed, should have batt or blown insulation and roof areas will allow less heat transfer if batt or rigid insulation is used. Any insulation should be fire-retardant. Insulation in masonry or concrete construction is a bit more difficult than for wood frame, but combinations of insulation and dead-air spaces will reduce heat transfer. Pipes and ducts should also be insulated to reduce heat loss. Density is not so important in the use of insulation, as the $R$-value (see Fig. 4-1) is based more on the amount and size of air bubbles entrapped in the insulation.

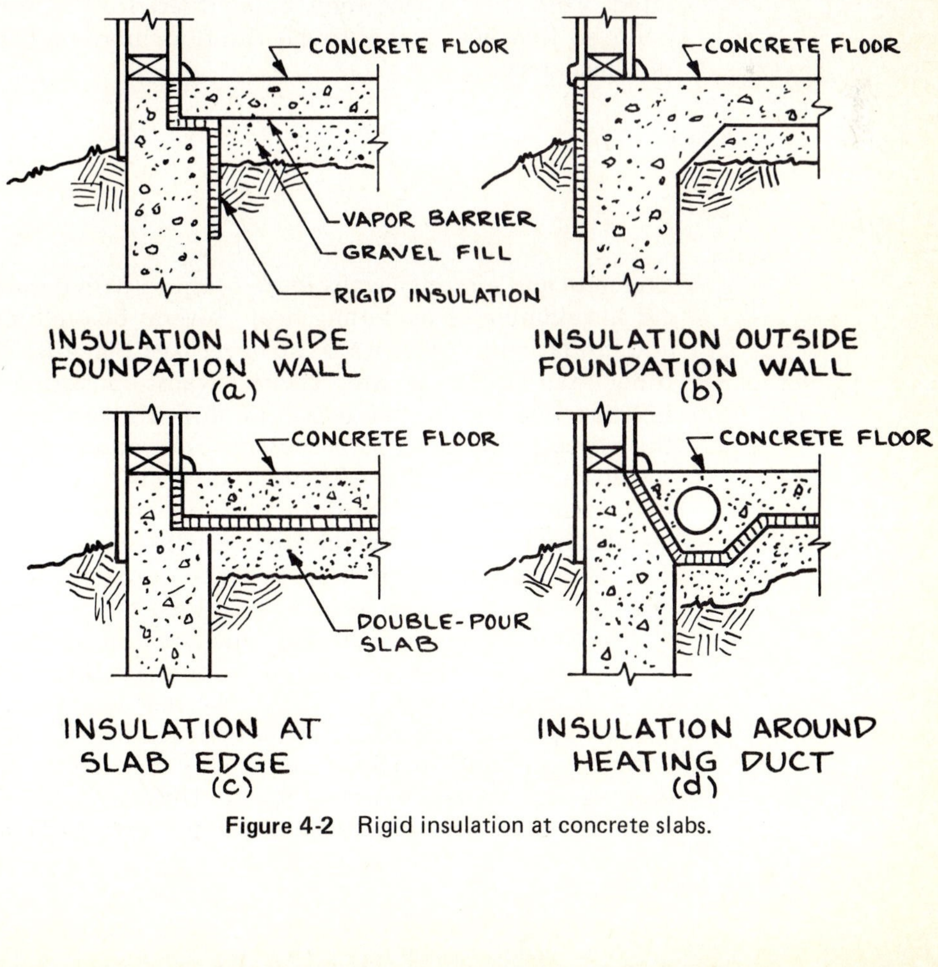


**Figure 4-2** Rigid insulation at concrete slabs.

## 4.3 FLOORS

Building floors may be separated into two types, depending on the material of construction: concrete or wood. These materials are for the structural floor and may also be the finish material if no other flooring is used. Concrete is usually used for floors below the earth grade or in all-concrete commercial construction. Slabs-on-grade in many localities are used for some residential building, but most residential and low-rise apartments have wood floors.

Concrete slabs-on-grade lose the most heat through the exposed sides of the slab and upper portion of the foundation. Rigid insulation board installed as shown in Fig. 4-2 will form a barrier to this leakage. If loss is expected through the slab, a double-pour as shown, in Fig. 4-2 with insulation between the pours, will provide a barrier. Floors above grade are more difficult to provide with intergral insulation, so in most conditions any insulation must be applied to the surface. For concrete walls, rigid board insulation may be attached to the inner surface, either by direct contact with adhesive, or over furring of wood strips or metal channels or studs.

Wood floors are constructed with a solid diaphragm of boards or plywood installed over a series of floor joists or beams. This is not a very dense combination. Insulation is usually batt-type, consisting of blankets of fiberglass installed between the joists and secured to them. In a few cases, rigid board may be used on top of the solid diaphragm, but this creates a problem if this is the walking surface. In some modern "heavy-timber" construction, floor diaphragms are constructed of nominal 2-in.-thick planks, but this does very little to increase the density. Wood floors at the first-floor level are exposed to the earth on the underside and, unfortunately, are seldom insulated. This crawl space may be ventilated by louvers or grilles required by code, so that heat loss through the floor may be high if this type of floor is not insulated.

## 4.4 WALLS

Concrete and masonry walls more often rely on density than on insulation to aid in retaining or excluding heat. Several possible methods may be used to add insulation to these walls, however (see Fig. 4-3). Brick walls in most areas, other than California, are "cavity" walls, consisting of two wythes or walls of brick with a space between. In California, earthquake requirements call for this space to be filled with cement grout and reinforcing, but in other areas the space may be filled with granulated insulation. On relatively smooth surfaces, rigid insulation board may be applied directly to the surface or may be installed over wood or metal furring strips.

Wood-stud walls are a bit easier to insulate using mineral batts or blanket material installed between the studs. Batts are available in 15- and 23-in. widths to fit the studs, 1 to 7 in. thick, and in lengths of 48 to 96 in. Blanket material is produced in rolls 50 ft long. Urethane foam may be sprayed in place or mineral pellets blown into open spaces. In many areas, rigid board is applied as a waterproofing barrier in addition to insulation, and may be composed of polystyrene, glass fiber, cane fiber, and other types of cellulose. This type of rigid board is often coated with asphalt or similar water-resistant material. Batts are usually enclosed in kraft paper or foil, on one or both

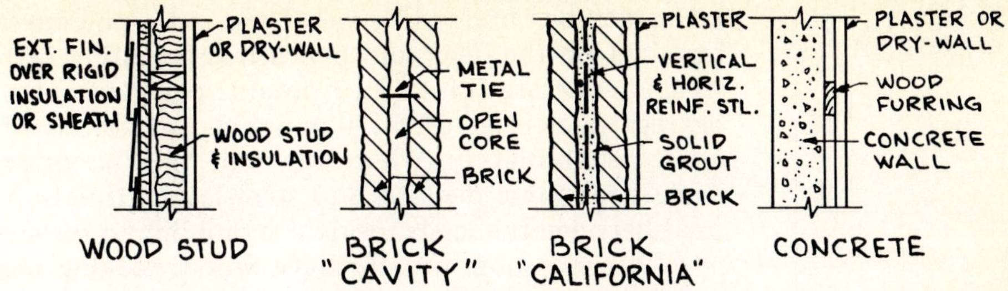


**Figure 4-3** Typical wall sections.

sides, and should be installed with this vapor barrier on the inside of the wall with no space between the batt and the exterior sheathing. *R-values* for various insulation materials are shown in Table 4-1.

**TABLE 4-1**
Thermal Values of Selected Building Materials

| | *R* | | *R* |
|---|---|---|---|
| Air space, ¾ in. | 1.01 | Hardwood flooring, 25/32 in. | 0.68 |
| Asbestos-cement board, ¼ in. | 0.06 | Insulation | |
| Brick | | 4-in. glass batt | 11.00 |
| Common, per inch | 0.20 | 6-in. glass batt | 19.00 |
| Face, per inch | 0.11 | 6-in. rock wool | 19.00 |
| Building paper | 0.06 | 1-in. rigid foam | 5.00 |
| Carpet | | 4-in. injected foam | 25.00 |
| Fiber pad | 2.08 | Acoustic ¾-in. tile | 1.89 |
| Foam rubber pad | 1.23 | Roofing | |
| Cement stucco, 1 in. | 0.20 | Asbestos-cement shingles | 0.03 |
| Concrete | | Builtup, ⅜ in. | 0.33 |
| Sand and gravel | 0.06 | Composition shingle | 0.44 |
| Sand and gravel, 8-in. thick | 0.64 | Wood shingle | 0.94 |
| Cinder, 8-in. thick | 1.72 | Wood | |
| Concrete bulk | | Softwood, per inch | 1.25 |
| Sand and gravel, 8-in. thick | 1.11 | Bevel siding, ½ X 8 | 0.81 |
| Lightweight, 8-in. thick | 2.00 | Hardboard, ¼ in. | 0.18 |
| Gypsum board | | Particle board, medium dense | 0.85 |
| ½ in. thick | 0.45 | Plywood, ⅜ in., DF | 0.31 |
| ⅝ in. thick | 0.56 | Plywood, ½ in., DF | 0.62 |
| Gypsum plaster, ½ in., lightweight | 0.32 | Plywood, ¾ in., DF | 0.93 |

Condensed from the *ASHRAE Handbook* 1981 Fundamentals.

## 4.5 ROOFS

The aesthetic design of a building roof often has little relationship to its value in connection with heat loss. In general, roofs are flat for commercial buildings and either flat or pitched for residences. If properly designed, they will shed rainwater and protect the building interior from water damage. Given their value in connection with heat loss, several items must be considered: (1) roof deck material, (2) insulation type, and (3) color exposed to the sun. None of these bear on the weathertightness of the roof but may vary according to building type, material used, strength of sun radiation, and type of roofing material used. The shape of the roof, flat or pitched, also affects the amount of overhang at eaves, which in turn affects the amount of shading of walls and windows.

Roof decks of concrete are normally reasonably flat owing to the inability of semiliquid concrete to "stay put" on slanted surfaces. Drainage pitch may be obtained by variations in the thickness of rigid insulation board installed on the deck and beneath the roofing material, or by using a lightweight concrete (85 to 110 pounds per cubic foot) with expanded shale as aggregate. Neither of these materials is classed as "roofing" but may be included as insulation. Where the roof deck is constructed of concrete placed over sheet-metal decking and steel joist, insulation batts or blanket may be used between the joists inside the building to obtain insulation values.

Pitched roofs usually have wood decking installed over wood rafters, and in most cases overhang the wall line. Insulation may be rigid board installed on the exterior surface of the deck or batts or blanket installed beneath the deck inside the building. Quite often, batts, blanket, or granular insulation is installed at ceiling level rather than at roof level. This method reduces heat loss through the ceiling into unused attic space and eliminates difficult placing of insulation at eave lines.

Material and color of the final roofing also affect heat loss or heat gain. Roofs that are constructed of several layers of saturated asphaltic paper combined with hot-asphalt moppings are termed "built-up" roofs and should be finished with pea-size rock granules as a final topping. If no rock granules are used, the finish surface will be dead black (asphalt color), will absorb heat from the sun, and will destroy the weathertightness by cracking and shrinking. Light-colored rock will act as a reflector while shielding the asphalt surface from the sun.

Although a built-up roof may be thoroughly watertight, it is rarely more than ⅜ in. thick and is not much of an insulator. Composition shingles are manufactured to overlap each other so that in most installations there may be two or three thicknesses of roofing and a finished surface of light-colored mineral granules to help insulate a building. Wood shingles and shakes are laid so that there are three layers of wood in most areas, but the darker exposed color of the wood, acting as a heat absorber, tends to offset any additional insulation value derived from thickness.

## 4.6 DOORS AND WINDOWS

Heat loss is greatest through doors and windows, mostly because they have little density and are difficult to insulate. Glass may be either ⅛-inch thick "sheet" glass or ¼-inch thick "plate" glass. In newer construction, *double-pane glass* is required by many state conservation laws. This type consists of two planes of glass separated from each other by about ¼inch with all edges sealed and with nitrogen introduced between the glass panes to reduce condensation. In cold-winter areas, *storm sash* may be installed in the window frame on the weather side of the window and separated from the window by an air gap. This provides two window sections with a dead-air space between and reduces heat loss.

Doors contribute equally to heat loss. Exterior doors are usually *solid-core*, with a solid wood or mineral core overlaid with veneer on each face. To reduce heat loss, storm doors may be installed in a manner similar to storm windows. Tight weather stripping at door and window edges will also aid in reducing heat loss. $U$ (transmission) values are shown in Table 4-2. Reflective glass, important in resisting heat gain from outside sun radiation, has little value in heat-loss calculations, as heat is lost from the inside and usually after the sun has set.

**TABLE 4-2**
Transmission Values for Doors and Windows

| | *U-value* | | |
|---|---|---|---|
| | *Winter* | *Summer* | *With storm door* |
| Doors | | | |
| 1¼-in. solid flush wood | 0.55 | 0.53 | 0.28 |
| 1¾-in. solid flush wood | 0.51 | 0.47 | 0.30 |
| 1¾-in. flush metal, mineral core | 0.59 | 0.58 | — |
| 1¾-in. flush metal, polyurethane | 0.20 | 0.18 | — |
| Windows | | | |
| ⅛-in. single-strength clear | 1.13 | 1.04 | |
| Double, ½-in air space | 0.60 | 0.56 | |
| Triple, ½-in. air space | 0.36 | 0.43 | |
| Storm window | 0.56 | — | |
| Miscellaneous | | | |
| Glass block, 8 X 8 X 4 | 0.56 | 0.54 | |
| Plastic sheet, ¼ in. | 0.96 | 0.89 | |
| Plastic skylight dome | 1.15 | — | |

## 4.7 SHADING

The shading of doors and windows on sun-facing walls is as important to heat transmission as is the shading of the entire wall. Sun and heat may be easily recognized on glass surfaces, but a solid wall is also subject to the same conditions, although it may not be so apparent. Adequate overhang of the roof is the easiest and most common shading device (see Fig. 4-4). In summer, when the angle of the sun is high, the roof overhang should be sufficient to shade an exposed wall to the earth. In winter, when the sun angle is lower, the overhang should allow the sun's rays to fall on the entire wall as much as possible. In general, the sun's heat from about noon until sunset is the warmest, so walls and and openings on the westerly side of buildings need special consideration for shading.

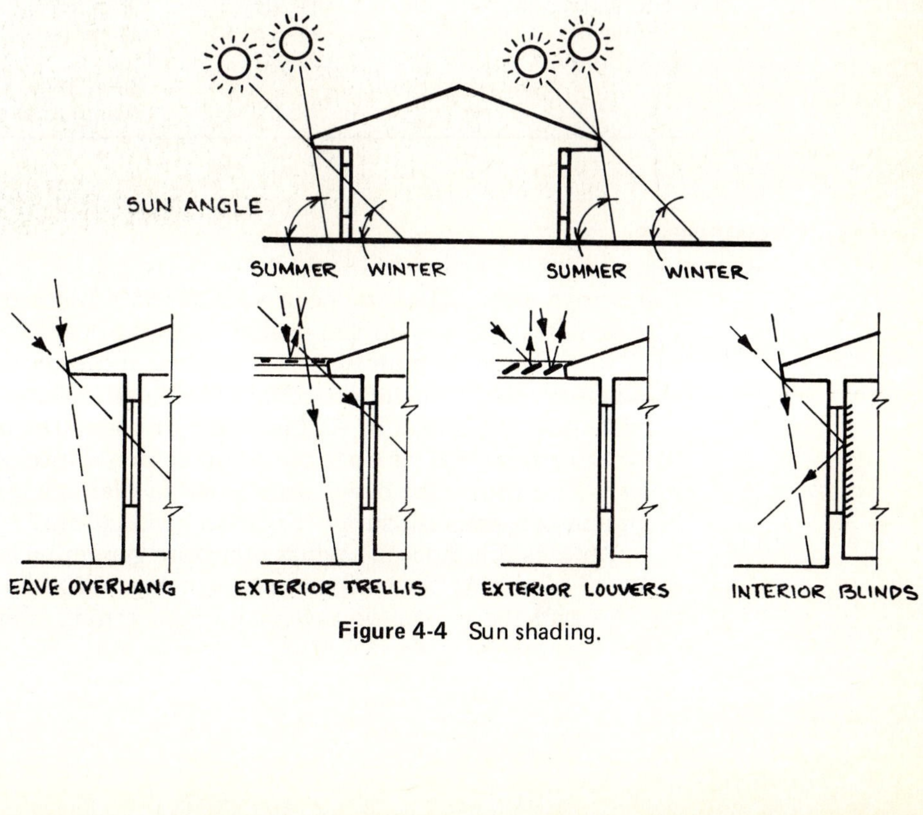


**Figure 4-4** Sun shading.

When it is not possible to provide proper shading with roof overhangs, other types of shading may be used. Proper landscaping and tree location may give some help when the trees are of reasonable size. Bushes planted along the wall may reduce part of the heat. Reflective glass and glare-reducing glass will also help against incoming heat gain but are not of much use in preventing outgoing heat losses. Interior solid shades, heavy drapes, and venetian blinds are a last resort but may be effective if no other cure can be found.

## 4.8 PAINT AND FINISHES

The color and texture of surfaces affect heat loss or gain as well as the illumination of a building. In general, a smooth surface reflects more than a rough surface and a light-colored surface reflects more than a dark surface. A rough concrete masonry wall painted a dark color such as brown, green, or black will absorb and hold heat while a similar wall with smooth-tooled joints and painted white or cream will reflect heat. Some light-reflection values are given in Table 4-3 and should be considered in establishing heat loss or gain and illumination requirements. Although not directly related to support-system needs, a smooth light-colored surface is easier to keep clean than a dark rough surface and often influences the use of some materials.

**TABLE 4-3**
Reflective Values of Materials

| *Material* | *Percent reflection* | *Material* | *Percent reflection* |
|---|---|---|---|
| Aluminum, polished | 85-95 | Paint | |
| Brick masonry, red | 20-35 | Black | 3-5 |
| Chrome plated | 70-75 | Cream, beige | 70-79 |
| Concrete, gray | 50-60 | Gold, light gray | 35-50 |
| Copper foil | 50-75 | Light green, blue | 70-75 |
| Glass | | Pink | 60-65 |
| Insulating | 71-78 | Olive green | 12 |
| Reflective | 24-51 | White, gloss | 85-90 |
| Paper, white | 60-70 | Wood | |
| | | Birch, pine | 35-50 |
| | | Walnut, mahogany | 5-15 |

## 4.9 ROOM ARRANGEMENT

Until now, very little has been said about room arrangement, but this is a very important factor in the design of support systems and especially in regard to economical plumbing installation. Exposed walls on the exterior of a building and the location of windows particularly affect heating and air-conditioning. Location of and amount of glass area in windows also affects the electrical system, as does the secondhand requirement for air-conditioning power. In quite a few cases, a little room planning can effect economics in labor and materials for support-system installations.

Perhaps foremost in savings are those generated by back-to-back arrangement of plumbing fixtures as shown in Fig. 4-5. Where fixtures are located on one wall the expensive sewer and vent piping is minimal yet adequate to

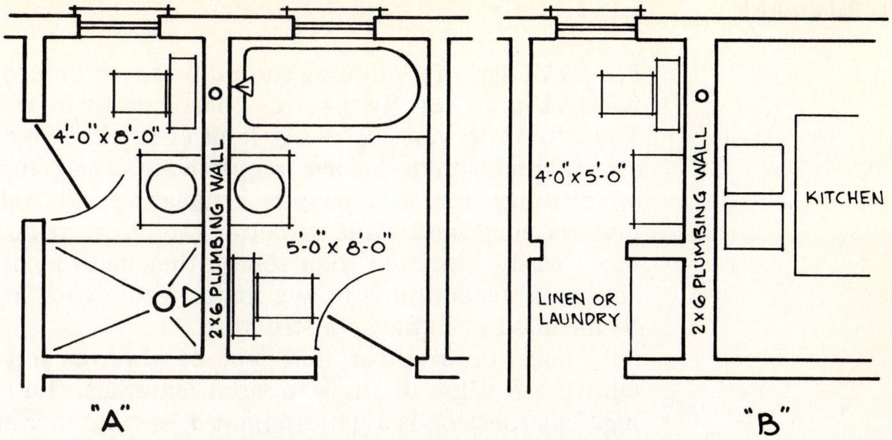


**Figure 4-5** "Back-to-back" common plumbing walls.

do the job and to provide full compliance with code requirements. Still further advantages are gained when kitchens or other baths are installed on the opposite face of the wall to reduce sewer, vent, and water piping. Although the floor area and entrance arrangement may be similar, careless consideration of fixture location may require additional piping of all types. In multistory buildings, economy is realized when bathrooms or kitchens are stacked one above the other in order to use a common drain, vent, and main water supply. Rather than having a long run from a main, supply-to-exterior hose bibbs should be made from a short extension from a bath or kitchen location. Location of the hot-water heater near the bath or kitchen should be more advantageous than location in the garage or a remote closet.

Location, type, and size of windows or doors will determine, to a great extent, the form of heating or air-conditioning system and the outlets. A rather continuous series of windows or window/door combinations may mean that wall or floor grilles or registers may not be used. A long series of openings also means a rather deep lintel or header over the openings, effectively eliminating over-the-head outlets. Grouped openings may force outlets and returns to be placed on inside walls and proper circulation may be "short-circuited," thus allowing the area adjacent to the windows to have little or no change of air. Flat roofs without space between the roof and ceiling offer another problem: what to do with required ductwork? Exposed installation on the roof looks like an afterthought and subjects the duct and equipment to the weather. Installation below the ceiling line is usually impossible and at best is unsightly. A similar situation exists when beams or other obstructions prevent sewer lines or other piping from having a uniform flow line.

One of the most serious mistakes made on multistoried buildings is the omission of adequate vertical chases or shafts for piping and electrical work. Quite often, no space at all is allowed for pipe or conduit to be installed from one floor to another higher or lower floor. This means that economical grouping is prevented and a less effective system may result, owing to increased bends or turns in ducts or fittings. Restricted space in walls or floors may also dictate the installation of several small pipes instead of one larger pipe. Electric panels and communication panels also require *working room*, usually 36 inches clear, in front of the panels, and this space may be interfered with unless careful planning is done.

## 4.10 SUMMARY

Proper design of a building should be based on proper function, but unfortunately this is not always true, particularly in relation to support systems. Too often the building designer has little knowledge of electrical, plumbing, or air-conditioning space requirements and may not provide adequate opportunity for this proper installation. A wall, floor, or roof of one material may lend itself to better retention or disposal of heat, and may be more easily insulated than some other method of construction. Proper and adequate insulation is now a major watchword in building design and must be included in all new construction.

Location, number, and type of windows and doors determine, together with the *R*-value of the structural materials, the heat loss or gain in a building. The location is a determinant of the number and location of radiators or registers and to a lesser extent the location of electrical outlets. Storm windows and storm doors reduce winter heat loss, but proper shading obtained by roof overhang is a year-round built-in feature that should be thoroughly studied for each building.

Certainly, the arrangement of rooms to take advantage of possible short runs for sewer, vent, and water should be considered. Often with only very minor relocation of fixtures, a considerable dollar savings may be realized. Multistory installations should be stacked to use minimum vertical piping, and proper chases are necessary to carry such work. Particularly in multistory installations, full consideration must be given to the amount of space needed for horizontal runs of pipe or ductwork, and obstructions that would interfere with gravity flow in sewer lines must be eliminated.

# 5

# Electrical Principles

## 5.1 INTRODUCTION

About 600 B.C., a Greek philosopher named Thales discovered by rubbing a piece of amber that it attracted light materials such as lint and feathers. Later, through his kite-flying experience, Benjamin Franklin discovered that lightning had electrical properties. In Italy, by means of a copper disk, a zinc disk, and a piece of paper between them wetted with acid, Allessandro Volta constructed what was probably the first battery. In Denmark, Hans Ørsted discovered that electricity has direction and magnetic properties. In 1831, a British scientist Michael Faraday and Joseph Henry, an American, almost simultaneously discovered that a piece of metal, moving in the field of influence of a magnet, could produce electricity. Edison, Thompson, Brush, and many others subsequently developed lighting, generators, dynamos, sound transmission, and the practical application of electricity to heating, cooling, and its multitude of other present-day uses.

Electricity exists in two major forms: static electricity (stationary) and current (moving). Static electricity occurs as lightning; attraction of various metals, cloth, or other objects created by friction; and in certain parts of electronic equipment in its "resting" stages. Through extensive experiments, theories related to the electrons and protons contained in an atom have been developed and a great number of standards for measurement discovered. Static electricity, however, is not very useful or of practical use, is of high voltage, difficult to control, and discharges in a fraction of a second. Static electricity, together with most phases of magnetism, are not therefore generally considered in major applications of electricity to building construction.

Electricity is an invisible, powerful energy force that may be used for lighting, power, transportation, sound, and heating and cooling when properly transmitted and controlled. Unlike many other types of energy, electricity must be used as it is generated and cannot successfully be stored for future use. It may be distributed over wires of nearly every size, through a wide variety of metals, water, earth, and air, and has a polarity factor that must be considered. At low voltage it may be used for the control of other voltages, for time-keeping, alarms, sound, and many other similar uses. Electricity may be transmitted at high voltage and reduced to workable voltage with transformers and it may be in the form of either alternating current (ac) or direct current (dc). In short, electricity is a very useful, adaptable, and reasonably inexpensive source of energy which may be handled in many different ways.

## 5.2 OHM'S LAW

Electricity may be compared to the flow of water. There are three major measurements involved in determining the usability of electrical energy: voltage, amperes, and watts. *Voltage* is the potential difference, or pressure, generated by the difference between the requirement at a negative location and that at a generating point or positive location. This is comparable to the force developed by opening a water faucet, which in turn requires a pump to provide pressure to push the water through the pipe. The *ampere* is a measure of the number of electrons that pass a given point in one second. This propagation of electrons is at a steady rate, so is comparable to the flow of water. The amount of resistance or impedance in a circuit is measured in *ohms*. When working with direct current the ohm is termed *resistance*; in alternating-current work it is termed *impedance*. Voltage is normally indicated mathematically by the letter $E$ or $V$ with the unit volts (V); amperes are denoted by the abbreviation A and mathematically by $I$ (current); and ohms are denoted by $\Omega$ and mathematically by $R$ (resistance). The relationship between units is represented by the formula volts (V) $\times$ amperes (A) = watts (W). One *watt* of electricity equals one volt times one ampere of current, and is the normal measurement indicator for lighting, heating, and other power requirements (thus watts are indicated mathematically by $P$). The relationships among these properties are credited to experiments of Georg Ohm performed in about 1826.

The diagram and table in Fig. 5-1 indicate various combinations and formulas used to determine the value of voltage, amperes, watts, and ohms. This information is easily derived, as several factors are almost always known for any electrical situation. Normal voltage for residential or commercial use is 110 to 120 V for lighting and convenience outlets. A voltage of 220 to 240 V is often required for motors, heavy appliances, and similar use. Wattage requirements are fairly easy to determine, as lamps are rated in watts, as are such heating appliances as irons, toasters, cooking tops, ovens, and unit heaters; standard wattage values may be assumed for convenience outlets and similar unmarked parts.

Using the diagram for Ohm's law, it is relatively simple to calculate the value of other electrical factors. As an example, assume that it is desired to find the resistance ($R$) in ohms ($\Omega$) for a motor rated at 2.0 A ($I$) at 120 V ($E$).

The formula is

$$\frac{E}{I} = R$$

$$\frac{120}{2.0} = R$$

$$120 \div 2.0 = 60R \quad \text{or} \quad 60\ \Omega$$

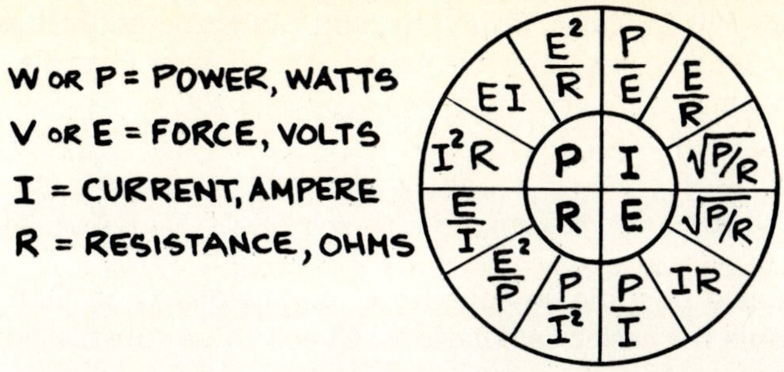


| TO FIND | DIRECT CURRENT | SINGLE PHASE A.C. | THREE PHASE A.C. |
|---|---|---|---|
| AMPERES IF KVA IS KNOWN | — | $\frac{KVA \times 1000}{E}$ | $\frac{KVA \times 1000}{E \times 1.73}$ |
| AMPERES IF HORSEPOWER KNOWN | $\frac{HP \times 746}{E \times \% COEFF}$ | $\frac{HP \times 746}{E \times \% EFF \times POWER}$ | $\frac{HP \times 746}{E \times 1.73 \times \% EFF \times POWER}$ |
| AMPERES IF KILOWATTS KNOWN | $\frac{KW \times 1000}{E}$ | $\frac{KW \times 1000}{E \times POWER\ FACT.}$ | $\frac{KW \times 1000}{E \times 1.73 \times POWER}$ |
| KILOWATTS | $\frac{I \times E}{1000}$ | $\frac{I \times E \times POWER\ FACT.}{1000}$ | $\frac{I \times E \times 1.73 \times POWER}{1000}$ |
| KILOVOLT AMPERES | — | $\frac{I \times E}{1000}$ | $\frac{I \times E \times 1.73}{1000}$ |
| HORSEPOWER | $\frac{I \times E \times \% EFF.}{746}$ | $\frac{I \times E \times \% EFF \times POWER}{746}$ | $\frac{I \times E \times 1.73 \times \% \times PF}{746}$ |
| WATTS | $E \times I$ | $E \times I \times POWER\ FACT.$ | $E \times I \times 1.73 \times POWER$ |

**Figure 5-1** Ohm's law.

A more commonly used formula is one to determine the minimum ampere rating required for the main safety switch of a residence. Again let us assume that we have 120 V service ($E$) and that the total load from all lighting, convenience outlets, and other connected electrical appliances is 8500 W ($P$). Determine the minimum amperes.

The formula is

$$\frac{P}{E} = I$$

$$\frac{8500}{120} = I \qquad 8500 \div 120 = 70.8 \text{ A}$$

With a minimum ampere requirement, proper design would probably indicate a main safety switch capacity of 100 A for safety to carry a slight additional wattage load.

## 5.3 POWER FORMULA

In addition to Ohm's law for electricity, there are several other laws or rules that might be used when power requirements are considered. Energy is often measured in *joules*, and one joule per second equals one volt-coulomb/second, also known as a watt. One watt equals one volt multiplied by one coulomb per second and one ampere is one coulomb per second, so a power formula may be written as

$$P = EI$$

where $P$ is in watts, $E$ in volts, and $I$ in amperes. Just as the formulas that express Ohm's law may be varied, the power formula may be rearranged as follows:

$$E = \frac{P}{I} \quad \text{or} \quad I = \frac{P}{E}$$

It is also possible to substitute expressions from Ohm's law into the power formula for easier calculations. Check these substitutions:

$$P = (IR)I \qquad P = IIR \qquad P = I^2R$$

Using the last formula, we may find the watts of power if the resistance is 100 Ω and the current is 4 A.

$$\begin{aligned} P &= I^2R \\ &= 4^2\ (100) \\ &= 16 \times 100 = 1600\ \text{W} \end{aligned}$$

## 5.4 FLOW OF ELECTRICITY

The *polarity* of electricity is what moves it from one place to another. In its more familiar form, a battery (see Fig. 5-2), the electricity is direct current (dc) produced by the action of an acid-water solution in direct contact with lead-calcium plates as in a "wet-cell" battery. This action is actually a process of regulated corrosion, as electricity or electrolitic action is always present when two metals or chemicals create a flow of electrons between them. The flow of current is assumed to go from the positive (+) pole to the negative (−) pole, and this theory is always used in electrical circuit design. The flow of direct current is steady and constant and becomes weaker only as the power source (battery) is depleted.

Whereas the direct current produced by batteries is used mostly in automobiles, flashlights, portable radios, and similar apparatus, the greater majority of lighting and power requirements in the world are satisfied by alternating current (ac). Alternating current actually means that the electrical

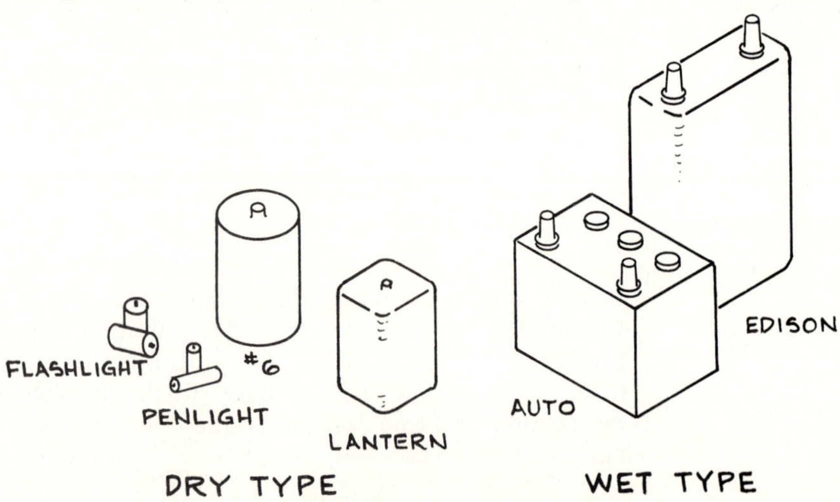


**Figure 5-2** Batteries

current alternates back and forth from one direction to the reverse at a given rate, at either 25, 50 or 60 cycles per second. This phenomenon is made possible by the use of rotating generators or alternators to generate the electricity from hydraulic power, steam, or nuclear power. Most electricity is now supplied at 110 to 120 V or 220 to 240 V at 60 cycles. When 25-cycle service is used, a noticeable flickering of lights is apparent and appliances require a longer time for operation or do not function at all unless designed for this service. The unit "cycles per second" has now largely been supplanted by the unit *hertz* (Hz). Electric power for motors and industrial equipment may also be available in 440 V and either single-phase or three-phase. These qualifications and requirements are reviewed further in Chapter 11.

## 5.5 TYPES OF CIRCUITS

Electric circuits of any sort, both alternating current and direct current, usually fall into one of two types: series or parallel. A current cannot flow unless there is a complete conducting path from the source of electricity (battery or generator) to and through various lamps or appliances and back to the source. If the conducting path is equipped with a switch, the circuit may be opened or closed at will, but the current can never flow unless the conducting path is complete, or closed.

In a *series circuit* (Fig. 5-3), the current is the same in all parts of the circuit. If a series of lamps is wired in a series circuit and one lamp burns out, the entire string of lamps will go out. A good example of this is one type of Christmas tree lighting, where the electricity illuminates one lamp after another as the current, almost instantly, passes through the filament of each consecutive lamp. If the filament of one lamp burns out, every lamp in the circuit must be tested to find the defective one. Some street-lighting systems are also on series circuits, with special bases used to bypass burned-out lamps. The principal disadvantage, in addition to the difficulty of locating defective lamps, is that the circuit resistance created by separate appliances is cumulative. This means that each resistance is added to the next one, which may result in a total circuit resistance that is higher than desirable, therefore necessitating more voltage or amperes.

### EXAMPLE

In Fig. 5-3, if the amperes are assumed as 3, the total resistance is $10 + 40 + 15 = 65\ \Omega$. $65\ \Omega \times 3\ \text{A} = 195\ \text{V}$.

$$\begin{aligned} 10 \times 3 &= \phantom{1}30\ \text{V} \\ 40 \times 3 &= 120\ \text{V} \\ 15 \times 5 &= \underline{\phantom{1}45\ \text{V}} \\ &\phantom{=}\ 195\ \text{V} \end{aligned}$$

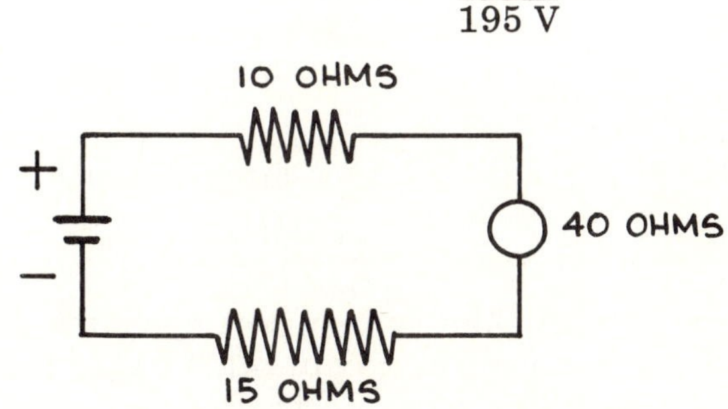


**Figure 5-3** Series circuit.

In a *parallel circuit*, each group of similar appliances or loads is connected separately across the same voltage. This causes the total current in the circuit to be the sum of the several branch circuits and is expressed by the formula

$$I = IA + IB + IC$$

Parallel wiring systems are most prevalent in building construction. Figure 5-4 indicates that each branch circuit may be calculated in accordance with Ohm's law, showing that the voltage remains the same in each branch, while the resistance or load varies, causing the current to be additive for the total load. If a connection inadvertently occurs between the "hot" wire and a neutral wire, the circuit is "shorted," the wires do not provide adequate resistance, and the fuse or circuit breaker opens the circuit. If the fuse or circuit breaker fails, the "short" may cause excessive heating and a fire may result. In electrical wiring design it is usually advisable to connect lighting separately from convenience outlets or power outlets. This is especially desirable when convenience or power outlets are used with moving machinery or heating, as in kitchens and home workshops. If a machine or cooking unit is then overloaded, causing the fuse or breaker to "blow," the lights will still be in operation, allowing the operator to see and therefore prevent accidental involvement with the overloaded or defective equipment.

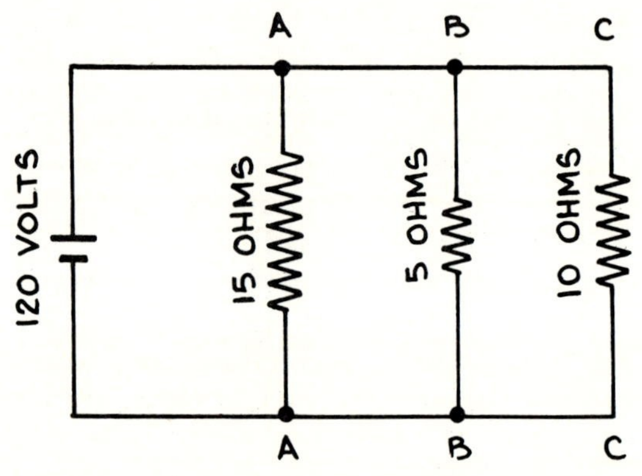


**Figure 5-4** Parallel circuit.

EXAMPLE

In Fig. 5-4, assume a 120-V power supply with resistance as noted.

$$\text{Current through } A\text{:}\quad \frac{\text{volts}}{\text{ohms}} = \frac{120}{15} = 8\text{ A}$$

$$\text{Current through } B\text{:}\quad \frac{\text{volts}}{\text{ohms}} = \frac{120}{5} = 24\text{ A}$$

$$\text{Current through } C\text{:}\quad \frac{\text{volts}}{\text{ohms}} = \frac{120}{10} = \underline{12\text{ A}}$$

$$44\text{ A}$$

A third system, which combines the two basic types of circuits, is the *series–parallel circuit*, often used where the actual size of resistance is important. In a series electrical circuit, the average rate at which a charge passes a given point must be the same everywhere. If two or more electrical circuit elements have one end joined, they must comply with *Kirchoff's current law*, which states that "the sum of all currents entering a terminal must be equal to the sum of currents leaving that terminal, and the sum of all rises in any closed electrical circuit equals the sum of all drops in voltage in the same path." Figure 5-5 shows a possible series–parallel circuit and illustrates that the rise in the battery of 12 V equals the drop in either circuit A-B or C-D. Series–parallel circuits may be separated into small units and solved by first applying Kirchhoff's current law, then computing the voltage by Ohm's law, and finally applying Kirchhoff's voltage law to find total voltage. This process may be used to find the resistance of conductors at different points from the voltage source, voltage across separate lamps in series–parallel, and solution of "ladder" circuits created by multiple sections carrying different currents.

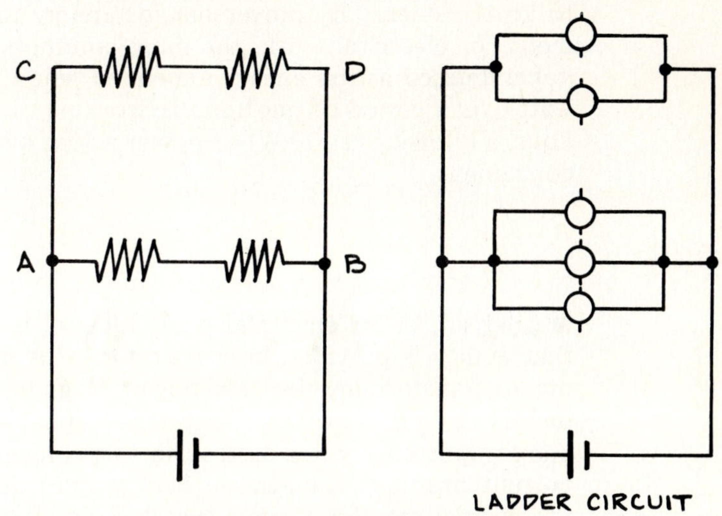


**Figure 5-5** Series-parallel circuits.

## 5.6 WORK, POWER, AND ENERGY

When considering electricity, the uninitiated often confuse the terms "work," "power," and "energy." *Work* applies to the accomplishment of doing some specific task regardless of method used or time required. A person may require 8 hours to dig a hole of a certain size. A modern backhoe may dig the same hole in 30 minutes. The process of digging the hole is "work" regardless of the time or method employed, and the amount of work is equal in each case. Work may be defined as doing a specific task where a given weight is moved a definite distance. In connection with electricity, a large amount of current may move the weight or load quickly, or a smaller amount of current may move the same load in a longer period of time. The work done is the same.

*Power* is the measure of the speed at which work is done. Power is a product of the distance an object is moved times the weight of the object. In mechanical calculations, an object weighing 100 lb lifted 10 ft requires a total of 1000 ft-lb. If the amount of power needed to accomplish this task is required, we must also know the time it took to raise the load. If the work was done slowly, it took less power than if the load was raised quickly. In electrical work, a moving current is a force capable of doing work. The pressure on the current is voltage, forcing the current through resistance to create a certain amount of heat or thermal energy. In electroplating, the electrical energy is transformed into chemical energy by conducting the plating particles from one node to another electrically. If electrical energy is used to rotate a pump motor, the electrical energy is changed into mechanical energy. Electrical power is therefore the amount of work that a specific current will do in a unit of time when there is a given electrical voltage behind it. To calculate this power is relatively simple using the formula

$$\text{electrical power watts} = \text{voltage} \times \text{amperes}$$

There are two principal types of energy: *potential energy*, the energy that something has due to its position; and *kinetic energy*, the energy that something possesses due to its motion. Potential energy may be converted to kinetic energy by movement, or energy must be used to stop such movement. In electrical work, the most common measure of energy is the *watt-hour*, defined as the energy expended when work is done at the rate of one watt over a period of one hour. In working with electrical power the *kilowatt-hour*, a kilowatt (1000 W) of power acting over a period of one hour, is more convenient.

## 5.7 MAGNETISM

Magnetism, as an electrical phenomenon, is not very useful by itself unless that ability is provided in connection with methods for measuring, generating, or transforming electrical power. Magnetism, the ability to attract ferrous metals or to function as a direction indicator in electrical circuits, is of primary importance when used as an electromagnet. A wire carrying an electrical current produces a magnetic field around the wire, regardless of how small the current may be. From a practical standpoint, a single wire through which a current is flowing does not constitute a useful electromagnet. However, if we wind that length of wire into an open coil (solenoid), the magnetic field is concentrated in a small space, and the force field is reinforced by the number of turns taken around the core. To further increase the efficiency and usefulness of the open coil, the wire may be wound around a core of iron or steel which may be firmly attached to the wire coil or may be movable.

The strength and usefulness of an electromagnet depends on the size and material of the core, the number of turns of wire in the winding, and the current (amperes) flowing in the wire. Assuming the current at a constant rate, increasing the number of wire turns will increase the strength in direct proportion (see Fig. 5-6). If we have a coil with 10 turns of wire at a 1-A current, we may assume that this electromagnet will lift 1 lb of magnetic weight. By increasing the coil to 20 turns of wire with a 1-A current, the electromagnet will lift 2 lb of weight. A coil of 100 turns at a 1-A current would be capable of lifting 10 times the weight possible with only 10 turns of wire. *Ampere-turns* is the product of current in amperes multiplied by the number of turns of wire in the coil.

Core size and material are additional factors that determine the strength of electromagnets. The strength of an electromagnet will increase rapidly with increased core size up to a saturation point, beyond which a further increase will have very little effect. Assuming a comparable number of turns of wire and the same current, generally a 1-inch diameter core will be able to provide a strength of pull twice that of a ½-inch core. Material for cores runs from low-grade iron with low properties through the entire ferrous (iron/steel) field to special nickel–iron alloys and other special combinations of

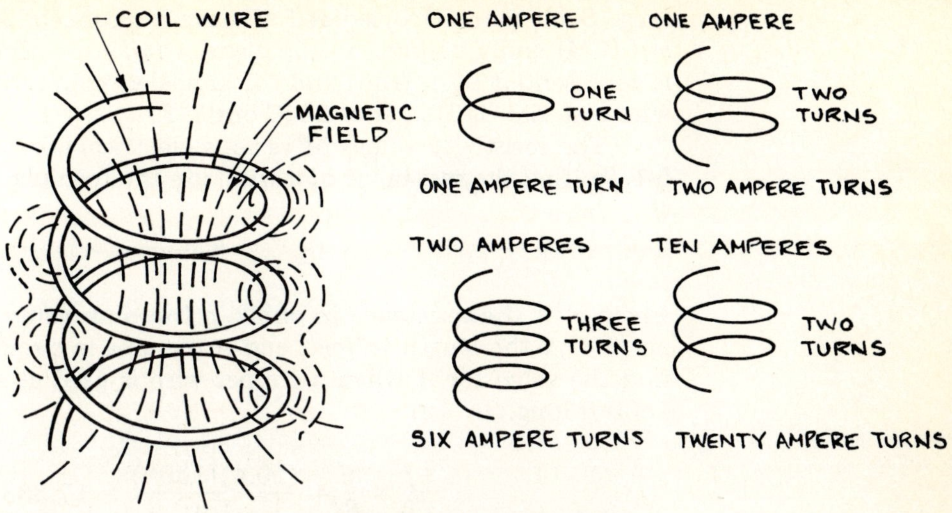


**Figure 5-6** Magnetic action.

very high grade. Obviously, the higher-grade material used as a core will reduce core resistance and provide more electromagnetic power from the same current than will a lower-grade core material.

The amount of current flowing through the wire coil is the third factor affecting the strength of an electromagnet. We know that an increase of turns of wire in the coil increases the lifting power of an electromagnet in direct proportion to the number of turns if the current in amperes remains constant. By the same token, if the number of wire turns remains constant, the electromagnetic power may be increased or decreased by increasing or decreasing the amount of current. A coil of 100 turns at a 10-A current may have strength to lift 1 pound. If the number of turns of wire is maintained at 100 but the current is increased to 20 A, the magnet strength will increase to lift 2 pounds. This means that an electromagnet may be produced with a great number of turns of wire to operate at low amperes, or with fewer turns of wire but designed to operate at a higher current of amperes. This possibility is very useful when the physical size of the coil is of considerable importance.

The use of electromagnets is so extensive that it would be almost impossible to itemize a complete list. Electromagnets are used to handle steel and iron scrap, in telegraphy equipment, telephone equipment, automobiles, locomotives, and innumerable other nonconstruction devices. In support systems for buildings, electromagnets are incorporated into elevator equipment, hundreds of types of motors, generators, door controls, sound equipment, relays for control systems, and other systems converting electrical power into mechanical action. In every case, however, the magnetic principles included here are not the major concern of the support-system designer but are only contributing to the design and function of equipment manufactured for inclusion in a complete building structure.

## 5.8 RESISTANCE

Every current-carrying electrical device has resistance to current. Three different sets of measurements are in common use: the British Engineering System (BES), the centimeter-gram-second system (cgs), and the meter-kilogram-second system (mks); The BES system is most commonly used in the United States, as is a standard wire with a diameter of 0.001 in. (1 mil) and 1 foot long. Such wire is considered to have a cross-sectional area of one *circular mil* (CM) and a volume of one *circular mil foot*. The resistivity of this unit is represented by $\rho$ (rho) and for annealed copper wire—the most common wire material—is 10.4 Ω at 20°C or 68°F.

The resistivity values of various electrical conductors are given in Table 5-1. To find the resistance of a wire, use the formula

$$R = \frac{\rho l}{A}$$

where $A$ is the resistance in ohms, $\rho$ the ohm value of wire in circular mils per foot, $l$ the length in feet, and $A$ the area in circular mils. As an example, find the value for $A$ when a copper wire is used at a 1.5-Ω resistance and is 1500 ft long.

$$A = \frac{\rho l}{R} = \frac{(10.4)(1500)}{1.5} = 17{,}333 \text{ CM}$$

**TABLE 5-1**
Resistivity of Electrical Conductors

| *Material* | *Ohms per circular mil foot* |
|---|---|
| Aluminum | 16.9 |
| Brass | 36.1-42.1 |
| Carbon | 21,000-42,000 |
| Copper | 10.4 |
| Lead | 132.0 |
| Mercury | 566.0 |
| Platinum | 66.2 |
| Silver | 8.85 |
| Tungsten | 33.1 |
| Zinc | 36.7 |

By reference to Table of Bare Copper Wire in the appendix, we find that the area of 17,33 CM falls between No. 8 wire with a circular-mil area of 16,510 CM and No. 7 wire with an area of 20,820 CM. Since most wire used in construction projects is available in even numbers smaller than No. 8, use No. 6 wire at 26,250 CM.

Heat is generated by resistance. The formula for power, except for motors, is $W = I^2R$, where $W$ is the energy in watts, $I$ the current in amperes, and $R$ the resistance in ohms; thus power lost equals $I^2R$. Heat has several undesirable effects: (1) heat increases the resistance of the conductor, (2) heat increases the heat of adjacent conductors, (3) heat causes a voltage drop, and (4) heat increases the cost of electricity that has no direct usage.

**EXAMPLE**

Using the formula $W = I^2R$, and given that line resistance is 0.0155 Ω, current is 200 A, and time is 1 hr, find the voltage drop and power (W) lost.

$$V = IR = (200)(0.0255) = 51.0 \text{ V (voltage drop)}$$

$$P = I^2R = (200)^2\,(0.0255) = (200)(51.0) = 10{,}200 \text{ W} = 10.2 \text{ kW lost}$$

Loss per hour, then, is 10.2 kW × 1.0 hr = 10.2 kWh.

$$1 \text{ kWh} = 3413 \text{ Btu, so } (10.2)(3431) = 34{,}812.6 \text{ Btu}$$

## 5.9 POWER

The watt is usually too small to be of value in determining requirements in building construction or power requirements. To simplify this measurement, multiply one watt by 1000 to obtain kilowatts (kW; 1000 W). In like manner, measure large amounts of resistance as kilohms (k$\Omega$; 1000 $\Omega$). Many machines are rated in horsepower (hp), so it is convenient to know the equivalent rating in kilowatts when determining electrical loads. Therefore,

$$1 \text{ kW} = 1.34 \text{ hp or } 1\tfrac{1}{3} \text{ hp}$$

$$1 \text{ hp} = 0.746 \text{ kW} = 746 \text{ W or } \tfrac{3}{4} \text{ kW}$$

**EXAMPLE**

Find the power requirements of an electric motor that requires 10 A at 220 V, by the following calculations.

$$\text{watts} = \text{volts} \times \text{amperes}$$

$$\text{power} = 220 \times 10 = 2200 \text{ W}$$

$$= \frac{2200}{1000} = 2.2 \text{ kW}$$

To find the horsepower of this same motor, the calculations would be

$$1 \text{ kW} = 1\tfrac{1}{3} \text{ hp or } \tfrac{4}{3} \text{ hp}$$

$$2.2 \text{ kW} = 2.2 \times \tfrac{4}{3} = 2.926 \text{ hp}$$

## 5.10 SUMMARY

The complete study of electricity and its many variables is not intended in this book. In addition to the information contained herein, there is a wide field, continuously expanding, in the research and application of electrical principles in radio, television, and electronic devices of all sorts, as well as improvements in equipment used in building construction. A general knowledge of the various formulas cited in this chapter will allow the architectural or construction student or general practitioner to better understand selection and installation of electrical equipment.

# 6

# Electrical Materials

## 6.1 INTRODUCTION

The generation, transfer, and control of electricity is often compared to the purification, transfer, and control of water, and this is a simile that should be kept in mind in the study of electricity. The generation and substation distribution of electricity is rather specialized, has relatively little to do with support systems for buildings except to provide required power, and is discussed in Chapter 3. The machinery, conductors, insulation, and supporting equipment for electrical generation is almost entirely different from that used in buildings. On the other hand, the conductor systems, controls, and equipment used for lighting, power, and related electrical usage in a building are essentially identical, except for size or capacity, in a residence or a major structure.

As in most other phases of mechanical equipment for buildings, the electrical material to be used and the installation are controlled by codes and local ordinaces. The *National Electrical Code®* (1981) and the Uniform Electrical Code, plus some state and local codes for special installations and elevators, control the material, size, arrangement, and other details of the various items used in an electrical system. For convenience of discussion, the major parts of an electrical system may be separated into the following groups: (1) conductor systems, including wire and conduit; (2) control devices, including switches, relays, circuit breakers, distribution panels, and other protection; and (3) electrical equipment, including lighting, motors, heaters, and other devices actuated by electricity.

## 6.2 ELECTRICAL CONDUCTORS

Electrical conductors that provide a path for the flow of electricity are wires of a number of different metals and sizes, bare or insulated, and installed in an exposed manner or carried in conduit or other duct types (see Fig. 6-1). Wire is drawn from a solid ingot and may be of aluminum, brass, copper, or iron, solid or stranded, one metal or combination, and in sizes to conform to the American Wire Gauge in the United States. Copper is the best and most commonly used material and is available in wire size from about No. 40 to No. 8 in solid wire and from No. 8 to No. 4/0 (0000) for most building use. The normal range of sizes for "line" voltage in buildings is from No. 18 to No. 16 for fixture wire, No. 14 to No. 8 for circuit wiring, and No. 8 to No. 4/0 for service wiring. Concentric lay stranded conductors from No. 250 MCM to No. 2000 MCM are also available for primary distribution work. All wire gauges increase as the wire size decreases, as shown in Table 6-1.

Solid wire is normally used for branch circuits for lighting, convenience outlets, and small appliances. Stranded wire is used for wiring fixtures and power wiring. Bare, noninsulated wire is used for grounding or neutral circuit wiring. Wire smaller than No. 18 may be used, bare or insulated, for low-voltage (6 V to 24 V) wiring designed for doorbell systems, low-voltage lighting controls, sound systems, and similar wiring, Wire smaller than about No. 30 is seldom used in electrical wiring circuits but is used primarily for winding of coils associated with transformers, relays, and solenoids, and may be bare or insulated copper or iron wire. Normal operating current for electrical systems is in the range 120/140 V to perhaps 408/440 V, but insulated copper wire may be used in sizes from No. 14 up to 2000 V and to No. 1/0 up to 35,000 V. Solid copper wire or bars may also be legally used as bus bars or as bonding conductors in a circuit.

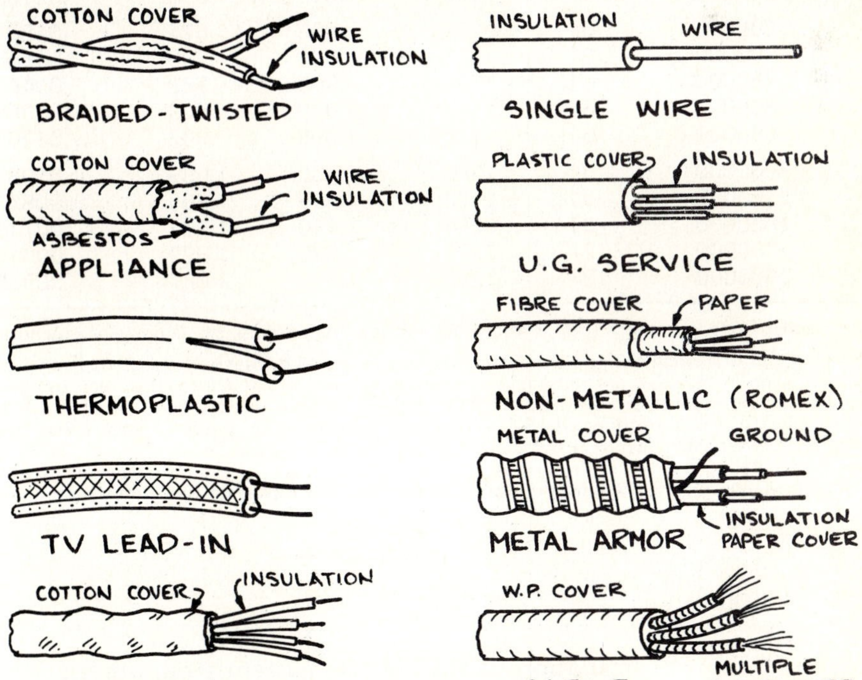


**Figure 6-1** Typical wire types.

**TABLE 6-1**
Properties of Conductors

| Size AWG, MCM | Area Cir. Mils | Concentric Lay Stranded Conductors: No. Wires | Concentric Lay Stranded Conductors: Diam. Each Wire Inches | Bare Conductors: Diam. Inches | Bare Conductors: Area[a] Sq. Inches | DC Resistance, Ohms/M Ft at 25°C, 77°F: Copper Bare Cond. | Copper Tin'd. Cond. | Aluminum |
|---|---|---|---|---|---|---|---|---|
| 18 | 1620 | Solid | .0403 | .0403 | .0013 | 6.51 | 6.79 | 10.7 |
| 16 | 2580 | Solid | .0508 | .0508 | .0020 | 4.10 | 4.26 | 6.72 |
| 14 | 4110 | Solid | .0641 | .0641 | .0032 | 2.57 | 2.68 | 4.22 |
| 12 | 6530 | Solid | .0808 | .0808 | .0051 | 1.62 | 1.68 | 2.66 |
| 10 | 10380 | Solid | .1019 | .1019 | .0081 | 1.018 | 1.06 | 1.67 |
| 8 | 16510 | Solid | .1285 | .1285 | .0130 | .6404 | .659 | 1.05 |
| 8 | 16510 | 7 | .0486 | .1458 | .0167 | .653 | .679 | 1.07 |
| 6 | 26240 | 7 | .0612 | .184 | .027 | .410 | .427 | .674 |
| 4 | 41740 | 7 | .0772 | .232 | .042 | .259 | .269 | .424 |
| 3 | 52620 | 7 | .0867 | .260 | .053 | .205 | .213 | .336 |
| 2 | 66360 | 7 | .0974 | .292 | .067 | .162 | .169 | .266 |
| 1 | 83690 | 19 | .0664 | .332 | .087 | .129 | .134 | .211 |
| 0 | 105600 | 19 | .0745 | .372 | .109 | .102 | .106 | .168 |
| 00 | 133100 | 19 | .0837 | .418 | .137 | .0811 | .0843 | .133 |
| 000 | 167800 | 19 | .0940 | .470 | .173 | .0642 | .0668 | .105 |
| 0000 | 211600 | 19 | .1055 | .528 | .219 | .0509 | .0525 | .0836 |
| 250 | 250000 | 37 | .0822 | .575 | .260 | .0431 | .0449 | .0708 |
| 300 | 300000 | 37 | .0900 | .630 | .312 | .0360 | .0374 | .0590 |
| 350 | 350000 | 37 | .0973 | .681 | .364 | .0308 | .0320 | .0505 |
| 400 | 400000 | 37 | .1040 | .728 | .416 | .0270 | .0278 | .0442 |
| 500 | 500000 | 37 | .1162 | .813 | .519 | .0216 | .0222 | .0354 |
| 600 | 600000 | 61 | .0992 | .893 | .626 | .0180 | .0187 | .0295 |
| 700 | 700000 | 61 | .1071 | .964 | .730 | .0154 | .0159 | .0253 |
| 750 | 750000 | 61 | .1109 | .998 | .782 | .0144 | .0148 | .0236 |
| 800 | 800000 | 61 | .1145 | 1.030 | .833 | .0135 | .0139 | .0221 |
| 900 | 900000 | 61 | .1215 | 1.090 | .933 | .0120 | .0123 | .0197 |
| 1000 | 1000000 | 61 | .1280 | 1.150 | 1.039 | .0108 | .0111 | .0177 |
| 1250 | 1250000 | 91 | .1172 | 1.289 | 1.305 | .00863 | .00888 | .0142 |
| 1500 | 1500000 | 91 | .1284 | 1.410 | 1.561 | .00719 | .00740 | .0118 |
| 1750 | 1750000 | 127 | .1174 | 1.526 | 1.829 | .00616 | .00634 | .0101 |
| 2000 | 2000000 | 127 | .1255 | 1.630 | 2.087 | .00539 | .00555 | .00885 |

National Electric Code, copyright 1983, National Fire Protection Association, Quincy, Massachusetts.

[a]Area given is that of a circle having a diameter equal to the overall diameter of a stranded conductor.

The values given in the table are those given in Handbook 100 of the National Bureau of Standards except that those shown in the 8th column are those given in Specification B33 of the American Society for Testing and Materials, and those shown in the 9th column are those given in Standard No. S-19-81 of the Insulated Power Cable Engineers Association and Standard No. WC3-1969 of the National Electrical Manufacturers Association.

The resistance values given in the last three columns are applicable only to direct current. When conductors larger than No. 4/0 are used with alternating current, the multiplying factors in Table 9 compensate for skin effect.

## 6.3 INSULATION

Insulation of the electric current from other systems is an absolute necessity if control of the electricity is to be maintained, and accidental grounding, fire possibilities, and physical damage to people is to be avoided. Fortunately, proper insulation of the wire will eliminate nearly all possibilities of accidents and contain the electrical current within the wire where it belongs. With the exception of low-voltage wire, which does not carry enough electricity to cause damage, all electrical wire is insulated unless specifically permitted otherwise by code. The *National Electrical Code*® (NEC) provides a complete listing of types of materials for different wiring locations together with minimum mil thickness and outer cover requirements. For general information and for normal situations, some of this information is given in Table 6-2.

Although insulated single wires form most of the circuits in any electrical system, the spacing of such wires side by side, each isolated from the other, would be space consuming, expensive, and inefficient. For many residential uses, combinations of two, three, or more wires, each insulated then collected into a bundle and further insulated as a package, may be used. This wire bundle, termed *nonmetallic sheathed cable* or *Romex* (a trade name), often includes a bare ground or neutral wire and in most cases is covered with a plastic outer sheath. Use of this conductor is normally limited to residential construction.

Identification of wire size, type of insulation, maximum rated voltage, and other information is stamped along the wire insulation or on the exterior cover of sheathed cable at intervals of not more than 24 in. In addition to this stamped information, insulation of wire is color-coded to simplify wiring. Neutral or ground wires are coded white, gray, or green. Black or red insulation is used to indicate "hot" wires. Blue insulated wire is used to connect special points in multiple circuits, as is yellow-colored insulation. The NEC also allows a number of exceptions "where the conditions of maintenance and supervision assure that only qualified persons will serve the installation." It is also possible to paint or "otherwise identify" by tags or tape a number of types of metallic covers.

## 6.4 SPECIAL WIRE AND CABLE

In addition to the wire and insulation already described, there are a number of other combinations common to electrical installations. Most prominent of these is metal-clad (type AC) cable, often called BX, which is a spirally formed metal sheath enclosing two or more color-coded wires and a grounding strip, all covered with a fiber or paper wrapping inside the flexible metal cover. Lamp or fixture cords are flexible, multi-stranded, usually with only two wires, enclosed in thermoplastic or braided-cotton fabric.

Heater and power cords are also special types, used on electric irons, toasters, waffle irons, vacuum cleaners, small power tools, and similar equipment. Heater cords, types HPD, HSJ, and HPN, are composed of two, three, or four wires insulated with asbestos, neoprene, or thermoplastic, with a cotton or rayon outer cover. Their use is limited to portable heat units. Type HPD cord is for dry locations and has a familiar black and green or yellow braid cover. A type of power cord termed a "hard service" cord (types SJ, SJT, SO, ST) is composed of two, three, or four No. 18 to No. 2 wires enclosed in thermoset oil-resistant thermoplastic and may be used in damp conditions with hard usage. Ranges and dryers use type SRD or SRDT cable, three or four No. 10 to No. 4 wires with thermopastic insulation.

Elevator power and communication control cable is composed of two or more conductors, wire size No. 20 to No. 14, type E, ED, or EN, insulated with thermoset plastic and covered with several layers of cotton and an outer layer of moisture and flame-retardant neoprene. Underground feeder or branch-circuit cable (type UF) is supplied of No. 14 copper or No. 12 aluminum to No. 4/0, insulated with moisture-resistant rubber and covered

**TABLE 6-2**
Allowable Ampacities for Insulated Conductors Rated 0-2000 V, 110 to 250°C[a]

| Size | 110°C (230°F) | 125°C (257°F) | 150°C (302°F) | 200°C (392°F) | 250°C (482°F) | 110°C (230°F) | 125°C (257°F) | 200°C (392°F) | Size |
|---|---|---|---|---|---|---|---|---|---|
| | Types AVA, AVL | Types AI, AIA | Type Z | Types A, AA, FEP, FEPB, PFA | Types PFAH, TFE | Types AVA, AVL | Types AI, AIA | Types A, AA | |
| AWG MCM | Copper | | | | Nickel or nickel-coated copper | Aluminum or Copper-clad Aluminum | | | AWG MCM |
| 14 | 30 | 30 | 30 | 30 | 40 | | | | |
| 12 | 35 | 40 | 40 | 40 | 55 | 25 | 30 | 30 | 12 |
| 10 | 45 | 50 | 50 | 55 | 75 | 35 | 40 | 45 | 10 |
| 8 | 60 | 65 | 65 | 70 | 95 | 45 | 50 | 55 | 8 |
| 6 | 80 | 85 | 90 | 95 | 120 | 60 | 65 | 75 | 6 |
| 4 | 105 | 115 | 115 | 120 | 145 | 80 | 90 | 95 | 4 |
| 3 | 120 | 130 | 135 | 145 | 170 | 95 | 100 | 115 | 3 |
| 2 | 135 | 145 | 150 | 165 | 195 | 105 | 115 | 130 | 2 |
| 1 | 160 | 170 | 180 | 190 | 220 | 125 | 135 | 150 | 1 |
| 0 | 190 | 200 | 210 | 225 | 250 | 150 | 160 | 180 | 0 |
| 00 | 215 | 230 | 240 | 250 | 280 | 170 | 180 | 200 | 00 |
| 000 | 245 | 265 | 275 | 285 | 315 | 195 | 210 | 225 | 000 |
| 0000 | 275 | 310 | 325 | 340 | 370 | 215 | 245 | 270 | 0000 |
| 250 | 315 | 335 | | | | 250 | 270 | | 250 |
| 300 | 345 | 380 | | | | 275 | 305 | | 300 |
| 350 | 390 | 420 | | | | 310 | 335 | | 350 |
| 400 | 420 | 450 | | | | 335 | 360 | | 400 |
| 500 | 470 | 500 | | | | 380 | 405 | | 500 |
| 600 | 525 | 545 | | | | 425 | 440 | | 600 |
| 700 | 560 | 600 | | | | 435 | 485 | | 700 |
| 750 | 580 | 620 | | | | 470 | 500 | | 750 |
| 800 | 600 | 640 | | | | 485 | 520 | | 800 |
| 1000 | 680 | 730 | | | | 560 | 600 | | 1000 |
| 1500 | 785 | | | | | 650 | | | 1500 |
| 2000 | 840 | | | | | 705 | | | 2000 |

Correction Factors

*For ambient temperatures over 30°C, multiply the ampacities shown above by the appropriate correction factor to determine the maximum allowable load current.*

| Ambient temp. °C | 110°C (230°F) | 125°C (257°F) | 150°C (302°F) | 200°C (392°F) | 250°C (482°F) | 110°C (230°F) | 125°C (257°F) | 200°C (392°F) | Ambient temp. °F |
|---|---|---|---|---|---|---|---|---|---|
| 31-40 | .94 | .95 | .96 | | | .94 | .95 | | 87-104 |
| 41-45 | .90 | .92 | .94 | | | .90 | .92 | | 105-113 |
| 46-50 | .87 | .89 | .91 | | | .87 | .89 | | 114-122 |
| 51-55 | .83 | .86 | .89 | | | .83 | .86 | | 123-131 |
| 56-60 | .79 | .83 | .87 | .91 | .95 | .79 | .83 | .91 | 132-141 |
| 61-70 | .71 | .76 | .82 | .87 | .91 | .71 | .76 | .87 | 142-158 |
| 71-75 | .66 | .72 | .79 | .86 | .89 | .66 | .72 | .86 | 159-167 |
| 76-80 | .61 | .68 | .76 | .84 | .87 | .61 | .69 | .84 | 168-176 |
| 81-90 | .50 | .61 | .71 | .80 | .83 | .50 | .61 | .80 | 177-194 |
| 91-100 | | .51 | .65 | .77 | .80 | | .51 | .77 | 195-212 |
| 101-120 | | | .50 | .69 | .72 | | | .69 | 213-248 |
| 121-140 | | | .29 | .59 | .59 | | | .59 | 249-284 |
| 141-160 | | | | | .54 | | | | 285-320 |
| 161-180 | | | | | .50 | | | | 321-356 |
| 181-200 | | | | | .43 | | | | 357-392 |
| 201-225 | | | | | .30 | | | | 393-437 |



[a]Not more than three conductors in raceway or cable based on ambient temperature of 30°C (86°F).

with a flame-retardant, moisture-, fungus-, and corrosion-resistant material, Special-purpose cords include a wide variety of multiple-wire, solid or stranded, shielded or unshielded, flat or round, plus cord manufactured for thermostats, TV lead-in, intercom systems, coaxial cables, computer circuits, and many other wiring systems necessary to complete buildings.

## 6.5 CONDUIT AND RACEWAYS

*Conduit* is a type of tubing or pipe which is used to enclose and protect electrical wiring, and includes rigid steel conduit, rigid aluminum conduit, thin-wall steel conduit (EMT), flexible steel conduit ("greenfield"), and PVC plastic conduit. *Raceways* include surface-mounted box-type units of metal or nonmetal, underfloor raceways, sheet-metal or expanded metal troughs (wireways), factory-fabricated busways, and several other surface-mounted wire enclosures. Both conduit and raceways provide an opportunity to install individual wires required for various circuits in accordance with the National Electrical Code at a location and time more convenient to the progress of the construction.

Rigid conduit or pipe is similar to ordinary water pipe, formed of steel or aluminum, in sizes from ½ to 6 inches, with threaded connections, and available in 10-foot lengths. Most rigid conduit is used for service entrances or connecting panels to other panels, either vertically or horizontally. Thin-wall steel conduit (electrical metal tubing), in sizes from ⅜ to 4 inches, is fabricated with thin walls to allow bending so is not threaded but connected with a variety of friction-set or screw-set fittings. Flexible steel conduit is similar in construction to BX but without the wires, is available in sizes from 5/16 to 2 inch, and is used in combination with conduit where sharp bends or difficult offsets are encountered. Some areas now allow conduit made from polyvinyl chloride (PVC) plastic, in sizes from ½ to 4 inches, in lengths of 10 to 20 feet, to be used in lieu of metal conduit.

Raceways are of two major types: (1) surface mounted and exposed, or (2) buried in floor construction and thus concealed. Surface raceways are for use in dry locations and are usually fabricated in two parts, the base and the snap-on cover. Sizes vary from the "wiremold" size of ¾ inch to boxes designed to carry several No. 4/0 cables. Many forms are available, including prewired systems, plug-in strips, and those designed for floors or ceilings. Most under-floor ducts are made of metal or grounded plastic and are buried flush in concrete floors with removable plates or recessed outlet covered with threaded brass disks. This type of under-floor system is much used to provide electrical or communication accessibility at regular intervals throughout office spaces or similar locations. For feeder wires between panels or from controls to machinery where a number of large cables are involved, a cable tray may be suspended from the overhead and the cables simply laid in the tray alongside each other and exposed.

## 6.6 BOXES AND FITTINGS

Conduit and wiring terminates either at a panel or in a "box" housing a switch, light fixture connection, convenience outlet, wire junction, or some similar end fitting (see Fig. 6-2). Boxes are made of galvanized steel, usually 14 gauge or nonmetallic (plastic), and are available in four primary shapes: square, rectangular, octagonal, or round. Circular cuts in the back and sides

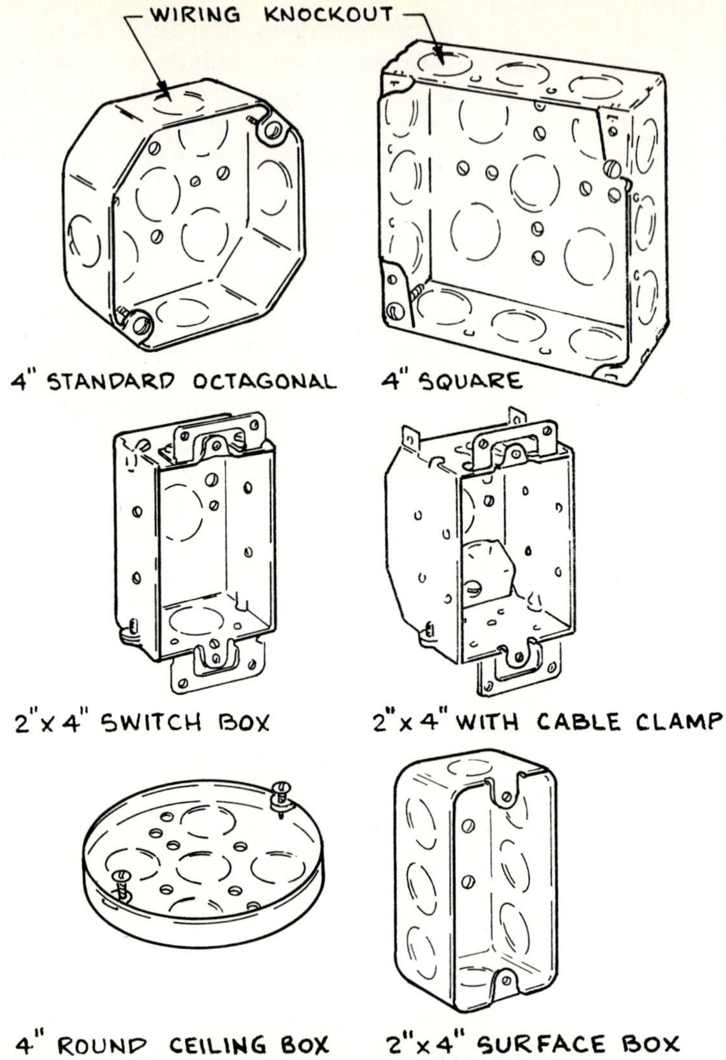


**Figure 6-2** Wiring boxes.

of boxes provide "knockout" holes when the enclosed disk is punched out, thus allowing conduit connectors to be attached to the box. "Ears" and punched holes provide possibilities for attachment to wall studs or other construction, and other threaded ears allow for attachment of switches, convenience outlets, and other devices contained within the box. Systems using steel conduit and steel boxes provide an automatic ground, but an additional grounding system is required when nonmetallic cable (Romex) is used or when boxes are plastic.

Switches of various types (see Fig. 6-3), convenience outlets, and miscellaneous other fittings are attached to, and enclosed by, the terminal boxes. A single-pole switch, with one "hot"and one neutral terminal, is used to control one light or other appliance from one location. Similar switches, with more terminals, are available for "three-way" circuits, in which two switches in different locations are wired to control the same outlet, or for "four-way"circuits, in which three switches in different locations are wired to control the same outlet. The most common switch type has a spring-actuated toggle to provide ON or OFF conditions, but similar switches may accomplish this by "touch plates" or silent mercury contacts. Switches are

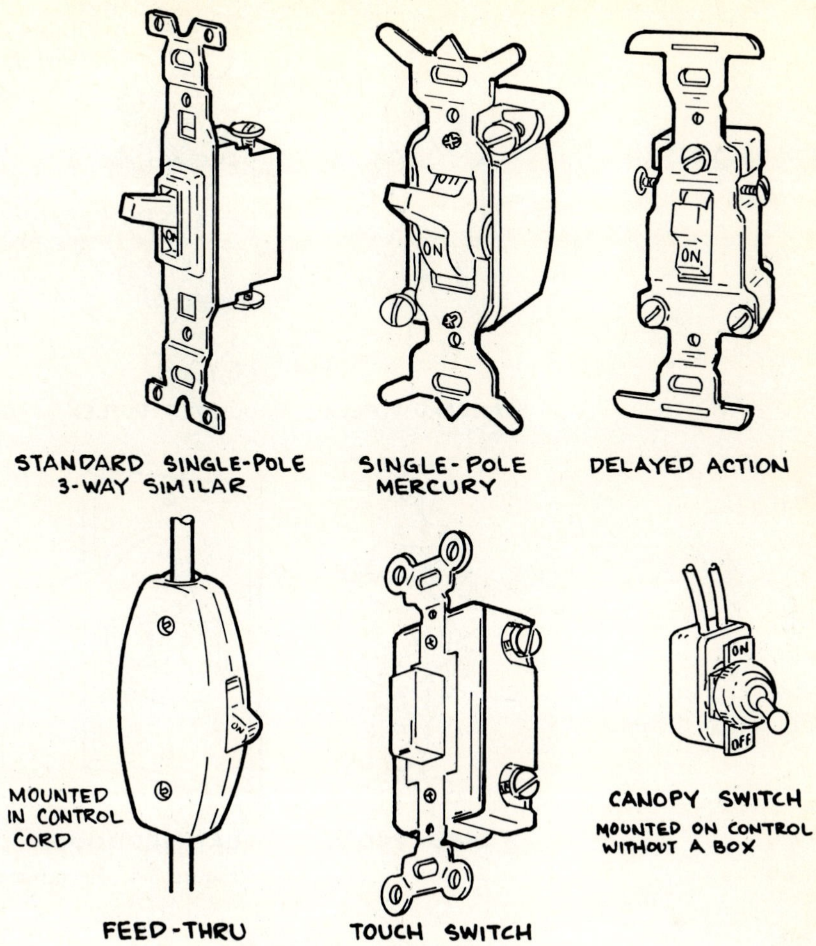


**Figure 6-3** Types of switches.

attached to the boxes with screws connecting the ears of the switch unit with the threaded lugs of the box.

Convenience outlets are also installed in boxes. These terminals are used to plug in portable equipment such as lamps, radios, television sets, kitchen appliances, and similar devices. The outlet is most often supplied with the capability to accommodate two plug-in devices (duplex outlet) and without a grounding connection. Where grounding is advisable as a safety precaution with power tools or some other appliances, a grounding plug is incorporated in the outlet to receive the grounding connection from the connecting cable. Outlets from 220- to 240-V systems are usually single receptacles with a number of different configurations, as shown in Fig. 6-4. Other special outlets are used for self-closing openings, weatherproof outlets, and connections for telephone or communication systems.

Cover plates for various single or combinations of switches, convenience outlets, blank covers, and dimmers are shown in Fig. 6-5. Plastic, either ivory or brown in color, is the most common material used, although stainless steel, ceramic, or plain galvanized steel is also available. Covers are attached with screws to the switch or outlet. Where lighting fixtures are attached directly to the box, no cover plate is required, but codes require that all other boxes be finished with a cover plate. Junction boxes (J boxes), which simply enclose wire connections, or junctions, must be mounted so that the blank cover plate is visible and readily accessible.

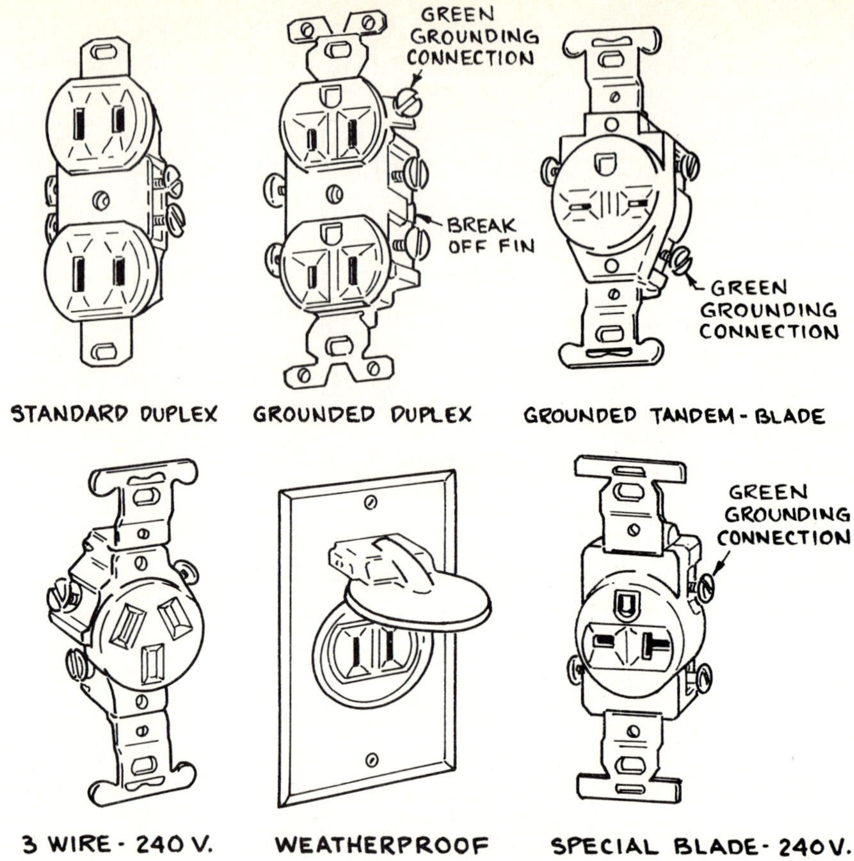


**Figure 6-4** Receptacles.

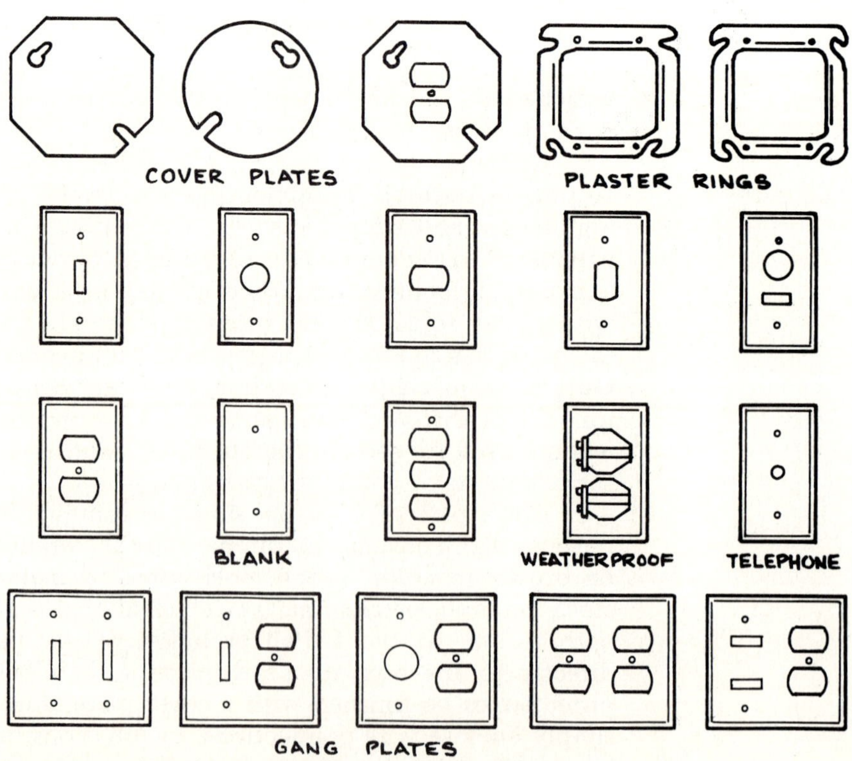


**Figure 6-5** Outlet plates.

## 6.7 PANELS

Panels, in their complete sense, cover a wide variety of control units that regulate or protect electrical circuits from entrance service through the distribution system throughout the building. Entrance service "panels" may be little more than a knife switch to complete the service connection or, more often, an enclosed and fused switch or circuit breaker. Depending on the type of building or occupancy, the following control panel may be of considerable size to distribute and control large quantities of electricity to other distribution stations, or may be simple multicircuit residential panels including separate fuses or circuit breakers. Switchboards, according to the National Electrical Code, are "large single panel, frame, or assembly of panels on which are mounted . . . . switches, overcurrent and other protective devices." Such equipment requires dry rooms for installation and handles initial distribution of electricity for lighting or power. Panels (load centers, panel box, fuse box, breaker panel) in their customary sense are for the protection of circuits and are of three configurations: (1) knife disconnect, (2) pull-out block disconnect, or (3) multiple breaker.

Service entrance panels are designed to shut off *all* electricity instantly. This condition is desirable in case of fire or when work is being done on the electrical system. Older "main disconnects" are often metal boxes enclosing a single-throw fused knife switch connected to an exterior lever or handle. When the switch is in the ON position, the "hot" service is connected by the knife blades, but pulling the lever to the OFF position disconnects *all* electricity. Newer construction may use a pull-out block disconnect, in which a fused block is physically pulled from the panel to disrupt the electrical circuit. A multiple-breaker panel (Fig. 6-6) is usually used to protect and dis-

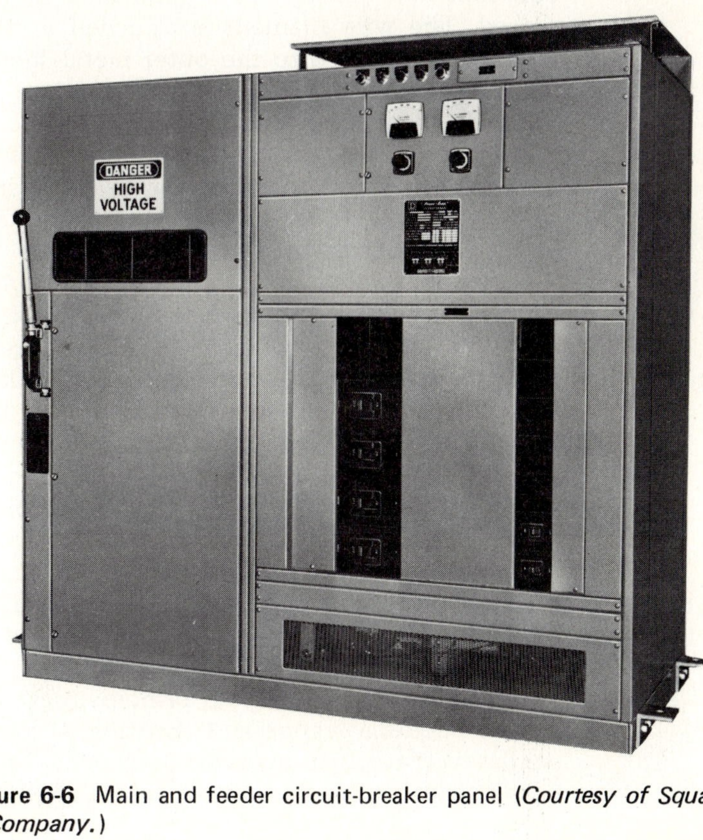


**Figure 6-6** Main and feeder circuit-breaker panel (*Courtesy of Square D Company.*)

tribute various circuits and may have a number of separate circuit breakers of different capacities to protect both 110- to 120- and 220- to 240-V lighting or appliances. Circuit breakers are separate units operated by internal springs and "fused" with a bimetallic strip that automatically disconnects the "hot" wire upon overload. After the trouble has been located, the switch toggle is flipped ON to restore power. Circuit breakers are ganged in distribution panels.

In many older or smaller residences, the National Electrical Code "rule of six" applies when not more than six circuits are used. This situation allows for a panel with six or fewer circuits and *no* main disconnect, on the theory that so few circuits could be manually disconnected in a very short time. Large installations or multistory buildings usually have a main-disconnect switch close to the transformer vault or "engine room," with a switchboard arrangement to divide the power into several systems serving different floors or facilities. Each floor, group of floors, or facility (kitchen, elevators, shops) then has a disconnect for *all* electricity plus circuit-breaker panels for various circuits. A somewhat similar situation may apply in large residences, where one or more subpanels are used to reduce the length of "homeruns" for circuit wiring.

## 6.8 LIGHTING

The principles of electrical lighting are discussed in Chapter 10, but a quick review of materials or types of lighting fixtures may be useful at this point. Three major types of lighting are available: (1) incandescent, (2) fluorescent, and (3) vapor type (see Fig. 6-7). *Incandescent* lighting is by far the most common and consists of a glass bulb in which a fine wire or filament is placed. The wire filament is attached to the base so that one end of the circuit is connected to the outer metal shell and the other end terminates at a contact at the center bottom of the base. Several types of screw bases are available, depending mostly on the bulb capacity (watts) and the use. The glass bulb is fused to the base and a partial vacuum created in the bulb. When current is applied, the filament heats to incandescence and a quantity of light is provided. Some clear glass lamps are still available, but "frosted" lamps give a better distribution of light.

Tubular lights are most often of the *fluorescent* type and are available in various wattage ratings from 10 to 96 in. long and from 3/8 to 1¼ in. in diameter. The most common type of fluorescent lamp is a straight tube, but circular tubes are also manufactured. The lamp is a glass tube coated on the interior with chemical phosphor compound, which will radiate light. Argon gas and mercury vapor replace the air contained in the cylinder, and each end is capped with a fitting, including an oxide-tungsten elment in the tube and contact pins outside the tube. When current is applied, a cloud of electrons is developed around each end electrode, and an electron arc develops. The electrons collide with the mercury and argon atoms, filling the tube with invisible ultraviolet rays, which in turn excite the phosphor coating to visible light. Current is controlled by a ballast and coil, which steps up the voltage and limits the current.

*Mercury-vapor* lighting, principally for exterior street and floodlights, is a bulb enclosing vapor and emitting a yellow tinted light. The reaction is similar to that which operates fluorescent lighting, and it reduces the amount of electrical current required and therefore operates at lower cost. The

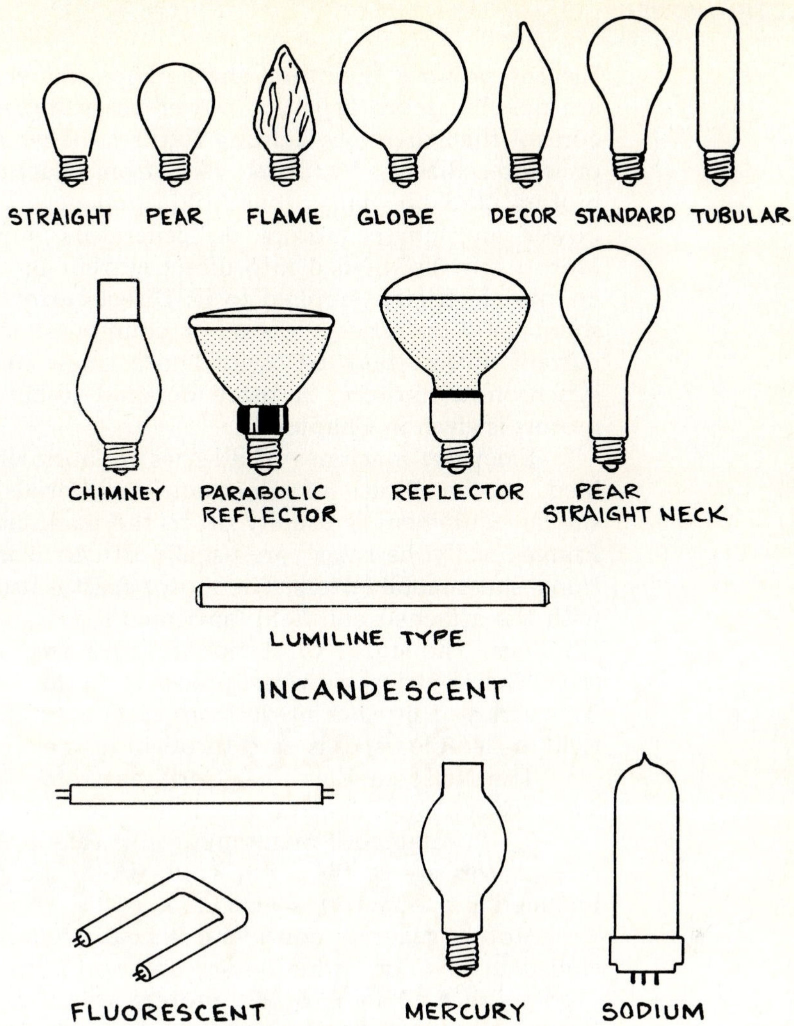


**Figure 6-7** Typical lamp shapes.

yellow-tinted light emitted may be a little strong until one's eyes become accustomed to the color. Lamps used for street lighting may also be wired differently, as much street lighting is of the series-circuit type to save wire, even though it has the disadvantage of opening the circuit if one lamp burns out. This disadvantage may be countered by automatic cutouts in the lamp.

Lighting fixtures, which hold or contain these various types of light-emitting units, are probably more numerous in type and design than any other piece of equipment installed in any building. The simplest fixture is a metal socket on a drop cord or a porcelain receptacle covering an outlet box, each accommodating a signle unshaded bulb. Fluorescent surface-mounted fixtures are generally used in kitchens, bathrooms, and similar locations in residences and in general lighting in offices and shops. Although fluorescent lamps do not emit much heat, the ballast or coil transformer requires adequate ventilation to prevent damage. Residential lighting fixtures are somewhat restricted as to size, weight, and number of lamps, but commercial fixture design goes from prosaic "workhorse" types to "the sky is the limit" and may include thousands of lamps in a single fixture. Selection of fixtures is based on certain specified models for each location or on an "allowance" in the construction contract which allows the owner to select desired fixtures within an agreed dollar amount.

## 6.9 ELECTRIC POWER

Electric power is the other, and perhaps greater phase of electrical use. There are possibly more types of power usage, types of motors, and methods of control than there are lighting fixtures, of which there are several thousand or more. Sizes of motors vary from fractional-horsepower single-phase motors used on many household appliances to heavy-duty multiple-horsepower multiphase motors on generators, pumps, and other equipment. Motors may be divided into direct current or alternating current, depending on the electricity supplied to it. Direct-current motors may be classified as shunt motors, series motors, or compound motors. Similarly, alternating-current motors may be divided into single phase, polyphase induction, or synchronous motors. A more detailed discussion of the various types of motors is given in Chapter 11.

Electrical machines of all sizes are normally supplied completely assembled, so the choice of construction materials for motors, generators, and similar equipment is usually up to the manufacturer, not the owner or user. Frames or "yoke rings" are usually of cast iron with integral covers or additional sheet-metal covers. The motor shaft is usually of precision-milled steel, with the attached coil field fabricated from laminated steel with copper-wire windings. The stator or stationary coils are manufactured of similar materials. The electrical power supplied to motors is transferred to the coil fields by a series of brushes made from carbon or metal, and insulation of various rigid or flexible types is used throughout the motor.

The National Electrical Code has some very specific requirements for the location and control of electric motors. The type of motor may be "open," "protected," or hermetically sealed. Most motors used in unit-type refrigerators are of the latter kind, where the motor and the compressor are included in one factory-sealed case.

Motors generate considerable heat when in operation and should be installed in positions where adequate ventilation and maintenance is readily accomplished. Where "open" motors are used, protection from sparks from the commutator must be provided. In severe conditions, motors may require complete enclosure to protect them from moisture or weather conditions. Enclosed pipe-ventilated motors may also require forced air or fans and in some special conditions may be enclosed in dust-tight rooms, properly ventilated and controlled remotely. Control of motors is discussed in Chapter 12; in general, however, the NEC requires that where one piece of equipment is "in sight from" the other piece (motor and controller), the maximum distance must not be more than 50 ft. All motors must be labeled, indicating the manufacturer's name, rated voltage and amperes, number of phases, rated speed, horsepower, and a variety of other data.

## 6.10 LOW-VOLTAGE MATERIAL

Low-voltage electrical systems, as mentioned previously, are generally considered to be between 6 and 24 V, direct current provided by batteries or alternating current provided by transformers. Transformers are fabricated from laminated sheets of steel that form a core around which copper wire is wound. Transformers for low voltage are relatively small, and the cases are customarily of pressed sheet steel.

Most other parts of low-voltage systems are also provided completely assembled. Pushbuttons, door bells, chimes, and other controls have metal or plastic cases, different metals for contacts or other parts, and of course, paint or plated finishes. Wire for low voltage is normally smaller than No. 18, bare or cotton-insulated, and installed without conduit or other protection. In a very few cases, emergency lighting may be low voltage, but such a system also does not require the same equipment as that required for "line" voltage lighting.

## 6.11 SOUND EQUIPMENT

The material used to install various types of sound systems, television service, telephones, and related equipment combines many of the materials outlined for use in other electrical installations. Each system is different, mostly because the manufacturer of the basic equipment designs equipment differently. Conduit is often preferable for wire runs even if not actually required, because a slight change may require more, or fewer, wires, or a different type. The simple sound system may be "designed" by the owner or the store providing the equipment, and may function properly or not. For more extensive systems, the manufacturer or authorized representatives need to be available.

## 6.12 DRAWINGS AND SYMBOLS

Electrical drawings are no different from those for other parts of a building. They show, in an abbreviated form and with the use of symbols (see Fig. 6-8), the general location, and perhaps the hookup, of the electrical service within the building. Symbols are used to indicate the various lamp locations, switch locations, panels, and other equipment. In commercial drawing, the conduit is usually shown with markings to indicate the size and the number of wires included within each conduit, and motors and the like are more definitely

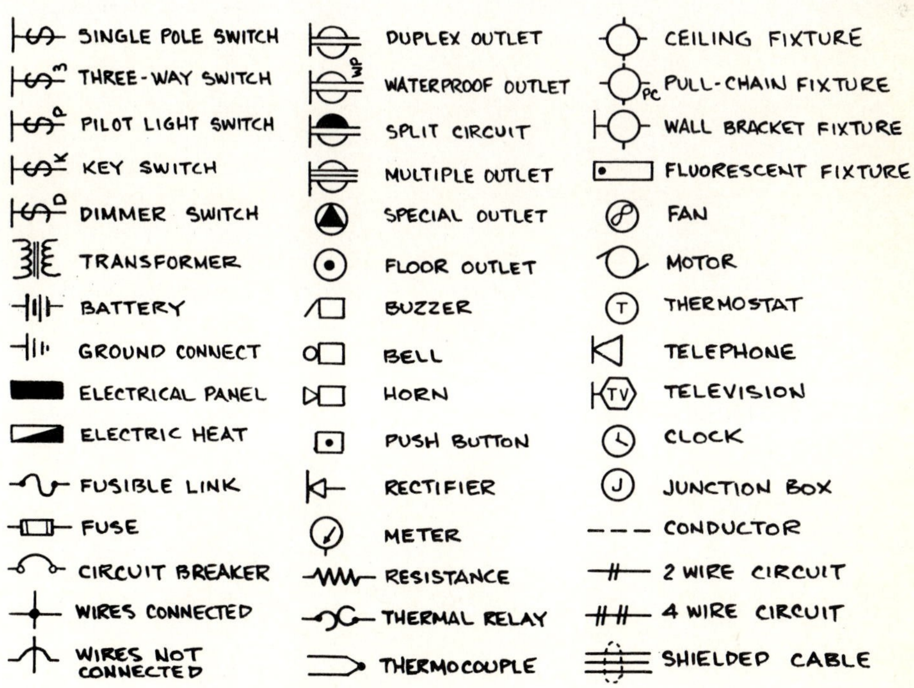


**Figure 6-8** Electrical symbols.

marked. Most residential electrical drawings simply indicate the location of the fixtures, switches, panel, and convenience outlets, and the actual wiring is installed in whatever manner the electrical installer elects—in compliance with the codes, of course.

## 6.13 SUMMARY

Much of the actual material used for electrical installations is familiar to people in a vague way because they see it every day even if they do not appreciate its function. Lighting and power controls turn lights and appliances ON or OFF, and kitchen appliances, heating units, radios, television sets, telephones, and a myriad of other electrical devices are used without much concern about the actual manner in which electrical power is supplied. But the average person knows little or nothing about the hidden parts of electrical installations. Conduit, wire, panels, and similar parts or materials are often not even very well known to construction personnel other than electricians.

As mentioned at the beginning of this chapter, practically all of the materials used in electrical work are controlled by codes. Little personal opinion regarding what material to use where is allowed to the electrical installer beyond the fact that certain materials or equipment may be used in residential work, while other materials or equipment must be used in commerical installations. A major factor that determines the material or construction of electrical devices is that of conductance or insulation. Electric current should be transferred as easily and efficiently as possible, mostly through certain metals, and needs to be protected from accidental grounding by good insulation of various kinds. Many materials and combinations provide these desirable and required characteristics.

# 7

# Electrical Generation

## 7.1 INTRODUCTION

Electricity is generated from two major sources: batteries and machines. Batteries may be dry cell or wet cell. In a *dry-cell battery*, a dry or moist condition exists to provide electricity, whereas in a *wet-cell battery*, a liquid (usually water–acid solution) is used to establish electrical properties. Batteries are used extensively to supply electricity for door bells, chimes, alarm systems, communication devices, and similar equipment included in support systems for buildings, and in larger sizes in submarines, locomotives, automobiles, and trucks. Batteries supply direct current (dc) for limited use and require considerable space. In addition to limited voltage, batteries are too expensive to be used for lighting, heat, or power, and either wear out or must be recharged periodically.

Generating machines, which may be used to provide direct current (dc) in greater quantity and higher voltage, are termed *dynamos*, or direct-current generators. Similar machines used to generate alternating current (ac) are generally termed *alternators*. Dynamos, as well as alternating-current alternators, have only a few essential parts: (1) a frame, (2) field magnets, (3) a rotating armature within the field magnets, (4) a commutator attached to the armature shaft, and (5) a system of brushes or slip rings to pick up the electricity from the commutator. An electrical generating machine produces an electromotive force (voltage) that is developed by electromagnetic induction by rotating an armature winding (coil) within a magnetic field. The resulting electric voltage is collected on a commutator and released to wiring through brushes or slip rings as the armature and commutator revolve. The entire generating principle is that the coils mounted on the armature core should cut the greatest number of lines of force of the field coil as possible. The greater the number of lines of force that are cut by the coils, the greater the voltage and current flow. Where direct current is required, the current generated by an alternating-current machine may be converted to direct current by use of the alternator-rectifier method or simply by rectification.

## 7.2 BATTERIES

Batteries are usually made up from a series of cells, each of which is composed of a positive and negative terminal and electrodes that are suspended in a chemical solution called *electrolyte* (see Fig. 7-1). Several chemical combinations are used for electrodes, such as a negative zinc plate with a positive carbon plate and with ammonium chloride used as the electrolyte, lead negative plate with lead peroxide positive plate and sulfuric acid as the electrolyte, or iron oxide negative plate with nickel positive plate and potassium hydroxide as the electrolyte. The electromotive force (EMF) is the same for any size of cell made of the same materials, so the size of the electrode is rather immaterial except for convenience in handling or usage. As the size of the electrodes, to a limited extent, determines the amount of current produced, electrodes with small areas may create a smaller amount of current than that created by those with larger contact areas.

Connection of the cells within the battery will determine the amount of voltage as well as the amount of current that may be drawn from the battery. In Fig. 7-2a, the cells are connected in series; that is, the positive electrode of one cell is connected to the negative electrode of the next cell. With this connection, the voltage is accumulative. If we assume that each cell produces 1 V, the total voltage will be 3 V. If the cells are connected in series, the internal resistance is also in series, so the amount of current (amperes) supplied is only the amount that one cell may produce. If the cells are connected in parallel as shown in Fig. 7-2b, all the positive electrodes are joined together and all the negative electrodes are joined together. This will provide only the voltage that one cell may produce, but the current will be determined by the total internal resistance divided by the number of cells, or one-third the total internal resistance, or three times the current one cell may produce. Where a larger voltage and a greater current are desired, a number of cells may be connected in series for greater voltage and these batteries connected in parallel

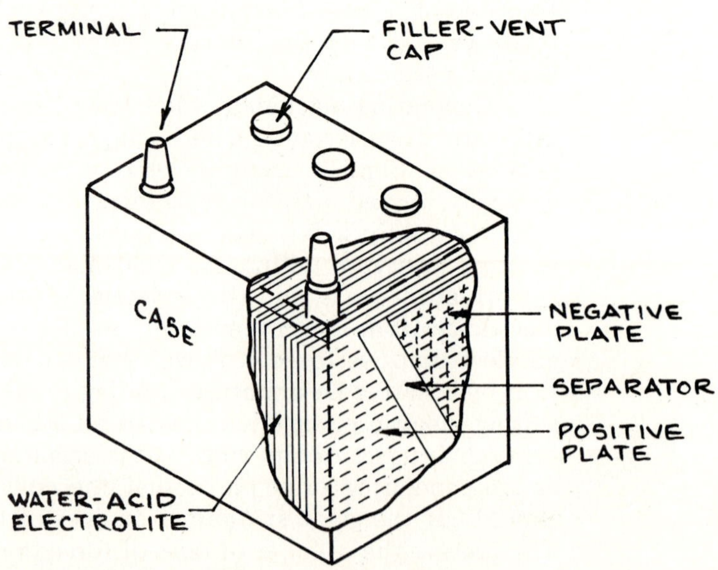


**Figure 7-1** Wet storage battery.

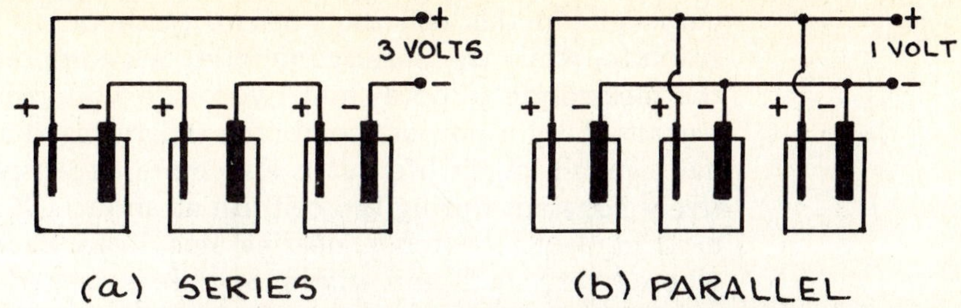


**Figure 7-2** Cell connection. (a) Series; (b) parallel.

for greater current. This type of arrangement is termed a *series-parallel battery*.

A serious disadvantage of a wet (voltaic) cell is that the electrolyte is a liquid and may easily spill. Since this electrolyte is also an acid-water solution, it will damage almost any material on which it is spilled. In addition, the weight and size of the plates, case, and electrolyte are considerable and may pose a problem in handling. Because they are cheaper and more portable, a type of dry-cell battery is produced that does not require a liquid and that comes in many sizes and shapes to fit different uses. Essentially, this battery (see Fig. 7-3) is a zinc can that is further contained by, and makes electrical contact with, an outer metal shell forming the negative terminal (anode). The zinc can contains a carbon rod surrounded by a paste or gel composed of ammonium dioxide, manganese dioxide, or zinc chloride-ammonium chloride plus carbon black and various fillers. The paste or gel is separated from the zinc can by a plastic film. The carbon rod is insulated from the external metal can and becomes the positive terminal (cathode).

When connected into a circuit, the dioxide gel loses some of its oxygen, thus becoming positively charged, and the zinc can is oxidized and becomes negatively charged. The carbon black in the gel acts as a conductor, thereby reducing the internal resistance of the battery. Regardless of the cell's size, the voltage will be approximately 1.5 V. The amount of current depends on the internal resistance, so a larger cell, with more active material, will function for a longer period than will a small cell.

If the positive element is of manganese dioxide and the anode (negative) element is zinc and the electrolyte is a paste of potassium hydroxide, the cell is an alkaline-manganese cell or simply an alkaline "battery." The voltage is approximately the same (1.5 V), but the internal resistance is much less, so the cell has greater life expectancy. A third type of primary cell is the mer-

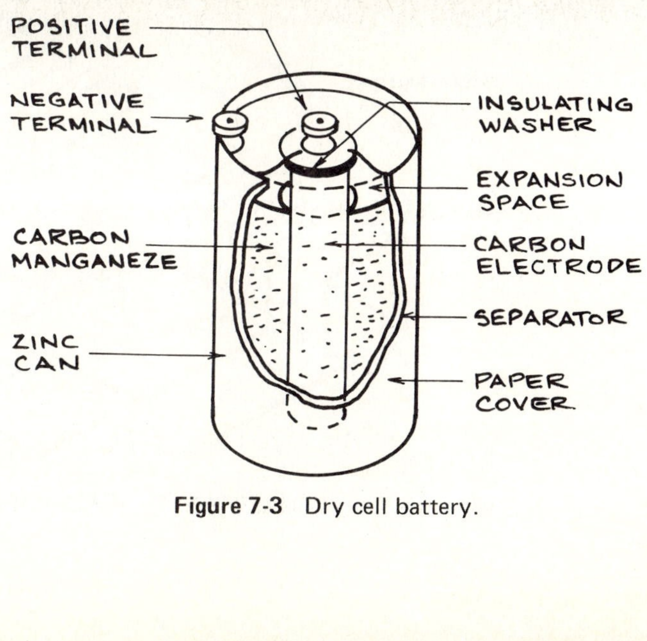


**Figure 7-3** Dry cell battery.

cury cell. In this cell the positive (cathode) is mercury oxide mixed with graphite, while the negative (anode) is a compound of zinc and mercury and the electrolyte is potassium hydrozide with zinc oxide. Passage of ions is permitted by a porous membrane separating the elements. This type of cell has a much longer life but is also more expensive. The voltage remains relatively constant during the cell life at about 1.3 V. Mercury cells are widely used in electric watches, hearing aids, heart pacemakers, and similar devices where a very small unit with a long life is desired.

## 7.3 ELECTRICAL GENERATING PRINCIPLES

An electric generator produces electromotive force (voltage) by electromagnetic induction, which involves the cutting of magnetic lines of force in a magnetic field by revolving coils across this field. This movement produces a potential difference that causes one end of the coil to be negative and the other end to be positive. When the rotating armature coil moves through one-half the cycle, voltage is induced and flows to one terminal or brush. As the rotation continues, the opposite magnetic fields are cut and voltage flows to the opposite brush. This creates an alternating current, which means that the flow of current moves first in one direction, then as the armature revolves, flows in the opposite direction. All machine-generated electricity is first alternating current but may be changed to direct current by the use of commutators. To understand this entire process fully, consider a very elementary type of generator, as shown in Fig. 7-4. The generator field is an electromagnet wound with a coil. In actual construction, this core may be many layers of thin sheet iron or steel, each insulated from the other, wound with many turns of wire. This construction reduces resistance and therefore reduces heat in operation. For illustrative purposes, consider the armature

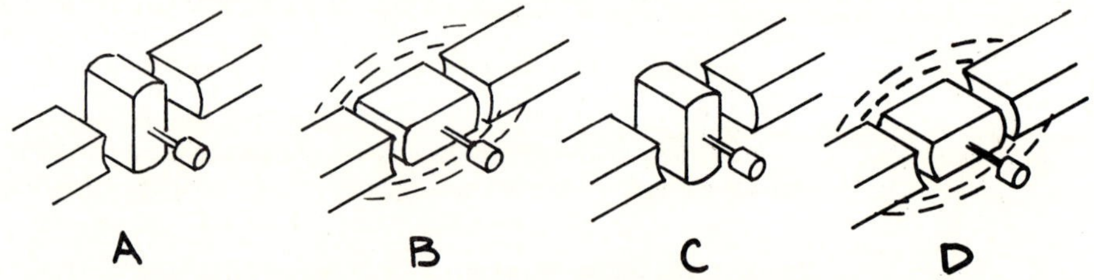


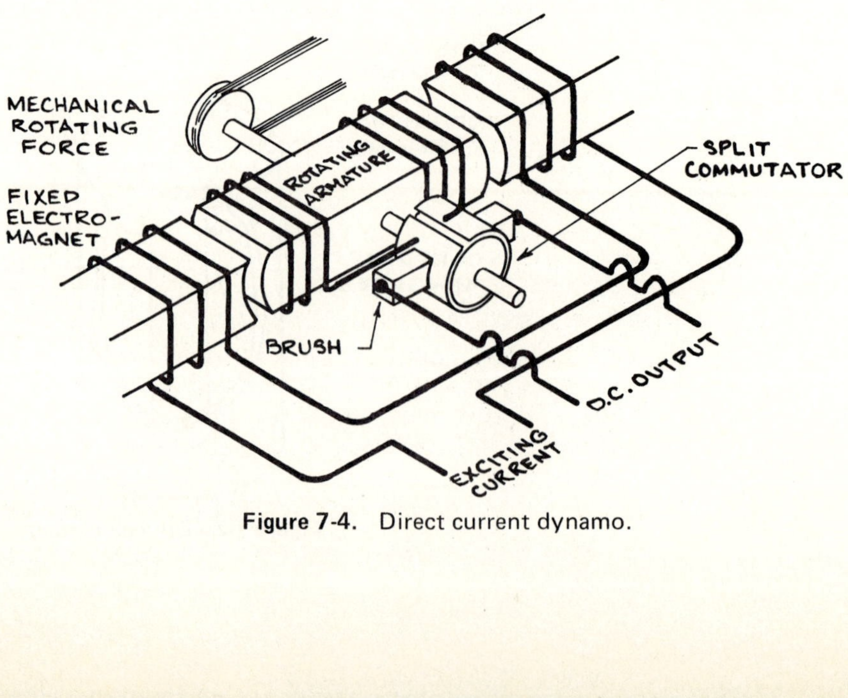


**Figure 7-4.** Direct current dynamo.

coil as a single wire shaped as shown. This is the armature and is designed to be mechanically rotated between the north and south poles of a field magnet. The end wires terminate in slip rings, which are insulated from each other and from the armature shaft on which they are mounted. Stationary brushes pick up the charge from the slip rings and transfer it to the wire circuit.

Assume that the armature coil is in the vertical position shown at Fig. 7-4a. In this position, no lines of force between the field coil poles are being cut, so no electromotive force is generated. As the coil turns 90° to the position shown in Fig. 7-4b, it cuts lines of force and the electromotive force travels through the armature coil to the collector ring or the negative terminal. When the armature continues to turn another 90°, the electromotive force is again reduced to zero. With another 90° turn, the electromotive force is again maximum, but note that now the flow is to the collector ring or the positive terminal. Another 90° turn brings the armature back to the original position and again to zero inductance.

Graphically, this can be shown as in Fig. 7-5. As the armature revolves, the force (volts) increases to the peak of the sine curve, then decreases as rotation continues, and finally returns to zero. Continuation of rotation again increases the force to maximum but to negative instead of positive, and then back to zero. The time scale can be related to the four quarters of time that are required for a full cycle. In actual practice, the rate of rotation is normally 60 times per second, termed 60 cycles or 60 hertz (Hz).

The armature must be turned mechanically by steam turbine, hydraulic power, or other source of energy. The field coil is excited by separate direct-current sources or self-excited by windings of the field coil called *shunt windings*. Voltage may generally be controlled by more windings, greater speed of armature rotation, or both.

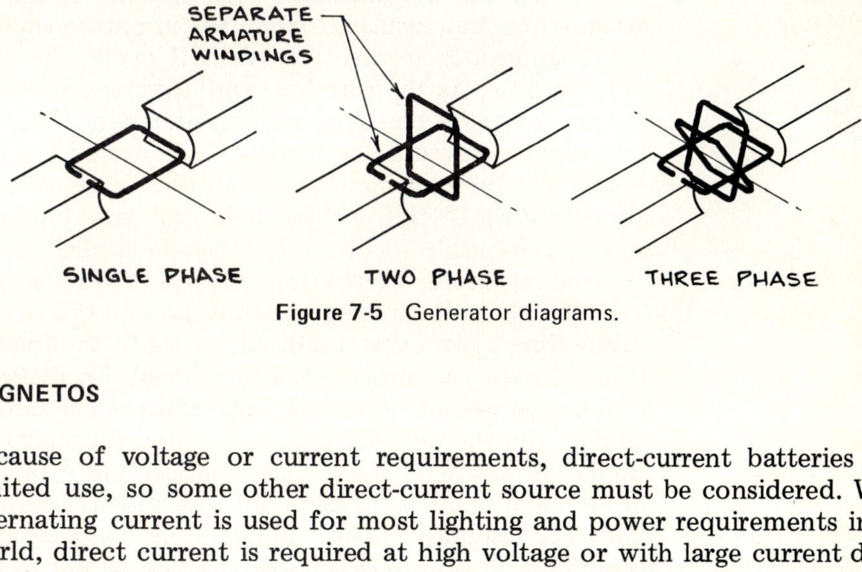


**Figure 7-5** Generator diagrams.

## 7.4 DIRECT-CURRENT MAGNETOS

Because of voltage or current requirements, direct-current batteries have limited use, so some other direct-current source must be considered. While alternating current is used for most lighting and power requirements in the world, direct current is required at high voltage or with large current draws for electroplating and various other processes. The *magneto* is an example of a fixed permanent-magnet machine which is capable of generating high voltage for ignition purposes in motorcycle, boat, and airplane engines. The magnetic field is composed of several permanent horseshoe-shaped magnets arranged side by side (see Fig. 7-6) so that all north poles are together and all south poles are together. Within the magnets, an armature of many turns of insulated wire wound on an iron core is turned at high speed through a system of gears. Electricity is picked off of this armature by a slip-ring collector and proper brushes.

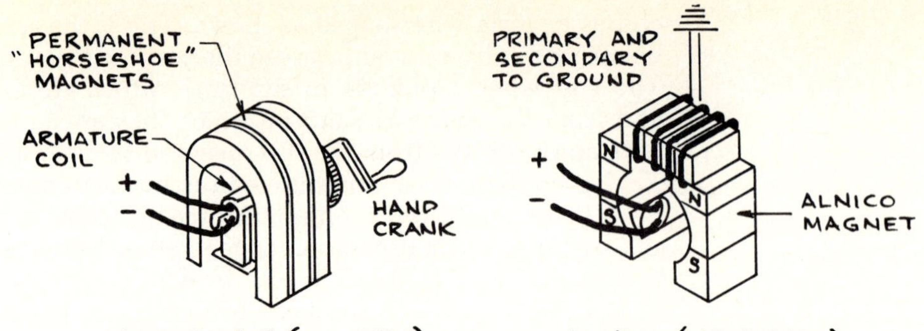


**Figure 7-6** Magneto.

### 7.5 DIRECT-CURRENT DYNAMOS

Where large amounts of current are required, as for lighting and operating machinery, another source of direct current is necessary. In these cases, electromagnets are used in place of permanent magnets to provide a stronger magnetic field and to provide more control of the electromagnetic field by adjusting the strength of the current passing through it. This control allows the voltage output of the dynamo to be varied as desired.

The direct-current *dynamo* makes use of an electromagnetic field within which a wire-wound armature rotates. As the armature rotates, it cuts the force fields of the opposing magnetic poles, creating a current of electricity that is transferred to the commutator, which is then further transferred by brushes pressing against the commutator surface. The commutator is a circular split ring divided into opposing parts, each of which is attached to one end of the armature winding. The entire commutator assembly is attached to, but insulated from, the armature shaft. When the armature and the commutator revolve for one-half cycle, the current is directed to the positive pole. As the rotation continues, the direction of current is reversed, but at the same time, the split commutator transposes the connections between the rings and the brushes to continue the flow through the positive brush. This process continues with each half-cycle, so the generated current always flows to the positive pole and into the circuit. This current flow, shown graphically in Fig. 7-7, is known as direct current, as it is always positive and does not alternate from positive to negative as does alternating current.

If only one winding or coil is used on the field and the armature, a pulsating direct current is produced, owing to the fraction of time in the revolution between cuttings of the force field. To overcome this, and to smooth out the pulsations, more windings and their commutator split rings may be added. By winding the armature coil continuously on a ring in the same direction, and with the ends connected and commutator connections made at an equal number of turns, it is possible to accommodate almost any number of commutators. This will cause the armature current to flow in opposite directions on opposing sides of the ring winding. The junction points of the windings are known as *neutral points*. The neutral points lie midway between the field poles and correspond to points on the armature windings where no lines of force are cut and thus no voltage is produced. The current in all cases flows to the negative neutral point through brushes set to contact at these neutral points.

A more common construction is the *drum armature* (Fig. 7-8), where the windings are formed in slots milled into the iron core. These coils lie flush with the adjacent surface of the armature and are connected in series.

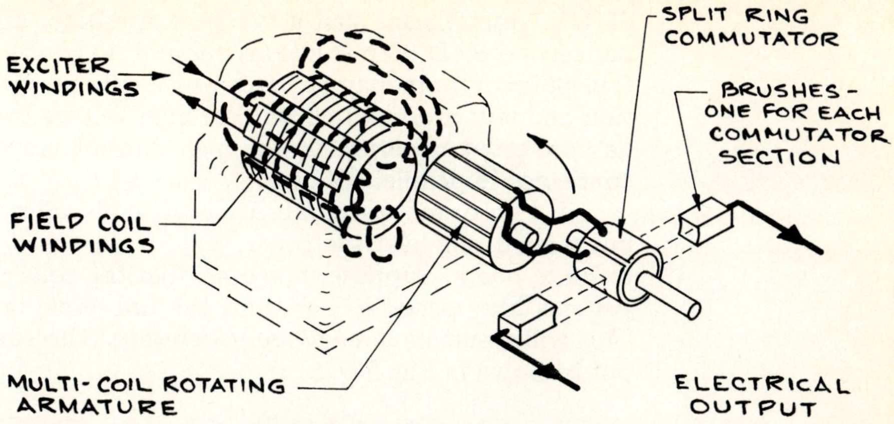


**Figure 7-7** Simple D.C. generator.

Each conductor to the commutator is made with two wires from adjacent coils. In this system the connections and operation of the commutators are similar to those of the ring armature, but there are no shielded portions. Because of the method of winding the armature coils, one-half of the windings are charged in opposite directions, and as the drum revolves, a coil is at the top and bottom neutral points. This allows a planned short circuit of the bars and eliminates sparking and excessive wear on commutator and brushes. An increased number of poles will reduce the size of the wire required to carry the current load. Generators for automobiles and similar equipment are usually two-pole, but those that handle larger current loads may have four to six poles. Despite the increased number of poles, a slight ripple may be apparent in the output produced, so filters consisting of capacitors and inductors connected across the positive and negative wires are often used to smooth out these fluctuations.

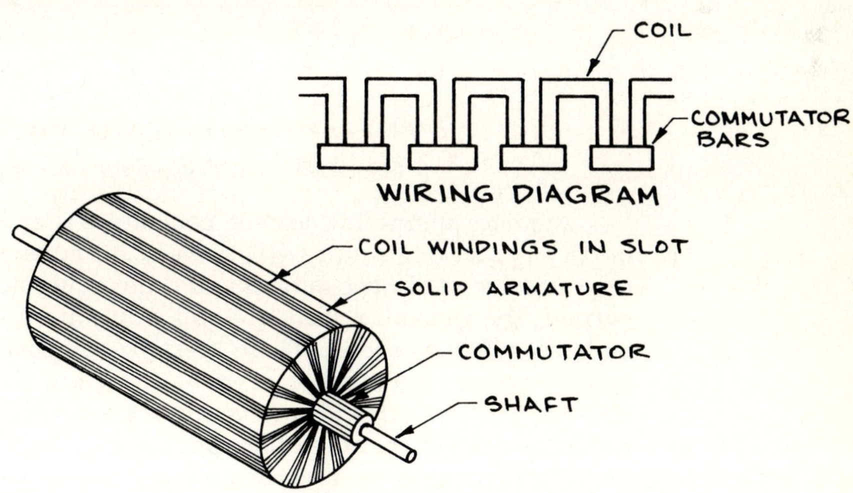


**Figure 7-8** Drum armature.

## 7.6 ALTERNATING-CURRENT GENERATORS

Alternating-current *generators* or *alternators* are constructed and operate in a manner similar to that described for direct-current dynamos except that the generated force is retained as developed, reversing each one-half cycle, or more often if more poles are used. Most of the alternating current used is at

60 Hz, which means that a two-pole alternator armature must turn 60 times per second or 3600 revolutions per minute (rpm). To reduce this high speed, four poles or more may be used, which reduces the rotational speed in proportion and still produces 60 Hz. For high voltage the stationary field coils may be connected in series, or for high current amperes the field coils may be connected in parallel.

If two or more separate armature coils are wound around the same core but wound one over the other, and are rotated together, the voltage is said to be *in phase*. Suppose, however, that the coils are wound as three separate systems but spaced at one-third the full circle or at 120° from each other. This will result in *three-phase generation*. The sine wave of three-phase output is shown in Fig. 7-9.

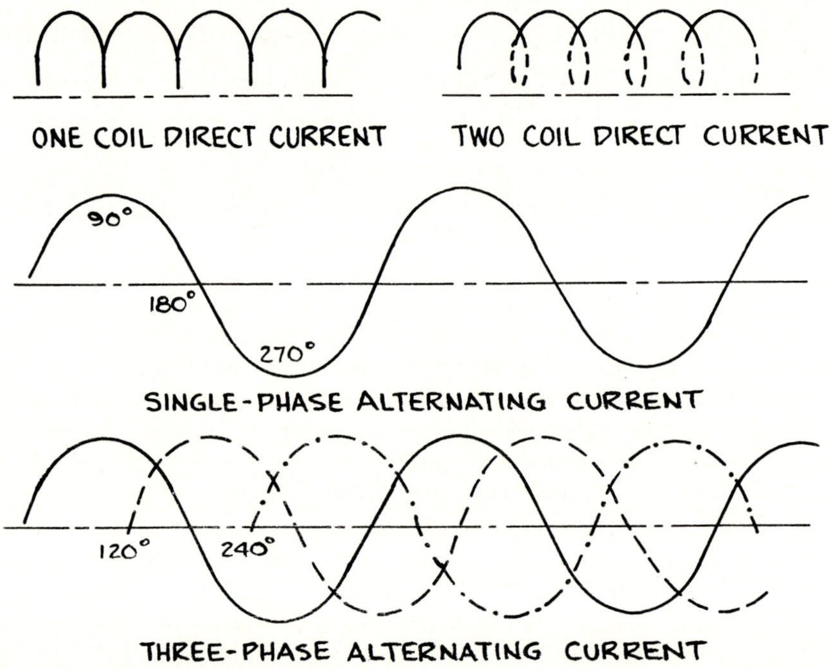


**Figure 7-9** Sine diagram for electric generation.

Modern alternating-current generators may have one set of poles [one north (+) and one south (–)], two sets, or more commonly three sets. Since the majority of electric motors and other equipment operates on three-phase current, the generator armature coil would have three coil systems with six ends, two for each coil. There are two variations possible to join ends of coils to reduce the number of wires and allow the current to be transmitted over only three wires. These variations are called *star* or Y (*wye*) for one method, and *delta* for the other method.

The three-phase wye connection develops from the fact that three separate generator coil systems are spaced 120° apart. In the wye connection, all three windings are tied together at point A, as shown in Fig. 7-10. This connection, shown diagrammatically, forms the letter Y. The voltage of a three-phase three-wire wye-connected system may be calculated by the formula

$$E_{\text{line}} = \sqrt{3E_{\phi}}$$

where $E_{\text{line}}$ is any of the line voltages (Fig. 7-9) and $E_{\phi}$ is any voltage to neutral. The equation for currents is $I_A = I_B = I_C = I_{\text{line}}$, which may be

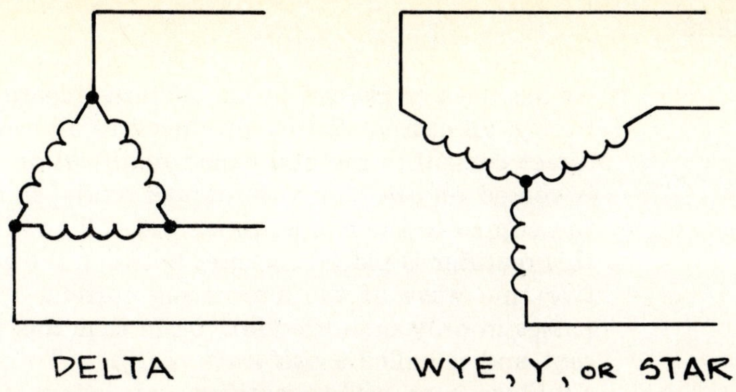


**Figure 7-10** Delta-Y connection.

rewritten as $I_{\text{line}} = I_{\phi}$. The lowercase Greek letter $\phi$ (phi) is used as identification for phase currents and voltage.

The three-phase delta connection also develops from the three coil systems used in the generator, but if the start of one winding is connected to the finish of the next for all three windings, the diagrammatic illustration looks as shown in Fig. 7-10. The resulting diagram looks like the capital Greek letter delta ($\Delta$), hence the name. In this method the voltages may be calculated by the formula

$$E_{\text{line}} = E_{\phi}$$

using the same values as in the wye system. The line currents and phase currents are calculated by the formula

$$I_{\text{line}} = I_{\phi}\sqrt{3}$$

The wye connection can operate as a four-wire system by the addition of a neutral (ground) wire. This is the system generally used for residential electrical distribution. In the delta connection, however, there is no point at an equal potential from the three-phase line voltages. Sometimes a ground neutral is taken off at midpoint between two points of the delta, but this is a dangerous practice and illegal in many locations. Some sample problems that use the formulas indicated follow:

**EXAMPLE**

If the line voltage of a three-phase four-wire distribution system is 2400 V, what is the voltage from line to neutral?

$$E_{\phi} = \frac{E_{\text{line}}}{\sqrt{3}} = \frac{2400}{1.74} = 1379 \text{ V}$$

The value 1.74 is the vector sum generated by two sides of the delta when current is drawn from the third side, and is a constant value, the square root of 3.

**EXAMPLE**

If the line current of a balanced three-phase delta system is 200 A, what is the phase current?

$$I_{\phi} = \frac{I_{\text{line}}}{\sqrt{3}} = \frac{200}{1.74} = 114.9 \text{ A}$$

## 7.7 RECTIFIERS

As has been explained in the sections regarding machine generation of electricity, all current is first produced as alternating current. To change this to direct current, a method called *rectification* may be used. If a copper plate is coated on one side with copper oxide, an electrical current will flow from the copper to the copper oxide but will not flow in the opposite direction. If this rectifier is placed in series between the generator and the load, the negative sine wave of the alternating current is damped out since the current passes in only one direction. Note that this produces a pulsating direct current and is called a *half-wave rectifier*. To overcome this pulsating current, a *full-wave* or *bridge rectifier*, consisting of several rectifiers in a diamond diagram, is used. This arrangement fills in the half-wave that is damped out and provides a current with only minor ripples.

## 7.8 NUCLEAR GENERATION

A considerable amount of the electricity used today is generated in power plants using nuclear energy as the prime mover. This is not a transfer system where the nuclear energy is immediately and directly turned into electricity, but rather a method that uses this energy to produce steam, which in turn rotates a steam turbine, which in turn rotates an alternating-current generator. Heat from splitting nuclear atoms is the basis for this generation process. A complete discussion of exactly how a nuclear plant works is not possible in this book, but a rather quick explanation will perhaps clear up some questions or misconceptions.

The fuel for a nuclear plant is uranium, the only substance found in nature that fissions (splits) and starts a chain reaction. The atom is very tiny (a million billion are required to cover a pinhead) but includes within each atom a core, termed a nucleus, around which electrons spin at tremendous speed. Within the nucleus are two kinds of particles: protons and neutrons. Protons are positively charged particles. Neutrons carry no electrical charge. As neutrons have no electrical charge, they can penetrate the nucleus of the atom. The splitting of a uranium atom starts a chain reaction as parts of the original atom fly off. In turn, their neutrons split other atoms, which then split, and so on. With each splitting, tremendous amounts of heat are generated, so much in fact that 1 lb of uranium completely fissioned will provide as much energy as burning 2,600,000 lb of coal. In the fission process, other fuel-producing compounds, such as plutonium, are produced, so as some fuel is used other fuels are manufactured.

A nuclear reactor core is an arrangement of rods of fissionable material, either uranium or plutonium. The reactor is usually controlled by moving rods in or out of the atomic pile, thus allowing free neutrons to be absorbed by the moderator or holding container, and bounced back to the fuel rods with a reflector. The moderator and fuel rods are contained in a reactor tank and are protected by a thermal shield. Cooling water enters the bottom of the tank, flows upward over the heat-producing fuel rods, and is drawn off at the top of the tank as superheated water. The coolant may be hydrogen, helium, fused salts, liquid metals, and a number of other materials, but water is usually more available, is cheaper, and may operate without physical change.

The simple diagrams in Fig. 7-11 show how the nuclear reactor generates heat, which is transferred to the coolant. This coolant then passes

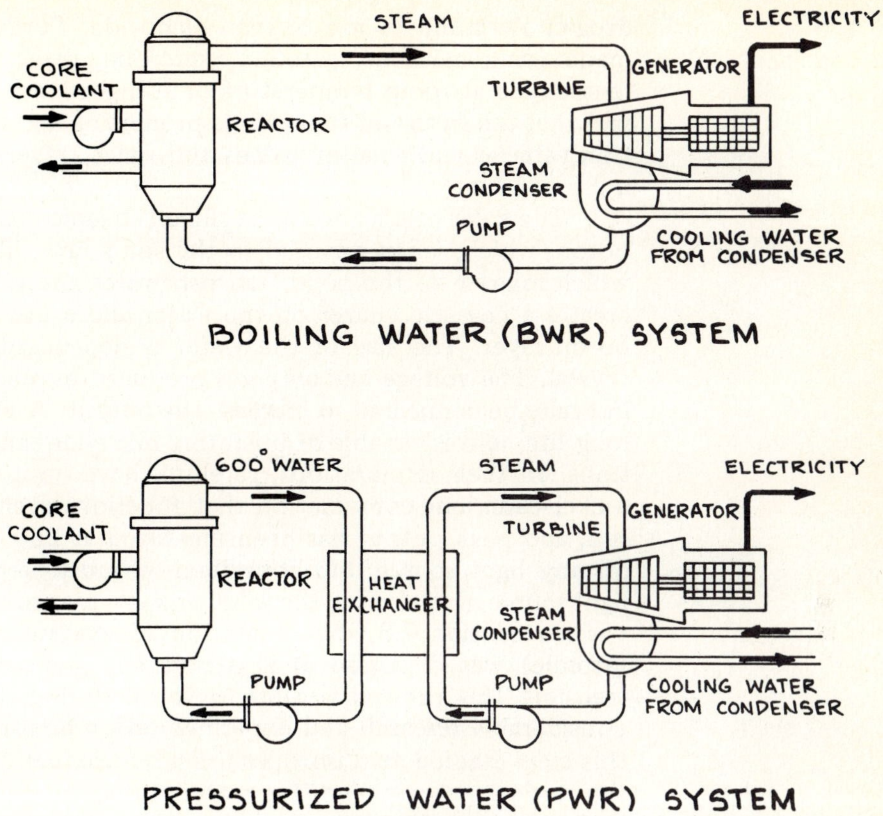


Figure 7-11

through a heat exchanger, which generates steam. The steam is then piped to a turbine, which is connected to an electrical generator that produces alternating current. The "used" steam is condensed back to water and returned to the heat exchanger to be reconverted to steam. If water is used as the reactor coolant, it may be dumped into an available large body of water and replaced, or it may be recirculated into the reactor.

## 7.9 MISCELLANEOUS ELECTRICAL SOURCES

In addition to the batteries, dynamos, and alternators already discussed, there are several additional sources of electricity. Among these are fuel cells, solar cells, and atomic cells. None of these will replace more established or more economical sources, but all have a potential use that may be accepted in the future. Almost every electrical source discussed previously has some form of wasted energy. Heat loss, friction in bearings and rotating devices, the high cost of chemicals, and similar difficulties slow up commercial use of these types.

*Fuel cells* were first constructed by an English scientist, William Grove, in 1839. The ingredients were hydrogen gas (fuel) and oxygen gas (oxidant), which were combined to form water, generating electricity in the process. In this process, hollow, porous carbon electrodes are suspended in an electrolyte of potassium hydroxide and water. Hydrogen gas is pumped into one cylinder and oxygen is pumped into the other cylinder. As the potassium hydroxide dissolves in water, it forms negative ions, which react with the hydrogen to form water, while the oxygen ions react to form hydroxyl ions. When the circuit is closed, electricity is produced by flow of the electrons at the hydrogen electrode to the oxygen electrode. Further experiments have been made using methane, gasoline, kerosene, gases of coal, alcohol, and formaldehyde at room temperature or at higher temperatures. Some authorities feel that the future of fuel cells is promosing and that their use in transportation vehicles and even in homes and space travel may eliminate undesirable noise and fumes.

The solar cell is a wafer of silicon-arsenic coated with a layer of silicon-boron, which, when exposed to the sun's rays, liberates excessive electrons, which migrate to the bond between wafer and coating. The light of the sun creates a negative charge on the wafer and a positive charge on the silicon-boron layer. The size of the wafer is dependent on the size of the silicon crystal. The voltage and amperes produced by such a solar cell are very small but may be connected to increase the output. A solar cell is rugged, light, has long life, and is capable of operating in a wide range of environmental conditions. In their explorations, satellites have been using solar cells made of a nickel-cadmium combination that functions even when not exposed to the sun, and peak output has been measured at 250 W. Solar cells may operate on any light source, but consideration must be given to the cost of such a light source in relation to the value or convenience of the electricity generated.

In Section 7.8, the generation of electric power with nuclear energy (atomic) was discussed as it is normally generated by a turbine and heat transfer. This process is quite wasteful during the heat cycle, however, so considerable research and experimentation has been undertaken to eliminate this step. Radioactive isotopes are a by-product of the fissioning of atoms in the nuclear process, and this material may be used to produce electrical current in a more direct manner. If a polystyrene capsule, an insulator, contains a small amount of radioactive material and the capsule is further enclosed in an aluminum collector shield, the electrons will penetrate the polystyrene wall and collect on the aluminum shield. If the aluminum shield is enclosed in a lead covering, the electrons will flow through the lead and collect as negative charges at a proper terminal. At the same time, the radioactive material acquires a positive charge through loss of electrons. This charge may be carried to a positive terminal. Although the power output from such a cell is small, the increase in atomic (nuclear) power plants provides a growing source of "used" material that continues to function for many years. Utilization of this "waste" material in multiple banks of small cells may be one of the answers to our growing demand for electric power.

## 7.10 SUMMARY

Several types of electrical power are available, but direct current, used for measuring many types of power and in portable equipment, is usually produced by batteries. This source is inadequate for lighting, power, and almost every common use of electricity. Dynamos and generators provide either direct current or alternating current, which may be supplied at different voltages or current. In this chapter these sources have not been treated in great detail, and the theories and advanced formulas have not been attempted at all. Many books cover the various aspects of electrical machinery design. By having a basic understanding of how different sources for electricity function, it should be easier to gain additional knowledge in regard to electrical support systems for buildings.

# 8

# Transmission and Distribution

## 8.1 INTRODUCTION

For electricity generated at a central power plant to be usable, there must be a system of transmission and distribution from that plant to the location where the electricity is to be used. Most generating facilities are located near a main supply of energy which can drive the engines or turbines that drive the generators. These locations may be at river dams, near coal or oil supplies, or near bodies of water used as coolants or condensers. Transmission lines from the generation site to the user location may be hundreds of miles long and are expensive to construct and maintain. In addition to the expense, such long lines cause serious loss of power due to ohmic and radiation losses during transmission. The actual weight of the transmission wires is also a serious consideration.

Electric power is the product of current times voltage ($P = I \times E$), where $P$ is the power in watts, $I$ the current in amperes, and $E$ (or $V$) the voltage. The same amount of power may be transmitted using a high voltage and low current or low voltage and high current. The ohmic power loss is a product of line resistance times the square of the current ($P = I^2 \times R$). By means of this formula it can be proved that the best method of electrical transmission is at high voltage but low current, and this is exactly what happens.

As indicated in Chapter 7, practically all electricity is first generated as alternating current, usually three-phase. By design, many plant generators are built to deliver about 18,000 V at low amperage. This initial voltage is then stepped up through transformers and transmitted over a three-phase four-wire system at voltage approximating 350,000 V (350 kV). Upon reaching the proper use location, the transmission line voltage is stepped down through transformers at a bulk-power substation and further distributed at lower voltage, through a feeder system (Fig. 8-1). The primary feeder may operate

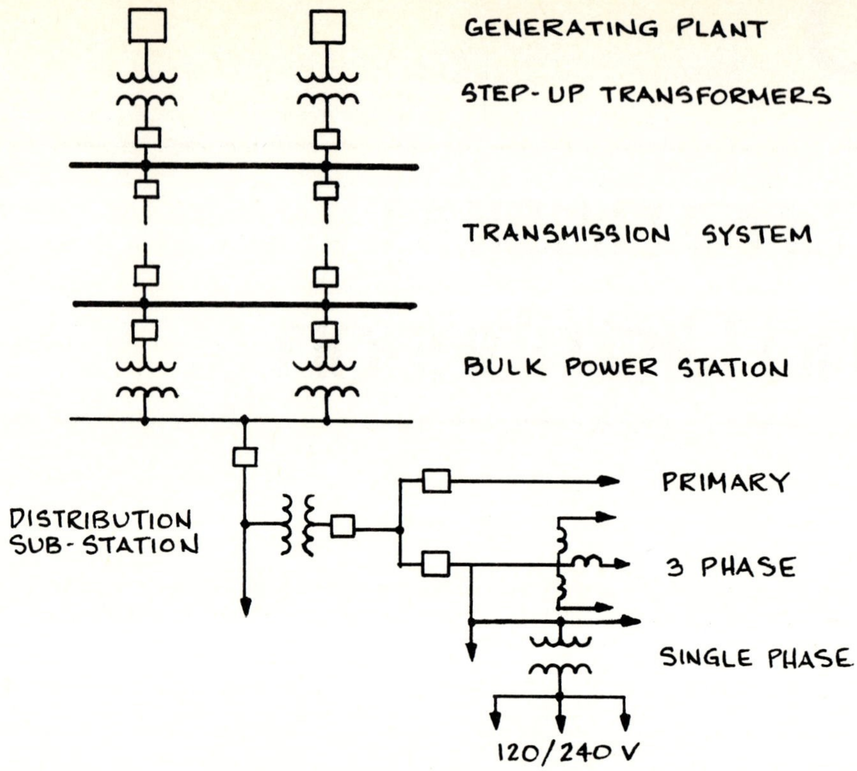


**Figure 8-1** Typical electrical system.

at 4, 13, or 35.4 kV, or combinations thereof. There is no standard combination for feeders. To complete the system, the feeder voltage is transmitted to additional transformers (at poles), where it is stepped down to consumer voltage and is then supplied to the user.

In recent years there has been considerable interest in transmitting power as high-voltage direct current. This system has both advantages and disadvantages. Since direct current flows in one direction continuously, only one wire instead of three is necessary to carry the current, and the neutral or ground wire is the same as for alternating current. This eliminates wire and costly insulators for two wires. Also, direct current does not give off electrical radiation as does the alternating-current method. Since most electricity is generated as alternating current, it must be changed to direct current before transmission, or stepped up to high alternating current and then rectified to direct current, before transmission. On the receiving end, the direct current must then be converted back to high-voltage alternating current and reduced to usable voltage with transformers. The cost and space requirements for rectifiers and converters have delayed general use of direct-current transmission.

## 8.2 TRANSMISSION LOADS

The optimum electrical transmission system depends on a number of factors: load density, area covered, rate of present growth, expected future growth, terrain, existence of rights-of-way, existing systems, total load, and other variables. *Load density* is usually calculated in required kilovoltamperes (kVA) per square mile, as shown in Table 8-1. This can be misleading, however. If a 1000-W toaster is considered as a measuring unit, the density would be 28,000,000 $kW/mi^2$. Open spaces such as parks, cemeteries, and undeveloped fields within an area to be calculated will increase the "local" load in direct proportion to the areas. In addition, a square mile of rural area is practically meaningless, as the total load may be from one set of farm buildings, whereas a square mile of highly developed downtown city property will include requirements for high-rise buildings as well as for less congested portions. Industrial loads must usually be estimated in very small areas, as the requirements for electricity vary widely with the manufacturing process.

**TABLE 8-1**
Range of Load Densities

| *Type of area* | *Load density ($kVA/Mile^2$)* | *Remarks* |
|---|---|---|
| Residential, low density — rural area | 10-300 | Number of farms or residences, and demand per farm can cover a broad range of values, e.g., 1 farm at 10 kVA to 150 farms at 2 kVA average each |
| Residential medium density — suburban area | 300-1,200 | Based on home saturation on 70 ft by 100 ft plots 20 percent of total area, with average kVA per house from 0.5 to 2 |
| Residential, high density — urban area | 1,200-4,800 | Based on home saturation on 70 ft by 100 ft plots 80 percent of total area, with average kVA per house from 0.5 to 2 |
| Residential, extra-high density — all-electric area | 15,000-20,000 | An estimated upper limit for total electric homes with heating and air-conditioning; also, high saturation of homes in total area |
| Commercial | 10,000-300,000 | Type of area and load includes such variations that more definitive breakdown is not practical; this range of values covers range of small shopping centers and commercial areas to downtown commercial areas of large cities; secondary network systems can be expected to be found in downtown metropolitan areas serving load densities in this range |

Westinghouse Electric Corp.

## 8.3 TRANSMISSION TERMINALS

The terminal electrical handling system is a series of *bus* (from "omnibus") *bars*, which may be bars, rods, tubes, or channels of either aluminum or copper. Aluminum may be favored because of its lighter weight and lower cost, but copper is a better electrical conductor. Bus bars (see Fig. 8-2) are simply oversized conductors that collect and distribute the generated electricity in an arrangement that allows various cross-connections between multiple generating units and different circuit feeders. This is desirable because multiple generating units provide opportunities for shutdown for maintenance and adjustment for high or low power demands in the feeder and transmission lines. Bus systems are equipped with protective circuit breakers ahead of, and after, connections to generators and feeders.

Bus systems may be installed indoors in enclosed cabinets or outdoors

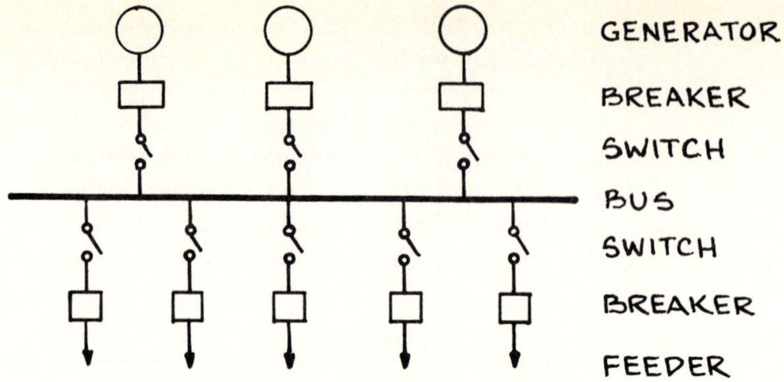


**Figure 8-2** Single bus diagram.

in larger metal lattice frameworks. Indoor installations are fairly compact and may be combinations of metal cabinets holding circuit breakers and concrete separators which carry the bus bars. Ventilation to control the generated heat is one of the primary problems of indoor bus systems. To eliminate damage caused by rats and mice, indoor bus bars are now carried at the tops of cabinets instead of at the bottoms. Outdoor bus systems use a much greater area but have the same function as indoor systems. Line-to-line and line-to-ground clearances are usually considerable, and the porcelain insulators used to support the lines are designed to withstand the pull of the magnetic field between adjacent bus bars. The design and proper selection of insulators is of great importance. Excessive amounts of dust, seawater mist, acid vapors, and similar material which may collect on insulators are a potential short-circuit hazard.

Circuit breakers used on generator high voltage are arc-interrupting devices immersed in a tank of oil. When the air surrounding a conductor is ionized (changed) by the passage of current in the conductor and flashes to the earth causing a short circuit, the circuit breaker automatically opens the circuit and reduces the ionization to zero. Reclosing must generally be delayed from 20 to 30 cycles before the breaker closes the circuit. Switches are used to isolate bus bars and breakers for maintenance and control and are used on both positive and negative connections. Metering, switching, and other control of the generated power takes place at a central station and is often combined with control of the prime-mover (hydraulic, steam) turbines or boilers.

## 8.4 PRIMARY TRANSMISSION LINES

Transmission lines connect the source of power generation with the area in which the power will be used. Poles for such transmission are probably familiar to most people and usually consist of latticed steel towers (see Fig. 8-3). Such towers have concrete foundations to which the steel structure is firmly bolted over a flat steel plate. Expanded steel poles made from refabricated I beams or H sections are also used in some locations. Wood poles, treated with creosote or other preservatives, are also used when the power load is less than about 70 kV and may be installed as single poles or in pairs with connecting horizontal crossarms. Crossarms are bolted to wood poles at the center of the crossarm and may be braced with diagonal metal straps to prevent displacement from the horizontal. When poles are installed in multiples, platforms for transformers or other equipment may also be in-

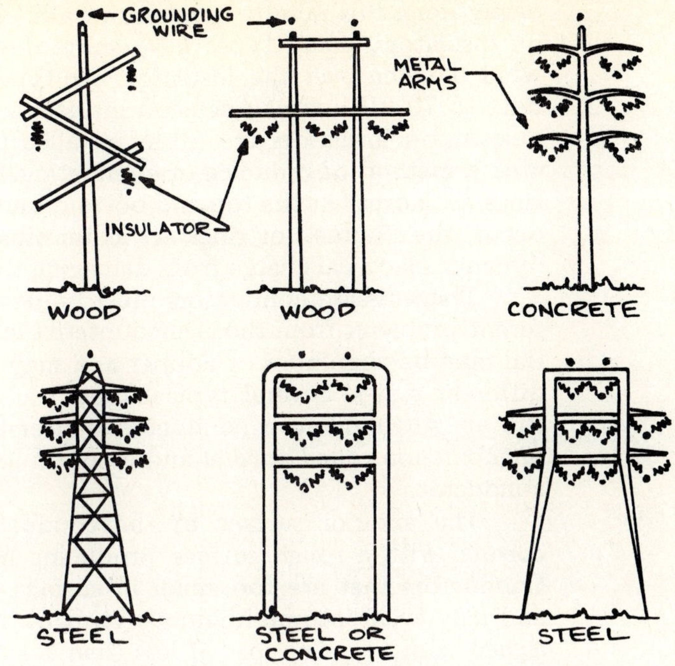


**Figure 8-3** Typical transmission line "poles."

cluded. A "wishbone" design to accommodate three-phase four-wire service may be fabricated from metal which will support three insulators and their connected wires, with the fourth (neutral) wire installed on top of the pole. In steel latticed towers, the fourth wire is often strung across the top of the towers in a similar manner.

Proper insulation between pole supports and the conducting wire is one of the most important and most costly items. Two principal types of insulators are in common use: suspension type or pin insulators (see Fig. 8-4). The design of individual insulators is also very important, as saltwater vapor, fine dust or grit, or other particles may be deposited on insulators and not be washed off by rain. If such films are not removed, a power "flashover" may

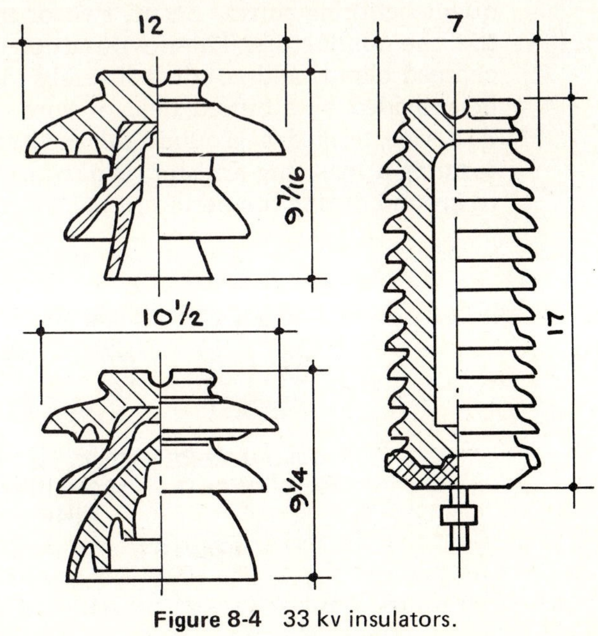


**Figure 8-4** 33 kv insulators.

occur along this minute conductor path. As winds also tend to deposit dust on insulators, many types have skirts designed to increase velocity of the wind blowing over the insulator. Pin-type insulators are generally used on lines to 70 kV, with suspension insulators used at voltages above that level. Suspension insulators are often installed in strings, and experience has created a method of reducing or eliminating flashover by the inclusion of metal rings or "horns" at the top and bottom units of the string. If a flashover does occur, the "horns" or rings act as terminals of a long air gap and keep the dynamic electrical charge from damaging the insulator.

Transmission conductors must be designed to overcome some very different problems from those encountered in building wiring. Conductor material may be aluminum or copper and may be manufactured in a number of different styles. Several typical cross sections are shown in Fig. 8-5. When the air surrounding a conductor is overstressed electrically, it produces a phenomenon called *corona* and a continuous electrical loss of energy in the conductor.

The exterior surface of the conductor has a considerable effect on corona, with a rough surface producing more loss than a smooth surface. Conductors that are too small in surface area also tend to vibrate noisily and may even become luminous at night. Most conductor lines are now designed with a corona loss of less than 1 kW/mi in a three-phase system. Vibration of the conductor is also a major problem in transmission lines between towers that may be spaced 600 to 1000 ft apart. Continued vibration caused by the wind would ultimately crystallize the metal and cause failure. To prevent this, the period of vibration must be changed by changing the tower spacing or increasing or decreasing the conductor tension between towers. A third method now used extensively is to place weights or wind vanes at or near insulators or at midpoint between supports. This method changes the natural period of vibration beyond the usual range of wind velocities.

Protection of towers and conductors from lightning is another important factor in the design of transmission systems. In areas where there are infrequent lightning storms, each tower is simply grounded through the base plate and rods driven into the ground, with the single neutral (ground) wire at the top of the tower connected to the tower. In areas where more frequent lightning storms occur, two or more ground wires are installed above the line conductors. These wires drain off any induced voltage in electrically charged clouds and conduct it safely to the earth. Additional grounding may be provided by a buried field of wire or metal plates attached to the tower. A device termed a ground-fault neutralizer, consisting of a series of tubes forming a lightning arrester, is also used to allow the line conductor to flash to ground during the period of lightning voltage.

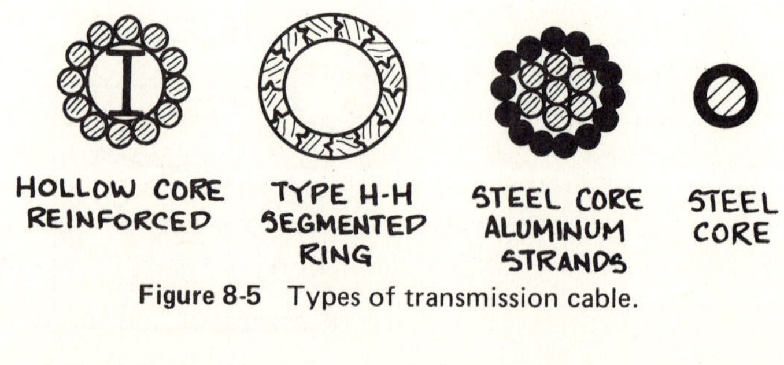


**Figure 8-5** Types of transmission cable.

## 8.5 SECONDARY SUBTRANSMISSION SYSTEMS

Distribution systems actually include all the poles, wires, insulators, and transformers required to convey electric power from generator to user. In the more specialized meaning of the term, however, distribution systems are the lines and circuits that carry the electric power from the step-down substation to the ultimate customer. Feeder voltage into this system is usually 440 to 480 V, with further transformer action to provide 220 to 240 V and 110 to 120 V. Distribution systems may be overhead or underground, a decision greatly dependent on the location of the user. Distribution in congested city areas is usually underground, as overhead wiring is both unsightly and difficult to connect. Overhead systems are widely used in residential and rural areas. Underground cables have a smaller amount of inductive reactance than overhead lines, but overhead conductors have a higher-rated safe current-carrying capacity. Costs for underground installation usually run 5 to 10 times those for overhead, but many municipalities or newly developed areas insist on underground installation. Changes in overhead are much simpler to make than changing underground ducts, manholes, and transformer vaults. On the other hand, the expected useful life of an overhead system is estimated at 25 years, whereas that of an underground installation is estimated at 50 years or more. Repair and maintenance should also be a major consideration. It is obvious that snow, ice, wind, and lightning affect overhead systems, but snow, water, or ice may also complicate underground repairs, and cable breaks are much more difficult to locate in underground systems. Comparisons of final costs for the two systems often reveal a standoff between the two when all factors of installation, maintenance, and life expectancy are considered.

When poles and overhead conductors are used, the system is similar to that for primary transmissions, with only a slight difference when corners are turned or service transformers are installed on poles or platforms attached to poles. Secondary distribution is usually either single-phase 120/208 V or three-phase 120/240 V, although some industrial distribution is 265/460-V four-wire wye. Poles are loaded by vertical action of wire, crossarms, and other equipment (see Fig. 8-6), or by transverse wind loads or change of direction. To counteract these forces, a crib brace may be bolted to the pole and buried when the pole is set, or the pole may be wire-guyed in the opposite direction of the pull. Increasing the depth at which a pole is installed below the surface may also increase its resistance to bending by as much as one-third. Tension on the guy wire may also be a factor to consider.

Conductors used in this type of distribution are either copper or aluminum, solid or stranded. Insulated conductors are used for underground installation, while either bare or insulated conductors may be used for aerial systems. "Weatherproofing" on wires, consisting of layers of asphalt-saturated cotton, will protect the wire from physical hazards but is not dependable when wet. Neoprene and polyethylene are now extensively used, as they both have excellent dielectric properties as well as weatherproofing values. The temperature at which a conductor operates is partially dependent on wind velocity, rain, and air temperature, and affects the current-carrying capacity of the conductor but is seldom a controlling factor in short lines. Underground cables are usually insulated with rubber compounds with an outer cover of braid, lead, or neoprene, and may be installed in wet or dry conditions.

**Figure 8-6** (*Courtesy of Pacific Gas and Electric Company.*)

## 8.6 UNDERGROUND CONSTRUCTION

Underground transmission requires an entirely different approach from that used for the more familiar overhead installation. The usual system consists of buried reinforced concrete vaults for transformers and terminal manholes connected with multiple ducts that contain the electrical cables. Of major importance is the location of other utilities, such as underground telephone, water, sewer and gas lines, storm drains, and perhaps steam piping. If all of these are of new construction, some adjustment in elevation may be made, but sewers and storm drainage are gravity disposal systems and thus may not be readily changed. The arrangement and location of manholes and transformer vaults as well as duct connections is endless in variety, depending mostly on location of the system within a city, concentration of electric load, subsurface earth conditions, and many other factors.

Cable for underground systems is either solid or stranded but differs from aerial cable in insulation and exterior protection. Single cable or multiple cables may be enclosed in one cable construction. Much underground cable is of the oil-filled or gas-filled variety, where the current-carrying wire is surrounded by oil or nitrogen to provide capability for high-voltage transmission. Pressure is put on the oil or gas and a pressure gauge in a manhole will indicate if there is a leak in the conduit. Small reservoirs located in manholes maintain the required pressure. Cables are routed from manhole to manhole via duct banks composed of multiple openings or single tubes. These units may be of precast concrete, clay, asbestos cement, fiber, or even wood, and are laid with concrete surrounding the installed lines as further protection. Where low demand warrants such construction, direct-burial cable may be installed in trenches 18 to 30 in. deep. Advantages of such construction are the considerable savings in cost of installation and the ability of the earth to dissipate heat generated in the cable.

Manholes at intervals up to 1800 ft are needed to provide space for connecting various laterals or interconnecting conduits and are normally of reinforced cast-in-place concrete with access through the familiar cast-iron manhole cover at pavement level. Transformer vaults (Fig. 8-7) are similarly constructed but require ventilation through grilles or mechanical means to carry off the heat generated by the transformer. Quite often, transformer vaults are constructed beneath sidewalks, where the top of the vault may be opened to facilitate repair or removal of equipment. Combination manhole-vaults are also possible. Ventilation should be calculated at about 2.5 to 3.0 $ft^3$ of internal vault volume for each kVa of transformer capacity, or about 1.5 to 2.0 $ft^2$ of vent area per kilowatt of heat-energy loss. The interior arrangement of transformers, cables, test equipment, lighting facilities, and miscellaneous gear requires careful consideration to provide adequate working space and safety clearance.

Final connection of an underground installation to the consumer's premises is normally through a street service box, a concrete box just large enough to accommodate low-voltage conduits. From this box, service is delivered to the consumer or to other boxes for feeder service. Boxes are usually set in the sidewalk and have a flush, removable concrete cover. Service to the electric meter of the consumer is through direct-burial cable or of the conduit-wire type.

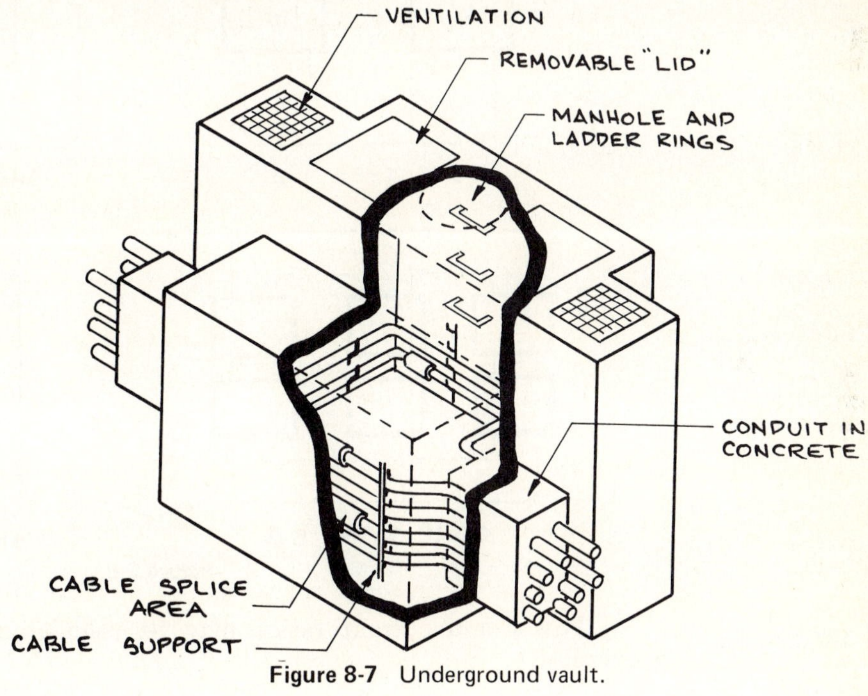


**Figure 8-7** Underground vault.

## 8.7 TRANSFORMERS

Transformers are electrical devices that may step up or step down the line voltage of an electrical circuit. Types of transformers range from those with a rating of 500 kVA or more (power transformers) through those with ratings of 3 kVA to reduce 110-V alternating current to 6-V direct current. The basic construction and function of almost all types are reasonably similar, but some types have special requirements.

A transformer consists of a laminated ring of steel or iron (core) upon which two coils of wire are wound. The coil connected to the source of supply is termed the *primary*, while the coil supplying power to the load is termed the *secondary*. The coils are insulated from each other and may be wound one on top of the other, or on opposite sides of the core. This arrangement of coils will cause an alternating current flowing in one coil to induce a voltage in the other coil. The transfer of energy from the primary to the secondary depends on the magnetic lines of force around the primary cutting across those of the secondary. Cores are sometimes made in the shape of a ring, but more commonly they are square or ring-shaped with a center crosspiece on which the coils are wound (see Fig. 8-8). The latter shape is called a shell core and is able to produce nearly 100% coupling. The alternating primary current tends to cause the core to change the polarity in step with the current frequency. This is an electrical loss and is called *hysteresis loss*. The other loss, termed *eddy currents*, is due to the electric current induced in the core by changing the magnetic fields of the coil.

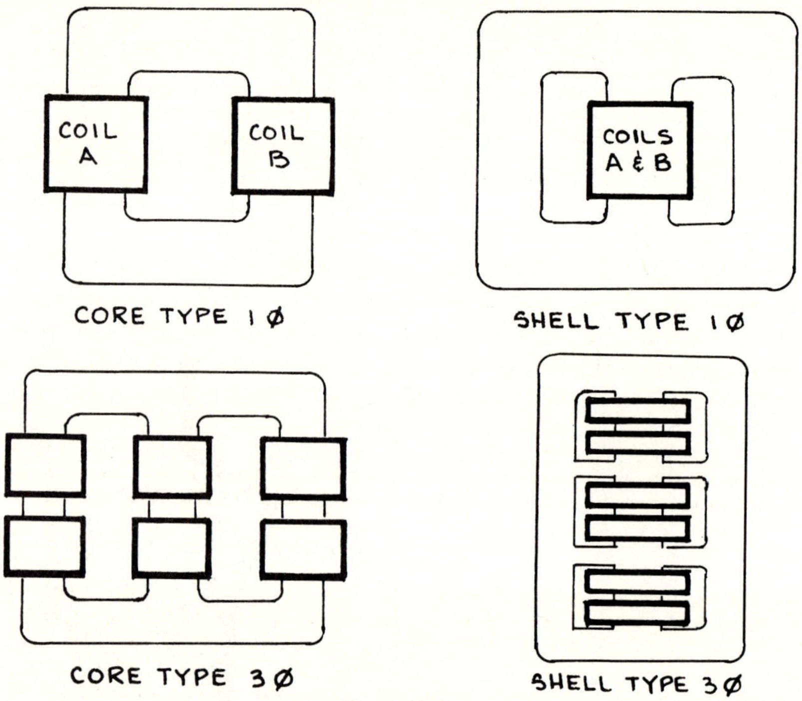


Figure 8-8

Power and current ratios may be calculated by using the following formulas:

$$\frac{E_1}{E_2} = \frac{T_1}{T_2}$$

EXAMPLE

Where $E_1$ is the primary voltage, $E_2$ the secondary voltage, $T_1$ the primary turns, and $T_2$ the secondary turns. With this formula, find the stepped-down voltage in $E_2$ if the primary windings ($T_1$) are 2000, the secondary winding ($T_2$) is 100 turns, and the primary voltage ($E_1$) is 5000 V.

$$\frac{5000}{E_2} = \frac{2000}{100}$$

$$E_2 = 250 \text{ V}$$

In this case the secondary voltage ($E_2$) will be 250 V. The voltage ratio of a close-wound transformer, then, is directly proportional to the turns ratio of primary to secondary coils. The current ratio of a transformer is the inverse of the voltage ratio, or

$$\frac{I_1}{I_2} = \frac{T_2}{T_1}$$

If $T_1$ equals 2000 turns of primary coil, $T_2$ 100 turns, and the input current is 50 A, the output current ($I_2$) will be 1000 A.

$$\frac{50}{I_2} = \frac{100}{2000}$$

$$I_2 = 1000$$

Distribution transformers (Fig. 8-9) are classified as liquid-filled or dry-type. *Dry-type transformers* are air-cooled and air-insulated. *Liquid-filled transformers* are filled with oil or a noncombustible liquid compound. Dry transformers are used primarily for industrial installations or where oil may be a hazard. Most pole transformers are oil-filled or are used in other outdoor locations. Overhead-installed transformers are termed *completely self-protecting transformers* (CSPs) if lightning protection and line-surge devices are included as part of the transformer. Single-phase transformers have a range of 2400 to 14,400 V on primary lines, with a secondary voltage of 120/220 or 240/480 V. Three-phase CSP transformers handle primary voltage from 2400 to 13,800 V, with secondary voltage at 208/120, 240, or 480 V.

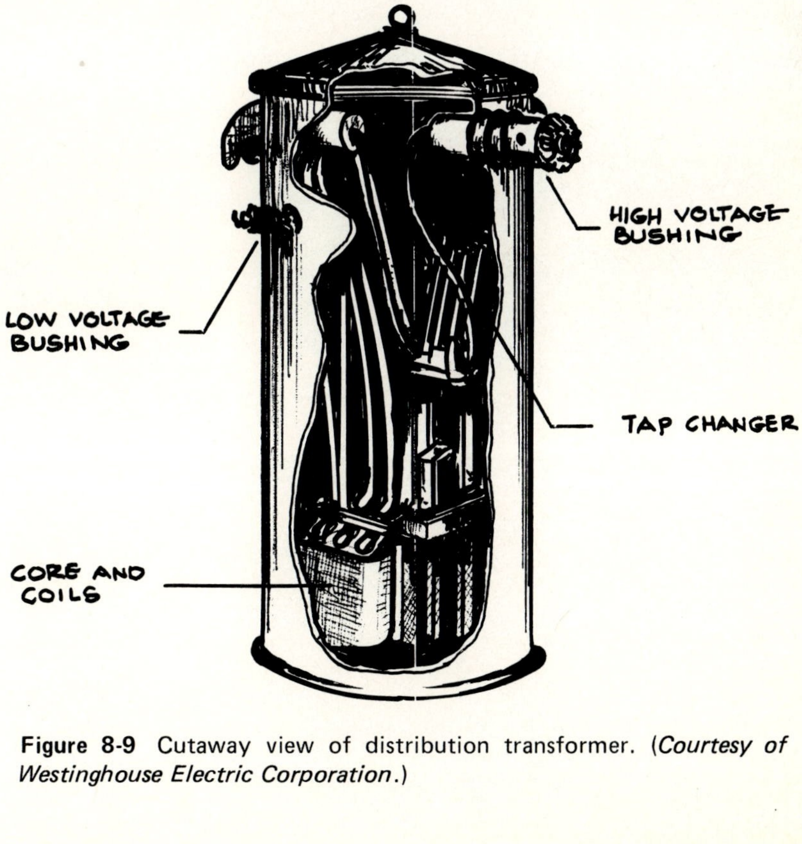


**Figure 8-9** Cutaway view of distribution transformer. (*Courtesy of Westinghouse Electric Corporation*.)

## 8.8 SUMMARY

The transmission of electrical power from generating point to user point is a major part of the entire system. A number of factors determine the type of transmission (overhead or underground), and the loads, demand periods, and overall economics contribute to the design. The distribution of electrical power is a highly specialized part of construction for energy. In addition to the design and construction of the generating plant, the transformer and busbar installations at each end of the transmission lines, and the local distribution overhead or underground systems, the safety and appearance of towers and cables is a continuing concern. The information included in this chapter is simply an overview, as specialization in each portion of the entire system fills many books; but this general knowledge should allow for further study of any aspect that creates interest.

# 9

# Service Systems

## 9.1 INTRODUCTION

Electricity may be generated and distributed by utilities in whatever capacity and manner that they, the utilities, elect if in compliance with applicable codes and other requirements, but such electricity is useless unless it is somehow provided to the building in which it is to be used. The portion of electrical distribution that provides a means to connect the utility service to the building is termed the *service connection* and is usually installed on the private property of the building owner. The exact connection is variable, depending on location, but may be overhead (service drop) from the utility pole to the building or underground (service lateral) from a vault or manhole in the street.

The choice between overhead or underground service is not always a simple one to make. Overhead distribution service by utility companies is 25 to 50% more economical than underground service but does require poles along a street or alley, multiple wires supported from crossarms on these poles, transformers exposed on poles, and guy wires. Overhead installation is more capable of handling higher voltages, especially on long runs, and is infinitely easier to maintain, replace, or to locate faults. On the other hand, many cities now require developers of new residential areas to install electric, telephone, and TV cable service underground to reduce the unsightly clutter of overhead service. Weather and terrain also influence the decision between overhead and underground service. Heavy snow, wind, and ice adversely affect overhead service, whereas rocky or extremely hilly country favors overhead service.

Underground utility service, even though generally more expensive, is more desirable from the standpoint of appearance, as well as being absolutely required under some conditions. In inner-city areas, overhead service constitutes a major problem because of the number and size of wires, poles, transformers, and other equipment, and the location and size of buildings. Underground service may be installed by direct burial cable or pulled through single- or multiple-duct systems, and service laterals to buildings are usually made by connection in an underground vault or buried manhole. Where long runs or many bends of underground service occur, additional problems and expense are incurred.

## 9.2 SERVICE TRANSFORMERS

An electrical transformer has coils wound on a core to provide step-down or step-up facilities, and the use of a transformer is almost always necessary to provide service to buildings. Transformers are used only on alternating-current systems, not on direct-current systems. Typically, a transformer is of the *step-down* type, reducing 2400-, 4160-, or higher-volt utility service to 480 V or 120/240 V for distribution in a building. Conversely, a transformer may be of the *step-up* type, boosting 120 volt utility service to 208, 240, 277, 440, or 480 volts, or higher. Step-up operation may be economical for transmission between points of connection or to provide proper voltage to special equipment or for special usages.

Transformers may be single-phase or three-phase, depending on the wiring (see Fig. 9-1). A single-phase transformer has several advantages over multiphase transformers: they are more economical to produce, smaller and lighter, and may be used in combinations to provide three-phase banks when connected properly. The physical size of a transformer is somewhat determined by the type of cooling used and type of insulation. Transformers may be oil-cooled, synthetic liquid-cooled, or dry air-cooled (see Fig. 9-2). Insula-

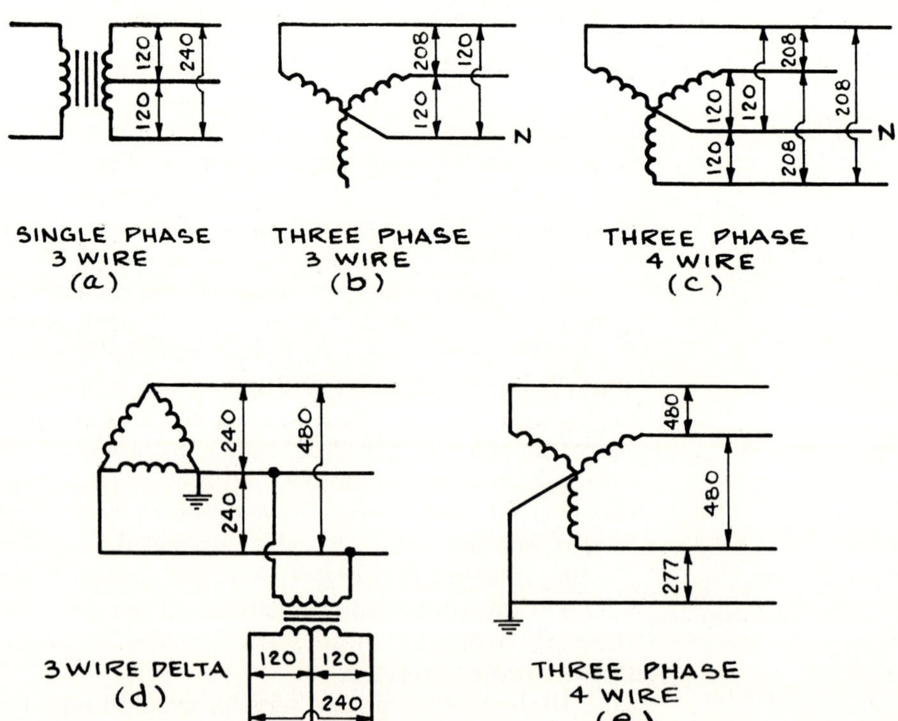


Figure 9-1

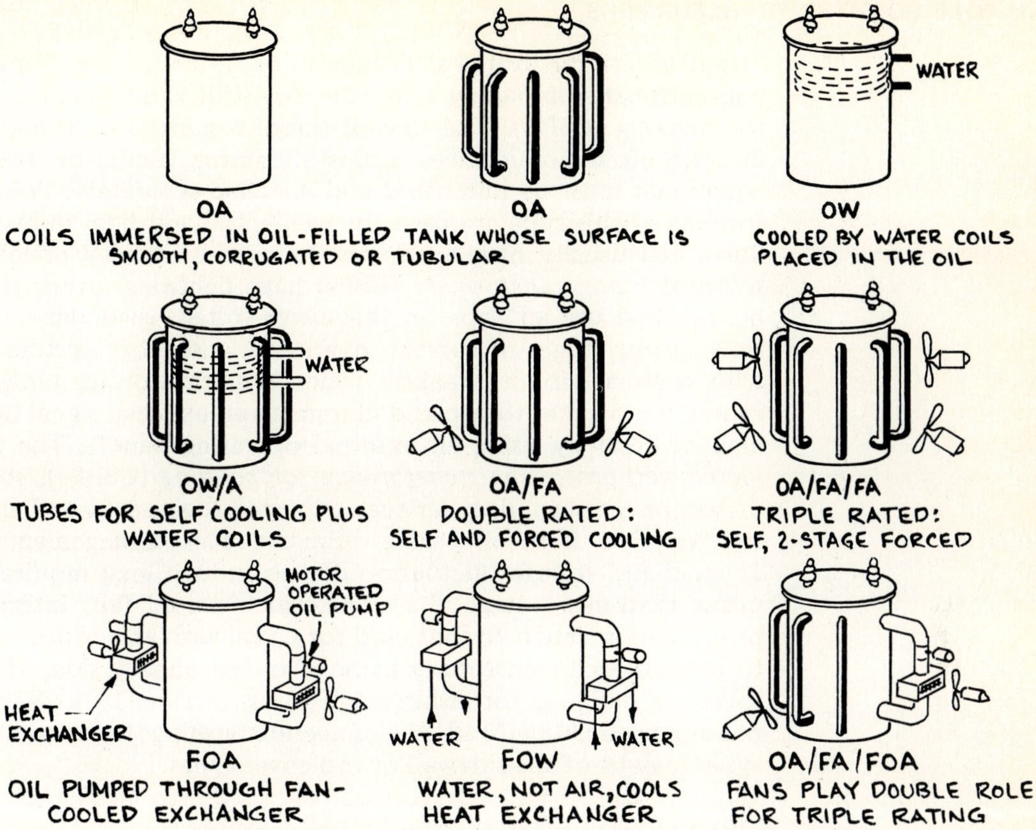


**Figure 9-2** Transformer cooling methods. (*Courtesy of Westinghouse Electric Corporation.*)

tion may be organic (type A), inorganic (type B) or silicone (type H). Heat generated by a transformer may be from 100 to more than 450°F, so the type of cooling as well as insulation are very important to consider. Smaller units usually run hotter than larger units.

Transformers are rated according to voltage, phase, and kilovoltamperes (kVA) of the unit. The voltage rating indicates the voltage of the *primary side* and the *secondary side.* A transformer rated at 2400/480 volts means that it steps down voltage from 2400 volts (primary) to 480 volts (secondary) for building use. If the transformer is a step-up type, it may be rated 120/2400 volt which means that "house" voltage is stepped from 120 volt (primary) to 2400 volts (secondary). A transformer may be either single-phase or three-phase, as indicated in an earlier section. Kilovolt-ampere (kVA) is a product of the voltage and the full ampere load. Voltages of primary and secondary windings are different, as noted above, and so are the ampere currents, but the kVA rating remains constant.

**EXAMPLE**

Given a 50-kVA transformer rated at 2400/480 V single-phase.

$$\text{Primary:}\quad \frac{50{,}000\ \text{VA}}{2{,}400\ \text{V}} = 20.83\ \text{A}$$

$$\text{Secondary:}\quad \frac{50{,}000\ \text{VA}}{480\ \text{V}} = 104.16\ \text{A}$$

## 9.3 POLE-MOUNTED TRANSFORMERS

Transformers for overhead installation are divided into three general types: conventional, completely self-protected (CSP) and completely self-protected for banking (CSPB). The first of these, *conventional transformers*, have no integral protection devices against lightning, fault, or overload, and such equipment must be purchased and installed separately. Pole-mounted transformers of this type are used for residential and light industrial/commercial duty, and usually have a primary voltage of 14.5 kV or lower. *Completely self-protecting transformers* (CSPs) have lightning arrestors on the primary busings and use air gaps on the lower voltage secondary. Overload protection against surges of current, overloads, and short circuits is accomplished with thermal circuit breakers inside the transformer tank. Excessive heat causes the breaker to trip and illuminate an external signal light. The breaker may be reset by using an external operating handle. The third type, *completely self-protecting transformers for banking* (CSPBs), consists of a series of sections of secondary service connected together in a "bank," all of which are supplied from the same primary. This arrangement distributes the demand and voltage fluctuation due to sudden load requirements and minimizes customer outage if a transformer should fail. Internal and external protection is similar to that used for CSPs, with an additional breaker system to protect each transformer bank of the secondary side. This type of transformer is designed for either single-phase service, 10 to 37.5 kVA, or three-phase service, 30 to 75 kVA. Connection bushings for these three transformer types may be of the sidewall or top-cover types.

## 9.4 UNDERGROUND INSTALLATION

Where underground service is provided, transformers may again be divided into three types: subway, underground residential, and network type. *Subway transformers* are generally located in underground vaults and are of two types, conventional and current-protected. Each of these may be installed in concrete vaults with access provided by manholes or full openings at grade or street level. The transformer is enclosed in a watertight case and connections are on top. The *current-protected* (CP) type has the same integral protection devices as those on CSPs except that the lightning arrestors are missing. Primary voltage ranges from 2400 to 14,400 V with secondary voltage ranging from 15 to 160 kVA. *Underground residential distribution transformers* (URDs) are essentially the same units as those used for overhead installation and may be mounted on grade-level pads in semiburied

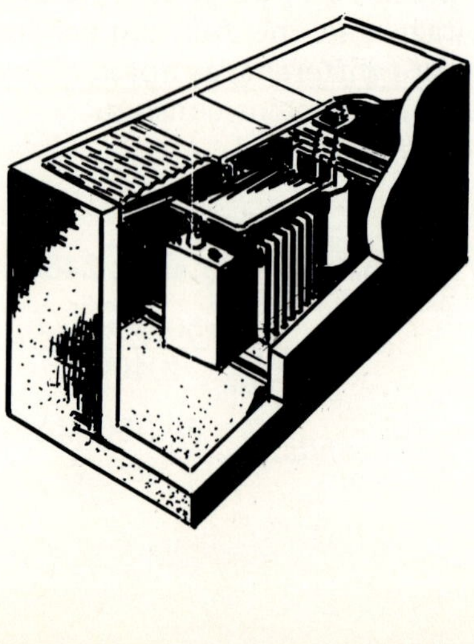

**Figure 9-3** Typical underground transformer vault. (*Courtesy of Westinghouse Electric Corporation.*)

housings. The latter type is often a pipe section of adequate size to accept the transformer and is capped with a sheet-metal cover or lid. Residential transformers range from 165 kVA to 14.4 kV and require adequate ventilation because of their semiburied condition. *Network transformers* provide low-voltage secondary electricity, and the disconnects, grounding devices, and other controls are mounted externally. Transformer units may be liquid-filled, ventilated dry type, or sealed dry type, with a primary rating of 150 to 2500 kVA and a secondary rating of 480 volts and higher. The transformer is generally located in a below-floor-level vault (see Fig. 9-3) but may also be installed in a basement or other indoor area. Sealed units are used where occasional submersion is expected or where a corrosive or explosive atmosphere may be present.

## 9.5 RESIDENTIAL SERVICE ENTRANCE

Service entrance equipment includes the wire, service head (pothead, weatherhead), anchors, conduit, meter ring, and main disconnect (safety switch) required to provide electrical service from the utility source to the building. In most residential and industrial areas, this is overhead and exposed outside the building. In areas where utility lines are underground, the entrance wiring is also underground, but the meter ring and main disconnect are usually above grade and outside the building.

Ordinary entrance service is from a pole line near the building. Utility service to the area is usually at high voltage (2400 to 4160 V), so a step-down transformer bank or single unit is also installed on a convenient pole. In older construction, only two wires, one "hot" wire and one neutral, are used to provide 110 to 120 V only. Equipment in the building, such as ranges, heaters, motors, and similar devices that require 220 to 240 volts or more, or multiphase power, will not operate on such service. Three-wire service, two "hot" wires and one neutral, is more common in newer construction and will provide 110 to 120 volts, as well as 220 to 240 volts when properly connected. In some commercial or industrial installations, this service may be 440 volts, three or four wires, and may be multiphase, depending on the requirements within the building. The size of service wire to be used is determined by the building electrical load and the rated ampacity of the conductor.

A typical residential service entrance is shown in Fig. 9-4. Required

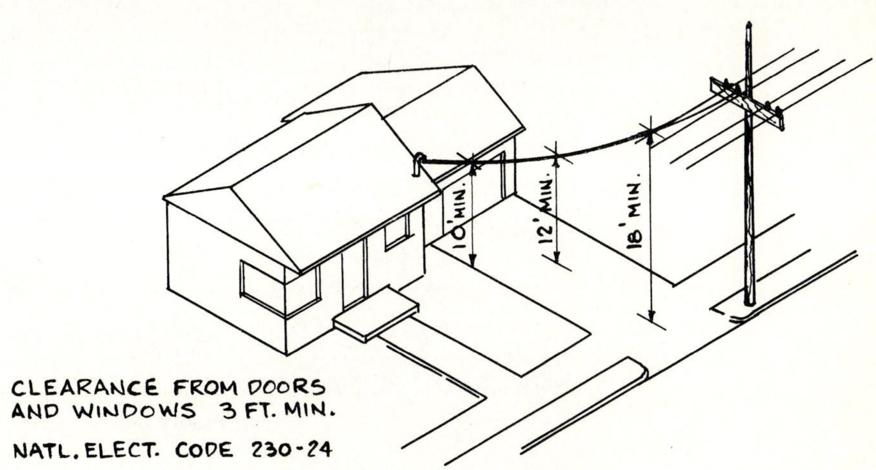


**Figure 9-4** Residential service.

clearances in accordance with the 1981 *National Electrical Code*® (NEC) are also indicated.

The two or three entrance wires may be entirely separated or may be grouped together in a spirally wound bundle and banded to a nonenergized carrying cable. Minimum wire size (NEC 230-23) may not be smaller than No. 8 copper or No. 6 aluminum but must have adequate ampacity to carry the load. In most residential cases, for 100 amp service this will require No. 2 or heavier with type RHW insulation. The individual wires or the carrying cable are anchored to the building, with screw-base standoff insulators, or to an insulated frame. At the building, the wires are collected into an entrance head (pothead, weatherhead) and water drips are formed in the wires to prevent rainwater from passing into the system.

The service conduit from entrance head to the entrance panel or meter ring is normally of rigid steel pipe, although in some localities electrical metallic tubing (EMT) is allowed. Figure 9-5 shows several types of overhead or underground installations. The meter ring is part of the electrical installation, but the actual meter is provided by the utility company, and most companies require that it be installed in a location where it may be read without entering the building. After passing through the meter, the circuit is interrupted by a main (safety) switch consisting of a fused manually operated knife switch or a circuit breaker of adequate capacity. In most residential work, this disconnect is 100 amp minimum but again is determined by the "house-power" requirements. A grounding connector is required from the main disconnect to an adequate "ground," usually a metallic water pipe.

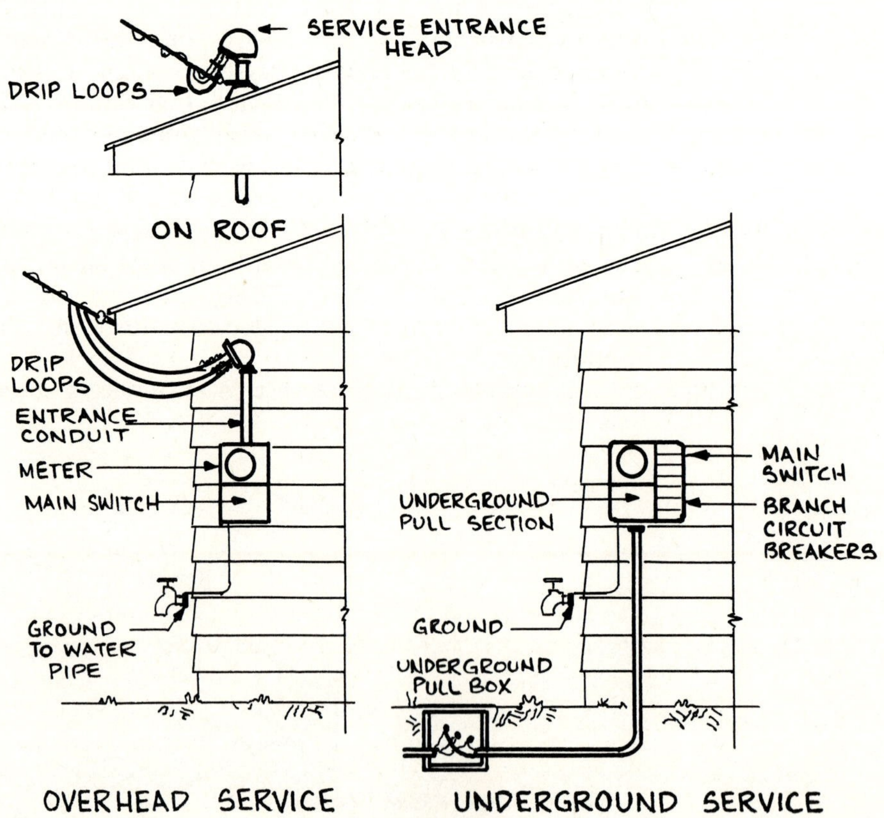


Figure 9-5

If the water system is installed in plastic pipe as allowed in some communities, a separate ground must be provided. In some residential or small electrical installations, the main disconnect may be a circuit breaker in the same panel as the circuit breakers for the various building circuits.

## 9.6 INDUSTRIAL SERVICE ENTRANCE

Electrical service requirements for industrial application are entirely different from most residential requirements. The voltage needed is greater, the phase is different, the characteristics of use are different, adequate grounding may be a problem, and there should be adequate comparison between utility-supplied energy and on-site generated energy. In addition to power loads, almost every industrial or commercial electrical system has a lighting load, may have a communication load, and should have some flexibility for current changes and future expansion or remodeling. All these needs require in-depth review and planning to arrive at proper service capabilities.

One of the first requirements is a *load survey* to determine the amount of electricity for lighting, motors, heating, special work (welding, heat treating), and miscellaneous use. Lighting is probably the most easy to determine. If the size of a building is known, the number of fixtures, lumens, or foot-candles required, the fixture mounting height and spacing, and the types of lamps needed are known or may be easily calculated. If these are not known, a rough "guesstimate" may be made by using about 5 volt-amperes per square foot. Outdoor lighting for service yards, parking, and similar areas should be estimated at about 10% of total interior lighting. Any special lighting at machines or other locations should be added.

The power load for machinery is a little more difficult to calculate. Tags on motors will give the voltage and ampere requirements, perhaps the horsepower-electrical requirements, phase, and related information. For new equipment, all this information may be supplied by the various motor or machine manufacturers. One must then determine the "special" power for welding or other nonmotor work. Together with this survey of power requirements comes the most difficult estimation of all—interval of use versus idle time. Some equipment may be operated 8 hours per day or more, whereas other equipment may be off and on at irregular intervals. Unless the electrical designer has a *thorough* knowledge of the manufacturing process, it is probably better to obtain this information from the plant engineer or other knowledgeable person.

Future load requirements are usually understated, either through lack of adequate foresight or because current installation of adequate equipment to handle a future load is considered too expensive. Completely ignoring future requirements obviates the necessity for planning for them, but often such inaction also means greater expense, and sometimes lack of space, when the situation finally does occur. To be safe, allow for double the present load and double the present electric service equipment space. Flexibility is a product of the electrical wiring system, although unit substations, plug-in bus ducts, and some types of floor systems allow considerable change as conditions differ. Here again some advance planning for floor space or connections to existing systems should be an advantage.

## 9.7 ELECTRICAL SUPPLY

As the demand load for electricity becomes known, the next step is to determine how to properly and economically provide the electricity required as well as considering the method of distribution within the plant. Two major sources for power are available: electricity supplied by a convenient utility company, or on-site generation. Where emergency power is required, as in a hospital, the utility service is usually augmented with standby generators to provide at least partial power if the utility service fails. In any case, space to accommodate the required equipment may again be a problem.

Since most industrial/commercial electrical energy is provided by a utility company, consider this alternative first. For small plants, installation requirements are relatively simple. Energy is supplied from a high-voltage utility line to a transformer or transformer pad, allowing both single-phase and three-phase service, normally at about 480 volts. From this transformer the energy moves to the normal meters and safety switch before being distributed further. Control systems within the project should be considered carefully for in-rush protection, overloads, grounding, and location in regard to central control or individual control at each machine. Distribution considerations should include possible use of conduit cable, bus duct, or other methods, and length of runs to reduce voltage drop. Where large amounts of energy are required, unit substations, which should be located so that future expansion is easily possible, may be used.

Private generation should be considered but is most advantageous and economical where large quantities of electricity are needed for heat treating, welding, and similar industrial processes. Generally, the small private plant is less able to cope with fuel and labor costs, and less adaptable when energy needs are increased. As fuel is a major cost, an on-site generating plant may be economical if waste material or reused heat is available, as is often the case in woodworking plants, cannery operations, and heat-treating plants. Once an on-site generating plant is constructed, it usually must be operated at nearly full capacity, as there is little opportunity to use only a portion of the equipment to provide for reduced requirements.

## 9.8 UTILITY SUPPLY

As mentioned in Section 9.7, service requirements for small plants may be satisfied with several step-down transformers mounted on utility company poles or grouped within a fenced area directly adjacent to the user's building. This equipment and its installation are usually provided by the utility company and the equipment remains the utility's property as well as its maintenance responsibility. Where greater amounts of energy are required, an array of small substations (2.4 kV) to large (13.8 kV and higher) may be required.

Ideal location of a substation (see Fig. 9-6) is directly at the plant load center, but as this is usually impossible, in most cases the location is at the perimeter of the plant. This may create some distribution and voltage-drop problems that need to be reviewed. Location outside the building is more normal than location inside the building and has several advantages and some disadvantages. Outside location reduces the costs of building construction

**Figure 9-6** Electrical sub-station. (*Courtesy of Pacific Gas and Electric Company.*)

to protect equipment, allows for cooler operation and circulation, is usually supplied overhead rather than underground and thus is easier to maintain, and is simpler to install or change since there are no surrounding obstructions. If excessive dust or air pollution, acid fumes, or heavy snows are a problem, these factors may deteriorate insulation as well as supporting frames and create more short-circuiting problems.

Primary substations are made up of one or more transformers electrically and mechanically coordinated with one or more switchgear assemblies. The transformer(s) are usually three-phase and probably oil-cooled or fan-cooled. The only major advantage in having three single-phase transformers connected to provide three-phase service is the possibility of continued operation on two single-phase transformers if one should fail. In addition to this transformer arrangement, most modern units are manufactured to provide tap control to maintain voltage at 3 to 5% above or below normal voltage. Even though this may be possible, it may be better to regulate voltage at each machine rather than to regulate voltage for the entire plant. Modern packaged substations are available for almost any condition and may easily be expanded. Several additional advantages are complete enclosure for positive protection, neat appearance, and a general savings in operating costs even if initial installation costs tend to be somewhat higher than those for open-site-constructed substations.

Proper grounding of any high-voltage system is essential and may be provided by an overhead grounding system paralleling the incoming energy wires, by a series of ground electrodes driven into the earth, or by a grounding bus system to which all equipment is attached. All metal, conduit, frames, raceways, and similar parts should be connected electrically to the preferred grounding system. Multiple-driven electrodes should be provided to eliminate the possibility of a fault developing in one grounding unit. The National Electrical Code has definite requirements for grounding conduit and raceways.

## 9.9 ON-SITE GENERATION

On-site generation is not feasible for the small plant for several reasons: the initial cost of installation is high; qualified engineers are required to operate and maintain the plant; flexibility is limited, as this is an "all-or-none" situation where the plant must operate for one machine just as it would if all the equipment were operating; and space requirements and air pollution may be major problems. For the large manufacturer, the on-site generator may be economical where considerable amounts of electricity are required for motors, welding, or heat treatment in addition to lighting, and especially if the manufacturing process is required 24 hours per day.

The most obvious method for on-site generation is the use of a steam boiler, fired by coal, oil, or gas, using the steam to operate an engine or similar device, which in turn rotates the armature of a generator to produce electricity. All of this equipment requires space, and costs money. The cost of fuel is a major problem, and adequate maintenance or replacement may be uneconomical. In almost every case, a building must be constructed to house this generating plant. If a breakdown occurs in any portion of the generating chain, the entire plant may be shut down unless auxiliary power is available. Despite all the disadvantages of on-site generation, and the growth of air-conditioning, there is increased interest in "total energy systems," where all required energy is produced from a primary fuel source.

## 9.10 TOTAL ENERGY SYSTEM

The *total energy system* depends primarily on the use of a gas turbine for generating shaft power for electrical generation and on the use of "waste" heat for water heating, process heat, air-conditioning, and other heat-energy operations. The essential configuration of this system is shown in Fig. 9-7, where the incoming fuel gas is first cleaned with filters and gas washers. After cleaning, the gas goes into a combustion chamber, combines with pressurized air, and is injected into the gas turbine. The turbine is connected by

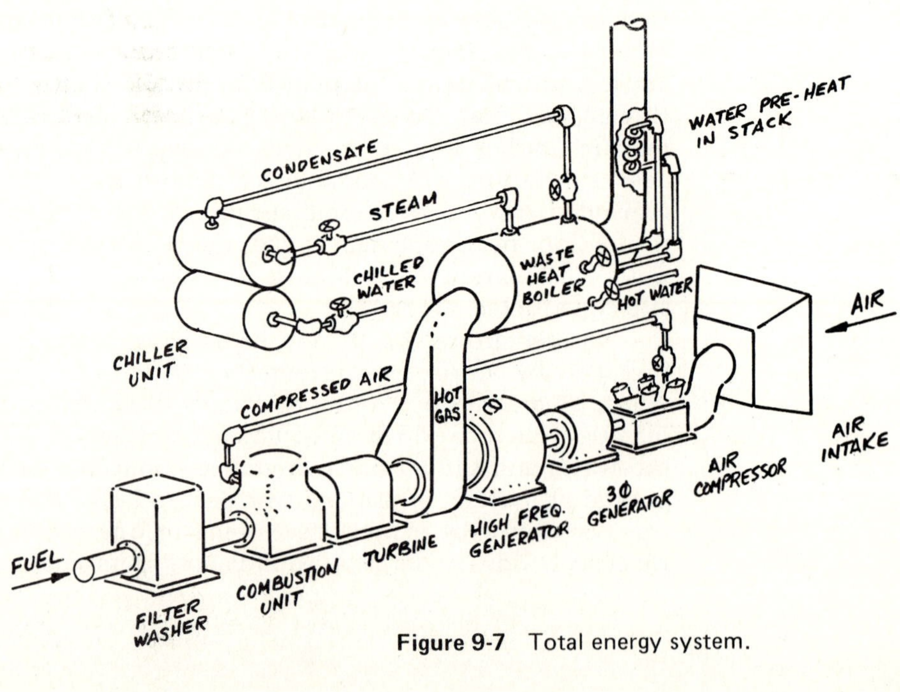


**Figure 9-7** Total energy system.

means of a shaft to the electrical generator and may also be connected by a shaft to the air compressor for combustion. After leaving the turbine, the heated gas goes into a waste-heat boiler, where water is converted to steam. The steam is then used to operate an absorption refrigeration machine to supply chilled water, or as a heat source for heating or winter air-conditioning. Additional takeoffs may be made between the turbine and the waste-heat boiler for process heat, between the waste-heat boiler and refrigeration equipment where steam is required, or from the hot-water side of the boiler for hot-water requirements. Coils arranged in the final exhaust may remove additional heat to preheat boiler water or may be used for other heating purposes.

This system is not without some disadvantages, however. The system is normally designed for the conditions at the time of installation, so changes in any one, or several, of the coordinated uses may be difficult to accomplish given the fixed parameters. Trying to match shaft power requirements (electricity) with heat requirements without loss of efficiency may be troublesome. Some of these difficulties may be overcome by using a number of units, which will also solve the breakdown problem but is more costly. Several such systems that have been in use long enough to provide accurate data indicate about 60 to 65% efficiency and a "payback" period of about 4½ to 5 years.

## 9.11 STANDBY EMERGENCY POWER

The requirement for "standby" electric power is generally applicable only to buildings or operations where an emergency may dictate a continuous noninterrupted supply of electricity. Hospitals and similar buildings where a very short interruption may mean the difference between life or death are most prominent. Planning for such a system should consider exactly what equipment is *absolutely* required to be operational in an emergency, and what equipment or lighting may be serviced without being connected into the emergency electrical system. This means that surgery rooms, recovery rooms, sterilizers, and similar areas have a priority over refrigerators, patients' room lighting and television, most laboratories, and general areas. Some planning in circuit design must include emergency power for continued operation of iron lungs, heart and kidney machines, and similar equipment. Power must be supplied at the exact voltage and proper phase.

Batteries, usually of the wet-cell type, are strictly a secondary low-voltage system and may be 6, 12, 32, or even 150 volts and are direct current rather than the alternating current normally supplied by a generator and converter. This type of emergency power is usually restricted to communication systems with manual control for switching. The use of direct current may prove to be a problem in that it requires equipment that can use both direct and alternating current, and a built-in charging system must be provided to keep the batteries fully charged.

## 9.12 SUMMARY

Electrical service is the connecting link between the generating facility and the user and varies from the low requirements of the single-family residence to the high requirements of a major manufacturer. In the former case, the service is almost always provided from curbside step-down transformers with

120/240-volt single-phase service. The average residence needs only 120 volts for lighting and rarely more than 250-volt single-phase for motors, air-conditioning units, or electric ranges. Three-wire service is available in all but very isolated areas, although there are still some locations where only 120-volt two-wire service may be available. Entrance wire size is usually No. 2 copper, so no special problem is presented.

Industrial service is completely different. Requirements include 120-volt single-phase lighting plus a variety of voltage requirements for motors and other electrical uses. Motor voltages may vary from 240-volt single-phase to direct line voltage of 4160 volts three-phase. Total loading may reach 15 kVA or more. Generally, the utility supply of electricity is most economical and most reliable. In planning a system, adequate attention should be given to the reliability of the utility regarding anticipated outage, voltage fluctuation, and of course estimated power costs. On-site generating plants are only for large manufacturers where power and lighting requirements are large enough to warrant the expense of initial installation and the increased cost of labor and maintenance necessary to keep the energy plant in operation.

In a few cases such as hospitals, elaborate standby systems consisting of instant-start generators, operated by oil- or gas-fired engines, are not only necessary but required by law. With some lesser systems, this standby equipment may consist of storage batteries to provide low-voltage temporary electricity. A careful review of the necessity for such equipment should be made, as additional costs include not only that for the major generator or battery bank but also those for switches, alarms, fuel tanks, pumps, and other devices, all of which will probably stand idle for most of the time but which must be maintained ready for instant duty.

# 10

# Wiring Systems

## 10.1 INTRODUCTION

Design of an adequate electrical wiring system is a matter of providing an arrangement of conductors and other equipment which will safely and efficiently supply energy from a source of power to various lamps, motors, and other electrically operated devices. Both residential and industrial installations have the same requirements, but the actual circuiting and installation in residential work is often left entirely to the electrical contractor. Most nonresidential systems are "designed" for the use intended; that is, they are each calculated for certain uses with reasonably accurate assumed loads from lighting, power, or other requirements, and are shown in more detail on installation drawings.

Every designed wiring system must provide not only the electrical energy needed at the proper location but also the flexibility for future additional loads or remodeling of loads, safety of installation and operation, and the physical characteristics of the building. Anticipation of future requirements by slightly oversizing wire, conduit, raceways, and switching equipment can help realize considerable savings both in the remodeling of electrical systems and in "downtime" of operation while such remodeling is going on. The safety of a system is a basic requirement and if the system is properly installed in accordance with the National Electrical Code there should be no problem. The physical size, shape, use, materials of construction, and possible hazards encountered or anticipated will also have considerable influence on the design and installation of any electrical wiring systems. A system that may be perfectly adequate and proper for a warehouse will usually be completely substandard for an office building and totally wrong for a hospital or school. The requirements for a single-family residence may be very similar to those for small apartment buildings, but even here the overall design must contemplate the additional loads for elevators, laundries, public-space lighting, communication and alarm systems, and perhaps some other heating or power installations. The reliability of the power supply source is another factor to be considered when designing a hospital or similar facility. If the power supply fails, even for seconds, will emergency power supply be required; and if so, how should it be distributed to the areas or devices that need it most?

## 10.2 VOLTAGE SYSTEMS

Requirements for an electrical wiring system depend a great deal on the voltage and current carried in the system. These voltages vary from 120 volts in the normal residential system to as much as 4160 volts for some large commercial installations. Within these parameters a considerable number of combinations should be considered.

The average residence is supplied with 120/240 volt, three-wire single-phase service, as indicated in Fig. 10-1. The utility service transformer is connected at each end of the low-side transformer winding and at midpoint to provide both 120 volts for lighting and similar loads and 240 volts for ranges, heating, motors, and similar devices. Small stores and other establishments where the load is not great and is used principally for lighting may have the same system. Each "hot" leg of this system carries line current, so total current load may be calculated by multiplying the load on each leg by 2 if the system is balanced. If unequal current is carried, the voltage times the current of each leg is calculated, then these sums are added together to obtain total kVA value of the system.

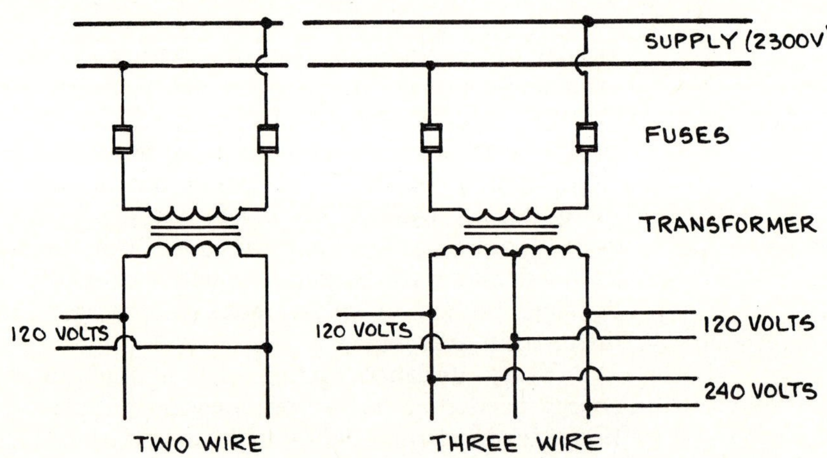


**Figure 10-1** 120/240 Volt service.

For small commercial installations that require 120/208 voltage a three-phase system is used with only two legs plus neutral. Fluorescent lighting is often 208 voltage, and this system may also be a part of a larger commercial/industrial complex where four-wire service is not required. The 120/208 volt three-phase four-wire system is most often used when a wider variety of power is needed to supply 120 volt lighting and single-phase motors smaller than ½ hp, 208 volts for fluorescent lighting and heaters on single-phase service, and 208 volt three-phase service for larger motors and other equipment.

Buildings that are multistory or spread out enough to require heavy feeder wiring are often wired as 480/277 volt three-phase four-wire. This allows 277-volt fluorescent lighting and 480 volt motors; and by using small dry-type transformers, the 480 volts may be stepped down to 120 volts for receptacles and incandescent lighting. The savings in installation is in smaller wire size and lighter switchgear, and tends to overcome the cost of the additional transformers required.

## 10.3 RESIDENTIAL WIRING

Residential electric wiring systems are most often simply shown on drawings as duplex outlets, light-fixture locations, switches, and occasionally as motors. The conduit, wire, or nonmetallic conductors are normally not indicated at all or only when some special connection is required. This means that the installing electricians actually design the various systems as they install them. Fortunately, a residence does not often have large loads, does not use multiphase power at high voltage, and does not require complicated wiring systems. This is not to suggest that electricians are not capable of proper installations, but that there is a minimum of deviation allowed in residential work and all such installations must conform to the standards of the National Electrical Code.

Service entrance requirements and installation for residential electricity are discussed in Chapter 9 and will be considered as properly adequate to supply the residence. The National Electrical Code (230-23) requires that service conductors "not be smaller than No. 8 copper, No. 6 aluminum, or copper-clad aluminum." Article 230-79(c) of the NEC also requires that the service disconnect (safety switch) "have a rating of not less than 100 A, three-wire," where the initial installation consists of six or more two-wire branch circuits. These two requirements cover the majority of residential-type service connections.

## 10.4 RESIDENTIAL CIRCUIT DETERMINATION

Before making any determination of the number of circuits to be installed, which in turn determines service requirements, there are a few fundamentals that should be considered.

1. Lighting load per square foot of floor area should be expected to be 3 watts [NEC 220-2(b)].
2. Maximum load on any circuit should be not more than 80% of branch-circuit rating if maintained for more than 3 hours [NEC 220-2(a)].
3. Divide the total watts by the ampacity of the circuit to obtain the required number of circuits.

**EXAMPLE**

Assume that the total floor area (enclosed living space, basement, porches, attached garage, and similar areas) is 3000 ft$^2$ and that 3 W/ft$^2$ is used.

$$\text{amperes} = \frac{\text{watts}}{\text{volts}} = \frac{3000 \times 3}{120} = \frac{9000}{120} = 75.0 \text{ A}$$

The 75 A obtained by this formula is the total load on the electrical service, but Article 230-79(c) of the NEC requires 100-A service, so the calculation serves only as a check on required ampere rating.

Normal wire size used in residential work is No. 14 American Wire Gauge (AWG) or No. 10 or 12 if "home runs" are long. NEC Table 310-16 shows that No. 14 copper wire has a maximum rating of 15 A. Therefore, the number of circuits for the residence is

$$\frac{75.0}{15} = 6 \text{ circuits}$$

Additional NEC requirements call for at least two 20-A circuits for each small-appliance circuit used to feed receptacles in the kitchen, pantry, dining room, and breakfast rooms. Another 20-A circuit is required when a clothes-washer circuit is added, and some spare circuits (15 A) for future additions should be included. Six 15-A circuits are required by earlier calculations, plus two small appliance circuits, plus one washer circuit, which equals nine circuits. As circuit-breaker panels are manufactured with an even number of opposing breaker locations, a minimum of 10 would be convenient, but two added spares will cost very little and provide greater flexibility. The minimum main switch (service entrance) amperage may then be refigured as follows:

**EXAMPLE**

First convert the three 20-A circuits to watts by multiplying.

$$3 \times \text{amperes} \times \text{volts} = \text{watts}$$

$$3 \times 20 \times 120 = 7{,}200 \text{ W}$$

$$\text{Add original watts} = \underline{3{,}000 \text{ W}}$$

$$10{,}200 \text{ W}$$

$$\text{Assume three spares at 1500 W each} = \underline{4{,}500 \text{ W}}$$

$$14{,}700 \text{ W}$$

$$\text{amperes} = \frac{\text{watts}}{\text{volts}} = \frac{14{,}700}{120} = 122.5 \text{ A}$$

However, a slight reduction may be realized by using 40% over 10,000 W, as allowed by code:

$$14{,}700 - 10{,}000 = 4700 \times 0.40 = 1880 \text{ W}$$

$$\text{amperes} = \frac{\text{watts}}{\text{volts}} = \frac{10{,}000 + 1880}{120} = 99.0 \text{ A}$$

A 100-A main will barely serve; and if all loads, plus some greater than 1500 W on spares, are included, it would be safer and only slightly more expensive to go to 150 A.

It should be noted that no circuits are included for 240-V ranges, furnaces, or similar equipment, which will occupy two breaker positions instead of one. Thus even more than 12 possible breaker positions may be warranted, with a corresponding increase in main protection size.

To make an even more accurate determination, a residential plan may be laid out for various circuits. In doing this, several factors should be considered. Lighting circuits should not be wired with convenience outlets, especially where electrical tools or appliances may be used. If the tool or appliance is overloaded and trips the circuit breaker, the lights will go out and people in the dark may be injured by moving appliances. Lighting circuits should not be calculated beyond 1500 W full load for each circuit. If convenience outlets are included at 150 W each, no more than 10 outlets should be on one circuit, assuming that they will not all be in use at the same time. Motors at furnaces and similar locations should be checked for horsepower and that figure reduced to watts (horsepower × 745.7 = watts). If an electric range is included, it will require about 8000 W and 240 V, No. 10 wire for the circuit. Some ratings for various residential appliances are given in Table 10-1.

**TABLE 10-1**
Typical Residential Appliance Wattages

| | | | |
|---|---|---|---|
| Air-conditioning | | Mixer | 150-250 |
| Room | 800-1200 | Oil burner | 250 |
| Central | 5000-6000 | Range | 8000-15,000 |
| Broiler | 1000-1500 | Cooktop | 4000-8000 |
| Can opener | 100-250 | Oven | 4000-8000 |
| Clothes dryer | 5600-9000 | Refrigerator | |
| Coffee maker | 1000-1500 | Frostless | 1000-1200 |
| Deep fryer | 1300-1500 | Standard | 750-1000 |
| Dishwasher | 1000-1500 | Shop tools, each | 1000 |
| Electric blanket | 200 | Television set | 250-400 |
| Electric iron | 1500 | Trash compactor | 1250 |
| Food processor | 200 | Vacuum cleaner | 250-800 |
| Freezer, frostless | 1000-1200 | Waste disposer | 500 |
| Frying pan | 1200-1500 | Washer, clothes | 850-1000 |
| Garbage disposal unit | 300-900 | Water heater | 3000-5000 |
| Lamps | | | |
| Fluorescent, per tube | 15-75 | | |
| Incandescent, per bulb | 25-250 | | |

## 10.5 RESIDENTIAL SPECIAL WIRING

Several electrical circuits are peculiar to residential construction and may occur only rarely in other types of buildings. One of these is the control of a lighting fixture from two or more switch locations, perhaps a yard light controlled from inside a residence and from the garage as well. The wiring system for this two-position arrangement is indicated in Fig. 10-2.

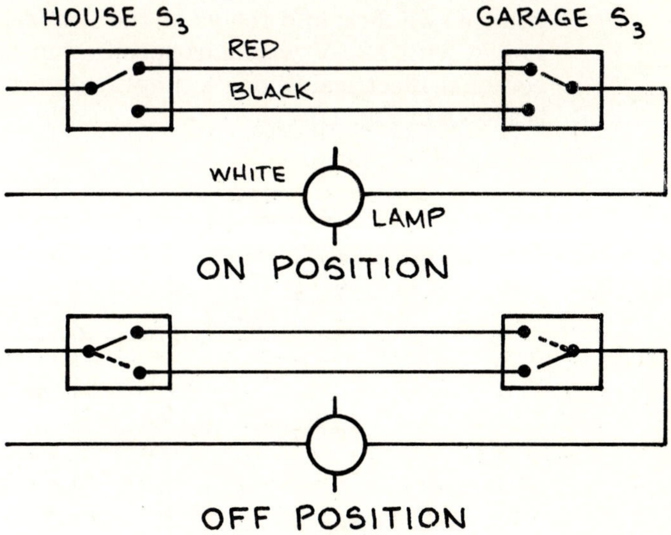


Figure 10-2

Another wiring "system" common to residential work is the split circuit for duplex convenience receptacles. This means that one outlet of the duplex receptacle is continuously "hot," while the other one is controlled by a switch. Normally, convenience outlets, except those above cabinets, are mounted 12 in. above the floor. If a room does not have a ceiling-mounted fixture, at least one wall receptacle must be controlled by a switch. The theory behind this requirement is that any portable lamp may be left ON, but the electricity to it will be controlled by the switch. The "hot" side of the receptacle is used for a television set, radio, or other devices that have their own integral controls.

Dimmer switches are fairly common in residential work and are usually simply inserted in the electrical circuit, as shown in Fig. 10-3a. Where the fixture is also controlled by two three-way switches, the dimmer replaces one of the normal three-way switches. If it is desired to use dimmer control on fluorescent lamps, special ballasts must be used to control the current in the lamp arc, wired as shown in Fig. 10-3b.

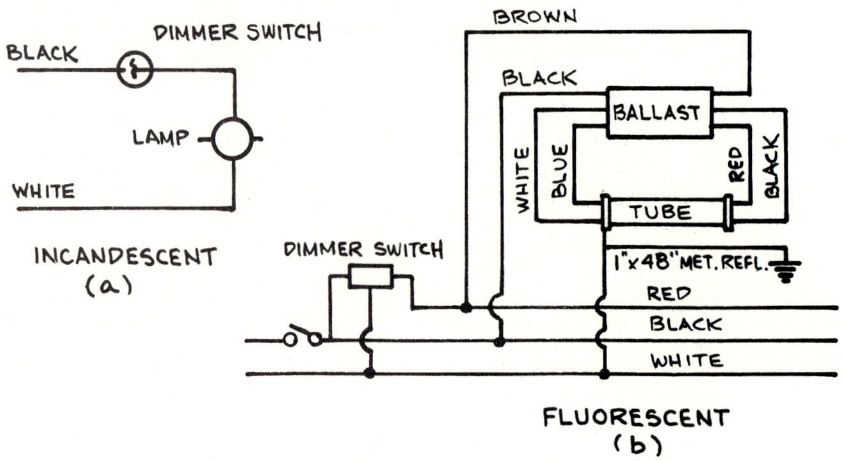


**Figure 10-3** Dimmer control.

Together with dimmer control of lighting has come the installation of low-voltage switching control. This method requires a step-down transformer, usually rated to 100 A, to reduce a 120-V supply to 24 V; special low-voltage switches; and relays at the fixture. Low-voltage wiring cannot be installed with 120-V wiring as specified in Sections 725-31 and 725-42 of the National Electrical Code. A simplified wiring diagram for low-voltage control is shown in Fig. 10-4.

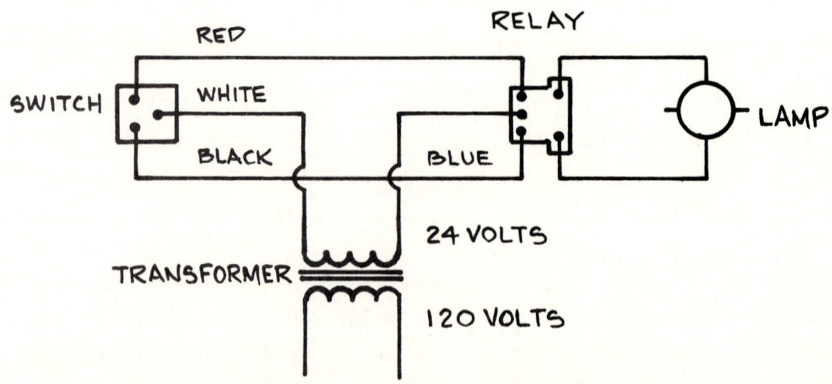


**Figure 10-4** Low Voltage control.

## 10.6 GROUNDING PROTECTION

All outlets, switch boxes, lighting fixtures, motors, panel boxes, and other devices should be electrically and physically grounded for safety. If thin-wall metal tubing and proper connectors are used as conduit for the wire and metal boxes are used, the entire system may be electrically grounded by a strap connection to a water supply system. If the system is wired in non-metallic sheathed cable (Romex), the enclosed ground wire should be attached to each metal box and to a water system. Often a bare copper wire is used for this purpose when not included in the cable. Special care must be used if plastic boxes or plastic water pipe are used, as these will break the electrical ground and must have additional "jumpers" or other connections installed to complete the system around the plastic portions. Grounding must *not* be to a fuel-gas system!

Electrical shocks are caused when the "hot" wire insulation breaks down, or a portion of a device fails, and the electricity may be transferred by contact with another grounded object, often a person. To reduce this possibility, a ground-fault circuit interrupter (GFCI) is required on all 15- and 20-A circuits in bathrooms where access is at grade level, at "readily accessible" garage outlets, and on all outdoor circuits (NEC 210-8). The GFCI in action provides a miniscule ampere difference when a fault occurs, which in turn is amplified sufficiently to trip the circuit breaker before anyone is injured or killed. The GFCI device is manufactured to actuate within $\frac{1}{40}$ sec for a ground fault of more than 0.005 A.

## 10.7 COMMERCIAL WIRING

One of the major differences between residential wiring and that used in a commercial installation, aside from voltage requirements, is the matter of distribution. Except for very large residences, apartment complexes, and motels and hotels, most residential systems have only one control panel (fused or circuit-breaker type). Each lighting fixture or group is controlled further by switches usually installed in the same room. The voltage throughout each system is the same except for a minor voltage drop in some longer runs. The voltage requirement is normally 120/240-V single-phase.

Commercial installation for factories, hospitals, office buildings, and similar occupancy usually has high incoming voltage, three-phase, three- or four-wire, which must be changed by transformers. Much of the lighting load may be controlled by circuit breakers at panels instead of individual room switches, and motor loads are protected at panels and controlled at machines or nearby locations. A possible 480/227-V, three-phase four-wire system is shown in Fig. 10-5. If wired in accordance with this diagram, all voltages—480, 240, 120, and 24—are possible, both single-phase and three-phase. The larger motors might require the most direct 480-V circuits, and each would be fused at the main distribution panel and again at the starter position. Other circuits may be fused at the main panel and again at a sub-panel and internally wired to provide 277 V for fluorescent lighting, the control for which may be 120-V or low-voltage 24-V provided by transformer or connected from another circuit. The incandescent 120-V lighting and 240-V power are obtained by transformers as in residential work, except that such transformers are usually a part of the in-plant system rather than a utility-supplied on-pole unit.

The actual installation of wire, conduit, and individual switches in small commercial plants and in office spaces in large industrial plants is little different from that used in residential work. Loads are essentially the same or

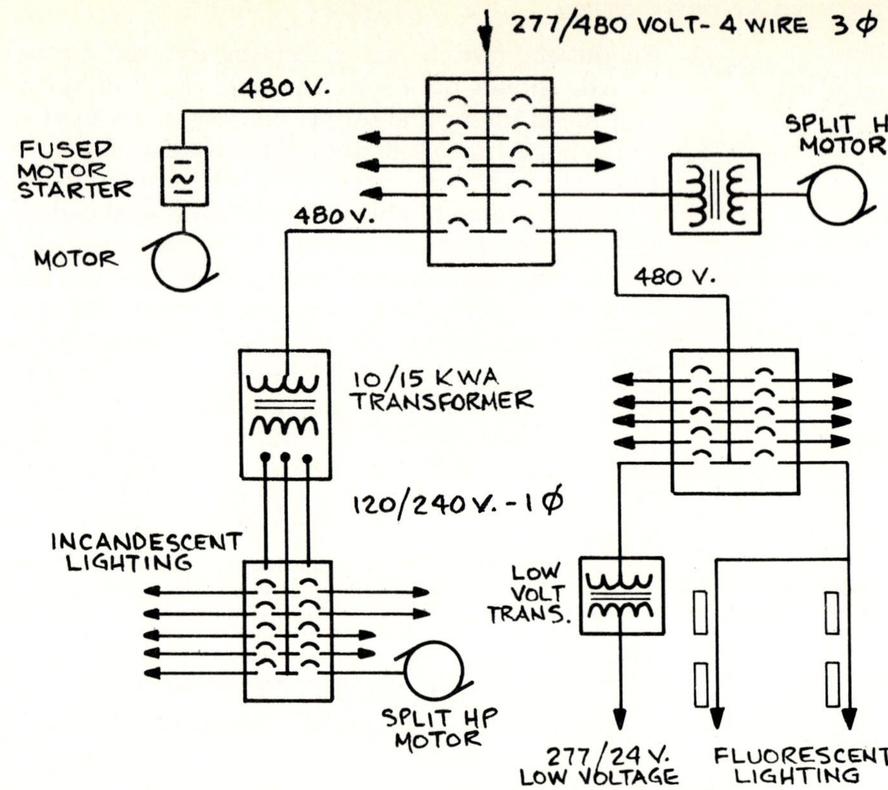


Figure 10-5

similar, and local in-room control is desired. If motors exist at all, they are usually fractional horsepower and thus may be accommodated on 120-V single-phase circuits. Larger commercial/industrial installations, however, use a great variety of conduit, raceways, cellular decking, and busways to carry the wire and an almost equal variety of panels and control devices. These are considered in the following sections.

## 10.8 DISTRIBUTION SYSTEMS

Conduit is probably the most familiar material used to carry electrical conductors from one location to another and to protect the wire from damage. Metallic conduit also acts as a grounding system and protects surrounding material from wire heat and possible fire. Conduit may be rigid steel, electrical metallic tubing (EMT), flexible metal conduit ("greenfield"), or nonmetallic conduit (see Fig. 10-6). Rigid steel conduit is heavy-wall steel pipe that may

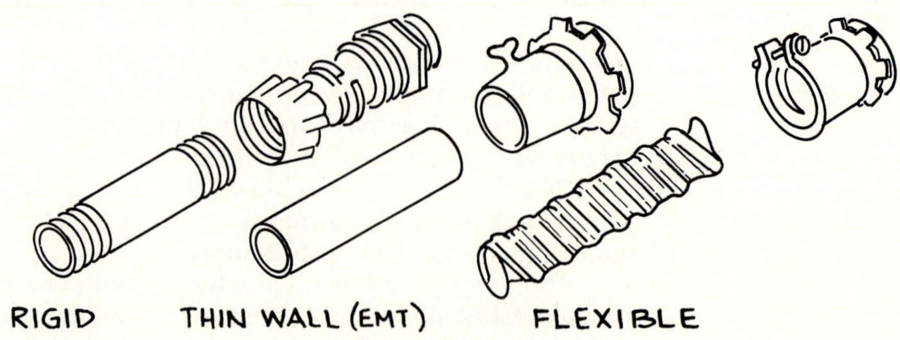


**Figure 10-6** Conduit.

be galvanized, sheradized, enameled, or asphalt covered, with required elbows, couplings, and similar screwed fittings for connection to boxes or individual lengths. Rigid steel conduit is usually used for the outside or initial service for residences but may be embedded in concrete or installed in walls or exposed locations. The more familiar electrical metallic tubing (thin-wall) is used in areas where no severe physical conditions exist, but is not allowed without special permission in hazardous locations, above 600 V, or embedded in concrete. Because of its thinner walls, this material may be readily bent and is connected by compression fittings. Where semiflexible protection is needed, as motor connections, vibration, or physical structural obstructions, flexible metal conduit is used. This material may be empty spirally wound casings or may contain wire (BX) and is available in both steel and aluminum. The classification of nonmetallic conduit includes a great variety of asbestos cement, plastic, and fiber and must be externally labeled if used for electrical distribution. Much of this type of material is used for underground and is commonly known by the trade names "Transite, Orangeburg, or PVC (polyvinyl chloride)." All nonmetallic conduit requires a separate ground-protection wire that must be included with the power wiring, as the material itself is nonconductive.

Raceways may be exposed or concealed and are most familiar when used in remodeling work or for locations where outlets are required at frequent intervals. Small sizes have a channel-form base with an interlocking top that supports the receptacles (wiremold), and is limited to 300 V. Larger sizes may be integrally divided to provide for power, lighting, or communication. Another form of raceway is a steel trough with hinged lid, often termed a *wireway*, which provides complete exposure of the enclosed wire or cable and allows taps and splices within the unit. In office buildings and similar occupancies, underfloor raceways or ducts may be incorporated to provide floor outlets at desk or equipment locations (Fig. 10-7). This type of installation is usually buried in the concrete floor slab with box outlets or flush plates at floor level. The wide variety of fittings and duct sizes provides flexibility, and these may be installed in single-level or multilevel fashion to

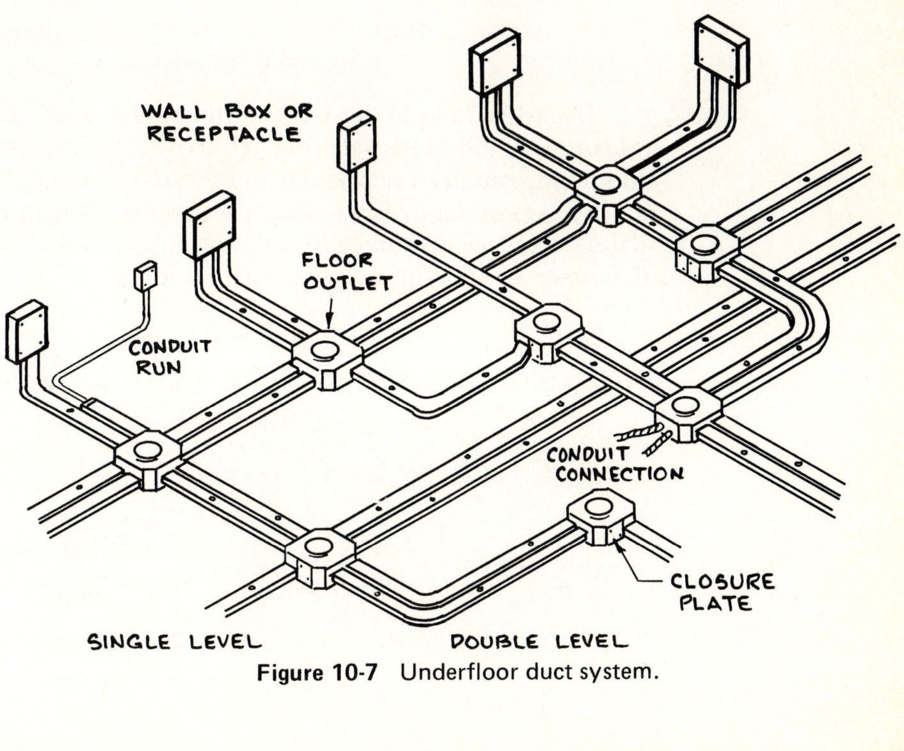


**Figure 10-7** Underfloor duct system.

distribute power, telephone, or other systems. In some circumstances, the grid may be installed below a floor but above a hung ceiling. This installation requires drilling the floor to accept the floor-level outlets but may have the advantage of greater maintenance accessibility.

Another form of raceway or duct is a cellular decking system. These systems, of either metal or concrete (see Fig. 10-8), serve both as structural elements and as ducts for underfloor wiring, and may feed either the floor above or the ceiling below. Metal cellular decking is a series of tubular spaces connected together which may be used with other tubular shapes or with standard metal decking and are usually used as formwork upon which the concrete floor slab is placed. One firm (H. H. Robertson Co.) is a major producer of this material and has trade-named it Q-Deck, which has resulted in its being called by that name generically. Another form of cellular construction is by use of precast concrete units. These may be joist sections nominally 24 in. wide to 10 ft long, or box girders in which continuous holes or tubular openings have been cast. With either metal or concrete cellular decking, a concrete topping at least $2\frac{1}{2}$ in. thick must be used and cross-headers for feeder wiring incorporated into the design. With careful planning, cellular deck systems may include heating and piping as well as electric wiring.

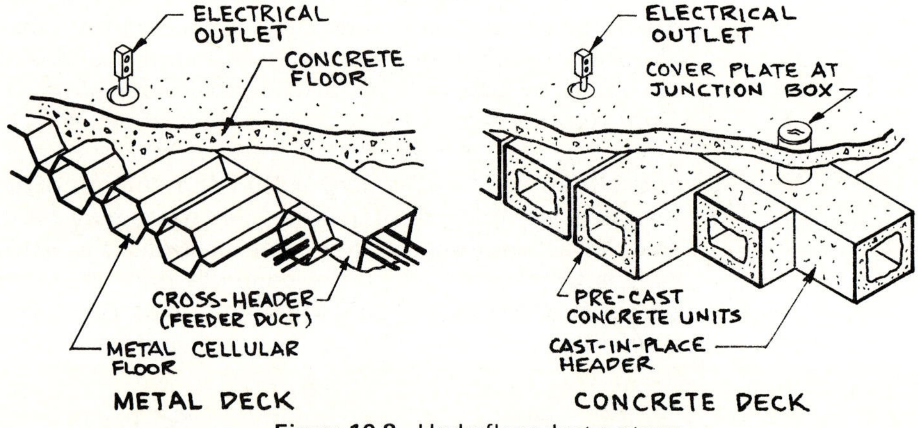


**Figure 10-8** Underfloor duct systems.

Electrical busways are still another method of distribution in commercial/industrial installations (see Fig. 10-9). In this system, solid bars of copper or aluminum, usually rectangular in section ($\frac{1}{4}$ in. × 2 in. or $\frac{1}{4}$ in. × 4 in.), are factory-assembled in an enclosing metal duct and connected with a variety of fittings to allow installation in almost any configuration. The current capacity of copper bus is figured at 1000 A/in.$^2$ of bus cross section. The bus duct

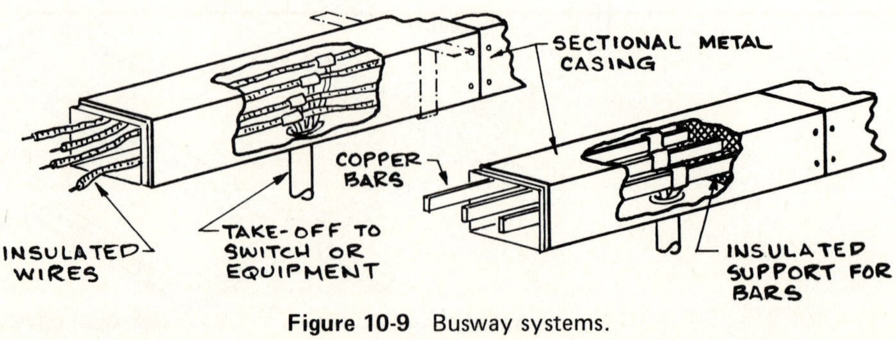


**Figure 10-9** Busway systems.

may include a number of outlets that allow devices to be directly plugged into selected circuits and is available in ratings of 100 to 4000 A and up to 5 kW. Another form of bus duct is termed cable bus and includes insulated cables rigidly mounted in a metal housing. Busways provide more compact and efficient systems (at about one-half the cost of conduit installation) but may be objectionable because of their bulky appearance.

## 10.9 WIRE AND CABLE

Without conductors there could be no electrical system, so selection of the proper size and type is of the greatest importance. Wire is sized in accordance with American Wire Gauge (AWG) standards; the larger the number, the smaller the size of the wire. For example, No. 8 AWG is larger in diameter and cross section than No. 14 AWG and is supplied as solid wire. Sizes No. 6 to No. 2 are 7-strand cable, and sizes No. 1 to No. 0000 are 19-strand cable. A 37-strand cable designated as 350 MCM has wire sized as 350,000 circular mils. Cable is available up to 2000 MCM.

Insulation of wire or cable is generally dictated by use and location, and some of the more common types and their location applications are given in Table 10-2. The NEC lists complete information in its Table 310-13.

**TABLE 10-2**
Applications of Common Wire and Cable Types

| *Type*[a] | *Material and characteristics* | *Application* | *Maximum operating temperature (°F)* |
|---|---|---|---|
| R | Rubber | Dry locations | 140 |
| RH | Heat-resistant rubber | Dry locations | 167 |
| RHH | High-temperature/ heat-resistant rubber | Dry locations | 194 |
| RHW | Heat and moisture-resistant rubber | Dry or wet | 167 |
| RVW | Moisture-resistant latex | Dry or wet | 140 |
| T | Thermoplastic | Dry locations | 140 |
| TW | Moisture-resistant thermoplastic | Dry or wet | 140 |
| THHN | Heat-resistant thermoplastic | Dry locations | 194 |
| THW | Heat- and moisture-resistant thermoplastic | Dry or wet | 167 |
| THWN | Heat- and moisture-resistant thermoplastic – nylon cover | Dry or wet | 167 |



[a]UF (underground feeder), USE (underground service), V, AVA, AVL, AVB (varnished cambric), and PFA (Perfluoro-alkoxy) are also used for a variety of locations.

There are six voltage classifications for insulation: 600, 1000, 2000, 3000, 4000, and 5000 V. All conductors are required by the NEC to have the first two numbers, which identify the maximum allowed voltage, stamped on the insulation covering. If there is no identifying number, the conductor may not be used for more than 600 V. This maximum voltage is that most often considered for residential wiring and in small installations and lighting systems.

Most wire tables indicate wire sizes by circular mils (CM), which is a circle 0.001 in. in diameter. To find the area of a round conductor, simply square the diameter. If the section is not circular, however, as is true with bus bars, the circular-mil area may be found by using the following formulas:

$$\text{square mils} = \text{circular mils} \times 0.7854$$
$$\text{circular mils} = \text{square mils} \times 1.273$$

**EXAMPLE**

Assume a bus bar ¼ in. × 2 in. What is the circular-mil equivalent?

$$\tfrac{1}{4} = 250 \text{ mils}$$

$$2 = 2000 \text{ mils}$$

$$250 \times 2000 = 500{,}000 \times 1.273$$

$$\text{CM} = 636{,}500$$

Resistance in a conductor is directly related to the cross-sectional area and the length of the run. All metals will act as conductors of electricity, with gold and silver being most conductive, but because of cost, most conductors are made of copper or aluminum. Resistance varies inversely with the cross-sectional area and directly with the length:

$$R = r\frac{L}{d^2} \qquad \text{or} \qquad R = r\frac{L}{A}$$

where $R$ is the resistance, $r$ the material resistance, $L$ the length in feet, $d$ the diamter in mils, and $A$ the area in circular mils. Copper has a resistivity of 10.4 per foot, and aluminum has a resistivity of 17.0 per foot.

The term *ampacity*, when applied to conductors, means the maximum current in amperes that can be carried in a conductor with safety, in free air. This rating is established by conductor material, type of insulation, and conductor size, at a temperature not over 30°C (86°F). Obviously, the temperature may differ, the conductor may be enclosed in conduit or other protective covering in actual installation, and seldom is one conductor installed alone. Allowable ampacity ratings for various sizes and insulations, together with derating correction factors for different temperatures above 30°C, are given in Table 10-3, taken from NEC Table 310-18.

**TABLE 10–3**
Allowable Ampacities for Insulated Conductors Rated 0 to 2000 Volts, 110 to 250°C[a]

| *Size* | *Temperature rating of conductor. See Table 310-13* | | | | | | | | *Size* |
|---|---|---|---|---|---|---|---|---|---|
| | *110°C (230°F)* | *125°C (257°F)* | *150°C (302°F)* | *200°C (392°F)* | *250°C (482°F)* | *110°C (230°F)* | *125°C (257°F)* | *200°C (392°F)* | |
| | *Types AVA, AVL* | *Types AI, AIA* | *Type Z* | *Types A, AA, FEP, FEPB, PFA* | *Types PFAH, TFE* | *Types AVA, AVL* | *Types AI, AIA* | *Types A, AA* | |
| *AWG MCM* | *Copper* | | | | *Nickel or nickel-coated copper* | *Aluminum or copper-clad aluminum* | | | *AWG MCM* |
| 14 | 30 | 30 | 30 | 30 | 40 | | | | |
| 12 | 35 | 40 | 40 | 40 | 55 | 25 | 30 | 30 | 12 |
| 10 | 45 | 50 | 50 | 55 | 75 | 35 | 40 | 45 | 10 |
| 8 | 60 | 65 | 65 | 70 | 95 | 45 | 50 | 55 | 8 |

**TABLE 10-3** *(cont.)*

| | | | | | | | | | |
|---|---|---|---|---|---|---|---|---|---|
| 6 | 80 | 85 | 90 | 95 | 120 | 60 | 65 | 75 | 6 |
| 4 | 105 | 115 | 115 | 120 | 145 | 80 | 90 | 95 | 4 |
| 3 | 120 | 130 | 135 | 145 | 170 | 95 | 100 | 115 | 3 |
| 2 | 135 | 145 | 150 | 165 | 195 | 105 | 115 | 130 | 2 |
| 1 | 160 | 170 | 180 | 190 | 220 | 125 | 135 | 150 | 1 |
| 0 | 190 | 200 | 210 | 225 | 250 | 150 | 160 | 180 | 0 |
| 00 | 215 | 230 | 240 | 250 | 280 | 170 | 180 | 200 | 00 |
| 000 | 245 | 265 | 275 | 285 | 315 | 195 | 210 | 225 | 000 |
| 0000 | 275 | 310 | 325 | 340 | 370 | 215 | 245 | 270 | 0000 |
| 250 | 315 | 335 | | | | 250 | 270 | | 250 |
| 300 | 345 | 380 | | | | 275 | 305 | | 300 |
| 350 | 390 | 420 | | | | 310 | 335 | | 350 |
| 400 | 420 | 450 | | | | 335 | 360 | | 400 |
| 500 | 470 | 500 | | | | 380 | 405 | | 500 |
| 600 | 525 | 545 | | | | 425 | 440 | | 600 |
| 700 | 560 | 600 | | | | 455 | 485 | | 700 |
| 750 | 580 | 620 | | | | 470 | 500 | | 750 |
| 800 | 600 | 640 | | | | 485 | 520 | | 800 |
| 1000 | 680 | 730 | | | | 560 | 600 | | 1000 |
| 1500 | 785 | | | | | 650 | | | 1500 |
| 2000 | 840 | | | | | 705 | | | 2000 |

Correction factors

| Ambient temp. (°C) | *For ambient temperatures over 30° C, multiply the ampacities shown above by the appropriate correction factor to determine the maximum allowable load current.* | | | | | | | | Ambient temp. (°F) |
|---|---|---|---|---|---|---|---|---|---|
| 31-40 | .94 | .95 | .96 | | | .94 | .95 | | 87-104 |
| 41-45 | .90 | .92 | .94 | | | .90 | .92 | | 105-113 |
| 46-50 | .87 | .89 | .91 | | | .87 | .89 | | 114-122 |
| 51-55 | .83 | .86 | .89 | | | .83 | .86 | | 123-131 |
| 56-60 | .79 | .83 | .87 | .91 | .95 | .79 | .83 | .91 | 132-141 |
| 61-70 | .71 | .76 | .82 | .87 | .92 | .71 | .76 | .87 | 142-158 |
| 71-75 | .66 | .72 | .79 | .86 | .89 | .66 | .72 | .86 | 159-167 |
| 76-80 | .61 | .68 | .76 | .84 | .87 | .61 | .69 | .84 | 168-176 |
| 81-90 | .50 | .61 | .71 | .80 | .83 | .50 | .61 | .80 | 177-194 |
| 91-100 | | .51 | .65 | .77 | .80 | | .51 | .77 | 195-212 |
| 101-120 | | | .50 | .69 | .72 | | | .69 | 213-248 |
| 121-140 | | | .29 | .59 | .59 | | | .59 | 249-284 |
| 141-160 | | | | | .54 | | | | 285-320 |
| 161-180 | | | | | .50 | | | | 321-356 |
| 181-200 | | | | | .43 | | | | 357-392 |
| 201-225 | | | | | .30 | | | | 393-437 |



[a]Not more than three conductors in raceway or cable based on ambient temperature of 30°C (86°F).

## 10.10 BRANCH CIRCUITS

Finding the requirements for branch conductors and the sizes for main service is somewhat similar to finding the required size of a building foundation. With foundation design, the calculations start with the roof, then the top floor, then intermediate floors, and finally the lower floor or basement, including in each case the loads of the construction material and anticipated occupancy loads. Electrical conductor sizes are determined in exactly the same manner, with branch circuits being analyzed, then feeders to which the branch circuits are connected, until finally the total loads of feeders are used to determine the service conductor sizes.

The National Electrical Code (210-3) classifies five ampacity ratings: 15, 20, 30, 40, and 50 amperes maximum. The ground conductor of a branch circuit is required to be continuous white or gray in color, except that grounding conductors for equipment are continuous green unless they are bare wire. In addition, the voltage between conductors on branch circuits for lighting, receptacles, or movable appliances for residential installation must not exceed 150 V. All residential circuits of 15 or 20 A, 120 V, must be provided with ground-fault circuit interrupters (GFCIs) for circuits in bathrooms and garages. Some of the minimum conductor sizes, ampere ratings, and other information is given in Table 10-4, which is taken from NEC Table 210-24.

**TABLE 10-4**
Summary of Branch-Circuit Requirements[a]

| | *Circuit rating (A)* | | | | |
|---|---|---|---|---|---|
| | *15* | *20* | *30* | *40* | *50* |
| Conductors (min. size) | | | | | |
| Circuit wires[b] | 14 | 12 | 10 | 8 | 6 |
| Taps | 14 | 14 | 14 | 12 | 12 |
| Fixture wires and cords | Refer to Section 240-4 | | | | |
| Overcurrent protection (A) | 15 | 20 | 30 | 40 | 50 |
| Outlet devices: | | | | | |
| Lampholders permitted | Any type | Any type | Heavy duty | Heavy duty | Heavy duty |
| Receptable rating[c] | 15 max. | 15 or 20 | 30 | 40 or 50 | 50 |
| Maximum load (A) | 15 | 20 | 30 | 40 | 50 |
| Permissible load | Refer to Section 210-23(a) | Refer to Section 210-23(a) | Refer to Section 210-23(b) | Refer to Section 210-23(c) | Refer to Section 210-23(c) |



[a]Type FEP, FEPB, RUW, SA, T, TW, RH, RUH, RHW, RHH, THHN, THW, THWN, and XHHW conductors in raceway or cable.

[b]These ampacities are for copper conductors where derating is not required, See Tables 310-16 through 310-19.

[c]For receptacle rating of cord-connected electric-discharge lighting fixtures, see Section 410-30(c).

Some wire sizes are established by the NEC. The code requires No. 14 as the minimum size for branch-circuit wiring, but good design would require No. 12 instead. Ranges and similar equipment are required to use No. 12 instead. Ranges and similar equipment are required to use No. 8 minimum, but No. 6 is better design. When aluminum conductors are used, they should always be at least one size larger than copper. If the ambient temperature exceeds 30°C (86°F), the current-carrying capacity of conductors should be derated as noted at the bottom of Table 10-4. Circuits that supply motors must be rated at 25% greater than motor load.

Branch circuits are designed by using several factors, as follows:

1. Determine the insulation type that suits the location.
2. Two- or three-wire circuits at any power factor:

$$\text{line current} = \frac{\text{volt-amperes of connected single-phase load}}{\text{line voltage}}$$

3. Three-phase circuits:

$$\text{line current} = \frac{\text{volt-amperes of balanced three-phase load}}{\text{line voltage} \times 1.732}$$

4. Select the conductor size from Table 10-3 (NEC 310-18);
5. Select the conduit size from Table 10-5 (NEC Table 3A) which will properly contain three wires or more.

**EXAMPLE**

Assume a 120-V two-wire single-phase lighting circuit, with 10 lamps of 100 W each to be installed in a dry location without heat problems.

1. Use type RUH or T insulation good to 140°F.
2. 1000/120 = 8.3 A. Use a 15 A breaker.
3. Table 10-3: Select No. 14 wire as the minimum size.
4. Table 10-5: ½-in. conduit will carry up to nine No. 14 wires. Better design would suggest No. 12 wire, which would still be accommodated in the ½-in. conduit.

**TABLE 10-5**
Maximum Number of Conductors in Trade Sizes of Conduit or Tubing

| | *Conductor size* | *Conduit trade size (in.)* | | | | | | | | | | | | |
|---|---|---|---|---|---|---|---|---|---|---|---|---|---|---|
| *Type letters* | *AWG (MCM)* | ½ | ¾ | 1 | 1¼ | 1½ | 2 | 2½ | 3 | 3½ | 4 | 4½ | 5 | 6 |
| TW, T, RUH, RUW, XHHW (14 thru 8) | 14 | 9 | 15 | 25 | 44 | 60 | 99 | 142 | | | | | | |
| | 12 | 7 | 12 | 19 | 35 | 47 | 78 | 111 | 171 | | | | | |
| | 10 | 5 | 9 | 15 | 26 | 36 | 60 | 85 | 131 | 176 | | | | |
| | 8 | 2 | 4 | 7 | 12 | 17 | 28 | 40 | 62 | 84 | 108 | | | |
| RHW and RHH (without outer covering), THW | 14 | 6 | 10 | 16 | 29 | 40 | 65 | 93 | 143 | 192 | | | | |
| | 12 | 4 | 8 | 13 | 24 | 32 | 53 | 76 | 117 | 157 | | | | |
| | 10 | 4 | 6 | 11 | 19 | 26 | 43 | 61 | 95 | 127 | 163 | | | |
| | 8 | 1 | 3 | 5 | 10 | 13 | 22 | 32 | 49 | 66 | 85 | 106 | 133 | |
| TW, T, THW, RUH (6 to 2), RUW (6 to 2) | 6 | 1 | 2 | 4 | 7 | 10 | 16 | 23 | 36 | 48 | 62 | 78 | 97 | 141 |
| | 4 | 1 | 1 | 3 | 5 | 7 | 12 | 17 | 27 | 36 | 47 | 58 | 73 | 106 |
| | 3 | 1 | 1 | 2 | 4 | 6 | 10 | 15 | 23 | 31 | 40 | 50 | 63 | 91 |
| | 2 | 1 | 1 | 2 | 4 | 5 | 9 | 13 | 20 | 27 | 34 | 43 | 54 | 78 |
| | 1 | | 1 | 1 | 3 | 4 | 6 | 9 | 14 | 19 | 25 | 31 | 39 | 57 |
| FEPB (6 to 2), RHW and RHH (without outer covering) | 0 | | 1 | 1 | 2 | 3 | 5 | 8 | 12 | 16 | 21 | 27 | 33 | 49 |
| | 00 | | 1 | 1 | 1 | 3 | 5 | 7 | 10 | 14 | 18 | 23 | 29 | 41 |
| | 000 | | 1 | 1 | 1 | 2 | 4 | 6 | 9 | 12 | 15 | 19 | 24 | 35 |
| | 0000 | | | 1 | 1 | 1 | 3 | 5 | 7 | 10 | 13 | 16 | 20 | 29 |
| | 250 | | | 1 | 1 | 1 | 2 | 4 | 6 | 8 | 10 | 13 | 16 | 23 |
| | 300 | | | 1 | 1 | 1 | 2 | 3 | 5 | 7 | 9 | 11 | 14 | 20 |
| | 350 | | | | 1 | 1 | 1 | 3 | 4 | 6 | 8 | 10 | 12 | 18 |
| | 400 | | | | 1 | 1 | 1 | 2 | 4 | 5 | 7 | 9 | 11 | 16 |
| | 500 | | | | 1 | 1 | 1 | 1 | 3 | 4 | 6 | 7 | 9 | 14 |
| | 600 | | | | | 1 | 1 | 1 | 3 | 4 | 5 | 6 | 7 | 11 |
| | 700 | | | | | 1 | 1 | 1 | 2 | 3 | 4 | 5 | 7 | 10 |
| | 750 | | | | | 1 | 1 | 1 | 2 | 3 | 4 | 5 | 6 | 9 |

If the ampere load had been greater because of additional loading, two alternatives would have been available: fewer load units to reduce the total load, or a larger wire size. Such a situation would also require a larger circuit breaker or fuse to protect the circuit.

## 10.11 FEEDER CIRCUITS

Most single-family residence electrical systems do not require "feeder" circuits, as the branch circuits are wired directly to the circuit-breaker panel. However, multiple apartments or condominiums often have main panels, which in turn have feeder circuits to subpanels to provide for entrance and corridor lighting, recreation rooms, public toilets, and other public spaces. Separate apartments may be connected in the same manner as single-family units or may be fed from a master main panel. In a similar manner, small commercial or industrial projects may not have feeder circuits, but larger installations will almost always require feeders to subpanels or transformers.

To calculate the ampacity of feeders, and thus the conductor size, the same procedure is required as for branch circuits, except that the ampacity required is simply a total of all the branch-circuit loads. Upon finding the total, a major decision regarding the type of system and the voltage required must be made. In locations with light to limited loads, the most common type is 120/208-V, wye three-phase four-wire (see Fig. 10-10). For commercial and industrial buildings with substantial motor loads, a 480-V three-phase three-wire distribution is used, with the 480 V carried to substations and directly to motors but to step-down transformers for lighting and receptacle circuits. Still another popular system is a 480/277-V wye three-phase four-wire (see Fig. 10-10), where the 480-V three-phase four-wire feeders go to each power location for motors, 277 V is used for fluorescent lighting, and 120/208-V single-phase is obtained from step-down transformers for incandescent lighting and receptacle circuits. For heavy motor loads, 4160/2400-V wye three-phase four-wire feeders may connect directly to motors and again, using load centers, to step down to 480 V for lighter motors and to other transformers for 120/240-V or 120/208-V circuits. The selection of feeder-conductor sizes depends almost entirely on the system used, which in turn depends a great deal on the number and size of motors, lighting units, and other connected devices throughout the building.

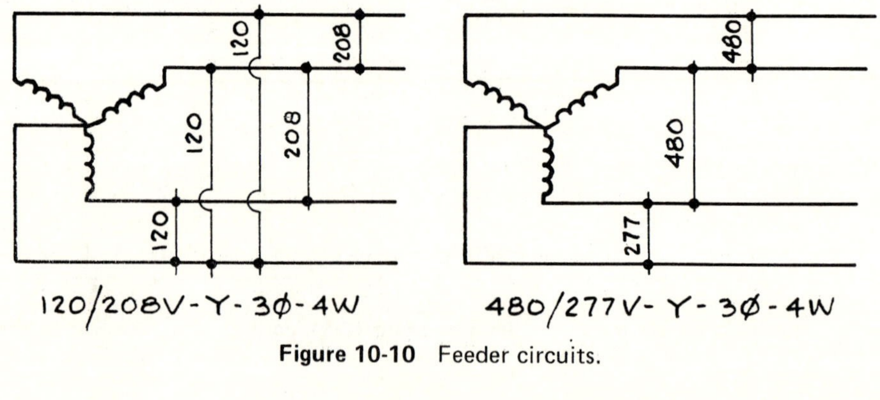


**Figure 10-10** Feeder circuits.

The ampacity of feeder conductors must not be less than 30 when two or more branch circuits are supplied and where the feeder carries the entire load of the service. Entrance conductors must have an ampacity not lower than that of the service conductors. Feeders using a common neutral may supply two or three sets of three-wire feeders or two sets of four- or five-wire feeders. Where a feeder supplies continuous loads and noncontinuous loads, the ampacity of the feeder must be that of the noncontinuous load plus 125% of the continuous load. In addition, demand factors may be applied to feeders to reduce required capacity, as shown in Table 10-6.

**TABLE 10-6**
Lighting-Load Feeder Demand Factors

| *Type of occupancy* | *Portion of lighting load to which demand factor applies (wattage)* | *Demand factor percent* |
|---|---|---|
| Dwelling units | First 3000 or less at: | 100 |
| | Next 3001 to 120,000 at: | 35 |
| | Remainder over 120,000 at: | 25 |
| Hospitals[a] | First 50,000 or less at: | 40 |
| | Remainder over 50,000 at: | 20 |
| Hotels and motels[a] — including apartment houses without provision for cooking by tenants | First 20,000 or less at: | 50 |
| | Next 20,001 to 100,000 at: | 40 |
| | Remainder over 100,000 at: | 30 |
| Warehouses (storage) | First 12,500 or less at: | 100 |
| | Remainder over 12,500 at: | 50 |
| All others | Total wattage | 100 |



[a]The demand factors of this table shall not apply to the computed load of feeders to areas in hospitals, hotels, and motels where the entire lighting is likely to be used at one time, as in operating rooms, ballrooms, or dining rooms.

## 10.12 FEEDER CALCULATIONS

Several factors are involved in commercial/industrial installations that are rarely found in residential work, although these factors may apply to multiple apartments or condominiums. Two factors, in particular, are in constant use: demand factor and diversity factor. *Demand factor* is the ratio of maximum demand to the total connected load, whereas *diversity factor* is the ratio of the sum of individual maximum demands of parts of the system to the maximum demand of the entire system. This factor varies between 1.00 and 2.00.

Another factor is voltage drop, which occurs in conductors with long runs. The size of feeder conductors should control this factor so that there is not more than a 1% drop in lighting loads and not more than 2% for power and heating loads. Allowance for load growth should also be included, both for future operation and current economy. Most power circuits are loaded to 50% capacity, while the NEC allows up to 80% limitation when circuits are in operation for long periods or are used for motor/lighting combinations.

Several formulas are used to find the current rating, power loss, and voltage drop, as shown in the following examples.

**EXAMPLE**

Assume a three-phase four-wire system at 480 V between one "hot" wire and the neutral wire, operating at 100 A.

$$I = \frac{\text{load (volts} \times \text{amperes)}}{K \times E}$$

where $K = 3.0$ for three-phase four-wire alternating current, and $E =$ 480 V.

$$I = \frac{480 \times 100}{3.0 \times 480} = \frac{100}{3.0} = 33.33 \text{ A}$$

NEC Table 310-18 allows up to 35 A for No. 12 wire when used in raceway or cable. Good design would indicate the use of No. 10 wire.

Power loss is calculated by

$$P = \frac{4 \times K \times L \times I^2}{\text{CM}}$$

where $P$ is power lost (watts), $K$ the resistance of the conductor (ohms/ft), $I$ the current in each wire (amperes), $L$ the one-way conductor length (50 ft), and CM the circular-mil cross section of No. 10 wire. Therefore,

$$P = \frac{4 \times 1.018 \times 50 \times 33.3^2}{10{,}380} = \frac{226{,}176.98}{10{,}380} = 21.789 \text{ W}$$

For voltage drop, first calculate for a two-wire single-phase system using the formula

$$\frac{V}{2} = \frac{2K \times L \times I}{d^2}$$

where $V$ is the voltage drop (volts), $R$ the resistance of conductor (ohms/ft), $I$ the current in the conductor (amperes), $d^2$ the circular-mil cross section, $K$ the resistance of the conductor (cir. mil-ohms/ft), and $L$ the one-way circuit length (feet). $K$ may be assumed as 11 for circuits loaded less than 50%.

$$\frac{V}{2} = \frac{2(11) \times 50 \times 33.33}{10{,}380} = \frac{3.532}{2} = 1.766 \text{ V}$$

If the allowed voltage drop is 2%, a drop of 1.766 V equals 0.38%, using the formula $(I \times Z)/V$, where $A$ is the total impedance (ohms).

## 10.13 SUBSTATIONS

The modern electrical system for large commercial/industrial buildings depends on substations or "load centers" for efficient operation. These units are made up of three primary sections; incoming alternating-current line voltage, transformers, and secondary control section. The incoming control may be circuit breakers, interrupter switches, or simply junction boxes. Transformers are usually step-down oil, askarel, ventilated or sealed dry type, but may also be step-up, depending on the voltage requirements. The secondary control system may be circuit breakers or motor control equipment.

The most common, and usually most economical type of substation is a radial system, where a centrally located load center supplies power to all parts of the building. This may be advantageous as long as the facility stays small and the motor-operated loads stay permanently located. If enough initial installation is made to provide about 15 V-A per square foot, this system may be able to accommodate minor changes.

Disadvantages are those commonly associated with only one of anything: the breakdown of conductors, switching, or transformers usually means a shutdown until repairs can be made. Much of this disadvantages may be overcome by using two smaller load centers, with a duplicate connection to the main source. The basic radial system is diagrammed in Fig. 10-11, as are two other systems. Single-end substations will generally satisfy most needs, but double-end units may be best where space is limited or power requirements vary considerably. If transformer characteristics in all load centers can be kept alike, the disadvantage of transformer failure may be rather easily overcome by having a spare transformer available for replacement.

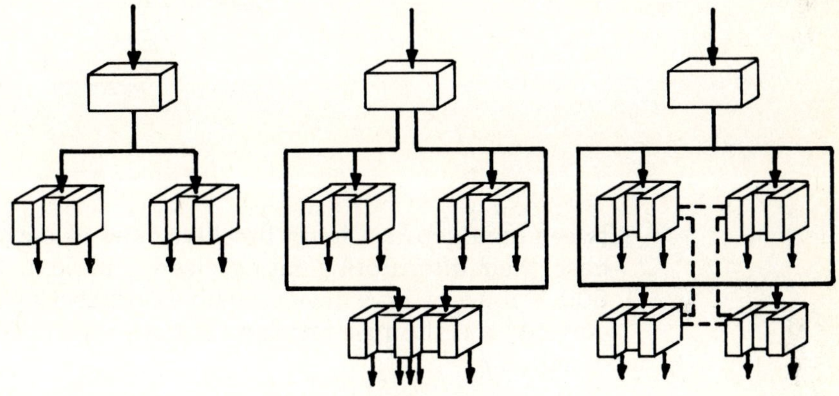


Figure 10-11

## 10.14 CIRCUIT PROTECTION

Protection of an electrical system is twofold: (1) protection of the equipment, and (2) protection of personnel operating the equipment. If the first is accomplished adequately, the safety of personnel is rather automatic except in special accidents. Damage to equipment, loss of production, and extra expense result from downtime caused by electrical faults. The location of typical protection devices is shown in Fig. 10-12.

Short circuits in a system occur when an accidental grounding is produced by failure of insulation, overloads, or physical damage to equipment. When a short circuit happens, the impedance is lowered and the protective device trips. This device may be a circuit breaker, fuses, or a combination, and are rated from 15 to 6000 A. Thermal cutouts, thermal relays, and similar devices are not approved protection units. The use of plug fuses (screw base) is prohibited where circuits operate at more than 125 V be-

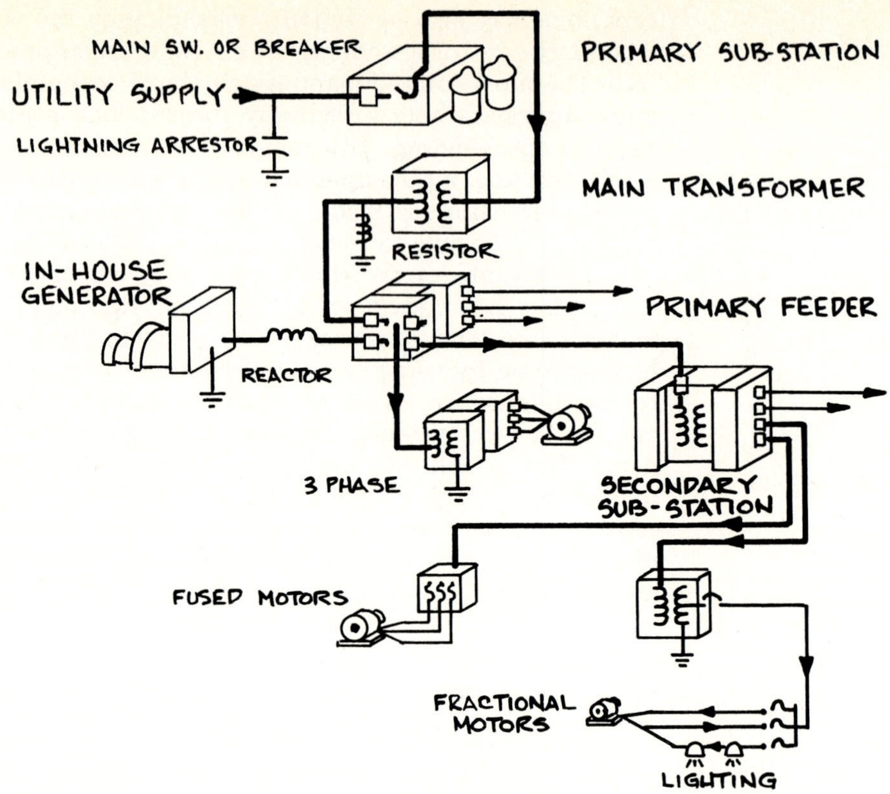


Figure 10-12

tween conductors. Circuit breakers must be capable of manual operation and have their interrupting rating clearly marked. Overcurrent protection over 600 V must be by a device capable of detecting and interrupting all values of current, and the fuse rating must not exceed three times the ampacity of the conductors.

## 10.15 SUMMARY

The primary function of a wiring system is to carry the electricity from the supplier to the user. To do this properly requires the correct type of wire, its insulation, and carrier for the wire. The two major factors involved in designing a system are the voltage and the current required, and the voltage may vary from 120/240-V single-phase to 4160-V three-phase four-wire. Rarely do single-family residences need more than the 120/240-V power, but commercial/industrial installations may require everything up to and including 4160 V.

Residential wiring systems are relatively simple, with nonmetallic cable or conduit as a carrier of the wire and probably with only one circuit-breaker control panel. A similar condition may exist for small commercial projects, but larger commercial/industrial installations will usually require a large main service panel, plus one or more substations or load centers convenient to major power uses. By using load centers, it is possible to provide "line" voltage, which may be 4160 V, or by transformers to provide 480, 277, 240, 208, 120, or even 24 V. Conductors may be carried in conduit, rigid steel tubing, raceways, underfloor ducts, or cellular decking, sized to include from 3 to 137 wires of different sizes or with different insulation.

Protection of electrical systems is accomplished by using circuit breakers, fuses, or switches at main panels, substations, load centers, individual lighting circuits, and at motors and other devices. Such steps are necessary to protect the equipment, the structure, and the personnel who operate the equipment or live in the building.

The design of wiring systems may be simple or very complicated, depending on the occupancy of the building and the lighting and power requirements. This chapter is only an overview of some possibilities but should provide an insight into the general characteristics of wiring for both residential and commercial/industrial electrical support systems for buildings. Complex requirements should be turned over to competent engineers.

# 11

# Lighting

## 11.1 INTRODUCTION

Lighting, or the phenomenon of illumination, is one of the major uses of electricity that is reasonably familiar to nearly everyone, yet is not well understood and frequently improperly designed. Radiant energy produced by electricity has a definite wavelength similar to those of radio, television, and x-rays and is directly related to temperature. When this wavelength is longest ($10^{-9}$ meter), a heat-absorbing body will become dull red, but as the wavelength decreases and more heat is absorbed, the color changes from red to orange to yellow to green and finally to blue. If the light source produces energy over the entire spectrum uniformly, the color white is produced.

When the color white is directed at a colored surface at an angle to that surface, the various color rays are absorbed by that surface and the ray of the surface color is reflected most strongly. If the color white is directed through a transparent surface, the same thing occurs, with the rays that match transparency being stronger. Reflection of a surface also affects a light. Light rays projected against a very smooth, polished, opaque surface will reflect at the same angle that they strike the surface and reflection will be about 80% of incidence value. If the surface is rough or irregular, the incident light is directed at many different angles by each irregular surface and appears uniformly bright from any angle. When both specular and diffused reflection are present, the surface will provide a bright background surface but will also mirror the light source. If the incident light strikes a transparent surface instead of a reflective surface, the rays will pass through the clear material slightly refracted but will emerge on the same angle as it entered, and at about 85% of the original incident light. Where the surface is of a diffusing material, the emerging light on the "back" side will be of uniform brightness due to the rays emerging at many different angles.

Four other factors must also be included when considering light and seeing. *Brightness* depends on the quantity of light falling on an object and the reflecting ability of the object. *Object size* is a second factor. Generally, the smaller the size of an object, the more light that is required to see it properly; more light is required to see a fly than to see an elephant. *Contrast* is a third important factor and is increased by differences in color, shade, or background. Objects moving swiftly through the field of vision require more light to slow them down for proper sight, so *time* is a fourth factor. Each of these, plus the reflection or diffusion, should be considered when designing building lighting systems.

## 11.2 LIGHTING LANGUAGE

A discussion of lighting must of necessity use words that are understandable. The following are some terms of practical applications.

*Candlepower:* A unit of light intensity based on the average of all candlepower in every direction about a standard candle.

*Cosine law:* Light is measured perpendicular to the light source. When this is not possible, the illumination is measured by the formula

$$\text{footcandles} = \frac{\text{candlepower} \times \text{cosine of angle from the perpendicular}}{\text{distance squared}}$$

*Distribution curve:* A plotted curve with all points equidistant from the light source, indicating the light intensity at various angles.

*Efficacy:* The ability of a light source to convert electrical energy into luminous energy, expressed in lumens per watt (lm/W).

*Efficiency:* The ratio of light output to energy input, expressed in lumens per watt.

*Footcandles:* The quantity of light on a square foot of surface; equal to lumens divided by area.

*Foot-lambert:* The unit of brightness directed at the observer, whether reflected or directly from the light source.

*Illumination:* The number of lumens falling on a surface measured in units of area. Illumination is denoted by the unit footcandle, which is equal to one lumen per square foot of surface.

*Intensity:* The candlepower value of a light source from one direction. Different light sources have different candlepower values, depending on the angle of direction.

*Inverse-square law:* A theoretical law in which illumination by a single light point decreases in proportion to the square of the distance from the light source (see Fig. 11-1). Illumination of one candlepower covers one square foot at a distance of one foot, but an area of one-fourth, or four times the area, at a distance of 2 feet.

*Lumen:* A quantity of light on a surface of one square foot a distance of one foot from a one-candlepower light source.

*Quantity:* Measured in lumens from a light source. One candlepower source equals 12.57 lumens, since a point one foot equidistant from a standard candle is a portion of a sphere with a radius of one foot.

*Reflected light:* Light that can be seen as reflected, not absorbed, from a surface. Foot-lamberts = footcandles × reflection factor.

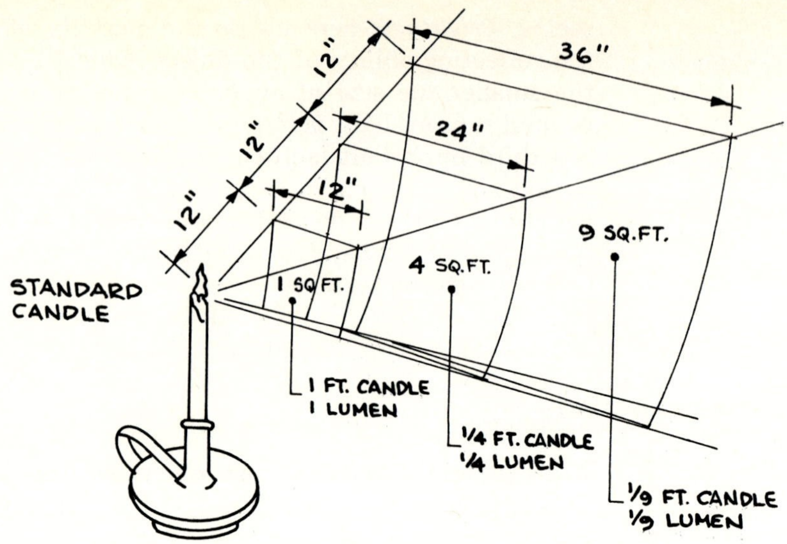


**Figure 11-1** Inverse square light law.

*Reflection factor:* The percentage of light reflected from a surface, expressed as a decimal.

*Spectrum:* The wavelength range given in angstrom units of visible light from red through violet where one angstrom equals $10^{-9}$ centimeter.

*Watt:* The unit of electrical power derived from multiplying one volt of pressure by one ampere of current. Most electrical devices, including lamps, are rated in watts.

## 11.3 LIGHT SOURCES

Lighting may be roughly divided into two major groups: the lamps providing the light, and the fixtures holding the lamp. The design of each has a definite effect on the quantity, direction, and efficiency of a lighting system. Principal artificial (electrical) sources of light are incandescent, fluorescent, and vapor types. Selection of a light source should be determined, or at least influenced, by three factors: light depreciation, life expectancy, and efficiency of the unit. Shape and size, operating temperature, operating voltage, frequency of starts/stops, and proposed location in the building also influence selection of the lamp type.

## 11.4 INCANDESCENT LAMPS

The incandescent lamp, constructed as a glass bulb enclosing a wire filament, is the oldest type of electrical light and still the most common. Compared with fluorescent or mercury-vapor types, the incandescent lamp is cheaper to purchase, easier to install, starts instantly, and is free of stroboscopic effects; but it also has a shorter expected life and lower efficiency. General lighting lamps range in rated wattage from $7\frac{1}{2}$ to 1500 W, with the majority in the range 25 to 150 W. Shapes have been determined by use and location, and typical shapes are shown in Fig. 11.2.

Lamp construction is basically similar for all types (see Fig. 11-3), with the exception of changes in bulb shape and arrangement of filament. The

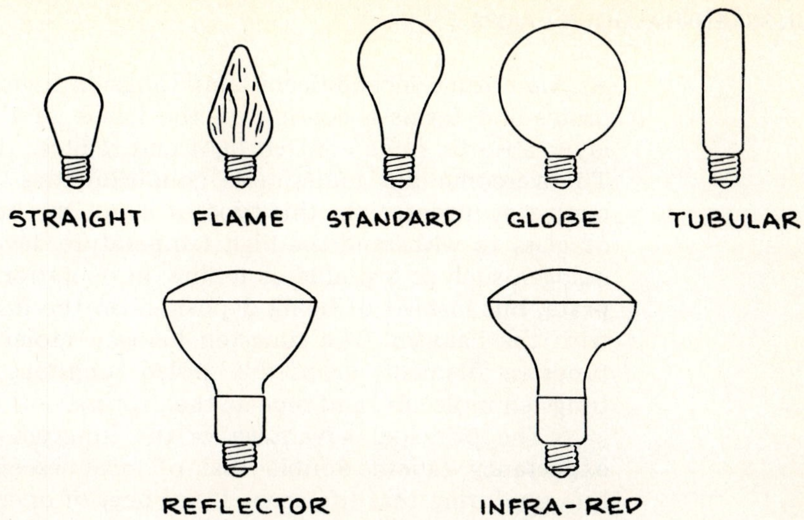


**Figure 11-2** Incandescent lamp shapes.

filament is a length of tungsten wire, usually tightly coiled, which is held by heavier wire supports. Special designs for the filament may be necessary when the normal operating position is "bulb up," or for special uses. In the manufacturing process, the air trapped in the glass bulb is removed to prevent the filament from burning up at its operating temperature of 3500 to 5000°F, and may be replaced with inert argon or nitrogen gas. Gradual evaporation of the tungsten will create a dark coating inside the bulb, reducing efficiency and eventually causing failure as the tungsten disintegrates completely.

Compared to fluorescent or mercury vapor lamps, the incandescent is low in original cost and easy to replace. Frequency of starts and stops has little effect on incandescent lamps, but excessive vibration may break the filament wires. Special R or PAR shapes may be used where light should be directed, as these shapes have a built-in reflector designed to focus the light away from the base.

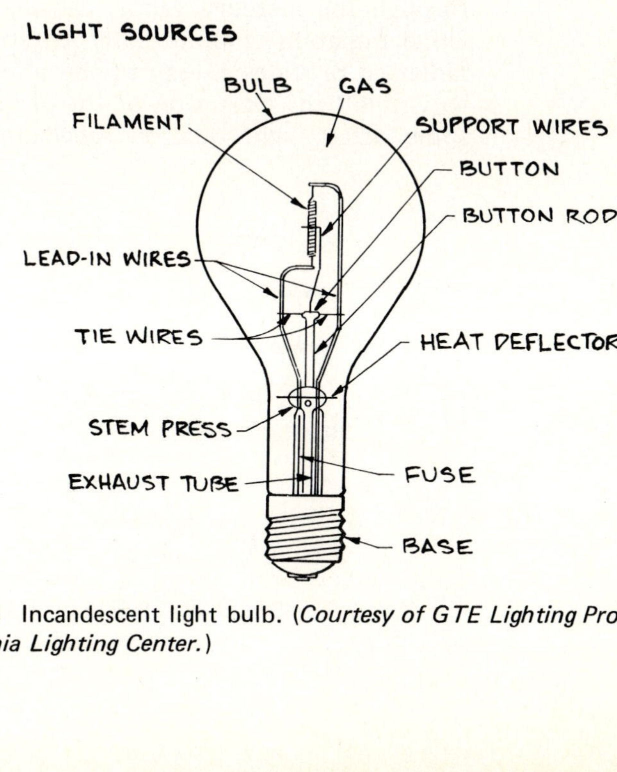


**Figure 11-3** Incandescent light bulb. (*Courtesy of GTE Lighting Products; Sylvania Lighting Center.*)

## 11.5 TUNGSTEN-HALOGEN LAMPS

As the normal incandescent lamp tungsten filament is heated, it slowly evaporates and forms a coating on the inside of the enclosing glass bulb. This layer absorbs some emitted light and reduces the light output of the lamp. To overcome this undesirable condition, the tungsten–halogen, or quartz, lamp was invented. In this type of lamp, the bulb is made of quartz instead of glass to withstand the high temperature developed. The lamp contains a halogen such as bromide or iodine. In operation, the tungsten tends to evaporate, but instead of being deposited on the inside of the bulb, it combines with the halogen. The tungsten–halogen molecule then returns to the hot tungsten filament, drops the cooled tungsten, combines with another hot tungsten molecule, and repeats the process.

The principal advantage of the tungsten–halogen lamp is greater life expectancy (about double that of incandescent nongas types), plus more lumens during that life span. If the heat of operation is reduced by dimming or ventilating, the rate of combining the tungsten and halogen molecules is also reduced, resulting in some tungsten deposits on the inner surface of the bulb and a reduction of lumen output. Another advantage is the smaller envelope size for the same wattage.

## 11.6 FLUORESCENT LAMPS

Whereas the light emitted from an incandescent lamp is generated by the high resistance caused by the tungsten filament in a complete circuit, the light created in a fluorescent lamp is created by an electrical discharge between two nonconnected terminals. The lamp body is a long glass tube (Fig. 11-4), from 12 to 96 inches long and about 1½ inch in diameter, coated inside with a phosphor coating. Each end of the tube is sealed with a filament, a drop or two of mercury, with some argon or krypton gas introduced to provide easy starting.

Light is produced by the electron stream from the electricity flowing through the mercury vapor, causing high-speed collisions that create ultraviolet radiation of about 2537-Angstrom wavelength. In turn, the ultra-violet radiation provides energy to the phosphor coating of the tube, converting it to visible light. Selection of the proper phosphor compound is important, as some affect light quality, especially where color matching is important.

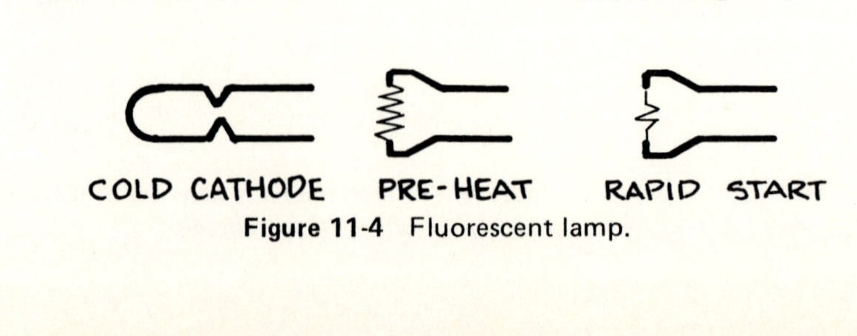


**Figure 11-4** Fluorescent lamp.

Three major types of fluorescent lamps are in common use: preheat, instant-start, and cold-cathode. Preheat lamps have bi-pin (two-prong) bases at each end, and each lamp has a starter designed to heat the filament to correct emission temperature. Instant-start lamps operate at higher voltage and require no preheating. Cold-cathode lamps operate at high voltage and low current and depend on a coated iron terminal to generate electrons, which in turn bombard the tube coating to produce light. All these types operate on alternating current, with a very few direct-current lamps for special uses.

Unlike incandescent lamps, fluorescent lamps require exterior ballasts and starters to control current to the lamp itself. Ballasts provide an inductive voltage for starting, but as the lamp heats the resistance decreases, so the ballast reduces the required current. Starters are of several different types, but all are designed to preheat the filament circuit for easier initial starting. Frequency of starts is of major concern with fluorescent lighting. With a burning time of 3 hours rated zero, a burning time of 12 hours will give approximately 60% longer life to the lamp, so in most cases a long burning period per start is most desirable. Ambient temperature and external humidity also affect the efficiency of the lamp, with a normal temperature of about 80°F giving best results. Undervoltage or overvoltage of more than about 7% will reduce both the life and the efficiency of the lamp.

## 11.7 MERCURY-VAPOR LAMPS

Low cost and high efficiency are two of the primary advantages of mercury-vapor lamps. In this type of lamp (Fig. 11-5), a small inner tube encloses an electrode at either end and the tube is filled with mercury vapor. Surrounding this arc tube is a glass bulb that contains support for the arc tube and the

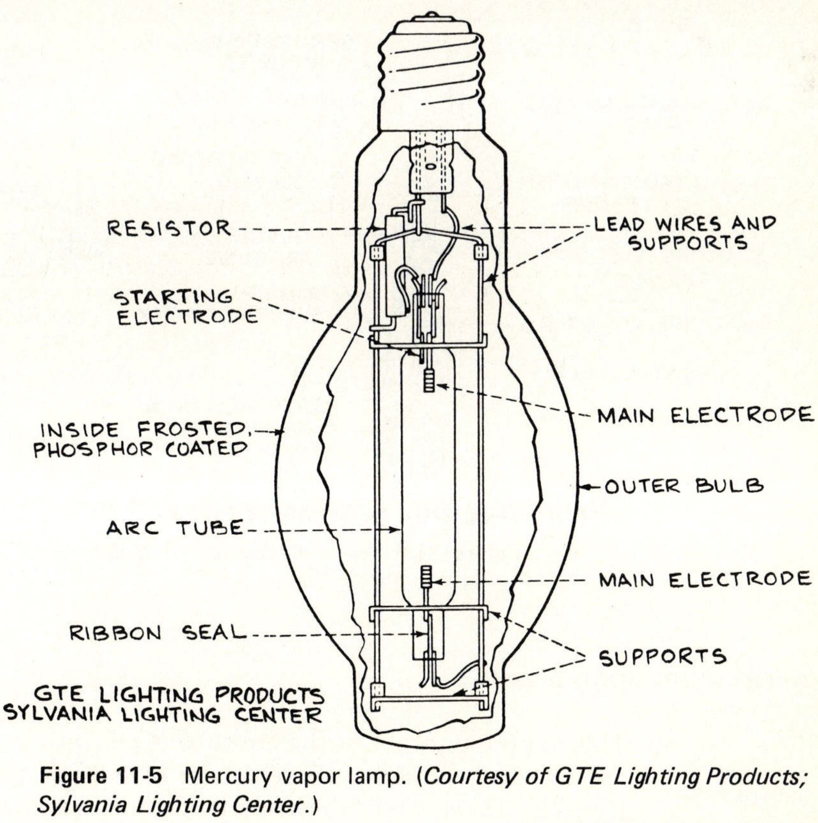


**Figure 11-5** Mercury vapor lamp. (*Courtesy of GTE Lighting Products; Sylvania Lighting Center.*)

inside of which is frosted with a phosphor coating. In starting, an auxiliary starting electrode sets up an electric field between it and the adjacent main electrode until the main arc is struck and the captured mercury is vaporized. As the tube fills with mercury vapor, a low-resistance path is created between the opposite electrodes at the tube ends, and the starter electrode is disconnected. This process requires from 4 to 10 minutes. If voltage drops or current is interrupted, the lamp will extinguish and must be allowed to cool before a new start it made. Ballasts to control voltage and current are necessary for mercury-vapor lamps.

Mercury-vapor lamps are most commonly used in heavy industry buildings where high mountings are available, as in warehouses, mills, and above crane-ways. The major objection to the use of these lamps is to the bluish color emitted, although this may be overcome by using color-corrected lamps or combinations with fluorescent or incandescent lamps. Initial and operating costs are less than for most other types of lighting, and lamp life is longer when operated at the proper voltage.

## 11.8 METAL-HALIDE LAMPS

To overcome the distinct blue color produced by the mercury-vapor lamp, a similar lamp (Fig. 11-6) has been devised which includes mercury, argon, neon, krypton, and some halide salts or iodides of metal in the arc tube. The additives provide the colors of red, yellow, and orange normally missing in the mercury arc. Burning position is important, as lamps designed for horizontal use have curved arc tubes, which should be installed with the curve uppermost.

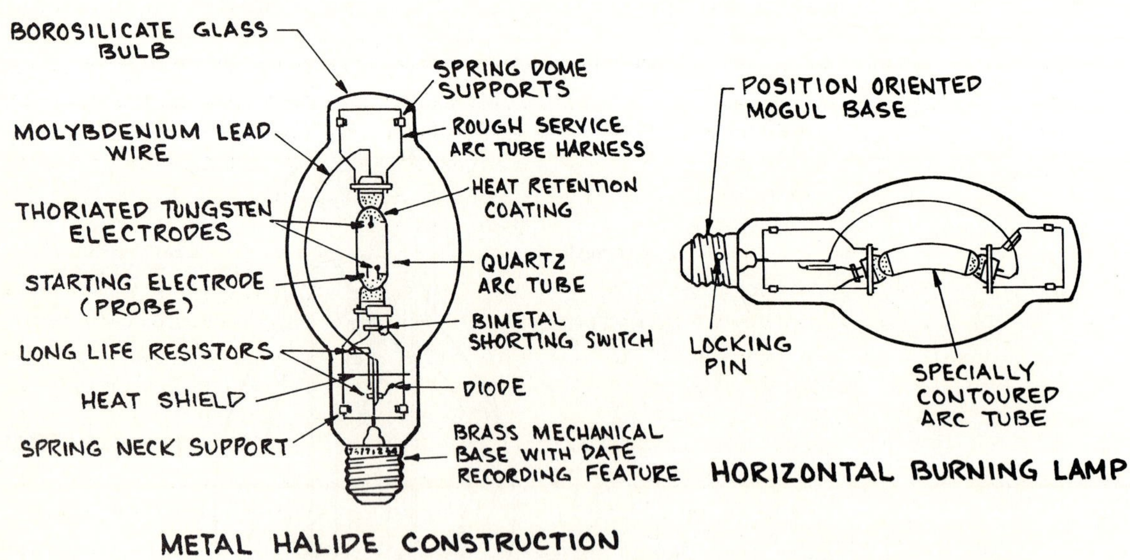


**Figure 11-6** Metal halide lamp. (*Courtesy of GTE Lighting Products; Sylvania Lighting Center.*)

## 11.9 HIGH-PRESSURE SODIUM LAMPS

This lamp is another of the arc-tube type, with a tube made of ceramic material to counter the corrosive action of sodium operating at high temperature (see Fig. 11-7). The tube contains xenon, mercury, and sodium operating at

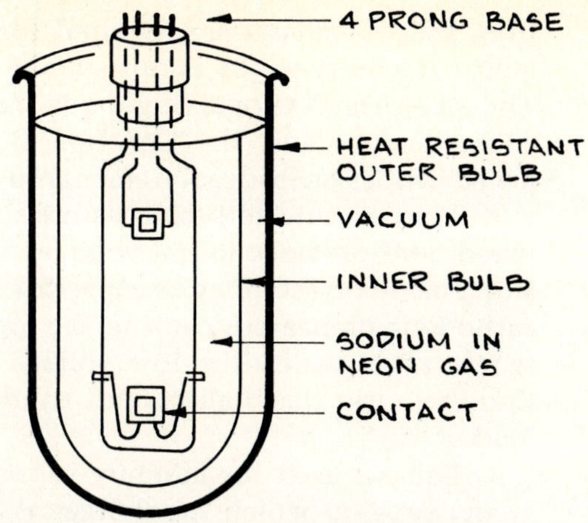


**Figure 11-7** Sodium lamp.

a pressure of 200 mm of mercury. This lamp produces light at all wavelengths, but the majority are concentrated in the yellow-orange portion of the spectrum. This creates a bright yellow light, which darkens most other colors.

This lamp requires a high-voltage (2500 to 5000 V) surge of energy to start it and requires an electronic starter to reduce the lamp resistance at start. There are no starting electrodes in the arc tube. The lumen output is very high and remains at about 80% even when the lamp has a life of 15,000 to 20,000 hours. Normal available wattages are 150, 215, 310, and 360. Dimming is limited to certain types.

## 11.10 NEW LAMPS

Manufacturers of lamps (bulbs) are presently about to produce several types that will give more light than the average incandescent with considerably less power consumption. Since a typical residential lighting system uses about 20 to 25% of total electricity for lighting, this is a breakthrough for the industry. High cost ($25 range) may slow use for quite a while, but the expected life of these new lamps is estimated at 5000 plus hours, which should make the cost per hour reasonable.

Recent sources indicate that the following will shortly be available. The Philips-Norelco SL-18 lamp puts out as much light as a 75 watt incandescent but uses only 18 watts and has a life expectancy of about 7500 hours. Westinghouse has a new lamp, the Econ-Nova, which has a 75 watt output using 25 watts. A U-shaped tube, produced by Interlectric Corp. and called a U-Lite, uses only 15 watts to produce 60 watts of light. The General Electric Co. has a lamp, first announced in 1982, which has the same shape as the familiar incandescent and is comparable in wattage to the others mentioned. Although the incandescent lamp is not the only lamp in widespread use, it is probably in more common use than is any other type, so these new units may have a definite effect on lighting design in the near future.

## 11.11 LAMP BALLASTS

Ballasts are required for all gaseous-discharge lamps and are referred to as negative-resistance devices. Their principal function is to provide starting voltage, limit current (resistance is zero), and provide power-factor correction. Proper ballasts are required for fluorescent lamps, low-pressure solium lamps, mercury-vapor lamps. metal–halide, and high-pressure solium lamps. The American National Standards Institute (ANSI) has set standards for light characteristics and the Certified Ballast Manufacturers (CBM) have established proper ballast requirements to meet the ANSI standards.

Fluorescent lamp ballasts are usually of the rapid-start type, in which a wired core provides initial voltage, which is increased by an autotransformer and phase-corrected by a capacitor. In two-stage or rapid-start ballasts, the cathode is preheated, causing the argon gas to ionize and decrease the resistance, after which the low voltage strikes an arc. The National Electrical Code requires that ballasts not overheat and that they perform safely under load.

Ballasts used for mercury-vapor lamps are usually of the inductive (reactor) type, in which the voltage is controlled over a low range of not more than 5%, plus or minus, with a line voltage of 240 and 277 V—except for lamps over 1000 W, which require a 480-V supply. Where voltage varies more than 5%, an autotransformer type is used which is more efficient, quieter, and operates satisfactorily on line voltage in the same way as a reactor type. Metal–halide ballasts may be autoregulated, regulating, or lead-peak autotransformer types, but these lamps will not operate on reactor ballasts. Ballasts for sodium lamps are larger, more complex, require auxiliary starters, require 2500 to 5000 V, and have a strike time of about 3 to 4 minutes.

## 11.12 LUMINAIRES

For correct and efficient lighting, the proper lamp plus the proper luminaire (fixture) should be combined. Many of the types of lamps discussed radiate light in all directions with varying intensity, and this light source needs to be directed and controlled. This may be accomplished by luminaires with properties of reflection, translucency, or transparency, depending on the material used. The location of the fixture also has a direct bearing on the design of the fixture, as it may be used for general low-intensity illumination, special spot lighting, or anything in between. Currently, most commercially available fixtures are of sheet-metal construction with glass or plastic covers or lenses, but special types may use wood frames or cases, and some "art" types are made from ceramics or are all-glass. Industrial fixtures for commercial or manufacturing are the most uniform and are available from hundreds of different companies. Residential lighting fixtures tend to be more varied and generally designed for candescent lamps.

## 11.13 LUMINAIRE CHARACTERISTICS

Reflection is one of the first methods to control light and it occurs in a number of different forms. Specular reflection occurs when a light source is reflected at the same angle as it strikes the surface. A spread reflection is similar, except that the light is scattered at the angle of reflection. A diffused reflection uses the cosine formula $I_\theta = In \cos_\theta$ from a matt reflecting surface. Where elliptical reflectors are used, the rays from the light source are typically returned as shown in Fig. 11-8. Each of the reflectors may be incorporated into a luminaire, depending on the requirement for a parallel or spotlight or a wide spread of the light.

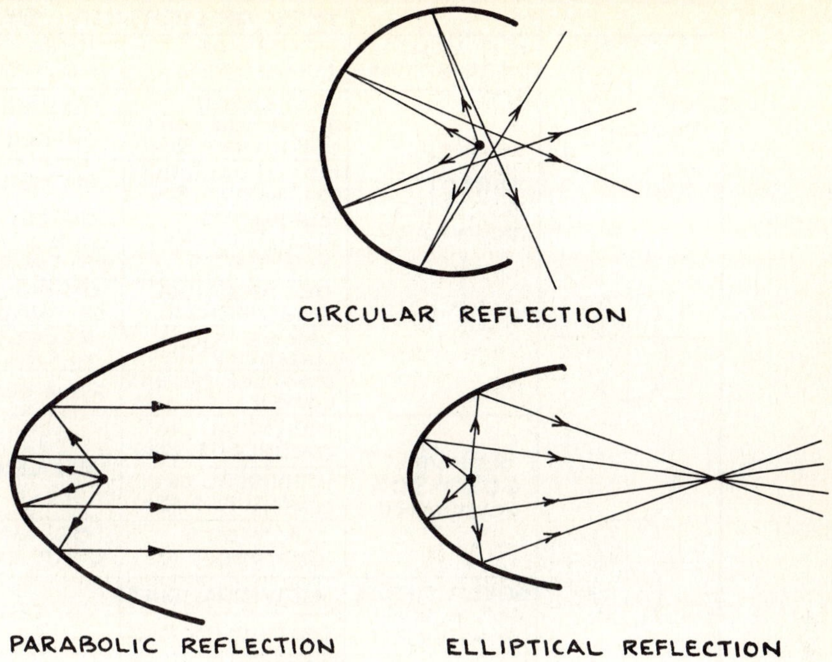


**Figure 11-8** Types of reflection.

*Refraction* is a change in the velocity of light, with the resultant bending of the light ray, as it passes through a more dense medium from a less dense (air) medium. As it emerges from the more dense medium, the ray again assumes the original direction but parallel to the direction of the entering ray. During the passage through the more dense medium, the ray is offset by an amount that may be calculated by formula and is dependent on the thickness of the material. In *absorption*, a surface may either change color by absorbing certain wavelengths of light or cause a change in luminance by neutralizing all wavelengths.

The *luminance* of a fixture represents the intensity of energy from a luminance in any particular direction. To determine this, the candlepower is measured at definite vertical angles, usually at 10° intervals, from directly below the light source, which is considered zero. These measurements are made at various planes taken around the fixture. This information may then be plotted to provide a *photometric curve* for each individual fixture and is usually supplied by the fixture manufacturer (see Fig. 11-9).

## 11.14 LIGHTING REQUIREMENTS

The proper amount of light, direction of light, and aesthetic appearance of the luminaire are three major problems that may be greatly affected by the location of the fixture and the type of use required. To some extent, particularly in residential installations, aesthetic appearance may be of greater concern than the amount of light supplied by a luminaire. In commercial or industrial installations, appearance is usually of minor importance compared to the amount and direction of the light and the cost of operation. Most installations indoors may be classified as either "closed-room, low-ceiling," as occurs in residences, stores, and similar areas, or "open-space, high-bay," which occurs in manufacturing plants, warehouses, and the like.

Although a great many factors affect the design of lighting systems,

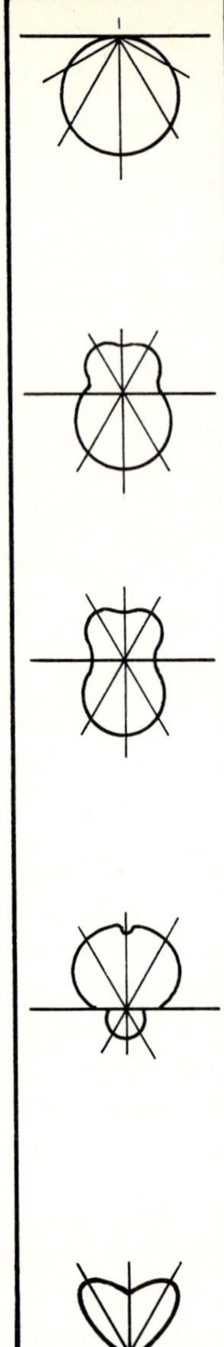

| TYPES OF LIGHTING SYSTEMS | | | |
|---|---|---|---|
| ICI* LIGHT DISTRIBUTION | TYPES OF LUMINAIRES | TYPE OF MOUNTING | LIGHTING SYSTEM CLASSIFICATION |
| DIRECT 90% TO 100% DOWNWARD | RECESSED | IN CEILING | RECESSED (INC. OR FL.) |
| | INDIVIDUAL UNIT CONTINUOUS ROW | ON CEILING | SURFACE MOUNTED DIRECT |
| | INDIVIDUAL UNIT CONTINUOUS ROW | ON CEILING | SUSPENDED DIRECT |
| | LUMINOUS ARCHITECTURAL ELEMENTS | ON CEILING, SIDE WALLS, OR COLUMNS | ARCHITECTURAL |
| | DECORATIVE ORNAMENTAL | ON CEILINGS, OR WALLS | ORNAMENTAL |
| | LARGE-AREA LOW-BRIGHTNESS | RECESSED IN CEILING | LUMINOUS ELEMENT |
| | DIFFUSER CEILINGS LOUVER CEILINGS | FROM CEILINGS | TRANSLIGHTED CEILING |
| SEMI-DIRECT 60% TO 90% DOWNWARD | INDIVIDUAL UNIT CONTINUOUS ROW | ON CEILING | SURFACE MOUNTED SEMI-DIRECT |
| | INDIVIDUAL UNIT CONTINUOUS ROW | | SUSPENDED SEMI-DIRECT |
| | LUMINOUS CORNICE | ON WALL NEAR CEILING | ARCHITECTURAL |
| GENERAL DIFFUSE 40% TO 60% DOWNWARD | INDIVIDUAL UNIT | ON CEILING | SUSPENDED |
| | CONTINUOUS ROW | | DIRECT-INDIRECT |
| SEMI-INDIRECT 60% TO 90% UPWARD | INDIVIDUAL UNIT CONTINUOUS ROW | ON CEILING | SUSPENDED SEMI-INDIRECT |
| | LUMINOUS COVE WALL BRACKET WITH DOWNWARD COMPONENT | ON WALL | SEMI-INDIRECT |
| | LUMINOUS ORNAMENTAL | ON CEILING | SUSPENDED SEMI-INDIRECT |
| INDIRECT 90% TO 100% UPWARD | INDIVIDUAL UNIT CONTINUOUS ROW | ON CEILING | SUSPENDED INDIRECT |
| | WALL URNS<br>COVE<br>PEDESTAL<br>CEILING<br>SOFFIT** | ON WALL<br>ON WALL OR IN CEILING<br>FLOOR OR COUNTER TOP<br>IN CEILING | INDIRECT<br>COVE<br>INDIRECT<br>INDIRECT** |

INTERNATIONAL CONGRESS ON ILLUMINATION APPROVED CLASSIFICATION OF LIGHTING SYSTEMS ACCORDING TO A RANGE OF DISTRIBUTION OF TOTAL LIGHT OUTPUT UPWARD OR DOWNWARD, IN PERCENTAGES AS SHOWN. A MORE DETAILED CLASSIFICATION, AS GIVEN HERE, HAS BEEN FOUND THROUGH PRACTICE TO BE MORE PRACTICAL.

** ACCORDING TO ICI PRACTICE, SOFFIT LIGHTING MIGHT BE CLASSIFIED AS "DIRECT". IN ACTUAL PRACTICE, IT IS NORMALLY REFERRED TO AS "INDIRECT", SINCE IT IS ACTUALLY A FORM OF COVE LIGHTING.

Figure 11-9

including exterior glare, reflection, absorption, room orientation, number and location of windows, and task requirements, two methods are in general use to determine the number of footcandles or lumens required. Using the lumen method, it is relatively easy to determine the number of footcandles required by means of the following formula:

$$\text{fc} = \frac{\text{lamp lumens} \times \text{MF} \times \text{CU}}{\text{area}}$$

where MF is the maintenance factor, CU the coefficient of utilization, and fc is footcandles. From this formula it is easy to solve from lamp lumens or other factors by shifting units.

The second method will give good approximations by means of the following formula:

$$\text{lamp lumens} = \frac{\text{fc} \times \text{area}}{\text{CU} \times \text{MF}}$$

Many of the requirements used in these formulas are fixed by some specification. Lamp lumens are fixed by the lamp and fixture selected. Room area is fixed by its dimensions. The maintenance factor (MF) is usually considered at about 75% to a low of 65%. The coefficient of utilization (CU) is most variable, but researchers have found 0.6 to 0.7 to be a good average. Required footcandles are usually determined by the location or task requirements or recommended lighting (Table 11-1).

The footcandles or lumens required, type of lamp, and type of fixture are rather easily determined. The next question is the number of fixtures and their location to provide the required light. For illustration purposes, assume a room with a 10 ft ceiling. Fixture spacing and mounting height should be nearly equal. The number of fixtures is determined by using the second formula above. Consider a room 30 ft by 60 ft with 75 fc required.

$$\text{lamp lumens} = \frac{\text{fc} \times \text{area}}{\text{CU} \times \text{MF}} = \frac{75 \times 1800}{0.7 \times 0.7} = \frac{135{,}000}{.049} = 275{,}510 \text{ lm}$$

**TABLE 11-1**
Recommended Footcandles

| | | | |
|---|---|---|---|
| Airplane hangars | 100 | Locker rooms | 30 |
| Art galleries | 30-100 | Machine shop | |
| Auditoriums | 15-30 | General | 75-100 |
| Bakeries | 50 | Fine-tool | 500 |
| Banks | 70-150 | Office, general | 100-150 |
| Bars | 30-50 | Printing | |
| Beauty shops | 100 | Press room | 70-100 |
| Brewery | 30 | Composition | 100 |
| Clothing manufacturers | | Electrotype | 70-100 |
| Cutting | 300 | Residential | |
| Sewing | 500 | Kitchen | 100-150 |
| Inspection | 1000 | Dining | 15-30 |
| Courtrooms | 30-100 | Bedroom | 50-70 |
| Dairy products | 100 | Living room | 50-150 |
| Foundries | 100 | Restaurants | |
| Garages | | Cashier | 50 |
| Repair | 100 | Dining | 30-50 |
| Parking | 10-50 | Kitchen | 75 |
| Hospitals | | School | |
| Corridors | 20-30 | Classroom | 75 |
| Examination | 50-100 | Drafting room | 100-150 |
| Surgery | 200-2500 | Laboratory | 150 |
| Laboratory | 100 | Service station, auto | 100 |
| Patient room | 30 | Shipyards, general | 30-50 |
| Hotels | | Storage, general | 5-50 |
| Lobby | 10-30 | Stores | |
| Corridors | 10-30 | Retail, circulation | 30 |
| Rooms | 10-50 | Showcase | 200-300 |
| Offices | 100-150 | Welding | |
| Industrial assembly rooms | 100 | General | 50 |
| Laundry | 50-70 | Precision | 1000 |
| Library | 70-100 | | |

If 40-W rapid-start fluorescent lamps (3150 lm) are used, two to a fixture, each fixture supplies 6300 lm. By dividing the total requirement, 275,510 lm, by 6300 lm per fixture, the answer is 43.73 fixtures. With a 10-ft mounting height, we can assume three rows of fixtures in the 30-ft room width, so we should perhaps require 14.6 (43.73/3), or 15 fixtures each row. If the room is exactly 60 ft long, rather than perhaps 59 ft 8 in., the 15 fixtures (each 4 ft long) will fit. If the room is short of 60 ft, the number of fixtures may have to be reduced to 14 in each row with a reduction of 6300 lm per row, or a total of 264,600 lm, or 72.03 fc. A selection of fixtures with three or four lamps may also solve the problem if the slight reduction in footcandles is unacceptable.

In the preceding example, the 75 fc would be delivered at floor level. For manufacturing, retail stores, and similar installations, the floor-level illumination is usually not important but the work-level lighting is important. This means that 75 fc was required at bench or counter height, say 4 ft, above the floor level and the 72.03 fc delivered at floor level could be acceptable as approximately 75 fc at bench height. This could be proven by reworking the formula to allow a mounting height of 6 ft and a respacing of fixtures if of vital importance, or the work surface could be augmented with local or spot lighting.

## 11.15 COST FACTORS

The cost of a lighting system must be based on initial costs plus operating and maintenance costs. An initial cost designed to provide the "most efficient lighting system" may sometimes be more economical, but it is usually more expensive owing to characteristics of color and heat, as well as light. Generally, a more efficient lighting device will cost more initially than one that is less efficient. On the other hand, the more expensive fixture is probably better manufactured and better designed. These factors tend to make maintenance costs minimal. Light loss due to dirt and surface deterioration are minimized with a good maintenance program.

Operating costs are determined largely by two factors: utility rates and time of use of the fixture. The utility rate is a very complex procedure, depending on the classification of the user and the ability of the utility to supply energy. Special discounts are often available for large users or within certain demand quantities of the utility. The usage of most commercial and industrial energy is controlled by circuit breakers at a panel which may control lighting some distance away. Thus in many cases, the theory of "turn the light off when not in use" is difficult if not impossible to accomplish owing to switching location.

In addition to the cost of different lighting systems, there are related costs that may be overlooked. One of the most important of these is the heat gain from operating lighting, which may adversely affect calculated air-conditioning loads. In a few modern installations, this lighting heat gain has been reused as a heat source for air-conditioning or industrial usage. The number of wall openings as well as their construction and orientation also affect the cost of operation. Full "window walls" on the southerly side of a building may provide adequate lighting without using lighting fixtures, thus reducing lighting operation costs, but may also increase air-conditioning costs because of the increased heat load. Overdesign of a system where excessive footcandles are used in minor spaces (75 to 100 fc in storage areas or corridors) may also increase operating costs.

## 11.16 EXTERIOR LIGHTING

Exterior lighting is generally concerned with three factors: safety, security, and advertising. In the first two, the desire is to provide a lighting system that will facilitate safe movement of people and equipment on roads, parking lots, and similar areas. Lighting used for "advertising" includes floodlighting of buildings, signs, and other devices to make the public aware of certain structures or conditions.

Selection of lamps is a major concern for exterior lighting. Incandescent lighting should be avoided because of its low efficiency and short life. Fluorescent lamps are quite sensitive to temperature changes, have poor optical quality, and may adversely affect colors. Low-pressure sodium lamps produce a yellow light which changes colors to shades of gray or brown, and although highly efficient, may become considerably less efficient when combined with a ballast. Mercury-vapor lamps have a low efficacy and optical control is poor. High-pressure sodium and metal-halide lamps have good optical control, high efficacy, and usually top efficiency. In general, the two latter types may be considered as first and second choice, with mercury-vapor lamps as a third choice. The choice of luminaires for lamps is limited but should provide low intensity 70 to 90° from nadir (directly below the lamp) and no intensity above 90°. Higher intensity at angles below 70° will provide more light where it is needed rather than spilling into the atmosphere or over areas not requiring light.

## 11.17 SUMMARY

Illumination is one of the major uses for electricity. The average residence uses the greatest proportion of electricity for lighting, and the same is true of a great many commercial/industrial installations. Just about every lighting condition is different thanks to the four primary governing factors: brightness of light, object size, contrast between object and background, and the movement or time the object is within the field of vision. The reflection of light rays or the absorption also affects the lighting requirements. The principal electrical lamp (light) sources are (1) incandescent, which relies on the resistance glow of a wire filament enclosed in a glass bulb; (2) fluorescent, which creates light by a discharge within a phosphor-coated glass tube between two nonconnected terminals; and (3) vapor lamps, which use mercury or sodium vapor to carry electrical current and provide light by action of an arc within the vapor.

Without proper luminaires, the lamp usually spills its light in all directions. To prevent this and to control the light, luminaires or fixtures are used. These vary from simple sockets in closets to ornate combinations of glass, metal, or other materials in residential installations, to dozens of patterns designed for efficiency rather than appearance for use in commercial/industrial projects. Luminaires are designed to provide reflection of the light source, refraction through lenses or glass enclosures, and absorption of undesirable light wavelengths.

The amount and type of light to be provided depends on a number of factors. In residential lighting, aesthetic appearance is often emphasized over good lighting, whereas the opposite may be true of commercial/industrial lighting. Lamp life, fixture and lamp maintenance, height of fixture installation, and similar factors determine the efficiency and operating cost of lighting. Selection of lamps and fixtures may also influence air-conditioning costs, optical color selections, and direction of the light. Although various tables give desired footcandles of light for certain rooms or tasks, each location and use requires its own type of lamp and luminaire for proper lighting.

# 12

# Electric Power

## 12.1 INTRODUCTION

Electric power is today's major key to industrial growth, and that growth represents the most significant use of this type of energy. Most of this energy is used as *shaft power:* electric motors turning a shaft, which in turn rotates other equipment. Motors range from small in size (¼ hp) to giant-sized (1000 hp); mostly alternating-current type; split-phase, single-phase, or polyphase; and provide operating characteristics to about 85% of industrial machines. A great variety of windings, frames, insulation, and other factors affect their design and use, as well as their operating voltage and ampere requirements.

Controls for the power units also vary extensively, depending on use, type of motor, speed, and economy. Simple on–off switches may be adequate for small motors with intermittent use, or controls may be complete load centers for larger equipment. If direct-current motors are used, different types of converts are necessary and may have advantages, particularly in regard to speed control. Motor starters are another control device that may be required, and protection from in-rush voltage, heat, overload, and stopping are needed for proper operation. The use of computers to control motor operation is of increasing importance, except perhaps in motors used in residential applications or small isolated equipment.

It is impossible here to discuss thoroughly all the advantages, disadvantages, design and manufacturing techniques, possible uses, and other attributes of electric motors. Many books have been written on the design and function of any one of the several motors discussed, or their auxiliary equipment, so the reader should be aware of the immensity of this field and consult specialized publications for additional information.

## 12.2 THE ELECTRIC MOTOR

A dictionary definition of *motor* is: "a machine that converts electric energy into mechanical power, causing, producing, or imparting motion." Mechanical features of a motor include its physical construction, rotating means, insulation, lubrication, and many special items. Electrical features include motor type, electrical supply, horsepower, and efficiency, as well as the operating conditions. Proper combinations of both mechanical and electrical features are necessary for the most satisfactory selection of an electric motor.

First consider the mechanical features in a general manner. The enclosure, the frame and case arrangement, is defined by the National Electrical Manufacturers Association (NEMA) as open or totally enclosed (see Fig. 12-1). *Open enclosures* may be simply defined as drip-proof, splash-proof, or weather-protected. A *drip-proof motor* is enclosed in a container or "box" and defined by NEMA as having vent openings so constructed that particles falling at less than 15° from the vertical, or liquids running on horizontal surfaces, cannot enter the motor housing. *Splash-proof motors* are designed so that particles or liquid falling on the motor at not less than 100° cannot enter. *Weather-protected motors* are completely enclosed but with ventilation provided by a fan blade attached to the motor rotor and circulating air from the enclosed case.

*Totally enclosed motors* are defined as pipe-ventilated, fan-cooled, or explosion-proof. A *pipe-ventilated motor* is completely enclosed except for ductwork that supplies incoming cooling air and removes exhaust heated air. A *fan-cooled motor* is similar to weather-protected motors in that it has an integral fan on the motor shaft to circulate air within its enclosed frame. An *explosion-proof motor* is designed to prevent explosion of any gas or vapor in or near the motor.

Second, consider the power supply, motor types, and performance expected. *Direct-current motors* are in common use despite advances in alternating-current design. These motors are available in six major categories: series-wound, shunt-wound, compound, split-phase, capacitor, and commutator type. Generally, the direct-current motor is used where control of speed is critical, and it operates economically at voltages from 250 to 900 V, depending on the horsepower required (500 to 1500 hp). Premium cost is a major deterrrent for extensive use.

*Alternating-current motors* include single-phase and polyphase (squirrel-

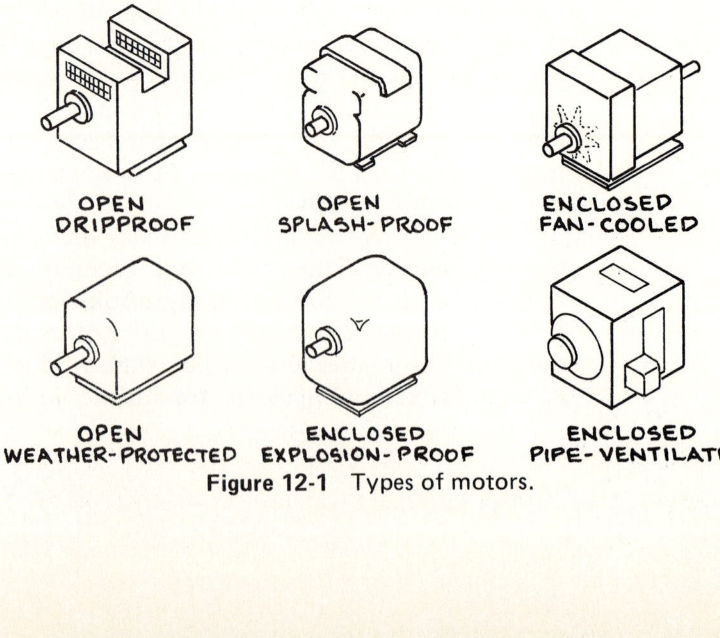


**Figure 12-1** Types of motors.

cage, wound-rotor, synchronous) and may range in horsepower and voltage requirements from 120 V–¼ hp to 4160 V–1500 hp).

## 12.3 DIRECT-CURRENT MOTORS

Modern electric power is seldom supplied as direct current, so convenience in the purchase and operation of an electric motor has had a great influence on the use of direct-current motors. High initial cost of the motor, together with the cost of necessary motor-generator sets or rectifying equipment to convert alternating current to direct current, have also been a deterrent. Despite these disadvantages, the direct-current motor is popular and is normally available in three major classes: series-wound, shunt-wound and compound (Fig. 12-2).

The *series-wound motor* is the simplest, with the armature and field in series, with the field wound with large enough wire to accept armature currents. This allows a nearly immediate change of speed, desirable in hoists, traction work, car dumpers, and gates, where a slow load speed is desirable but a rapid no-load speed is necessary. Positive connection through gears or couplings is necessary, as a series motor without load may race to high speeds and to ultimate destruction if not controlled.

*Shunt-wound motors* have their armatures and fields wired in parallel, thus giving constant field strength and motor speed. Speed is variable from about 5 to 10% from load to no-load and is adjusted by using a rheostat in the field circuit. This type of motor is used primarily with elevators, fans, and centrifugal pumps.

The *compound motor* is a combination series and parallel wound arrangement, where high starting torque and constant running speed are desired. Here the shunt field is connected in parallel with the armature, while the series field is connected in series with the armature. Compound speeds may vary as much as 25% from load to no-load, and this type is usually used on pumps, presses, and geared elevators.

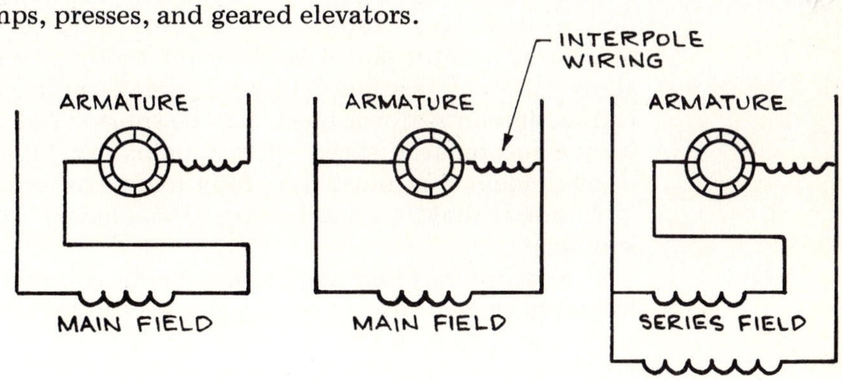


**Figure 12-2** Direct current motors.

## 12.4 ALTERNATING-CURRENT MOTORS

Alternating-current motors are available in two major types: *single-phase* and *polyphase*. Older and smaller motors tend to be single-phase for several reasons: the smaller motors are quite often used on residential equipment (refrigerators, disposal units, pumps, home shop equipment), which is 120/

240-V single-phase 60-Hz, and some alternating current is still supplied by utilities as 120-V 133-Hz, which operates single-phase motors more efficiently. Most single-phase motors are capacitor-start/capacitor-run or split-phase/ repulsion-start and require more auxiliary starting means. The *split-phase motor* is the simplest and consists of a *stator* or field fixed in the stationary frame, and a *rotor* or armature mounted on a drive shaft and rotating within the stator field. In the split-phase motor, the two stator windings are physically 90° apart and intermediate starter windings are located between the main windings. When the motor is started, both windings are in parallel, but as the motor reaches about 75% full speed, a centrifugal switch cuts out the auxiliary windings. This type of motor is most often connected to small fans, electric heaters, and similar equipment. A capacitor-start motor has a separate starter winding in the stator which holds the starting current out of phase with the running current until about 80% of operating speed has been reached, at which point a centrifugal switch cuts out the starter current. If the capacitor remains in circuit during the running period, the motor becomes a two-phase motor which runs more quietly and with less vibration. A repulsion-start motor has a series of brushes bearing on a commutator attached to the rotor shaft. A mechanism short-circuits the commutator bars before the brushes are raised during the start and the motor operates on the stator windings. Applications for this type of motor are those with heavy starting torque, such as compressors, stokers, and pumps.

Heavy-duty motors may be of squirrel-case, wound-rotor, or synchronous type. The *squirrel-cage motor* is most common and is usually of the induction type, three-phase, with each separate winding electrically connected to a separate leg of the supply line. When the power is switched on, the magnetic field induced between the three different stator windings and the self-contained rotor winding causes the rotor to turn in much the same way that an electric generator operates, but in reverse order. The motor voltage is whatever the supply voltage is and may vary from 240 to 4160 V, all three-phase.

*Wound-rotor motors* have a rotor winding with the same number of poles as the stator and with the rotor winding connected to slip rings, which allow different resistance values to be set up on each ring. By this method various amounts of resistance may be shorted out to allow maximum starting torque and reduced speed when running. This type of motor is well adapted to equipment that must start fully loaded or requires high starting torque at low current draw, such as compressors, plunger pumps, reciprocating pumps, and fans.

*Synchronous motors* have a magnetic stator core and windings connected to alternating current to supply the rotating magnetic field. Rotor windings are connected to a direct-current source, which interlocks with the stator field when operating. An auxiliary exciter is direct-current and supplies magnetizing current to the field windings. Motors are rated from 700 to 1200 hp with speeds of 360 to 1200 rpm.

## 12.5 MOTOR STARTING

All alternating-current motors may be started if connected directly across the electric supply lines, but this throws a heavy load on the electrical supply and develops a high torque impact. As the motor picks up acceleration, the current required will eventually balance with normal requirements, but the initial ampere inrush may be noticeable whenever a motor is started. Manual starters close the line contacts with a great variety of switches and are still used in many installations. However, magnetic control devices operate by a small current from the stator feed and have the advantage of remote location for control by pushbuttons or similar devices. Across-the-line starting is inexpensive and generally satisfactory where motors are started only a few times daily and remote control is not necessary. NEMA standards limit manual starting to 440 V three-phase 7½ hp or to 5 hp for single-phase.

An *autotransformer* starter (see Fig. 12-3) reduces motor voltage through transformer action coupled with higher starting torque for the same current inrush. Two or three leg starters are available, and the ratio of current to torque depends on the location of the transformer taps, which may control from 25 to 65%. Wound-rotor motors are started either manually or automatically by introducing resistance into the circuit and gradually reducing it as the motor accelerates, thus reducing inrush shock and high starting torque. Manual starters may consist of a resistor of three legs, each tapped at intervals, with a manually operated contactor cutting in various resistances. Actual start is by a pushbutton after all resistances are in the circuit. An automatic starter uses a series of relays and resistances that are removed by short-circuiting as the motor gains speed. Drum controllers operate in a manner similar to other manually controlled devices, except that all resistances are energized by the main control until it reaches the OFF position. Another starting method is by use of *impedance resistors*, where the incoming voltage is lowered. Since voltage drop varies with current drawn, the motor voltage increases as current pull reduces as the motor speed increases. First-step current is about one-third of normal across-the-line starting current. Acceleration control is accomplished by adjusting resistance values and relay timing.

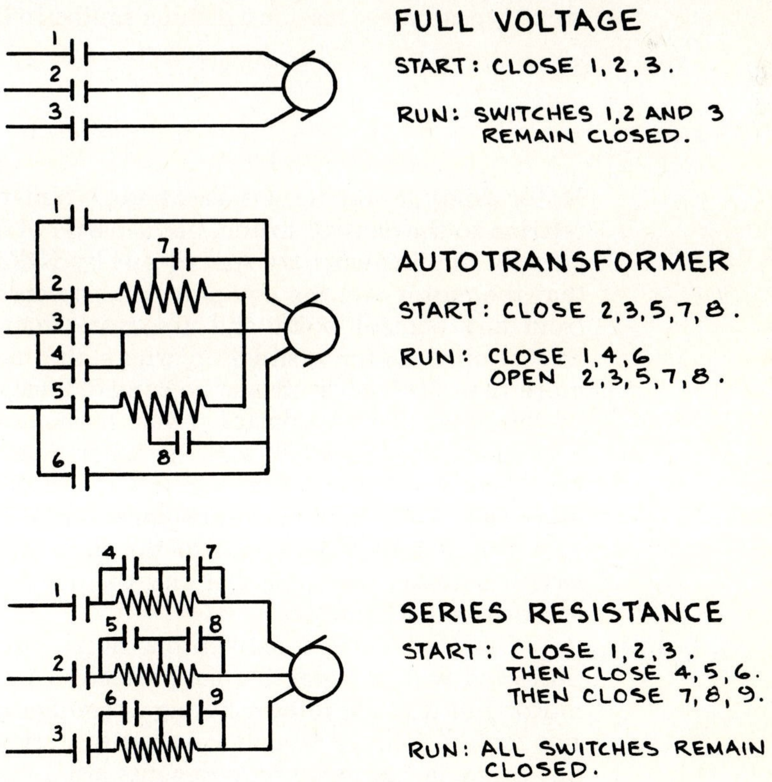


**Figure 12-3** Starting methods for alternating current motors.

## 12.6 PILOT DEVICES

*Pilot devices* are elements inserted into a control circuit of a motor which energize or deenergize the power circuit. The simplest is a manually operated *pushbutton* with one or two sets of contacts for normally open (NO) or normally closed (NC) function. A few features of pushbutton control are ease of identification, easy mounting, safety, locking possibility, and interchange of basic parts. *Limit switches* may be used to start, stop, or reverse equipment when a part being manufactured reaches a certain point, or a machine reaches a predetermined number of revolutions or number of parts. Floats in tanks holding liquids may actuate *liquid-level devices*, which in turn operate switches to energize pumps or drains. The actual switch may be sliding contacts or a mercury tube tilted by the level float. Another type is a *pressure switch*, consisting of a chamber and a diaphragm connected to a contact switch. As the diaphragm is forced from a preset value by gas or liquid pushing against it, the attached switch is used to control pumps, blowers, and similar equipment. In the same category are *temperature switches*, which use the expansion of gas or metal to open or close switches that control other devices.

*Relays* and *fuses* may also be classified as "pilot" devices when used to control circuits, although fuses are usually considered as protection against overloads. Control relays may be used as voltage amplifiers or as controls for interlocking or sequence operations. Relays may be constructed with coils, contacts, and return arms, or may be magnetized "reeds" of thin metal moved by exterior magnets. The fact that relays can be adjusted for time sequence makes them especially valuable for computers, motor starting, proximity switches, and similar uses. When fuses are used as pilot devices, their main function is to break a circuit because of overload or heat and in some types may be reset in a manner similar to that used with circuit breakers.

## 12.7 MOTOR SAFETY

Motor safety, as the term is used here, is a matter of location of the motor in relation to the control switch, not a matter of design of the equipment. Three primary requirements are spelled out by NEMA standards. The basic rule is that the motor and the equipment it powers must be visible to the operator from any control point and not more than 50 ft away from the motor. Exception A is for a situation where the motor and driven machinery are more than 50 ft away from the control location or in a location out of sight of the controller. A control device is required between the motor and the disconnect power switch, and the latter must be capable of being locked in the open position. Exception B is when the motor and driven machinery are more than 50 ft from, or out of sight of, the controller. In this case a disconnect switch is required ahead of the control device, with another disconnect switch between the controller and the device and within sight of the motor and driven machinery.

Overload devices for motors over 1 hp but under 1500 hp and not protected with integral thermal protectors are required to be set for 125% of motor full load for motors having a temperature rise of not more than 40°C. For motors 1 hp or less, automatically started, out of sight of the starter, or not portable, the same requirements are necessary. Portable motors of 1 hp or less, manually started and within sight of the starter, or started by plug-in at an outlet, must be protected by adequate circuit breaker or fuses at the distribution panel. For stationary motors of ⅛ hp or less, the branch-circuit overcurrent device serves as a disconnect.

## 12.8 RESIDENTIAL ELECTRIC HEAT

Electric heat may be classified as residential or industrial. In the former case, electric heat may be as simple as a heat lamp screwed into a light fixture, a wall-mounted resistance unit, radiant resistance wiring installed in a plastered ceiling, or as a heating unit in an air-conditioning unit. The use of electricity to heat boilers is usually not economical, and electrically "fired" domestic water heaters are not classified as heaters.

*Unit heaters* are most often installed in bathrooms as auxiliary heat to provide more comfort while washing or bathing, or in isolated locations such as basements, garages, or home workshops. Units are generally recessed into the wall, operate on 120-V single-phase, and may or may not have an integral fan to provide better distribution of heated air. Wattage loads run from about 1000 to 5000 and are usually manually operated with a visible switch. *Baseboard units* are contained in a metal case from 36 to 96 inches long, 750 to 4000 W, and 2560 to 13,660 Btu/hr. This system requires minimum space at room walls, is usually totally visible, economical to install, and may or may not incorporate a fan. The general provisions for motors of ⅛ hp or less apply if fans are included, and any thermostat control is accepted in lieu of a disconnect switch if normal grounding is provided. *Resistance cable* heating is installed in ceilings or walls and covered or enclosed with plaster to form a large radiation surface. The cable is available in rolls to about 1500 lineal feet, 400 to 4500 W, and from 1350 to 15,500 Btu/hr. Cable is usually rated at 2.75 W per lineal foot. Installation must provide at least 6 in. of clearance from walls and at least 8 in. from electrical junction boxes or outlet boxes. Calculations for electric heat are made in the same manner (figuring heat loss, heat gain, insulation of space) as for any other type of heating.

## 12.9 INDUSTRIAL ELECTRIC HEAT

Industrial electric heat for manufacturing processes is on the increase. It is used to heat liquids, gases, and solids; to forge, roll, and form metals; and to heat-treat a wide variety of products. Infrared heaters solve drying problems and difficult space-heating problems. Dielectric heating is used for processing wood, ceramics, and plastics. Although not a direct heating process, welding also requires large amounts of electric energy in manufacturing processes. Also, there is always the possibility of, or necessity for, industrial temperature control in shops and yards where other types of heating may not be possible. Industrial heat use is based on electricity's quick response, automatic operation, wide temperature range, easy distribution, and safety.

Three basic methods for generating electrical heat are by resistance, by induction, and by dielectric heating. When a material *resists* the flow of electric current, heat is produced which may be transferred by radiation, conductance, or convection. *Radiation* (infrared) is obtained by heating a substance to the point where it emits radiant energy, which can be directed against an object to provide heat. When a resistance coil is embedded in a material that will absorb the heat developed and in turn pass it along by contact or air movement created by the heat, it is called *conductance* when in direct contact and *convection* when not in direct contact. Electric arc welding is another form of resistance heating, where an electric arc is struck between a piece of metal and the welding electrode with sufficient heat to partially melt the base-metal parts together.

*Induction heating* uses alternating current to charge a magnetic coil with high frequency, thus setting up a magnetic field around the coil. Any work to be heated is placed within this magnetic field as a short-circuited secondary coil. Heat transfer is rapid and the magnetic field will penetrate any nonmetallic material, so work may be easily handled without physical contact with the coil. Extremely high temperatures are possible.

*Dielectric heating* is effective where other methods may not be possible. In this process, nonconductive materials are heated as positive or negative electrodes with low current but high voltage. Material placed between the plates is uniformly heated from 2 to 15,000 V at frequencies of 10 to 30 million cycles per second. To produce different shapes, only the plates need to be formed to the shape required and heating needs.

Electric heat is so versatile that new uses for it are constantly being developed. Strip heaters are used to heat liquids in tanks, and in forced-air systems, ductwork, ovens, and all sorts of dryers. Infrared heaters are used in paint-baking ovens, for drying paper products, for curing resins and metal parts, and in all phases of food preparation. Induction furnaces are used for the vacuum melting of metals and other processing of metal parts. Many liquids in tanks or railroad tank cars are heated to increase flow by means of resistance heaters. Other heat uses include processes of brazing, shrink fitting, off-peak water heating, and plastic manufacturing and forming.

## 12.10 SUMMARY

Power has by now taken first place in the consumption of electric energy and its use will probably continue to expand as more applications for it are found in the future. Foremost in the use of this power are the thousands of installations of electric motors, together with their necessary auxiliary equipment. Whereas, in the past, direct-current motors have been relegated to a secondary position, primarily because alternating current was more readily available, new convertors and rectifiers are making these units more popular because of their adjustable speed and high efficiency. In general, alternating-current motors may be divided into two types: low-horsepower single-phase motors or large-horsepower polyphase motors. The first of these will probably always be a standard for motors rated 1 hp or less for residential and small industrial/commercial equipment, while motors to 1500 hp will be developed further to provide a better ratio of weight and size to horsepower.

Together with the increase in better design and efficiency of motors will come more and different uses for electric heat in both residential buildings and industrial processing. Shortages of plant expansion space and demand for more production at greater economy will force manufacturers to rely more heavily on electric heat to produce their wares. As fossil-fuel shortages grow, electricity generated by hydropower will be more economical and will also hasten the use of electric heat in industry.

# 13

# Special Electrical Systems

## 13.1 INTRODUCTION

In addition to the relatively normal use of electricity for lighting and power, a great number of special uses have come about in the past few years. Prominent among these special systems are the several types of heating, communication wiring, alarm systems, and emergency power. Heating systems and emergency power are usually "line" voltage (120/240 V), whereas most communication and alarm systems are usually low voltage (6 to 24 V).

## 13.2 ELECTRIC HEATING

Electric heating is available in several different forms: ceiling-embedded radiant heating, baseboard heating, unit heaters, and heat lamps. In addition, heat pumps and air-conditioning systems may also be considered as "electrical." These systems are usually controlled by thermostats but may also be controlled by normal switching.

Embedded radiant heating is usually installed in ceilings, although heating panels may also be placed in walls. The heating wire is factory-assembled and connected to nonradiant power leads, and thus should not be cut or altered in any way. Most systems operate on 120, 208, or 240 V and are embedded in plaster or laminated gypsum board, which functions as the radiator. Radiant wiring should be kept at least 6 in. away from a wall–ceiling intersection. The requirement for heat loss is calculated in exactly the same manner as for any other heating method (Chapter 23). The formula used to determine the required wall spacing (WS) is

$$\text{WS} = \frac{12(L-1) \times (W-1) - 12A}{U}$$

where $L$ is the length of the room, $W$ the room width, $A$ the total unusable area, and $U$ the length of the heating wire in feet.

**EXAMPLE**

Assume that a room 10 ft × 15 ft requires 1.8 kW of heating wire 500 ft long. Unusable ceiling area over cabinets and the like is 10 ft$^2$

$$\text{WS} = \frac{12(15-1)\times(10-1)-12(10)}{500} = \frac{12(14)\times 9 - 120}{500}$$

$$= \frac{1392}{500} = 2.784 \text{ in., say 3 in.}$$

Wire should not be spaced closer than 1½ in. Average heat dissipation is about 2.75 W per foot of cable length. Baseboard heating and wall units are used as sole heat sources or as auxiliary heating units. Baseboard heaters are designed to provide the maximum amount of radiation for transfer of heat expressed in watts per square inch and to provide natural or forced air over the heat fins into the room. Sections are available from 1 to 10 feet with ratings from 100 to 3000 W. Metal enclosures vary from 4 to 10 inches in height and 1½ to 3 inches deep. Most baseboard units are thermostatically controlled and operate on 120-V or 240-V single-phase.

Wall radiant units are manufactured as a complete unit with a rated resistance heating unit installed in a metal base (see Fig. 13-1). These units are used in bathrooms and similar small spaces where instant concentrated heat is desired. Units are available from 17 in. × 11 in. × 3½ in. in size and from 500 to 1250 W in power. Units may be thermostatically controlled, controlled with a wall time switch, or most usually by an on–off switch incorporated in the unit. Most units are designed to be recessed into the wall between studs and operate on 120-V single-phase current. Forced-air wall heaters are similar in construction and range from 7 in. × 14 in. × 2 in. to 21 in. × 13 in. × 4 in. in size, from 1000 to 4000 W, and up to 13,500 Btu/hr. All forced-air heaters are equipped with fans and may be of updraft or downdraft type.

Electric heating by resistance coils in connection with air-conditioning is another method but is usually considered as part of the air-conditioning system rather than as a separate form of heating. In bathrooms or similar small areas, infrared heat lamps of various types may be installed in standard lamp sockets to provide local heat.

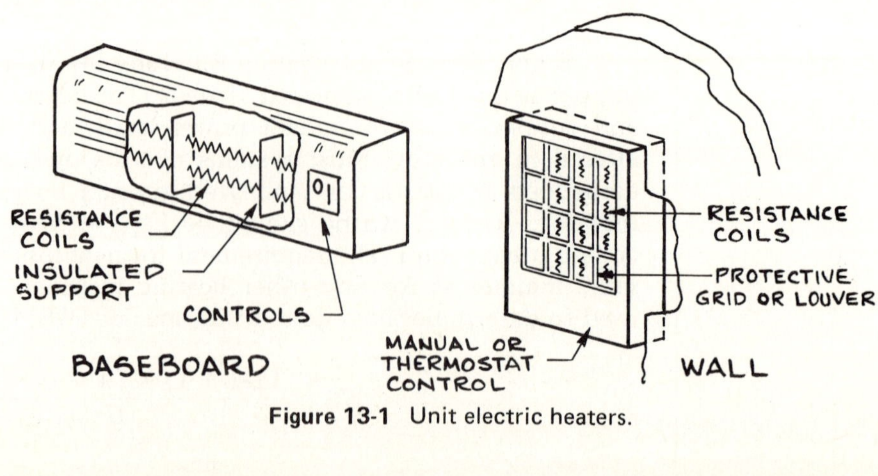


**Figure 13-1** Unit electric heaters.

## 13.3 SHORT-METHOD HEATING CALCULATION

To find the exact amount of heat required for a building is a rather long and sometimes unnecessary calculation, so an approximate heat requirement, more easily done, can serve for estimating both heating values and operating costs. In Fig. 13-2 the heating requirements are expressed in kilowatts for various sizes of residences with fully insulated frome on concrete slabs on grade in areas of 0°F design temperature. To use Fig. 13-2, several factors of a local nature must be known or determined. The outdoor design temperature of the location must be obtained, the total square foot area of the building must be determined, and the total area of all doors and windows must be known.

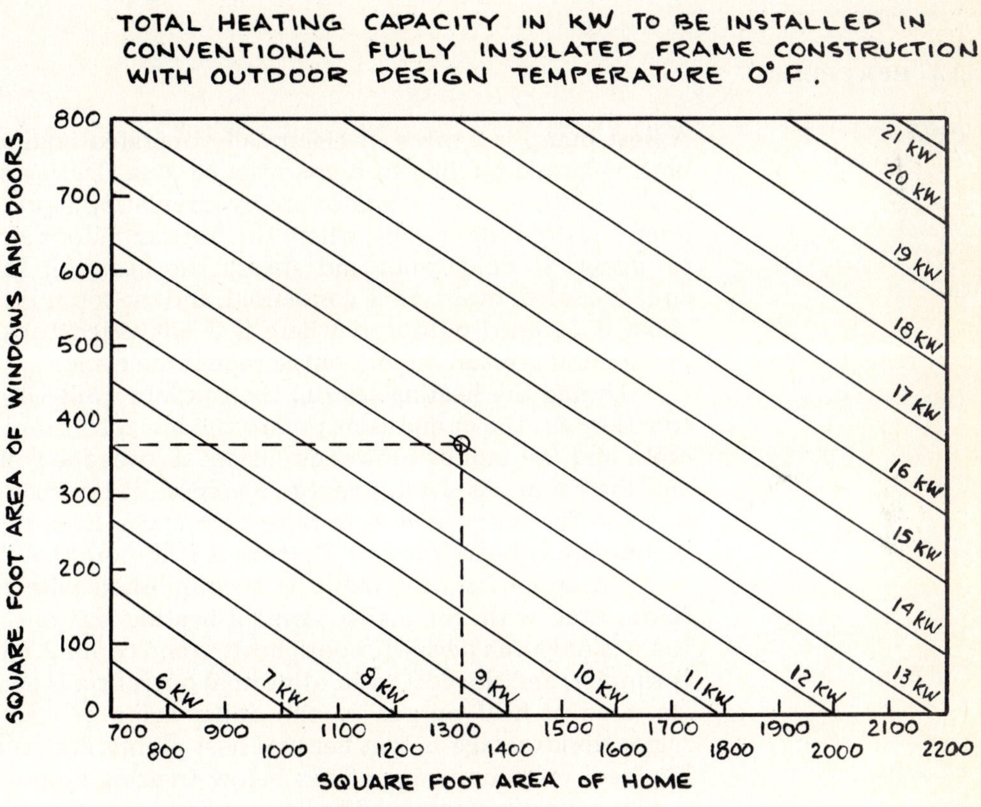


| | WATTS | | | | | WATTS | | | |
|---|---|---|---|---|---|---|---|---|---|
| SQUARE FT. | ZONE 1 | ZONE 2 | ZONE 3 | ZONE 4 | SQUARE FT. | ZONE 1 | ZONE 2 | ZONE 3 | ZONE 4 |
| 50 | 360 | 500 | 600 | 750 | 300 | 1650 | 2300 | 2750 | 3450 |
| 75 | 500 | 700 | 850 | 1050 | 325 | 1750 | 2450 | 2950 | 3675 |
| 100 | 650 | 900 | 1075 | 1350 | 350 | 1900 | 2650 | 3175 | 3975 |
| 125 | 750 | 1050 | 1275 | 1575 | 375 | 2000 | 2800 | 3350 | 4200 |
| 150 | 900 | 1250 | 1500 | 1875 | 400 | 2150 | 3000 | 3600 | 4500 |
| 175 | 1050 | 1450 | 1750 | 2175 | 450 | 2375 | 3300 | 3950 | 4950 |
| 200 | 1150 | 1600 | 1925 | 2400 | 500 | 2675 | 3700 | 4450 | 5550 |
| 225 | 1300 | 1800 | 2150 | 2700 | 550 | 2975 | 4000 | 4800 | 6000 |
| 250 | 1400 | 1950 | 2350 | 2925 | 600 | 3175 | 4400 | 5300 | 6600 |
| 275 | 1550 | 2150 | 2575 | 3225 | | | | | |

**Figure 13-2** Short-method calculator.

**EXAMPLE**

Assume a building of 1500 $ft^2$ of floor area, with 380 $ft^2$ of exterior door and window area and a design temperature (exterior) of 20°F. What is the amount of electricity in watts that is required?

1. Using Fig. 13-2 find 380 $ft^2$ on the vertical scale at the left of the graph.
2. Draw a vertical line upward from 1500 $ft^2$ on the scale at the bottom of the graph.
3. Find the intersection at approximately 13.0 kW.
4. Multiply 13.0 $\times$ 0.72 for temperature correction, which equals 9.36 kW.

## 13.4 HEAT PUMPS

A heat pump is a piece of electrically operated equipment that can provide both heat and cooling to a conditioned area. During warm summer periods, heat is removed from indoor air by circulating liquid refrigerant from outdoor coils to indoor coils, where the warmer indoor air causes the refrigerant to change to cold vapor and absorb the heat. Air flowing over the indoor coils is cooled, moisture is condensed, and the room is cooled. The refrigerant vapor is pumped outdoors, where it is converted back to liquid by the compressor and is returned indoors to repeat the cycle.

During the heating season, the outdoor coils absorb heat from the surrounding air, the compressor pumps the hot refrigerant vapor into the indoor coils, and the indoor blower circulates air over the coil, thus warming the air and the room. As the hot vapor condenses, it returns to a liquid state and returns to the outdoor coils to repeat the cycle. Even though the exterior temperature may be as low as 0°F, there is still more than 75% of heat at 100°F.

Control of a heat pump is accomplished with a two-dial low-voltage thermostat, with one dial to control heating and one to control cooling. As there is only one device for both heating and cooling, mechanical maintenance is simpler, and the cost ratio of normal operation is about four times the dollar value of heat output to one dollar value of electricity (heat) input. A major disadvantage is that because heat pumps are usually sized for the cooling load, outdoor temperatures below freezing do not have enough heat and auxiliary heating may need to be supplied for room comfort.

## 13.5 EMERGENCY POWER

For many occupancies, emergency power is not a real problem, but for hospitals, some industries using continuous heating processes, and locations such as theaters where panic might occur, it is a necessity. Such systems are designed to supply illumination and/or power automatically to areas where disruption of normal electrical service might endanger the safety of human life.

Sources of power may be storage batteries, generator sets, uninterruptible power supplies (separate service), or by emergency service connected ahead of normal service disconnect. Emergency current must be supplied within 10 seconds after failure of normal supply (NEC 700-12). Battery installations must be capable of supplying 87½% of total emergency load for at least 1½ hours, must be connected to a "compatible" automatic charger, and all lead-acid battery cases must be transparent to facilitate checking water levels. Automobile batteries are not approved for use.

Generator sets must be driven by a prime mover, which must start automatically upon failure of normal service. Fuel for internal combustion engines must be provided on-site and in sufficient quantity to operate the engine for not less than 2 hours on full demand.

Separate service, where approved by local authority, must be a complete installation, with a separate service drop, safety switch, and "widely separated electrically and physically" from the normal service. For connection ahead of service disconnect, the emergency service bypasses the normal main disconnect, and thus all emergency-system wiring must be carried completely separated from any other wiring.

## 13.6 SIGNAL SYSTEMS

Signal systems are of many types but should not be confused with sound or communication systems. The former usually means that a signal such as a bell, light, or similar device is actuated when heat, smoke, water, or other substance reaches a high temperature, low level, or other state. These systems are not usually designed as communication (voice) systems and are used to alert supervision personnel to malfunctions. Residential signal systems are most often confined to door chimes, smoke detectors, and perhaps low-level alarms for fuel in the heating system. Industrial signal systems may include fire alarms, sprinkler systems, smoke detection, watchman's reporting, security alarms, and elevator location systems.

Signal systems are governed by *National Electrical Code*® Articles 720, 725, 760, and the several related articles referenced in these. Of major interest is Article 725, which defines the wiring requirements of the several systems. Generally, all signal wiring systems must comply with the requirements of the NEC. There are two classifications for the systems.

> *Class 1 systems:* Supply source of not more than 30 V and 1000 A, remote control not in excess of 600 V, minimum wire size No. 14.
>
> *Class 2 systems:* Conductors must be isolated from power or light wiring, transformer supply must be rated at not more than 100 VA, overcurrent protection is required, and the line side must conform to general requirements for power wiring.

Residential door bells or chimes are one of the oldest of signal devices and most often consist simply of a pushbutton, low-voltage step-down transformer, and the bell or chimes. If used at more than one entry location, the bell or chimes may have a different tone for each location. Whem combined with electric door locks in apartment buildings or condominiums, the buzzer or bell for each apartment is actuated by its own pushbutton, and the apartment, in turn, has a pushbutton circuit to release the electric lock. In place of a bell or buzzer some systems may have an intercom to allow visitors to announce their name to apartment occupants. A simple diagram for a two-location signal system is given in Fig. 13-3.

Fire alarms and smoke detector systems are not yet common in resi-

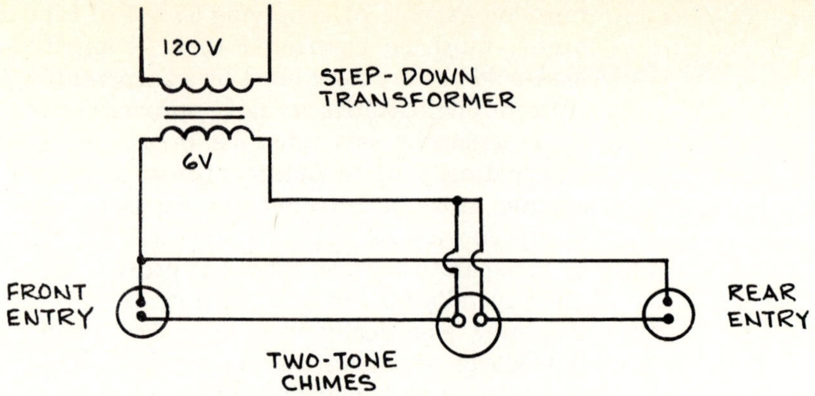


**Figure 13-3** Residential door signal.

dences except as battery-operated units, but are almost a necessity in commercial/industrial projects and in multiunit apartments and condominiums. When used in the latter locations, the alarm system is usually connected to a fire-station system and may be a general alarm that indicates only that there is a fire in a certain building or may be coded in various manners to indicate the exact location by floor or other space. The indicator may be a bell or buzzer arranged to "tap out" the signal or may be a drop annunciator with accompanying sound device. In a few installation types, a recording machine may be added to the receiver end to record the location of the fire. A simpler system uses manually operated reporting stations (the common "break-glass-pull-lever" type) that may be connected in the same manner to a central station. Whereas fire alarm detectors are made operational by temperature, the smoke detector is actuated by smoke and is installed in the same manner. The primary advantage of the smoke detector is that actual fire may not be needed to give the alarm. In many locations such as manufacturing plants, public assembly rooms, and other situations where people may be concentrated, a fire sprinkler system is required. This may be a wet system with pipes filled with water at all times, or a dry system with no water in the pipes until admitted by automatic valves or by connection to fire department equipment at the street level of the building (via a "siamese" connection). In either case, the sprinkler head spews water as temperature melts a plug, and the alarm is actuated by a drop in water pressure or by automatic opening of a valve. All of these alarm systems may be manufactured to turn off automatically after a preset interval or by manually switching.

Many systems in industry use similar alarm circuits to indicate low or high levels in fuel tanks or other compartments containing liquids or granular solids. A device somewhat similar to a smoke detector may also be used to indicate the presence of liquids where only dry conditions should exist. Some manufacturing plants, retail stores, bonded warehouses, and other locations use a human watchman who is required to make regular trips through the facility during his watch. Many have recording devices initiated by a pushbutton or keylock at each station to ensure that the watchman is actually making his rounds. Where large areas are patrolled, an annunciator or visual indicator in a central office may be incorporated to show the location of the watchman as he moves from one station to another. More sophisticated installations sometimes use television cameras connected to in-house circuits to supervise all areas of a plant.

## 13.7 CLOCK SYSTEMS

Clock systems may be just that, a system of clocks used to indicate the correct time, but more often they include some sort of signal system. The latter setup is most common in schools, where preset programs are announced by a bell or buzzer. Clock systems employ a master clock, which in turn transmits electrical impulses to each separate clock and correction impulses at the hour and daily by synchronous motors. One system requires separate circuits to each clock from the master. Another system uses the local power supply to operate the clocks, but correction signals are transmitted over the power wiring. Still a third system, familiar to children of school age and their parents, is a minute impulse system where the secondary clocks "jump" every minute instead of moving continuously. Widely used in schools, the secondary clocks are connected to the master in a simple two-wire circuit, which transmits a direct-current impulse every minute. A more modern system uses electronic high-frequency waves from the master to the secondary clocks to correct on the minute or hour, and the individual clocks are powered from adjacent light or power service.

When the clock system is also used as a signal device, an additional piece of equipment for programming must be included. This instrument usually consists of a series of drums or wheels which provide control by the hour and minute over a 24-hour period. By inserting plugs or pins in the drums or wheels, it is possible to arrange a contact with relays, which will in turn provide contacts for ringing bells or other signals.

## 13.8 TELEPHONE SYSTEMS

Prewiring for telephone service is almost a "must" for both residential and commercial/industrial projects today. This system should be carefully coordinated with the telephone company, which will normally supply the required cable and accessories, while the electrical installer supplies the boxes, conduit, and other fittings. Cable may be installed in wood-framed buildings without conduit and buried in concrete when provided with a special watertight covering. Entry into residence buildings is usually done in a manner similar to that used for electrical service, but much commercial/industrial service is provided underground from telephone manholes. Underground telephone cables may be installed in the same trench or raceway as that used for electric service but must be separated in different ducts.

For new construction, it is definitely recommended that the building designer and the electrical designer confer with the telephone company during the design stage. The telephone company will have much vital information regarding the average number of instrument stations that various occupancies have required, the size and type of wiring they can supply, the space needed for distribution panels, and a great deal of other information. Installation of spare conduit, ducts, or wire may also be recommended for future expansion or change of use of the building. A little extra thought and material at an early stage may provide better service and eliminate unsightly exposed wiring at a future date.

Entrance space for small residences is minimal to nonexistent, but for apartments, hotels, condominiums, and other locations where a number of telephone circuits are required, space must be provided for distribution panels. For small installations, a wall space in a closet or similar location of 4 ft by 6 ft is usually adequate. For large installations, a separate room or cupboard is needed to accommodate the larger distribution panel, relay racks, power supply, and accessories. Utility companies often require at least 3 ft of clear working area in front of such equipment.

## 13.9 TELEVISION

The ability of television to change as rapidly as it has in the past few years makes it almost impossible to plan very far in the future. Expanding even more rapidly than radio reception, television is now one of the foremost distribution systems for news and entertainment. Four primary "antenna" systems are available: self-contained "rabbit ears" on portable units, outdoor antennas for single-family homes, systems for multiple connections, and cable systems. Of these, the first, rabbit ears, may be eliminated from consideration as the antenna is a part of the receiver and may be used for only one particular unit. In somewhat similar fashion, the last method, cable TV (CATV), may be eliminated, as about the only building requirement is the installation of a coaxial cable about ¼ in. in diameter to each station.

An individual antenna for a single-family residence is quite familiar in form and provides the incoming television signals through a coaxial cable. This cable may be approximately ½ in. wide by ⅛ in. thick or may be about ½ in. in diameter. Where more than one or two outlets are desired or necessary, it is advisable to install a master antenna, amplifier equipment, and outlets in multiple locations. A simple diagram is shown in Fig. 13-4 where a single antenna serves a number of apartments. Cable television is installed in a somewhat similar manner but with more amplifiers to increase the picture and voice reception at the television set.

Information on television installation would not be complete unless closed-circuit television (CCTV) was included. This is a complete in-house installation with one or more cameras and one or more receivers. Such systems may be used extensively for security surveillance in industry, utility maintenance, video schedule displays, education, and a myriad of other uses. Older systems used vidicon vacuum tubes for camera pickup and produced black-and-white pictures.

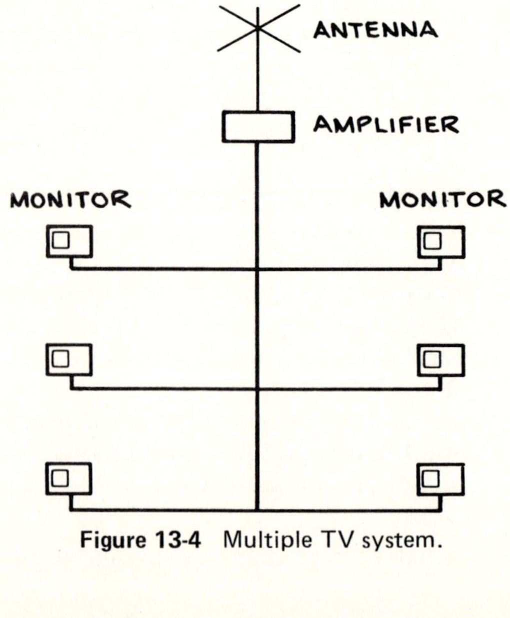


**Figure 13-4** Multiple TV system.

The second phase consisted of transistorized cameras and receivers. The third stage, now in use, includes solid-state cameras and better amplification. The clumsy size of early cameras has been replaced with hand-held units and one such camera is as small as 6 in. X 9 in. X 3 in. thick. A typical circuit, shown in Fig. 13-5, uses several cameras and monitors (receivers), as well as recorders and switching gear.

Transmission lead-in from antenna to receiver must be provided by using a coaxial cable consisting of two parallel wires of matching impedence. This impedence is typically 300 ohms. If other types of wiring are used, the wire will pick up extraneous sounds and noises and will not provide acceptable sound or video. About 500 lineal feet of standard coaxial cable is all that can be used without amplification before serious reception loss is encountered. Lead-in through building walls or beneath windows must be carefully made to eliminate possibility of shorting the circuit. Cable television (CATV) may be directly connected to the receiver or may be installed to several wall outlets with coaxial cable. With this type of installation, the receiver is connected to the circuit with a double-pin plug in much the same manner as with a convenience outlet.

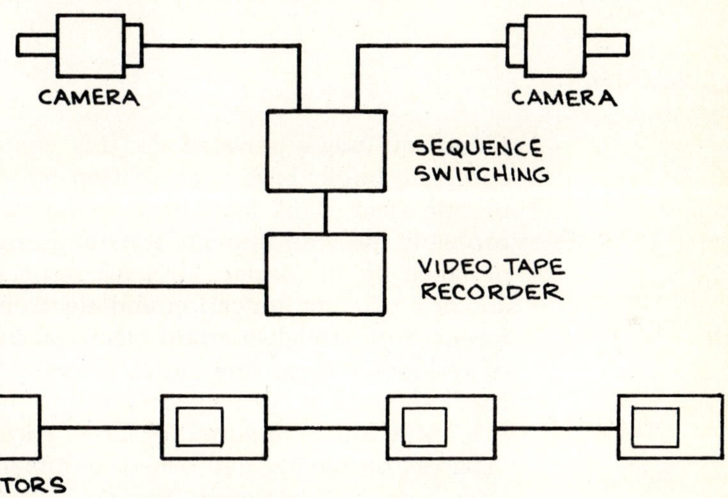


**Figure 13-5** Closed-circuit television.

## 13.10 SOUND SYSTEMS

A paging system generally uses a voice announcement to locate someone, who may in turn use a pushbutton indicator, return-voice acknowledgment, or a physical return to the caller's location. Perhaps the best-known systems are those used in hospitals to locate personnel who are known to be somewhere in the building, or nurse calls in patients' rooms to indicate that the patient needs attention. The first of these, as well as with systems at airports, large industrial or commercial establishments, and some schools, may be auxiliary uses for an intercom or music system and simply break into the steady flow of music for each announcement. In situations such as this, an acknowledgment to the caller is not normally possible over the same circuits, so some other signaling or voice device must be used.

Several factors determine the design of paging and intercommunication systems. The first factor is the occupancy and size of space that is expected to benefit by the installation. Factories and similar occupancy require the greatest amplification, or up to 100 W for 40,000 to 50,000 ft$^2$, while 30 W is adequate for most other indoor installations of about 8000 to 10,000 $ft^2$. The number and location of speakers may be determined by having a few high-output types or a greater number of low-output types. Acoustically, "live" areas will generally function better with low speaker sound levels and all speaker–microphone relationships should be carefully considered for feedback created by reverberation.

Wiring for communication systems is, in general, about the same as for power wiring, except that a smaller-size wire is used. Wire may be No. 14 to No. 20 for most circuits, single or twisted pairs, insulated, and enclosed in conduit. The length of speaker cable is important in preventing excessive power loss and may vary from about 1200 ft for No. 20 wire to 4000 ft for No. 14 wire on 140 V (50-W amplifier). Speaker or microphone wires should not be run in the same conduit with power wiring, and all amplifiers, cabinets, relay racks, and similar equipment should be properly grounded. Many different equipment units are used in communication systems, including AM/FM radio, TV, cassettes and other type tapes, automatic timing devices, microphones, and alarms. This diversity suggests that each installation must be based on a definite need, location, and combination of units, and this in turn usually means a special wiring arrangement.

## 13.11 SUMMARY

Electrical usages covered in this chapter are so diverse and varied that complete books have been written by experts in one field or on one system in one field. This fact suggests that although the information herein will probably provide enough *general* information, any installation should be designed by an "expert" in each respective field. This holds particularly for all sorts of communication and electronic systems, as these are basically different from straightforward electrical design. Further, although many types of residential alarm and indicator systems will never require expansion, space for it should always be provided for commercial/industrial installations. The few extra dollars required for larger wire size, more wires, larger conduit, and "spares" on panels, and other equipment may be returned with interest at a later time. Special care is also necessary when adding to an existing system. Almost without exception, it is both economically and "audibly" expedient to provide equipment that is of the same type, brand, and function than to try to redesign the entire system.

# 14

# Plumbing Principles

## 14.1 INTRODUCTION

Human beings have been using one system or another of plumbing for thousands of years. Early use included underground water ducts, hot-water systems, sewers, and storage facilities. The Romans had all these systems, at least in buildings belonging to the upper classes, and numerous types of buried piping have been unearthed. Many so-called primitive tribes and cultures used open ditches, bamboo troughs or pipes, and a number of other devices to carry water from a distant source. Unfortunately, the disposal of waste, did not keep pace with water delivery, and even at this late date barely keeps abreast.

The average person is familiar with plumbing only to the extent that when a faucet is turned on water comes out, or perhaps that a plumber, who has been called to clear a sewer stoppage, does something beneath the house and charges too much for whatever he did. Modern plumbing includes a number of systems, and most plumbing work is concealed behind walls or ceilings. In addition to the supply of potable water, plumbing work also includes supply and installation of building sewers, extension of building sewers to municipal (public) or private disposal systems, venting to prevent infiltration of sewer gas, fire-protection systems, irrigation systems, and installation of gas and other liquid-fuel piping. In addition, special piping may be required for commercial or industrial processing, glass piping for chemical transfer, piping for long-distance transfer of oil or gas, and to a varied extent, hot-water and steam piping in connection with some types of heating systems.

As more and more plumbing problems arose in connection with human living, plumbing regulations also increased, and today two major codes are used, the National Plumbing Code and the Uniform Plumbing Code. These codes are very specific in requirements for sewers, venting, water supply, and their relationship to the health and safety of people. The use of various materials, particularly pipe and fittings, is limited or allowed by local authorities and ordinances. Except for local differences, most of the pipe, fittings, fixtures, and related items are fairly common in the United States, as well as in other parts of the world. Plumbing in general is based on gravity flow for sewer systems and venting, but requires pressure for water and gas systems. Each of these aspects will be discussed in more detail in later chapters.

## 14.2 WATER SUPPLY

Although the disposal of waste is very important, and critical population growth has increased this importance, there would be little waste to dispose of unless adequate water was available first. Most water requires treatment to make it fit for human consumption, or *potable*, and is obtained from lakes or rivers (surface water) or from wells, springs, or underground streams (groundwater). Impurities in natural water are many and varied. All types of matter are common in surface water, as animal and vegetable wastes fall into the water and are absorbed or disintegrated by the water action. Underground water is contaminated by minerals or may be a carrier of bacteria from human waste when private or public disposal systems are not properly located. Normal water sources are not under pressure, so any building water supply must be pressurized as well as decontaminated before use.

Since water supply is a pressure system, pipes may be installed horizontally, vertically, or in almost any position, and most municipal water supply is provided at a pressure of 50 to 60 pounds per square inch (psi). In a few locations, where supply pressure is higher, a pressure-reducing valve must be used on the main building supply pipe to prevent pressure greater than 60 psi, which may cause damage. Most supply installations have a water meter and main shutoff valve to measure the amount of water supplied at each location. In areas where freezing is a winter problem, both water meter and valve may be installed below the frostline or inside the building.

Water is important not only to plumbing but is a major requirement for some forms of heating, air-conditioning, and power supply. In heating systems using hot water or steam as a heat-transfer agent, the water must be of adequate purity to prevent fouling of the system, in sufficient quantity, and properly installed. Water in connection with air-conditioning systems is most often used as a coolant for the condenser operation and may be used in very large quantities in some commercial applications. In low-humidity conditions, water in small quantities may be needed to adjust the humidity to a comfortable level. Steam, obtained by heating water, is used in many power-generating plants and again must be of good quality to prevent boiler scale, and in sufficient quantity to replace vapor lost through leaks, exhaust valves, and condensation. Although a great deal of this work is installed by "steam fitters," the work is essentially plumbing but subject to higher temperatures or pressures than in normal circumstances.

## 14.3 SEWAGE DISPOSAL

The disposal of waste is probably the most difficult problem that faces the world today. Our everyday living produces a seemingly unlimited amount of cans, bottles, paper, and other wastes. A great deal of this type of waste is a by-product of the food we eat, which in turn produces human waste, which is detrimental to human health if not disposed of properly. Two primary systems are used to handle this waste or sewage: (1) private disposal systems, and (2) public disposal systems.

Private systems are necessary when habitation is remote from municipal systems or when such systems cannot be connected into public systems owing to adverse terrain or lack of sewer-main extensions. In conditions where only a few people live, or where the ground level may be very irregular, disposal units may be simply holes in the earth which are backfilled when they become full. With more modern "inside plumbing" conditions, the fixtures, sewer, venting, and water supply are approximately the same as for a municipal installation, but the sewage disposal system is usually a septic tank and absorption field. Permission to install such facilities must be obtained from local health authorities. Adequate tests to determine the absorption and percolation of the proposed disposal area must be made and approved. Location of the septic tank and its disposal field in relation to the residence and water supply are also vital to prevent contamination.

Public sewer systems are seldom given a thought by people using a connection into those installations. The average person recognizes the existence of a system by the fact that waste from fixtures goes somewhere "down the sewer," or that manhole covers in the street provide access to the sewer, but beyond that, most people know little about the actual system. Plumbing inside the building connects all fixtures in a definite arrangement of pipe types and sizes, dictated mostly by the plumbing code. The number of fixtures, their type, arrangement, and accessibility are all determined by building codes, health department requirements, handicapped statutes or ordinances, and function of the building. Perhaps the most important single problem in connection with sewage disposal is the fact that practically every system operates as a gravity system. This means that a minimum "fall" or pitch of pipe must be maintained to cause the sewage to flow. It also means that careful planning of any sewer is necessary to prevent intersection with water, gas, or electrical piping. Building installation may pose a problem when long runs are necessary, as the total pitch may force the pipe to be exposed beneath a ceiling or otherwise interfere with the finished appearance.

## 14.4 VENTING

Bacterial action on sewage creates a gas (methane) which at best may have a disagreeable odor or at worst may be unhealthy and dangerous. To prevent this gas from entering the building, water-filled traps are required on all fixtures to seal off the sewer line and its possible gas. These traps alone are not adequate to prevent entrance of gas into a building, as a rapid drop of a volume of sewage in a multistory building could easily create a vacuum and siphon the water from each trap. Vents, piping connected to each fixture, are required to maintain atmospheric pressure throughout the plumbing. These vents are attached to each fixture on the sewer side of the trap, and the system is "vented" above the roof of the building into the free air. Each fixture may have a separate vent, or individual vents may be connected together to form various combinations of "circuit" vents. Municipal sewers in the street may also be vented, with pipes equipped with ventilated caps installed at the curbs, or simply by holes in sewer manhole covers. Vent systems in buildings are generally considered as part of the sewer installation.

## 14.5 PLUMBING FIXTURES

The term *plumbing fixtures* is applied to those parts of a plumbing system that are usually exposed to public view. Quite a list could be made if *all* fixtures were included, but those most common and easily recognizable are bathtubs, lavatories, water closets, and kitchen sinks. Water heaters are often included in this list as well. Less common fixtures are bidets, floor sinks, custodial (slop) sinks, shower units, laundry trays, drinking fountains, and similar units. Fabrication materials are ceramic, stainless steel, enameled cast iron, pressed steel, plastic, and about every combination that might be imagined. Although not strictly a fixture, the faucets, spouts, and other "brass" required to complete the installation may be included in this list.

Fixtures have only one function, to provide a container, connected to both water and sewer lines, which can be used essentially for human cleanliness. Some fixtures are mounted directly on the floor, others are hung on walls, and some are included in cabinetwork or other supporting devices. The use, residential or commercial, usually dictates the design and the method of installation. Location and type of support or attaching methods are provided by the manufacturer's "rough-in" diagrams for each type and size of fixture.

## 14.6 GAS SUPPLY

Piping for liquid or gas fuel may be installed either as part of the plumbing work or as part of the heating system. In either case, the work and material required and the principles used are the same. Gas piping should be black iron, not galvanized steel. The zinc content of the galvanizing process deposited on the surface, both inside and outside, of steel pipe may combine with some fuel-gas elements or may chip off with the passage of gas or oil if the coating is not firmly secured. The fittings used with black-iron pipe should also be black iron and connections are generally threaded, although welded connections are also acceptable and more common on larger pipe.

A great deal of fuel pipe is buried in the earth, either from a street main to the building or from a fuel tank to the building. This pipe should be buried below the frostline as below-freezing temperatures may cause contained moisture to freeze or at best to form ice crystals that could plug valves or nozzles. The pipe should be wrapped or completely coated with asphalt compounds to control corrosion. Another method often used on commercial installations is to provide an anode–cathode electrolitic arrangement in which the pipe is protected by deliberate destruction of metal poles of opposite value buried in the earth and connected to the piping system. Fuel piping inside the building is installed almost exactly in the same way as water piping, except that fuel piping should *never* be used as an electrical ground in the way that water piping is used.

## 14.7 IRRIGATION

Piping for landscape irrigation may also be installed as plumbing work but is most often installed by the landscape contractor. In earlier installations, the pipe and fittings were of galvanized steel, primarily because that was the only material available. When copper pipe and flexible copper tubing came into more common use, they began to replace the steel pipe, since it could be easily bent and was available in long lengths as tubing. Today, plastic pipe and fittings have just about replaced all other types for initial installation, as well as for many repair jobs. Plastic pipe is durable and noncorrosive, easy to fabricate with a minimum number of tools, and usually more economical. The location of pipe runs and sprinkler heads is usually indicated on the drawings and the pipe is buried below the frostline to prevent freezing in winter. The capacity to drain the system completely should also be included if freezing in winter may be a problem. Fixed heads spray the water in a full circle or preset quadrants of a circle, whereas moving heads provide a stream or modified spray. Manual or automatic supply valves are used to control the installed system.

## 14.8 DESIGN PRINCIPLES

Plumbing design is perhaps less complicated than most other support systems for buildings but does require adherance to a few basic principles. The sewage disposal system is almost always a gravity-flow operation and requires a minimum pitch of ⅛ in. per foot to keep the flow going. Each fixture requires a trap to prevent sewer gas from entering the building, and the venting system must be arranged to provide atmospheric air pressure throughout the system. In multistory construction (see Fig. 14-1), the disposal system must be sized to accommodate the increased sewage flow as additional fixtures are added floor by floor. Venting in multistory plumbing usually constitutes a separate system, but in a few cases the sewer line may also serve as a vent.

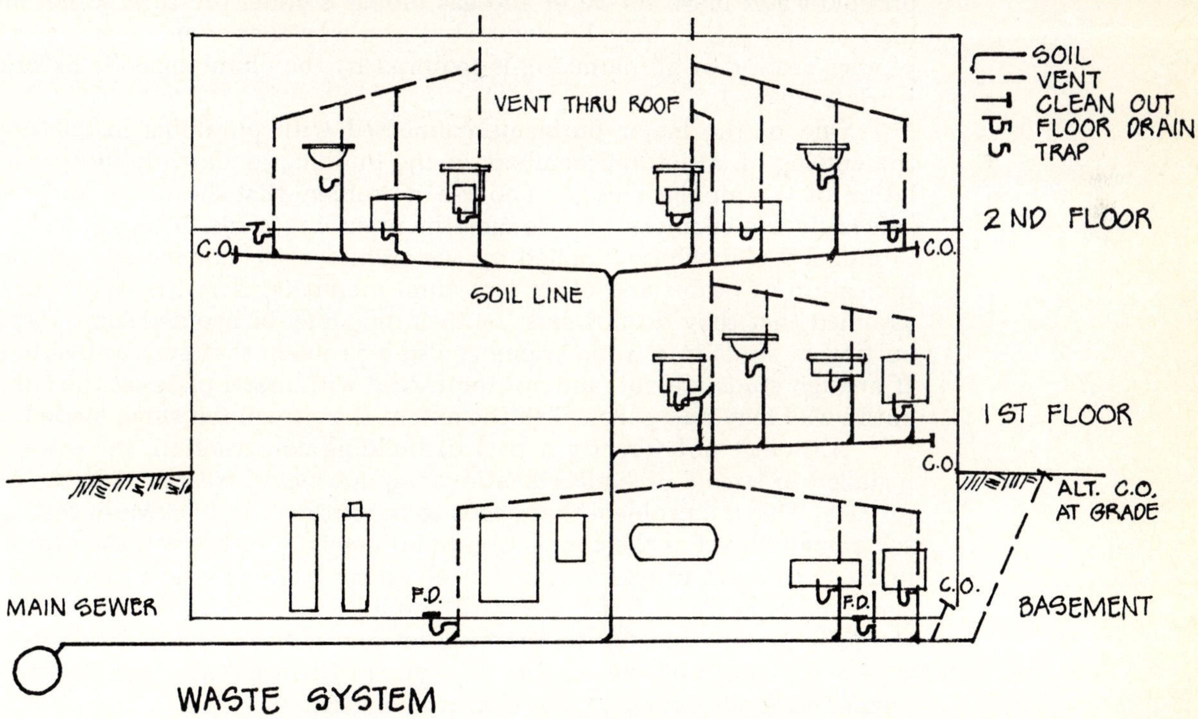


**Figure 14-1** (*From the Architectural Draftsman's Reference Handbook.*)

Water is required at all plumbing fixtures. Cold water is used at all fixtures, but hot water is normally supplied only to bathtubs, showers, lavatories, kitchen sinks, bidets, laundry washers, and perhaps custodial sinks (see Fig. 14-2). For proper installation, each waterline should be controlled by a stopcock ahead of the fixture. Insulation of hot-water lines is advantageous,

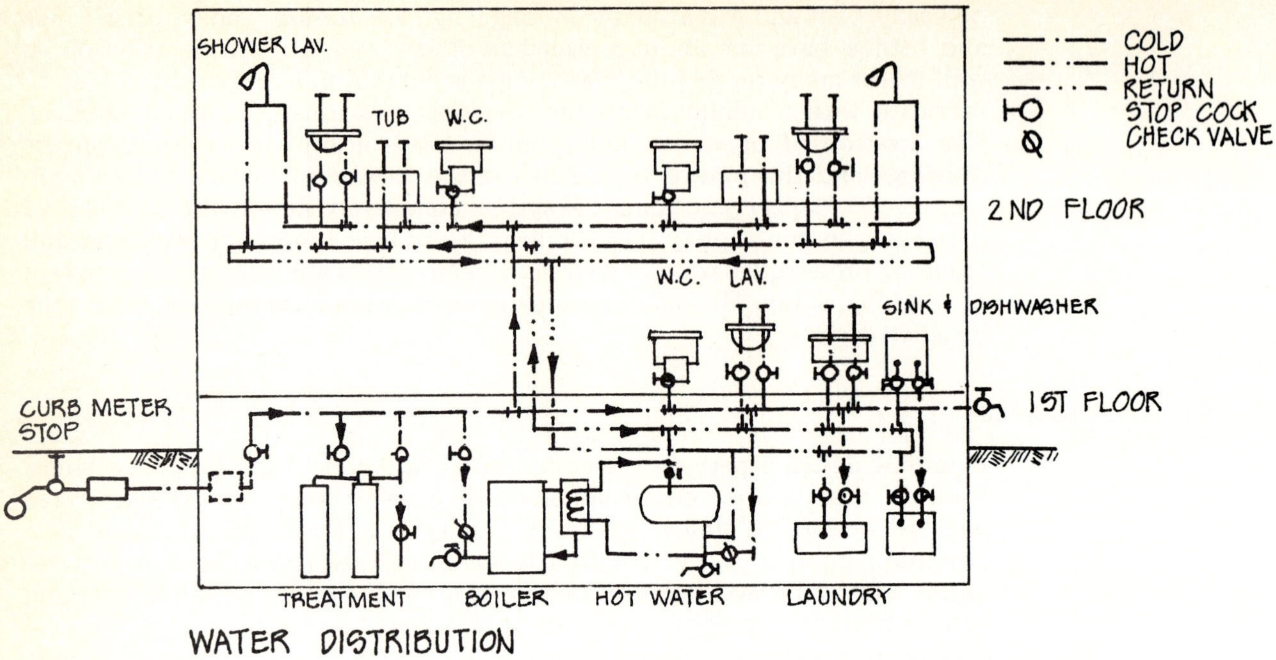


**Figure 14-2** *(From the Architectural Draftsman's Reference Handbook.)*

but only in high-humidity conditions is the application of insulation desirable on cold-water lines. As water and gas piping is under pressure, sometimes as high as 150 psi, it may be connected in a way that avoids sewer and vent piping. Testing of all plumbing is required by the plumbing code before any work is covered.

One of the major problems connected with plumbing installations is the cutting of structural members of the building to allow the proper installation of the piping. This is, of course, a problem that should be, and can be, solved by the engineer who designs the plumbing work. Often, however, the "dummies" (drawings supplied by the architect) do not include the proper indication of beams and other structural members, with the result that it is assumed that they do not exist or that they may be avoided some way. The cutting of studs in a wood frame is also a problem that may be overlooked. If enough studs are cut, and not reinforced with metal plates at the cuts, the entire wall may be weakened to the extent that it will fail when loaded.

Although not directly a part of building construction, the sewer lines installed in streets or similar locations require special treatment in the design process. This is a problem that needs to be worked out backward; that is, the sewer-main invert at the connection point needs to be known, then the sewer laid out to drain properly into that elevation from all points "upstream." If the building sewer invert is arbitrarily established without this consideration, the flow may be too flat to function or may even be uphill. Friction in the pipe and fittings and the number and type of fittings that retard flow in both water and sewer lines is also of primary concern.

## 14.9 HYDRAULICS AND PNEUMATICS

A basic knowledge of hydraulics and pneumatics is necessary in the design of water supply and drainage systems. A cubic foot of water weighs 62.4 lb, and a 12 in. × 12 in. × 12 in. cube of water exerts a static pressure of 62.4

lb on the 12 in. × 12 in. base area or 0.433 lb on each square inch of the bottom surface. For a depth of water different from 12 in., use of the following formula will provide the pressure at any depth in pounds per square inch.

$$P = \frac{62.4}{144} = 0.433 \times D$$

where $P$ is the pressure per square inch and $D$ is the depth of water in feet.

**EXAMPLE**

Assume a water tank 80 ft high and a house with its floor level 25 ft below the tank-base level. What is the pressure at the base of the tank and at the floor level of the house?

$$P_1 = \frac{62.4}{100} \times D \text{ of tank} \qquad P_2 = \frac{62.4}{100} \times D \text{ of house floor}$$

$$= 0.433 \times 80 \qquad = 0.433 \times (80 + 25)$$

$$= 34.64 \qquad = 45.46 \text{ psi}$$

Pressure *head* is measured by the depth of a column of water above the point where the measurement is taken. This quantity is often used instead of pounds-per-square-inch psi. If the pressure head is known, it may be converted to psi, or vice versa, by using the formula

$$H = 2.31 \times P$$

where $H$ is the pressure head in feet and $P$ is the pressure in psi.

**EXAMPLE**

What is the pressure head required to create a pressure of 50 psi?

$$H = 2.31 \times 50$$

$$= 115.5 \text{ ft}$$

## 14.10 FRICTION LOSS

Friction in a pipe is a major consideration in the flow of liquids. A smooth pipe interior creates little friction, whereas a rough inside surface and multiple fittings create friction and velocity loss. Lists of various fittings in different materials are given equivalent lengths of straight pipe, and graphs indicate the friction loss of pipe by size and capacity. For both drainage and water supply, the entire piping system is reduced to equivalent lengths of straight pipe. Knowing the length of pipe, the size, and the estimated flow, the required pressure of water may be calculated. Drainage-pipe sizing depends primarily on the number of fixtures connected and the proper pitch of the pipe.

Water-supply-pipe size has been established for most fixtures, so the selection of the size alone is not a great problem. Loss of pressure due to friction at various fittings has also been adequately determined and is given in Table A8 of the Appendix. Knowing the number and type of fittings and the actual lengths of connecting straight pipe, it is relatively simple to calculate the friction loss in a pipe system. The flow rate is determined by the amount of water, in gallons per minute, (GPM) that is required to serve adequately all the fixtures connected to the water piping.

**EXAMPLE**

Assume a 1-in. galvanized-steel pipe installed as shown, with a flow rate of 5 gal/min.

| | |
|---:|---:|
| Total length of pipe (50 + 10 + 25 + 3 + 2) | 90 lin. ft |
| 4 90° elbows × 3 ft each | 12 lin. ft |
| 1 90° tee × 5 lin. ft | 5 lin. ft |
| 1 globe valve × 25 lin. ft | 25 lin. ft |
| Total equivalent length of pipe | 132 lin. ft |

Friction loss in 1-in. pipe has been established at approximately 0.03/100 lin. ft when the rate of flow is 5 gal/min. Different values may be obtained from a variety of tables, depending on the different rates of flow, condition of pipe, and so on. Complete the calculation by multiplying the equivalent linear feet of pipe by the head loss per 100 ft as follows:

$$\frac{132 \text{ lin. ft}}{100} = 1.32$$

$$1.32 \times 0.03 = 3.96 \text{ lb friction loss}$$

Using a combination of pressure loss due to height above the main water supply plus friction loss through pipe and fittings, it is possible to determine the required initial pressure to supply-water operating pressure to any fixture or group of fixtures at any floor level or elevation above that of the main supply pipe.

## 14.11 SUMMARY

Plumbing is now a necessity for all people living in a modern society. Perhaps the only exceptions are those people who live in remote areas away from potable water and proper disposal systems, or those who do not have access to these facilities owing to religious beliefs, excessive costs, or inadequate public provision. Despite the availability of potable water and approved disposal methods, millions of citizens of the United States and many more millions in other parts of the world live without these benefits.

The various aspects of plumbing are covered in the following chapters. Not all plumbing work is identical, but the installation of certain groups of fixtures or separate units is controlled primarily by building and plumbing codes and is reasonably identical, or at least very similar. In addition to code requirements, plumbing installation is also determined by the friction loss in pipes, temperature and pressure variations, type of building and occupants, location and number of fixtures, and the availability of both potable water and sanitary sewers. All these factors are important to proper design and installation of plumbing.

# 15

# Plumbing Materials

## 15.1 INTRODUCTION

Most people are reasonably familiar with certain plumbing materials simply because they see them or use them daily, but this probably includes only the actual fixtures that are exposed and does not include the concealed piping, valves, traps, and other parts necessary for an operating system. Fixtures (tubs, sinks, lavatories, water closets, etc.) are manufactured from a number of materials in a great number of sizes and shapes, each designed to help satisfy requirements for special use or location and to allow a cost range for economic reasons. Pipe is also available in a wide variety of materials, sizes, connections, and strength to satisfy use or the plumbing code requirements. Finishing "brass" includes faucets, valves, spouts, drains, showerheads, and similar parts that are installed with the fixtures and may be obtained in a number of finishes. Towel bars, robe hooks, paper holders, and other "bath and toilet accessories" are *not* normally a part of plumbing work but are considered specialty items or supplied as part of "finishing hardware." In addition to these three major categories, there are a myriad of other materials required to install piping, caulking, connect fixtures, or to simply cut and assemble concealed work.

## 15.2 FIXTURES

Plumbing "fixtures" are generally considered to be those units that are exposed to view and used for baths, washing, collection and disposal of human waste, and related functions. Before we investigate each of these units, we should be sure of their proper names and some of their uses. Misuse of proper names may cause confusion, misplacement of units, or misunderstanding of code requirements or manufacturer's information. Following is a list of proper fixture and/or location identifications.

*Bathroom:* A room containing facilities for taking a bath; not a *toilet room*. Usually found in residences, apartments, and hotels.

*Toilet room:* A room containing water closet(s), urinals, and lavatories; not a *bathroom*. Usually found in restaurants, bars, and offices.

*Shower room:* A room containing various types of shower units. Usually found in schools, gyms, and other recreational construction.

*Bathtub:* A unit designed for bathing. May include a shower fitting.

*Lavatory:* A unit for washing hands or face. Not a "bathing" unit. Usually installed in connection with water closets and/or urinals.

*Water closet:* A unit designed for the collection and disposal of solid and liquid human waste. Often misnamed a "toilet."

*Urinal:* A floor or wall-mounted unit designed for the collection and disposal of liquid human waste.

*Sink:* A fixture designed in a great number of sizes and shapes for kitchen use in the preparation of food and cleaning of dishes.

*Laundry tray:* A unit similar to an oversized sink made of cast iron or cement and used in connection with laundry machines.

*Service sink:* A wall or floor-mounted unit designed for janitorial use to accommodate buckets, mops, and similar equipment.

*Floor sink:* A small drain receptacle, often used as an indirect waste outlet for refrigerators, bars, and similar installations.

Bathtubs (Fig. 15-1) are manufactured in lengths of 4′-0″, 4′-6″, 5′-0″, and 6′-0″, with 5′-0″ the most common, in widths from 27 to 32 in., and heights at the rim of about 14 in. Good-quality tubs are formed from molten iron (cast iron), then covered with powered enameling material called "frit," and finished in high heat to melt the frit into a homogeneous glossy coating. Bathtubs may also be formed from sheet steel by a drop hammer, then trimmed and several coatings of sprayed enamel (paint) applied as a finish. Tubs or tub-shower units are also made of plastic with integral wall

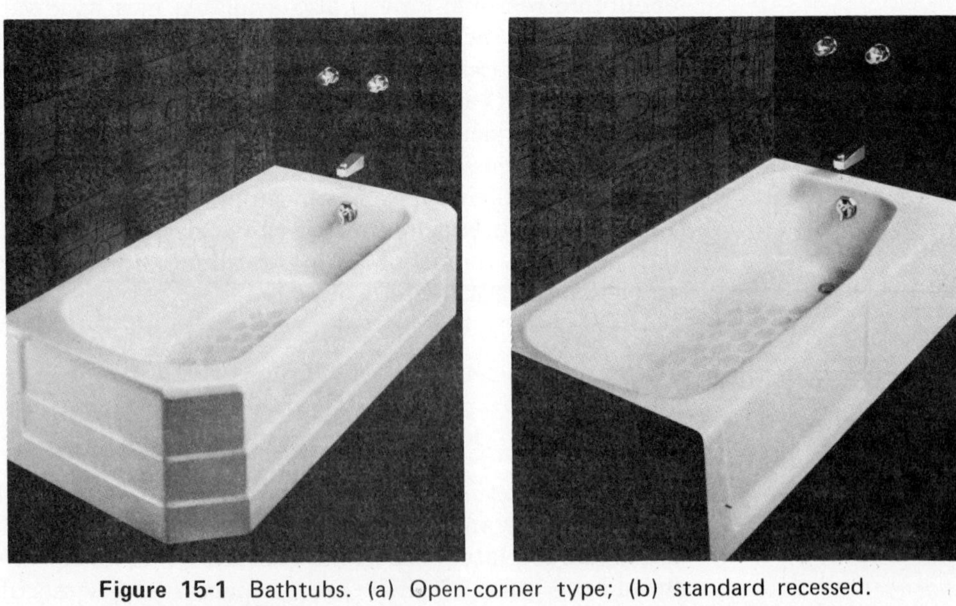

**Figure 15-1** Bathtubs. (a) Open-corner type; (b) standard recessed. (*Courtesy of Eljer.*)

surfaces of approximately 5′–6″ high and without seams or joints. Pressed steel or plastic units are lighter in weight and lower in cost, but to be installed properly should have blocking or solid insulation placed between the bottom of the tub and the floor to eliminate flexing or cracking of the tub bottom.

Lavatories may be installed on pedestals, in cabinet tops, or wall-hung (see Fig. 15-2) and will average about 18 to 20 in. deep and 20 to 24 in. wide for wall-hung units. Countertop lavatories may be designed to fit on top of ajdacent countertops or may be designed to fit flush or below the actual top, with chrome trim covering any exposed edges. Lavatories are normally of enameled cast iron or vitrified clay (china) but may also be obtained in pressed steel with enamel finish or of "cultured marble" plastic. Wall-hung units may be supplied with a turned-up "backsplash" of 4 to 6 in. or may simply butt against the wall. Holes for the finishing faucet sets are usually made by the manufacturer, but some lavatories for countertop installation require space in the countertop for these fittings.

**Figure 15-2** Bathroom lavatories. (a) Wall-hung; (b) counter in-set; (c) pedestal. (*Courtesy of Eljer.*)

Water closets, often misnamed "toilets," are of several types but almost without exception are manufactured of vitrified glazed clay (china). The exterior appearance of water closets is more or less limited to two types (see Fig. 15-3): those with a close-coupled flush tank, normally used in residences, and those without a tank but equipped with a *flushometer* or flush valve, as installed in more public facilities. The tank-type unit uses about 12 to 15 gal of water, at practically zero pressure, to flush the unit. The flushometer installation requires a minimum of 15 psi of water pressure and a minimum pipe size of $1\frac{1}{4}$ in. to provide adequate water during the short flushing period of about 30 seconds.

The interior arrangement of a water closet is such that it includes the trap required on all sewer waste lines. This is accomplished by one of the two major configurations of passages and baffles shown in Fig. 15-4. The cutaway section in Fig. 15-4a shows a unit termed a "washdown" water closet, in which the water enters from the rear and rim, spills over the leg of the trap, and causes siphon action on the bowl contents. The section shown in Fig. 15-4b is a "siphon jet," in which the bowl contents are accelerated by a jet stream of water into the leg of the trap at rear of the fixture. The

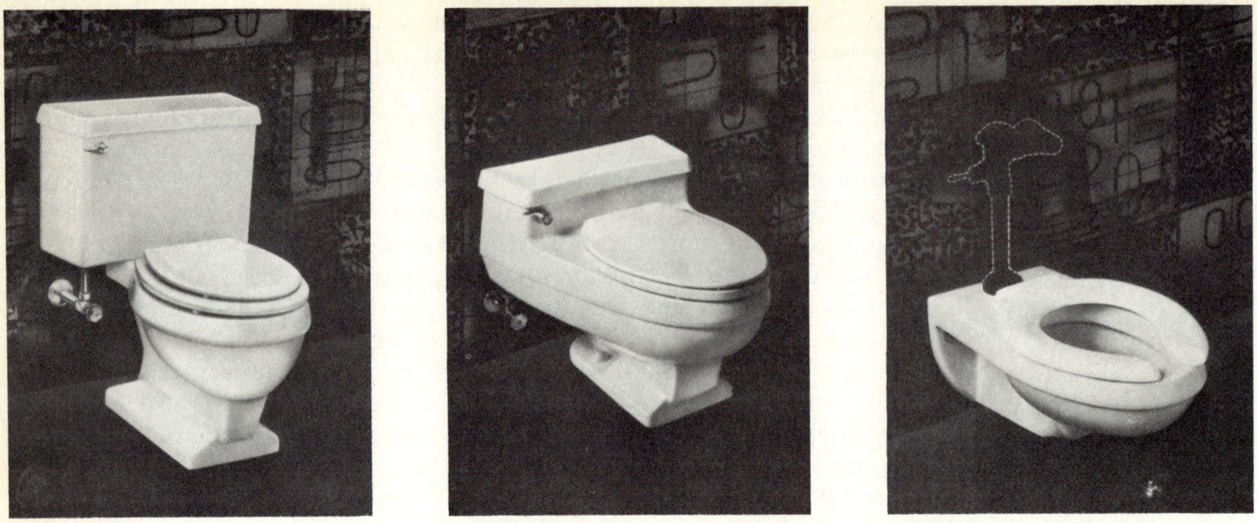

**Figure 15-3** Water closets. (a) Standard floor-mounted; (b) water economizer; (c) wall-hung flush-valve. (*Courtesy of Eljer*.)

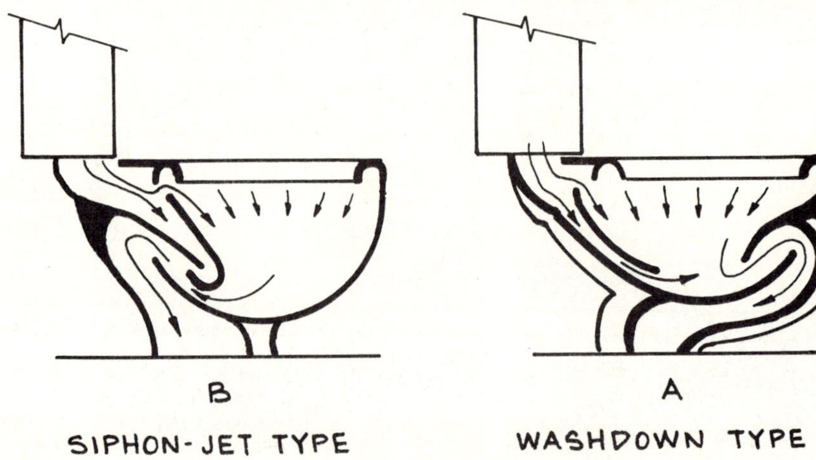


**Figure 15-4** Types of water closets.

waste or soil line of a water closet is usually 3 in. in diameter and often determines the minimum size of sewer lines or house drains allowed in construction.

Urinals are not normally included in residential plumbing installations but are very common in most public toilet rooms and usually used by men, although for several years in the 1950s one fixture manufacturer produced a Sanistan model for women. Urinals may be floor-supported stall type, wall-hung, or freestanding (see Fig. 15-5) and may be enameled cast iron or vitrified clay. Flushing water is provided by a controlled steady-stream valve or by flushometers. In a few areas, where quantity and water pressure are not adequate for flushometer use, urinals may be cleared by installing wall tanks with pull chains similar to those used for water closets.

Sinks are probably available in more sizes, shapes, materials, and various combinations than any other plumbing fixture. Designed almost exclusively for installation in counters or cabinets, sinks have a rim that allows countertops to be installed beneath the rim, flush with the rim, or over the rim (see Fig. 15-6). Material of manufacture includes enameled cast iron, vitrified clay, enameled pressed sheet metal, stainless steel, "cultured marble" plastic, and even glass. The depth of the sink basin is usually about 8 in.,

**Figure 15-5** Urinals. (a) Single wall-hung; (b) multiple wall-hung. (*Courtesy of Eljer.*)

**Figure 15-6** Residential sinks. (a) counter in-set; (b) roll rim. (*Courtesy of Eljer.*)

but there are combinations available with one compartment of standard depth and another compartment 8 to 24 in. deep. With the advent of garbage disposal units, most sinks are now provided with a drain opening of proper size for installation of a disposal unit, and various combinations of openings are available for different types of faucet and spout trim.

Laundry trays or sinks are manufactured from cast iron, cement compounds, or in a few cases from sheet metal. Usually available in single- or double-compartment configuration, these utilitarian units are approximately 24 in. × 24 in. each and about 24 in. in depth, and may be wall-hung or supported on metal legs. Service sinks are not normally installed in residential work but are essential wherever janitorial service is provided. These units are similar to laundry sinks, manufactured in cast iron or china, pedestal or wall-mounted, and designed to accommodate the equipment of the cleaning services. Floor sinks are miniature sinks, about 6 to 12 in. square, generally of cast iron and provided with a perforated cover. Installed complete with a separate trap, they serve as a waste receptacle for drains from refrigerators, bar sinks, and so on, and provide an air gap to prevent siphonage by waste or drain lines from such units.

Henry Ford, the automobile manufacturer, is reported to have once said, "You can have any color Ford car you want . . . as long as it's black." A similar statement might be made in regard to plumbing fixtures, since white is the predominant color regularly supplied. A wide range of other colors are available from some manufacturers, however. Colored fixtures cost 10 to 25% more, colors of different units may not exactly match because of the different types of finishes, and darker colors show water stains and dust more readily. Enameled cast-iron fixtures, procelain-enameled formed steel fixtures, and vitreous china fixtures are all produced in accordance with standards of the American National Standards Institute (ANSI) and the Uniform Plumbing Code.

## 15.3 PIPE

Plumbing pipe is made from a wide variety of materials: cast iron, wrought iron, copper, steel, stainless steel, brass, lead, clay, cement, glass, and plastic. Some are allowed for waste lines only, some for vents, some for water, and some (glass) are used for distilled water or acids. Codes and local ordinances specify which pipe, or combinations, may be used, so what may be approved in one locale may not be approved in another. For many years, the drain, waste, and vent (DWV) piping was installed only with cast-iron pipe and fittings, and water lines were installed with galvanized-steel threaded pipe and fittings. Today this has changed to the extent that in addition to these old standbys, iron pipe, copper, brass, and plastic may be used for general purposes indoors and outdoors and clay and cement pipe are extensively used outdoors.

Cast-iron pipe is used for soil or waste lines and in older plumbing installations for the vent system. Available in "standard" weight or "extra heavy" weight, cast-iron pipe is produced with a number of jointing possibilities and in a great number of fitting types and sizes. The most common connection is the bell and spigot, shown in Fig. 15-7 which allows the spigot

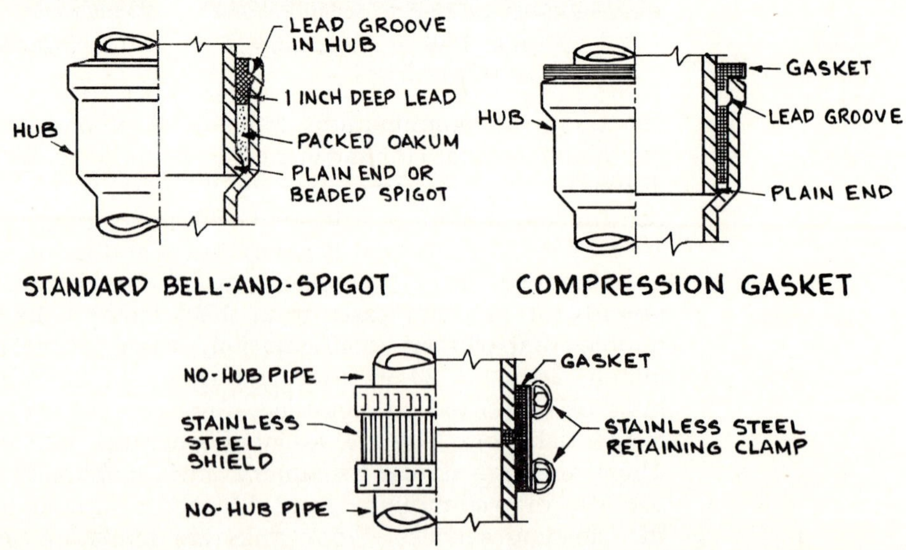


**Figure 15-7** No-hub cast iron pipe connections with gasket and clamp.

end to be caulked into the bell end of another pipe with oakum and molten lead. Another type has flush squared ends and joints are made with a metal band secured by tightening screws. In large pipe, 8 to 48 in. in diameter, used in waterworks installations, the joint is usually made with flanged ends and connecting bolts, and a rubber gasket joint, either mechanical or push-on, is easily assembled and can be watertight.

Galvanized-steel pipe and black-iron pipe are manufactured to the same general standards, are assembled and joined with threaded fittings (see Fig. 15-8), and are normally available in 40-ft lengths. Pipe is sized by the inside diameter of the pipe and may be obtained from ¼ through 12 in. for most uses in plumbing work. Galvanized pipe was formerly used extensively for water pipe, but "hard" water tends to deposit contained calcium on the interior of the pipe, thus reducing the actual opening until the pipe may be completely closed. Black-iron pipe (not galvanized) is used for gas piping.

Copper pipe or tubing is usually the most costly to install for a water-supply system in a building but is also the least likely to give trouble if properly installed. Rigid pipe in sizes from ¼ to 8 in. is available. Copper tubing is fairly flexible and soft-tempered, so that is can be bent cold to accommodate itself to irregular offsets. Heavy-wall type K is used for underground service, interior systems, in either hard or soft temper. Medium-wall type L is also available in hard or soft temper and in rigid pipe or tubing. The various dimensions and capacities of copper pipe and tubing can be compared by reference to Appendix Table A-4. As the inside surfaces of copper pipe and tubing are smoother than galvanized steel pipe, primarily because of the manufacturing process, it is sometimes possible to install copper pipe that is one size smaller than galvanized-steel pipe would be if used in the same installation. Although not usually used for soil, drain, or vent piping, copper may be used for these installations, especially when prefabricated systems called "Christmas trees" include these items. The cost of copper pipe and tubing is higher than that of steel, but the greater ease of installation and faster connections tend to offset this higher cost to make it comparable to steel-pipe installations.

The newest pipe material on the market that is accepted by code authorities for plumbing systems is plastic, in a number of different compounds.

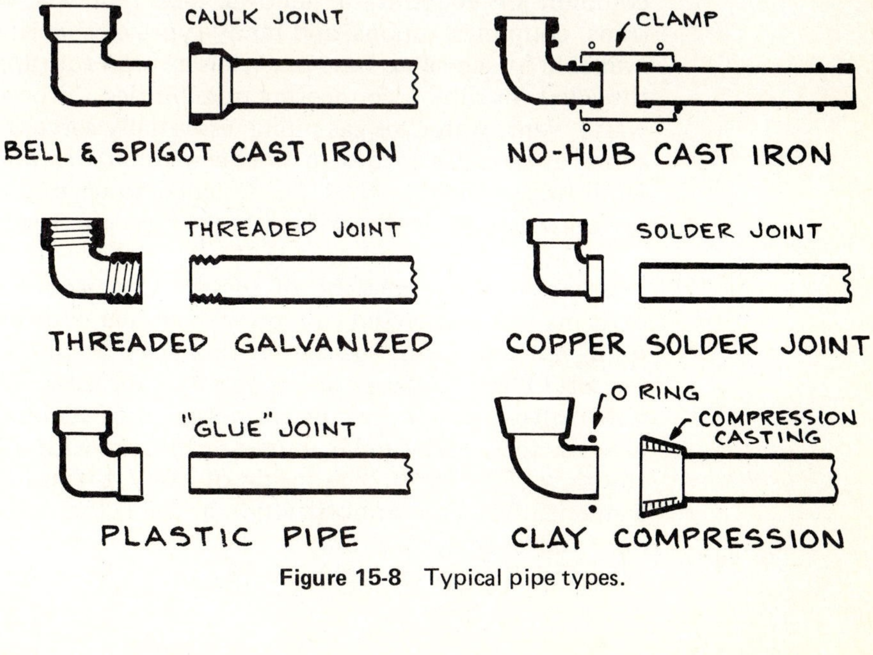


**Figure 15-8** Typical pipe types.

Among the several types of approved plastic pipe are acrylonitrile–butadene–styrene (ABS), which is used extensively for drain–waste–vent (DWV); polybutylene (PB) pipe and tubing approved for hot-water piping and water service; polyethlene (PE) for natural-gas yard piping; and polyvinyl chloride (PVC) for natural-gas yard piping, DWV use, and irrigation work. Polyvinyl chloride types PSM and PSP are also used for sewer installations. Most plastic pipe is available in the same sizes and lengths as steel pipe, but only one or two types with extra-heavy walls use threaded connections. Unfortunately, most plastic pipe does not have the inherent strength to span the same distances between supports as steel pipe does and thus must be supported more often. Cutting may be done with a hacksaw or other sharp tool, and connections are simply made with proper fittings and liquid adhesive.

In addition to the major types and materials of pipe, lead pipe, asbestos cement pipe, clay pipe, brass, bituminized fiber, and glass pipe are also available. For the most part, these types are reserved for special uses. Lead pipe is probably used most often as a connection to water closets as a "closet bend," but this is strictly regulated nowadays. Asbestos-cement pipe is used for municipal water supplies, and "cement" (concrete) pipe, either lined with steel or unlined, may be used for municipal water, sewer, or storm drain lines. Clay pipe, in many sizes and lengths, is used for exterior sewer lines, septic-tank disposal fields, and similar nonpressure application. Brass pipe is used only occasionally in connection with copper or brass fittings. Fiber pipe, usually treated with bitumen, is used for exterior drain lines for sewers or septic-tank disposal fields, irrigation systems, or when perforated, as collector piping for various drainage installations. Glass- and tin-lined pipe is restricted to use where acids, distilled water, or other special liquids are to be handled.

## 15.4 FITTINGS

In addition to the various types of piping that may be used in a project, there is an almost unbelievable variety of "fittings" required to connect the lengths of pipe together or used to connect pipe to fixtures. Among the most common are 90° and 45° elbows, tees, reducers, caps or plugs, Y's, offsets, traps, couplings, unions, and many types of special fittings. These fittings are available in the same base materials as used for pipe, and with flush, flanged, threaded, or slip-on connecting possibilities. Although fittings used for drain, waste, vent, water, or gas piping essentially serve the same purpose, there are some differences in design between drain/waste and water or gas fittings. Drain/waste fittings must be designed to accommodate the gravity flow or sewage waste, while those for water or gas are designed for pressure systems.

Drain, waste, and vent (DWV) fittings are usually made of cast iron, or in some localities may be of plastic. Cast iron formed with bell-and-spigot ends may be connected to pipe with similar ends by means of caulked joints, or if ends are flush, the connection may be made with sleeve or strap connectors. Plastic fittings are generally fabricated for a slip-on joint secured with solvent. DWV fittings are designed to facilitate the flow of sewage, so elbows, tees, and similar units are formed with a sweep in the direction of the flow (Fig. 15-9). The inside of DWV fittings is also designed to provide a smooth flow free of obstruction at the actual joint.

Galvanized, steel pipe is usually connected by malleable iron galvanized threaded fittings (see Fig. 15-10) and is used principally for water piping,

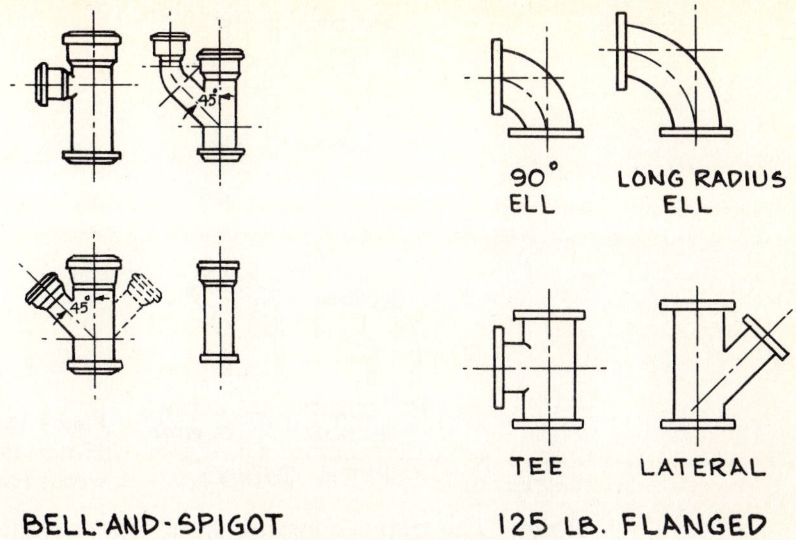


**Figure 15-9** Cast iron pipe fittings.

although in some locations galvanized pipe may be used for vent piping. Measurements for threaded fittings are usually taken from manufacturers' catalogs, which give the distance from the face of the fitting to the center of the fitting, and allowance for pipe threads must be made for connecting pipes. Threaded fittings are screwed to a tight fit, but because the formation of the last four or five threads from the pipe end is incomplete, the normal connection will show these incomplete threads when assembled. Male fittings are threaded on the outside; female fittings on the inside.

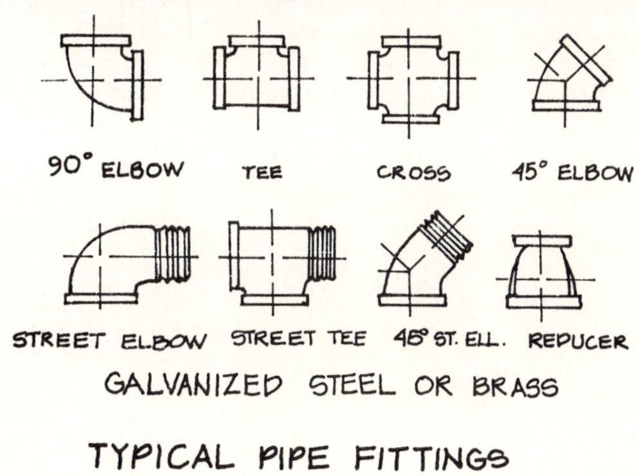


**Figure 15-10** Typical galvanized pipe fittings. (*From the Architectural Draftsman's Reference Handbook.*)

Fittings for copper and brass pipe are of two types. If the pipe is brass or type K or L copper, the fittings may be threaded in a manner similar to that required for galvanized steel. A more common practice, however, is to use soldered or "sweated" connections. Soldered fittings fit over the pipe of same size in a slip joint, so an obstruction is eliminated by offsetting the ends as shown in Fig. 15-11. Solder is applied at edges of the fitting or flowed into a groove in the fitting, thereby effectively connecting fitting and pipe. Copper tubing may also use solder fittings or may use a flared fitting.

Traps are required on all plumbing fixtures unless they have indirect waste (into another fixture) or have a built-in trap, as is the case with water

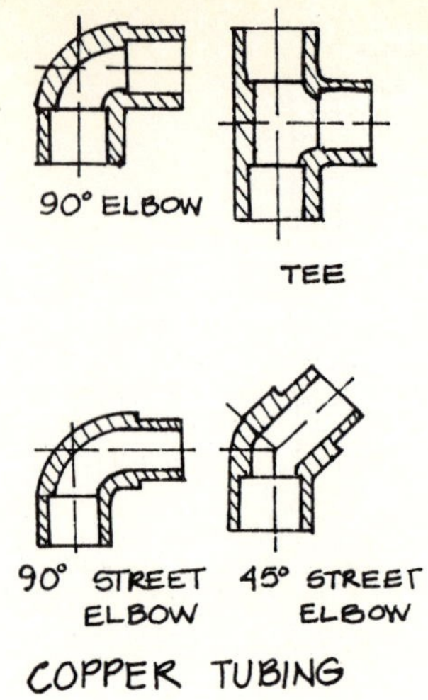


**Figure 15-11** Typical copper tubing fittings. (*From the Architectural Draftsman's Reference Handbook.*)

closets. The trap is constructed of pipe material, usually in several removable sections, in the shape of the letter P or S (Fig. 15-12) and is designed to retain liquid as a seal to prevent sewer gases from entering a room through the fixture waste opening. Cleaning is possible by unscrewing slip joints or by a threaded plug. Drum traps are also used in some installations but prohibited in others. Influent through the bottom of the drum may become obstructed by solids to the extent than in remote cases a solid "tube" is formed instead of a liquid seal, thereby allowing free passage of sewer gas.

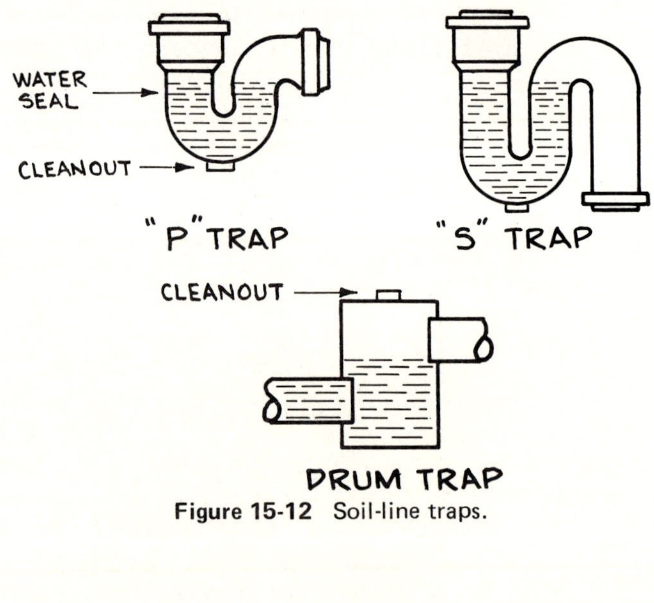


**Figure 15-12** Soil-line traps.

## 15.5 CONTROLS

The flow of liquids in a plumbing system must be controlled by some type of valve to regulate the amount of liquid to be removed, to provide the capability to change volume or direction, or to stop or start flow. Sewer lines are usually devoid of control equipment, as the flow is by gravity, not under pressure, and may only occasionally require a check valve to prevent backflow from a main sewer into a house drain line. Water and gas systems, however, use valves of various designs quite extensively. Most valves are manufactured with cast-iron bodies or brass bodies, and with stems and other parts of compatible material. The type of valve and material used are somewhat dependent on size and use, but various types of valves are available from 1/4 to 48 in. Special valves are manufactured for large pipes or unusual situations.

Gate valves and globe valves are the two types most commonly used to control flow in supply lines and are shown in Fig. 15-13. The gate valve depends on a solid disk or plate (the "gate") that operates between guides perpendicular to the line of flow to block the passage of liquid as controlled by a threaded stem and hand wheel. The globe valve uses a disk or ball (globe), controlled by a similar stem and hand wheel, to regulate the amount of flow through a horizontal opening parallel to the line of flow. Gate valves designed to allow full flow (full-way valve) when open are required on all water-supply lines by the Uniform Plumbing Code.

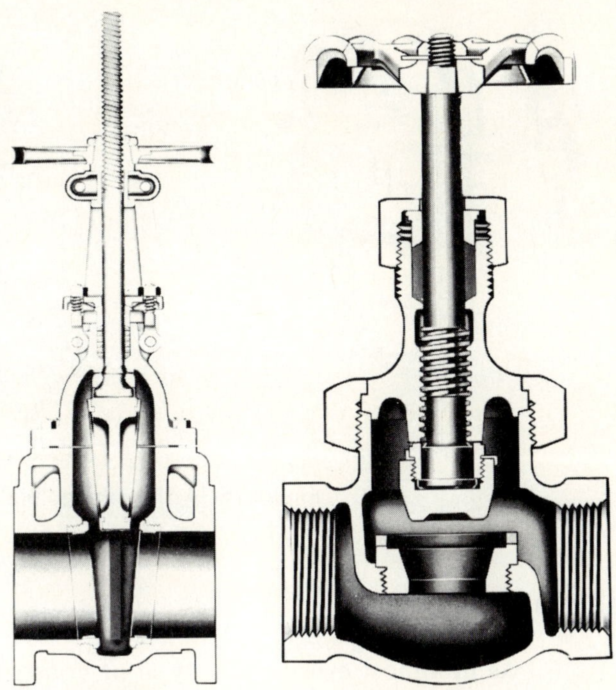

Gate valve Globe valve

**Figure 15-13**

Sink and lavatory faucet sets may be composed of separate valves to control hot and cold water and connected beneath the fixture to the supply spout, or may be a one-piece unit combining these controls and surface-mounted on the fixture (see Fig. 15-14). Gooseneck water-supply controls may be deck-mounted and have integral valve control. Modern drinking fountains for schools and commercial installations are most often supplied with a jet unit and control valve as a part of the unit. Hose bibbs for exterior hose connection have an integral valve and outlet threaded for attaching a garden hose. Bathtubs may be supplied with holes to accept water control valves and spout, or may be designed with no openings but require wall-mounted valves and spout above the tub rim. Similarly, showers may be controlled from a simlple cutoff valve integral with the water spout or by wall valves installed above the tub or shower receptor. Dial-type (Moen) water temperature and volume controls are usually wall-mounted and controlled by a knob that is turned to change water temperature and by push-pull action to increase or decrease volume. All water lines, both hot and cold, should be equipped with a simple valve to shut off water supply to the fixture if the normal faucet set, valve, or fixture needs repair or replacement.

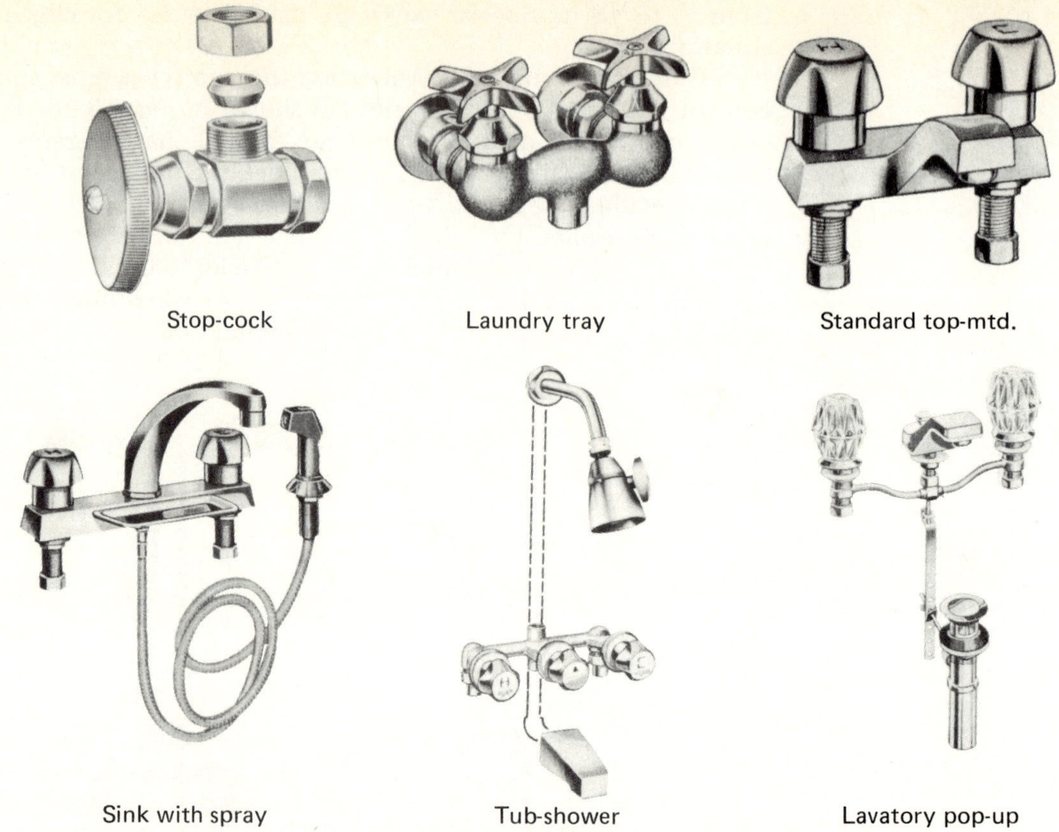


**Figure 15-14** Plumbing "brass." (*Courtesy of Eljer.*)

## 15.6 PLUMBING SYMBOLS

As in almost every other type of construction diagrams, symbols specific to plumbing work must be used on drawings. The symbols shown in Fig. 15-15 are typical for indicating pipe runs, fixtures, controls, and related work, and should become reasonably familiar to those who make or read such drawings. Piping runs are also shown in a diagrammatic location, which may not be the exact location of installation, so variations for valves, connections, and fittings must be considered when electrical work, heating ducts or pipes, and structural elements are also shown in the same diagrammatic location, or possibly not at all.

## 15.7 SUMMARY

Such a variety of plumbing materials exists that it is entirely possible that no two installations are ever alike simply because of the material used. Fixtures represent the largest cost in most plumbing systems and are often selected for appearance rather than function or upkeep. Since most fixture items of similar function operate about the same, the size, shapes, material, or color may be the determining factor from the owner's standpoint.

The type of pipe, fittings, valves, and insulation, if any, and the other "concealed" materials necessary for a complete plumbing installation are often left to the discretion of the installing plumber unless some agreement

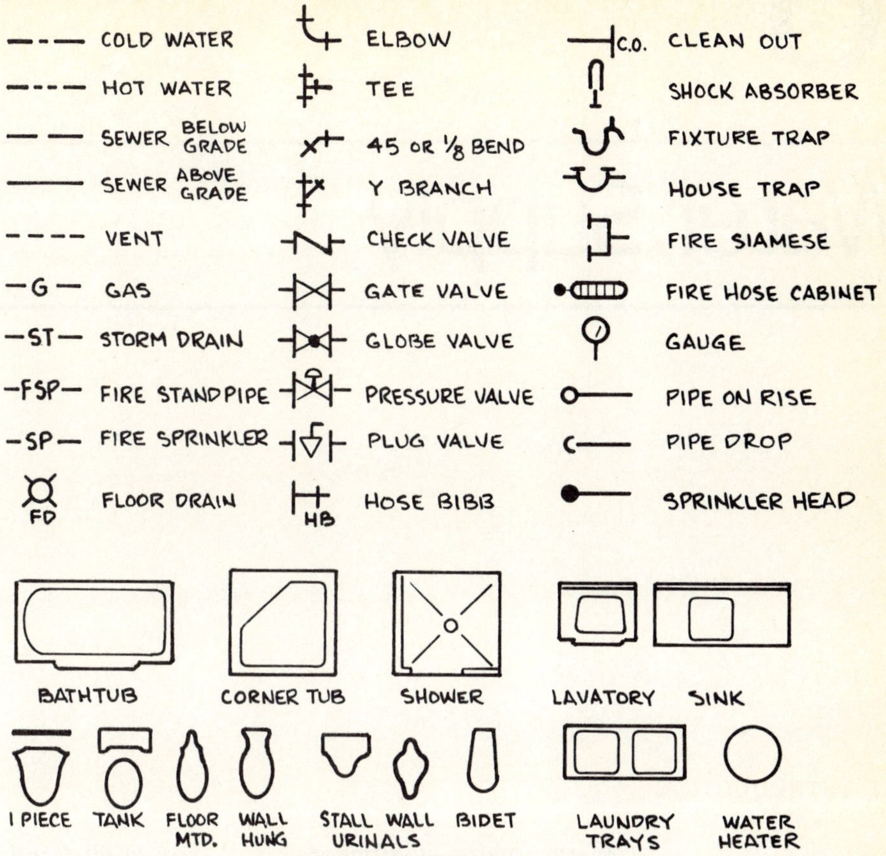


**Figure 15-15** Plumbing symbols.

is reached beforehand. Most pipe and fittings should provide adequate service for 25 to 50 years. Unfortunately, the pipe installed for either the water-supply or waste system, or both, is often the minimim size, or smaller, and eventually the entire system fails to perform adequately and needs repair or replacement. The Uniform Plumbing Code, like most codes, is a minimum code, so a statement that "it meets the code requirements" is not necessarily a guarantee that the system will function properly.

Although control valves are not usually required on waste systems, they are always required on water-supply systems. A pressure valve may be necessary at the street service connection to assure that adequate, but not excessive pressure is supplied to operate the system. Each water line, hot, cold, chilled, or other type, should be supplied with a stop valve at each fixture as well as at other branch locations to allow shutoff in case of repair. For commercial or similar locations, each floor or area feeder should also be valved for complete control.

While standard plumbing symbols are in common use to indicate the fixtures required for many residential installations and some small commercial/industrial projects, it is not common to see pipe locations, fittings, and valves shown. These items should be carefully considered in making plumbing layouts for major work, because two single pencil lines on a drawing indicating pipe or fittings are very different from job-site conditions if the equipment is larger than perhaps ½ in. Lines representing two 6-in. pipes and their valves may look fine in a drawing but may not even fit in the space left to them in the construction, and relocation or replacement may be very costly and difficult.

# 16

# Water Supply

## 16.1 INTRODUCTION

"Water, water everywhere . . ." We are all familiar with this phrase and its conclusion ". . . but not a drop to drink." This is not necessarily true, although there is water everywhere in some form or other, at some point or other, even in desert areas. Water is a primary requirement for human existence. Water, as we usually consider it, will keep people alive even if food is not available for quite a while, but deprived of water a person or animal will not live for more than a few days under normal conditions. In addition to the use of water by human beings, water is essential for most domestic and wild animals, crops, and vegetation of all kinds. It affects the erosion of land, changes temperatures of areas, and provides a system for transportation, as well as many other uses. This book is concerned primarily with the availability, sources, and uses of water for human consumption.

Practically all water on the earth is in its oceans, and all water eventually returns to these reservoirs. Because of the heat of the sun, water is vaporized to form clouds. As clouds are carried over land areas by the wind, cold air currents cause the clouds to lose their moisture in the form of rain or snow. As the rain or snow falls, it may pick up contaminants suspended in the air, but it is not until it makes contact with the earth that the major contamination starts. About one-third of total rainfall returns to the ocean by rivers or other surface drainage, but two-thirds is absorbed into the earth or evaporates back to clouds. Water coming into contact with minerals or metals in the earth may dissolve them and these compounds are carried along with the water. "Suspended solids," consisting of mud, clay, sewage, and industrial waste, may be added to the water as it flows and some gases may be formed.

Unfortunately, water may also contain bacteria that can cause people to become sick or can create epidemics in communities. Small communities without adequate water-treatment facilities are more prone to be affected by diseases carried by water than are larger communities with newer or better plants. Increases in drinking-water problems since about 1955 caused the United States government to formulate the Public Health Service Drinking Water Standards in 1962 and the Safe Drinking Water Act (SWDA) of 1977. These regulations state only the desired results and do not indicate how they are to be accomplished or enforced. One of the requirements of the SWDA is that water suppliers must notify their customers when water systems are in violation of requirements, but since most consumers have little or no idea of the requirements, this is a useless gesture. As it stands now, states and local communities make and enforce most of the requirements for safe drinking water.

## 16.2 NATURAL SOURCES

Water is composed of two major elements, hydrogen and oxygen, in the proportions of two atoms of hydrogen to one atom of oxygen ($H_2O$). In addition to these components, water also contains minute quantities of other chemicals, dust, and other materials picked up along the way. In some forms it occurs as snow or ice, and may also be saturated in various degrees with saline compounds (salts) or minerals from rocks (limestone). Human beings further contaminate relatively usable water with sewage, industrial waste, storm drainage, chemicals from crop production, and general misuse. Water to be used for human consumption must be free of contaminants, although different types of water or water sources may contain some chemicals. Water suitable for drinking is usually termed *potable* water.

Most of the domestic water used in the world comes from rivers, lakes, or other freshwater surface sources. Such water may contain a minimum of contaminants or may be heavily loaded with sewage or industrial waste. Even water obtained directly from a spring or similar source may contain minerals dissolved from rocks below the surface. Where surface water is not available, or municipal systems do not exist, the most common domestic water source is from wells drilled into the earth to tap a water table or subterranean river. The amount of water available from wells depends on the depth of the water table below the surface, the quantity of water below the surface, and amounts of seasonal rain to replenish the water table. Other sources for domestic water are a bit more complicated and include (1) evaporation from salt-laden water, and (2) recycled water from sewage. Each of these requires much equipment and expertise and results in a high cost per gallon, but with dwindling water supplies these are sources that must be considered for the future.

## 16.3 QUANTITY AND QUALITY

The amount of water used in present-day living is astronomical, and is increasing with every additional person and with each use in households and industry. Several years ago, the amount of water used by one person in one day was reasonably estimated at 50 gallons per day. In the past few years, this has increased to approximately 80 to 100 gallons per day and in industrial usage may be as high as 500 gallons per day per person. This high figure includes water actually used in a product, but a great majority of industrial water use is for cooling of equipment and facilities in which a product is being manufactured or for cooling or quenching heat generated in a product during the manufacturing process. Most of the water used in industry is lost to further use, except condensed steam in a closed circuit, and is dumped into the nearest river. Some firms have now realized the economic loss of such dumping and are providing lakes or ponds adjacent to their operations to save and reuse cooling water or water that has not been polluted.

Water considered appropriate for human use may be obtained from a variety of sources, but in every case certain requirements must be met in our present-day living mode. In many countries rivers are the source of water for drinking, bathing, laundry work, and sewage disposal—often all at the same point. Streams in mountains or coursing through pasture lands may look clean, smell clean, and even taste clean but may also be contaminated by animals, rotted vegetation, undesirable minerals, or other pollutants. The concentration of impurities in water determines its usability. Visible or physical characteristics are readily discernible in the form of color, taste, odor, or in the presence of silt or microorganisms. On the other hand, chemically pure water is not palatable. Small amounts of gases or minerals are therefore desirable to provide water that is pleasant in taste and acceptable in appearance. The cleanliness of water may be measured by the amount of impurities in terms of *parts per million* (ppm) or *milligrams per liter* (mg/liter), which are approximately equivalent. Rainwater, before it is contaminated by roof coatings or other contacts, may have as low a contaminant count as 1 ppm, while normal seawater will have about 35,000 ppm. River water will contain from 300 to 500 ppm, while raw sewage contains above 1000 ppm.

Three major systems or types of treatment may be necessary to provide water that is usable. The impurities in the parts-per-million ratio may be from several sources and obviously are not all alike. One of the most common treatments to which raw water is subjected is filtration, which removes most objectionable color, turbidity, and bacteria. Another treatment is *deferrization* and *demanganization*, which removes dissolved ferrous metals (iron) and manganese and converts them to insoluble flocs by oxidation. A third, and probably the most commonly used, is softening of the water by removing excessive amounts of scale-producing, soap-consuming calcium and magnesium ions by the addition of lime and soda ash, which precipitates calcium as a carbonate and magnesium as a hydrate. Further treatment of water may include passage through activated carbon or application of chlorine to kill any remaining bacteria and remove any final odor or color. After passage through sedimentation beds and sand filters, water may be further purified by being allowed to trickle through beds of purified coke. Depending on the amount of undesirable pollutants and their type, a water treatment plant may use all these methods or only one or two to provide potable water.

## 16.4 WATER-TREATMENT PLANTS

Treatment of raw water may be required regardless of its origination, but the water generally obtained from rivers or lakes for municipality use almost always requires purification and proper distribution. Even cities that obtain their water from deep wells must, in most cases, treat that water for color, hardness, mineral content, or other impurities. Most private water supplies, such as driven wells, private lakes, or similar sources are not purified, or at most have a small filter and perhaps a chlorine doser.

If possible, raw water should be impounded in a lake or reservoir before being drawn into the treatment plant. This storage will allow some of the solids that may be contained in the water to settle to the bottom and will simplify the control of flow at the intake. Figure 16-1 shows a diagrammatic scheme for a conventional water-treatment plant which may be increased by multiple units or decreased by eliminating units as the quality of raw water varies. Major impurities in the raw water supply are (1) floating objects such as logs, floating cans and cartons, and fish; (2) small suspended solids, clay, silt, and leaves; (3) dissolved salts and compounds; and (4) dissolved hydrogen sulfide and carbon dioxide gases. In addition, there is usually the problem of color and odor caused by some of these impurities.

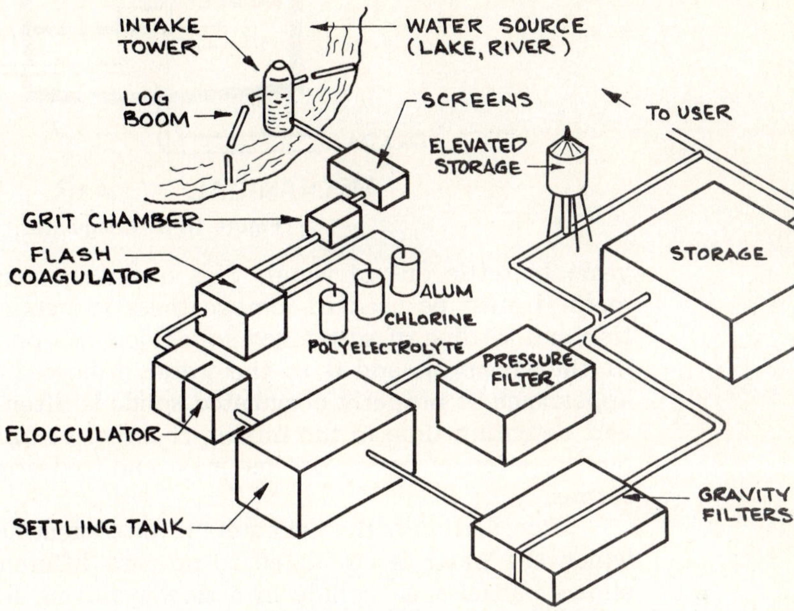


**Figure 16-1** Typical water plant.

The intake for raw water is usually housed in a fenced area or protected with floating booms to exclude large objects such as logs or floating trees, wooden crates, or other objects larger than about 3 to 4 inches. Further screening may then be necessary to remove twigs, leaves, paper, and other trash that may pass through the larger mesh. A common type of secondary screen is a mesh drum of fine weave that is rotated in the raw water supply and which is continuously cleared of trash by scrapers or brushes. Such screening will remove algae and larger microorganisms. Algae in ponds and reservoirs are a major problem, as these plants tend to grow rapidly and absorb oxygen from the water and produce undesirable odor and taste as plants die.

*Coagulation* is a process for combining small particles carried in raw water into larger particles for easier removal (see Fig. 16-2). Water is treated chemically to make suspended particles "sticky" or unstable so that when the water is stirred rapidly, small particles will adhere to others and form a larger unit which may be removed by filtration. Alum is a common additive used in coagulation and acts to form a gelatinous coating on particles as well as absorbing positively charged polymers. Alum is used as crystals or liquid. Large or relatively heavy particles of clay, rock dust, or similar material will precipitate out of water rather easily. However, if a rock of approximately 0.5 square inches of surface area is pulverized to colloidal size, the total surface area of particles is about 7 acres and would require about 65

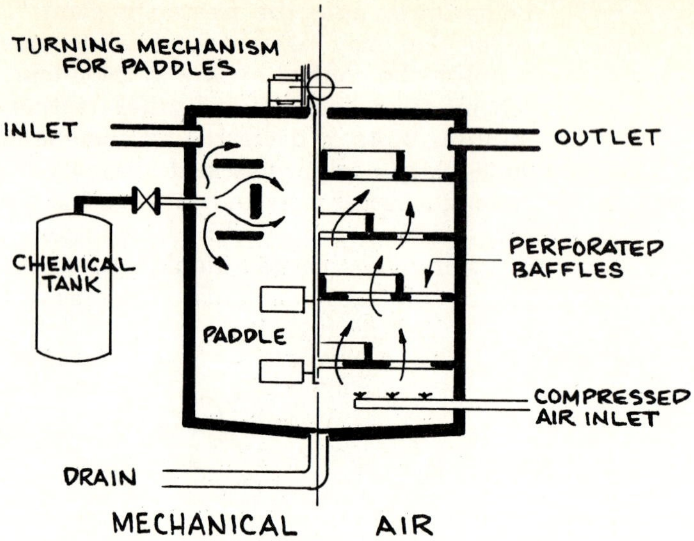


**Figure 16-2** Coagulator.

years to settle out of water. This obviously means that some more rapid method must be used to remove these suspended particles. Good coagulation and settling of suspended solids depends on rapid and thorough mixing of the chemicals added to the water, followed by slow flocculation. The appearance of properly coagulated solids is often described as "spongelike," and detention time in the mixing chamber of only a few minutes is important, as a longer period of time may tend to destroy the flocs that have been formed.

*Flocculation* is the next step in water treatment and consists of tanks in which the water is circulated to provide different velocities, which in turn allows particles to collide in a slower-moving liquid (see Fig. 16-3). Two major types of flocculation tanks are in common use. The nonmechanical type is generally a concrete or metal tunnel in which horizontal or vertical baffles alternate from top and bottom or side to side. As the water passes these baffles, a different velocity of passage occurs in the center of the stream, along the edges, or as eddies at bottoms of baffles. This change in velocity causes particles (floc) of different gravity to group together. A mechanical flocculator is a tank with a variety of different paddles or agitators which create different velocities as the water passes through the tank.

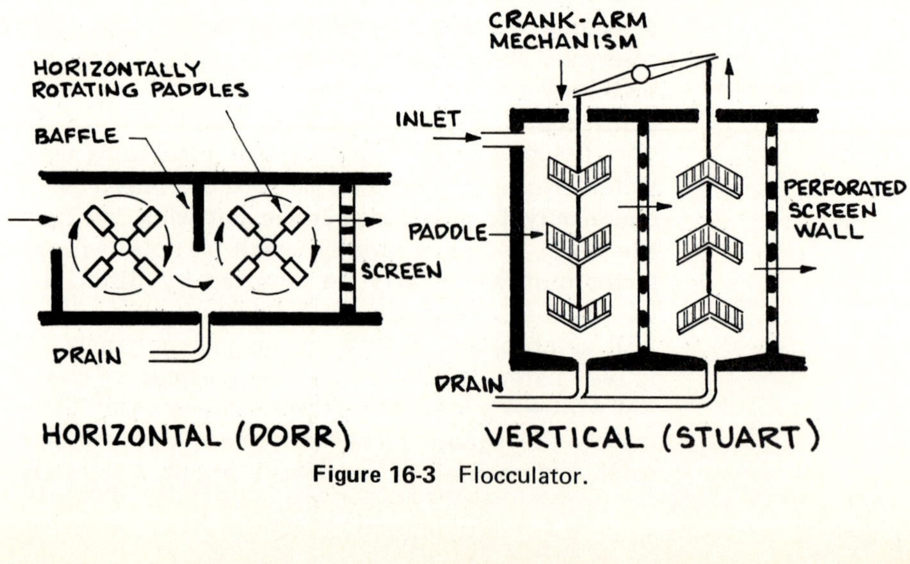


**Figure 16-3** Flocculator.

Flocculation time is generally recommended to be from about 20 minutes to not more than 60 minutes.

From the flocculation chamber, the water passes into a settling chamber or basin. Settling chambers may be long basins equipped with sludge scrapers, small square or round tanks equipped with hopper bottoms, or tube groups. In addition, sedimentation tanks fall into two groups related to time: conventional tanks, which provide about 2 to 8 hours' detention, and rapid-flow tanks using upward-flow principles which provide 1 to 1½ hours' detention time. Most basins are designed for about 4 hours' detention time. Rate of flow should be about 1 to 2½ ft/min. To cope with variable water-use requirements, settling basins may be designed to operate in parallel, which also allows downtime for maintenance. Mechanical sludge removal is almost a necessity, as manual removal is a thankless time-consuming task and sludge may lead to deterioration of water quality. The inlet end of a settling tank must provide some form of baffle to slow and evenly distribute water flow so that a greater quantity of sludge will accumulate at the inlet weir end of the tank than farther along. Rectangular basins (see Fig. 16-4) are usually designed with side-water depth of 15 to 18 ft for 100-ft long tanks and approximately 30 ft wide. Circular tanks are constructed in a manner similar to those used in sewage disposal systems with a diameter of 125 to 150 ft and a water depth of at least 10 ft. Influent water is fed into a center vertical diffuser, which directs the water in a tangential current. Sludge precipitates to the bottom and is removed by rotating scrapers, while the clarified water overflows into weirs at outer edges of the tank.

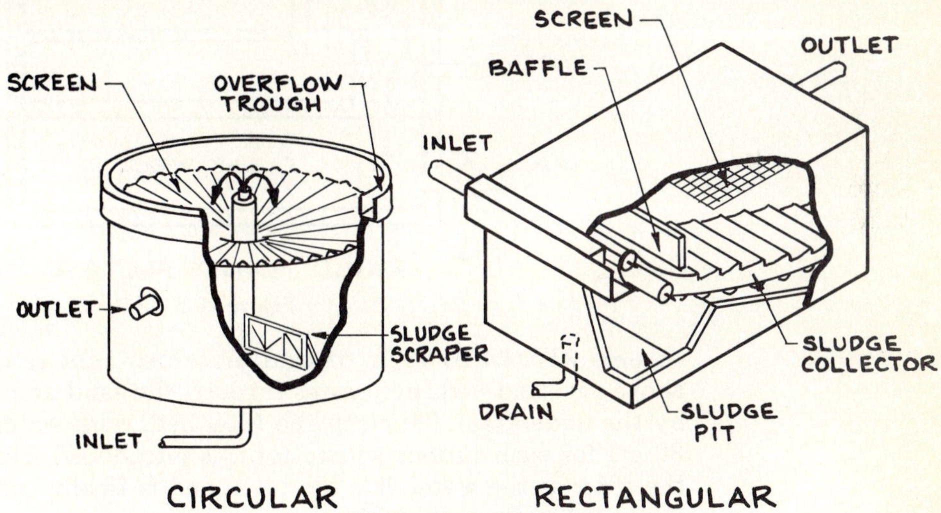


**Figure 16-4** Clarifiers.

Removing suspended solids by passing water through a porous medium is primarily a mechanical action and is the next step, called *filtration*. Filters (Fig. 16-5) may be of the gravity type, in which the water passes by gravity through a sand-gravel bed, or of the rapid-rate pressure type, where the water is forced through the filter bed under high pressure. Water that has been pretreated by coagulation and settling tanks has a large amount of the suspended solids removed. This in turn reduces the filter load, and the coagulating material coats the filter surface, which reduces the amount of finely divided particles that penetrate the filter bed. Filters are tanks made from concrete, steel, or wood, in the bottom of which is installed a collecting system for the cleaned water and a piping system for backwash water. Over this underdrain system, a layer of stone, gravel, or anthracite coal is installed and

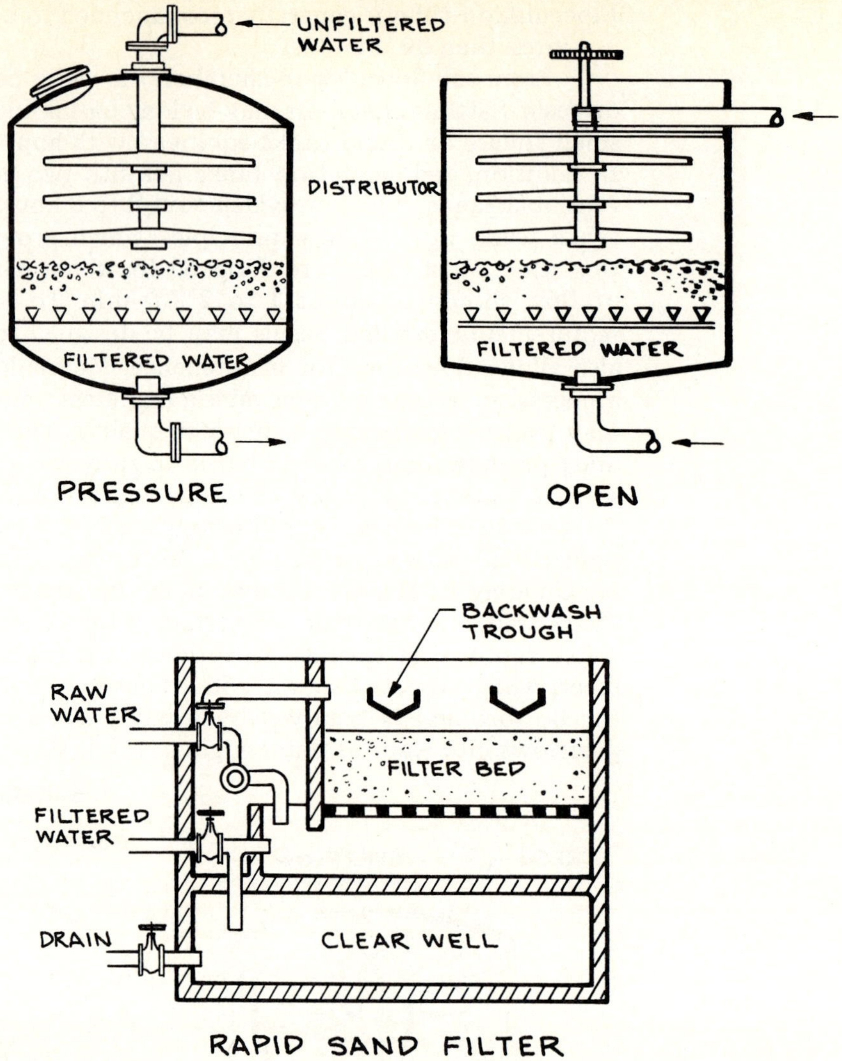


**Figure 16-5** Filters.

covered with 24 to 36 in. of sand. Water from the settling tank is introduced above the sand bed, percolates through the sand and gravel, and is collected by the underdrain. To clean the filter bed, wash water, approximately 15 to 30 gal for each surface square foot, is introduced from the bottom, washes the filter as the water fills the tank, and is finally drained off as the residue is scraped or flushed off the top.

Pressure-type filters are essentially closed tanks, either horizontal or vertical, that contain the filter material. Water forced through the filter bed at medium to high pressure is removed from beneath the filter bed. Vertical installations are generally used where small amounts of water are needed, while horizontal units normally provide greater volume. The depth of the filter bed in any type varies from 18 to 36 in. Vertical filters vary from 1 to 10 ft in diameter, while horizontal units vary from 6 to 8 ft in diameter and up to 25 ft long. Working pressure is from 65 to 100 psi. Pressure filters are used primarily for industrial water treatment because the tanks require less space than municipal-type equipment and water systems are usually under pressure in industry anyhow.

Final chemical treatment of water may or may not be needed. If the filtered water is free of color, odorless, and free from harmful bacteria, there is probably no reason to introduce chemicals into the water. Chlorine is used to kill any bacteria and reduces or eliminates color and odor. The chlorine may be introduced as crystals or more often as a gas or liquid and is distributed into the water by gravity feeders or under pressure and in accurately measured quantities. Lime or other chemicals may also be added to the final water to control corrosion or algae that may develop during water storage. Chlorine dioxide is used in place of chlorine in many foreign countries and destroys phenols, persists longer than chlorine, and does not react with ammonia nitrogen. Ozone is the most powerful oxidant but is not used in the United States. Cities obtaining water from heavily polluted rivers, however, could hardly refrain from using ozone to remove objectionable taste, odor, or color. Unfortunately, ozone is expensive and residuals last only a few minutes, so chlorine is usually required in addition to ozone. Although chemicals may be added at this point to help control hardness, retained metallic compounds, or other undesirable particles, this control is better accomplished as described in the following sections.

## 16.5 SPECIAL SYSTEMS

In earlier sections, only minimum consideration was given to the special treatment required for deferrization and/or softening. These are somewhat different from the normal water treatment and may be required together or separately, depending on the raw water source. Removal of injurious ferric compounds from groundwater wells or springs requires an aerator unit composed of a bed of purified coke and a system to allow the raw water to pick up oxygen from the air. The water is sprayed into the air, takes up oxygen from the atmosphere, and trickles through the coke bed, where carbon dioxide is removed and soluble iron is oxidized into insoluble floc. The floc and water pass into a settling basin, where the floc is removed and the water is purified further by passing through a slow sand filter.

Water softening is particularly important where water has a high calcium carbonate ($CaCO_3$) content and is used as feedwater for boilers. When used to generate steam, "hard" water precipitates a crust on the inside of boiler tubes, eventually reducing the efficiency of the boiler. A similar condition occurs in domestic water systems when hard water is distributed through galvanized-steel pipes. In this condition the action is slower, but the end result is the same. Hot water will precipitate scale more rapidly than cold water and water heaters are especially vulnerable. When water is softened by a treatment plant operation, lime and soda ash (sodium carbonate, $Na_2CO_3$) are added to the raw water. This solution is thoroughly mixed and the water passes through a reaction basin similar to a flocculator and then into sludge-settling basins, where the water remains from 2 to 4 hours. Quantities of sludge are formed and cleaning should be done from a minimum of 15 minutes to a maximum of 4 hours, using scrapers and backflushing. The filtered water is then stabilized with carbon dioxide before being stored.

The softening process indicated in the preceding paragraph is too cumbersome for individual industrial plant use or for residential use, so an *ion-exchange* process is used (see Fig. 16-6). Ion-exchange processes include sodium-ion exchange (zeolite), hydrogen-ion exchange, or anion exchange. Sodium zeolite is the most used method and consists of passing water through a bed of material capable of removing calcium and magnesium from the water and replacing them with sodium or potassium. The exchange is very rapid, so water passing through the bed is almost completely softened. As

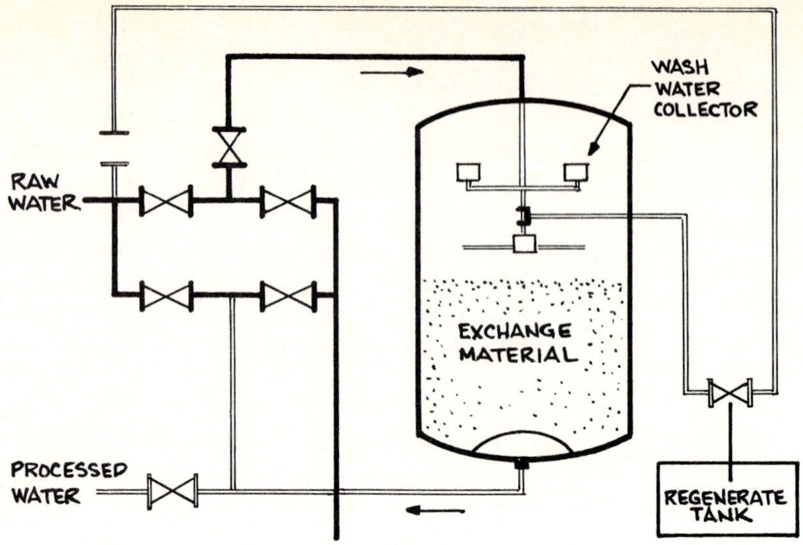


**Figure 16-6** Ion softwater exchanger.

water passes through the zeolite bed, calcium and magnesium chloride are converted to sodium chloride, changing the surface bed to calcium and magnesium zeolite. This process continues until the bed must be replaced or recharged. This is accomplished by treating the bed with a concentrated solution of sodium chloride (table salt), which reverses the chemical action, and calcium and magnesium ions pass into the brine solution and are washed out as waste. In use, the purified water may carry some of the salts to about 4 ppm before regeneration is required.

## 16.6 WATER DISTRIBUTION

Water may be distributed in open channels or in closed piping either by gravity or by pumping. In actual practice, treated water is normally distributed in closed piping and by a pressure system. Raw water, irrigation water, or wastewater may be distributed in canals by gravity. Canals are cut into the earth or blasted through rock, carried over depressions in flumes, and through hills or mountains in tunnels. In cases where flumes or piping will interfere with surface traffic or other action, closed pipe siphons or pumped sections may be used to conduct water underground beneath a canyon or other surface obstruction. Tunnels through rock are usually concrete-lined to reduce the friction of water passing through the tunnel. *Grade tunnels* are simply tunnels that continue the gravity grade through obstructions, while *pressure tunnels* are built to carry water beneath rivers or similar surface conditions or as entrance piping from collecting intakes, and require pumping or pressure to continue the water flow.

Water distributed from a treatment plant must of necessity be made in a pressurized piping system, or there would probably be no reason to clean the water. This pipe is termed a *trunk main* and connects the plant with a storage reservoir or tower. Water reservoirs are concrete basins with a wood or metal cover, not under pressure, and often at grade or below grade. Since these facilities are not under pressure, water distributed from such reservoirs must be pumped into further piping to provide operating pressure to plumbing fixtures. Water storage may also be made in steel tanks which may resemble "oil tanks" that are not elevated above ground level. These tanks may be 30 to 50 ft in diameter and several stories high. Still another type of reservoir is the elevated tank in its many shapes and heights. Elevation of water reservoirs is very important because water will provide an automatic static pressure at the rate of 0.433 psi. This means that since water weighs 62.4 $lb/ft^3$, the static pressure beneath a 12-in. cube of water is 62.4/144 = 0.433 psi, or inversely 1 psi of pressure will sustain a static (stationary) column of water 1/0.433 = 2.3 ft in height. Pressure of waters in service mains is usually maintained at 50 to 70 psi, so pumping is required from nonelevated tanks, or all or a portion of required pressure may be obtained by height of tank. As an example, if a pressure of 60 psi is desired, it will be necessary to provide elevation of 138.5 ft. This may be possible with some elevated tower tanks, but pumping will be required to augment the static pressure.

Secondary mains distribute water from the service reservoir to the street service mains under pressure and in various patterns (see Fig. 16-7). The loss of pressure (head $H_1$) is dependent on several factors, such as pipe friction ($f$), diameter of pipe ($d$), velocity of flow ($v$), length of pipe ($l$), and the acceleration of gravity ($g$), and may be calculated by the Darey-Weisbach formula:

$$H_1 = \frac{flv^2}{2gd}$$

Secondary mains do not usually have laterals from consumers tapped into them but form the major system for supply to street mains, which in turn serve ultimate consumers.

Secondary mains usually follow the topography of the area. That is, they are uniformly buried below the surface so as to go up when the land rises and go down when the land declines. At high points, air collects in the line and an air lock may result unless an automatic air valve is installed to allow entrapped air to escape. Similarly, the low spots in the waterline may require a valve or cleanout for the removal of scale and other waterborne debris. Another major consideration for waterlines is the elimination of

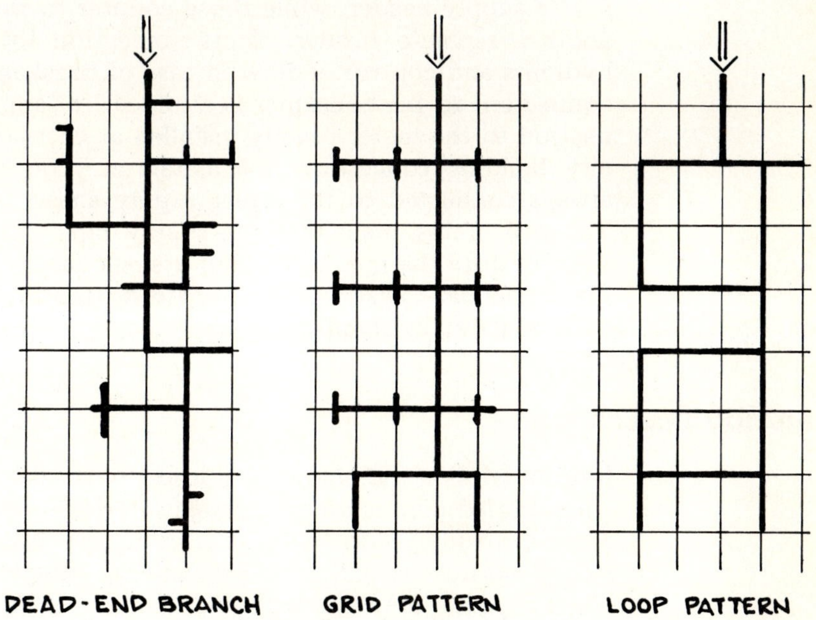


**Figure 16-7** Water distribution.

possible *water hammer* or pipe collapse due to vacuum created by loss of pressure. Water hammer is noise and a resultant excessive jolt of water velocity caused by the rapid opening or closing of valves or when pumps are started or stopped. This phenomenon is controlled by the use of surge tanks or vertical standpipes at the ends of lines. In this method, a limited quantity of water is topped by an enclosed air chamber. As the surge in the pipeline admits more water, the entrapped air is compressed, and as the surge passes, the increased air pressure pushes the excess water back into the line until normal operation is again reached. Vacuum conditions are also possible from the same causes. If a waterline breaks or a large amount of water is drawn off rapidly, the pipe area formerly occupied by the water may become a vacuum and cause the pipe to collapse. Automatic air valves or surge tanks may be used to fill the empty spaces quickly and prevent damage.

In calculating the size of pipe and hydraulic capacity of a water system, it would seem that domestic demand, industrial demand, and fire-hazard demand would have to be combined. This is seldom necessary, however, as peak demand of all three rarely occurs at the same time, and new pipeline construction would normally consider possible increases in demand by a growing community. Fire-hazard demand is commonly calculated by the *standard fire stream* of an engine's "red line" 1⅛-in. hose at a pressure of 45 psig. When this amount is translated into the 2½-in. standard hose, the friction resistance is about 15 psi per 100 ft of hose and makes the standard fire stream effective to about 70 ft in height and horizontal carry. Increases in pressure and capacity in gpm are affected by the fire-engine pumping unit and may reach 1000 gpm. Extra care must be taken, therefore, to see that cross-connection or loop systems of piping will allow this increased demand and not reduce the pressure in the mains below 20 psig, or vacuum failure may result.

*Street mains* are usually laid out in patterns that follow the street patterns. In normal grids, the pipe systems installed in all the streets are connected at all corners, and proper valves enable the supply of water to be reduced or eliminated as occasion demands. In a grid system of dual mains, the lines are not connected, but those parallel in one direction are connected into a supply header, while those counter to these lines are connected into another separate header. Cross-connection by valves is supplied for fire hydrants and control of flow in case of breakage or for maintenance. Final connection to the consumer is attained by tapping the street main and connection to the meter usually installed at curbside. In some cases of central-city building congestion, the meters may be installed in the building but always connected to the street supply ahead of any building distribution. The size of the main building supply pipe varies with the type of installation, as does the meter size, but a single-family supply pipe of 1 in. is normally adequate. Distribution of water within the building structure is covered in more detail in Chapter 18.

## 16.7 FRICTION AND CORROSION

Interior friction and pipe corrosion are two of the major problems encountered when buried pipe is used. A rough interior pipe surface or poor joints with edges protruding into the central cavity are two of the principal reasons for loss of pressure (head) in water handling. Most commonly used materials for water pipe are (1) cast iron with flange ends, (2) welded steel, (3) spun concrete, (4) asbestos cement, and (5) plastics. Cast iron is brittle, heavy, and available in fairly short lengths. Welded steel is relatively light and flexible, has a smooth interior surface, but is subject to corrosion. Spun concrete is usually reinforced with mesh or rods, so cutting is difficult and the sections are heavy. Asbestos-cement pipe is dense, smooth, and has high corrosion resistance, but is a bit brittle. Plastic pipe has most of the desired characteristics. It is light, smooth, available in many sizes and lengths, easily joined, and practically corrosion-proof. Plastic pipe, however, has not been entirely satisfactory for installation outside buildings because of its lack of impact resistance and defense against accidental puncture by repair equipment or gnawing rodents.

Corrosion is caused by minerals in the earth, which may set up electrolytic action with metal pipe or metal joint fittings, chemicals carried in the water which attack the pipe from inside, or the scouring action of the water flow in rough-surfaced pipe. Electrolytic action may be reduced or suppressed by using a cathodic protector that uses galvanic or directly impressed electric current to arrest the corrosion. Chemical corrosion is best controlled by using a film or coating of cement or coal tar. Calcium carbonate ($CaCO_3$), which may be deposited as a coating from the treated water, may also be used but may eventually build up enough to restrict water flow. Interior rough surfaces may be reduced by mechanical grinding or application of a cement liner. The subjects of pipe friction and corrosion are of great importance in the study of liquid transportation but are too extensive to be considered further here.

## 16.8 GROUNDWATER

Sources for water are generally twofold: *surface water*, obtained from rivers, lakes, and similar catchments; and *groundwater*, obtained from wells, springs, or other underground galleries. Underground water may not be as free of pollutants as often thought. Water that courses through limestone may pick up calcium compounds. Wells near the sea may allow salt water to intrude into the fresh water. Sewage may percolate into the underground water, the water table may not remain at its customary level or quantity, or ordinances may restrict the amount of water that may be pumped. The volume of flow is dependent on the amount of rainfall or percolation flow from surface water, but no underground water source is inexhaustable, and most underground water is contaminated with some form of undesirable pollutant.

Practically all underground water must be pumped. Water may flow by gravity from infiltration galleries into pump wells, and springs and artesian wells may provide gravity flow if they lie high enough above the ultimate user. When the water level is less than 25 ft below the surface, standard suction pumps may be used, but wells deeper than about 25 ft require deep-well submerged pumps. The size, number, and arrangement of wells are determined by the quantity of water required, amount and depth of water available to be pumped, method of pumping, and the hydraulics involved. The cost of well construction is also a major factor, as wells in sand require extensive casings, wells in rock require slow drilling or blasting, and wells in compacted gravel or soil may be subject to cave-ins before casings are installed.

Wells are generally constructed in one of four accepted ways. *Dug wells* (see Fig. 16-8) are excavated by hand with a diameter of up to about 6 ft to allow workers to work as the well deepens. In most cases, dug wells are lined with masonry, timber, or concrete and should be dug deep enough to supply water faster than the water can be removed. *Driven wells* (Fig. 16-8) are used

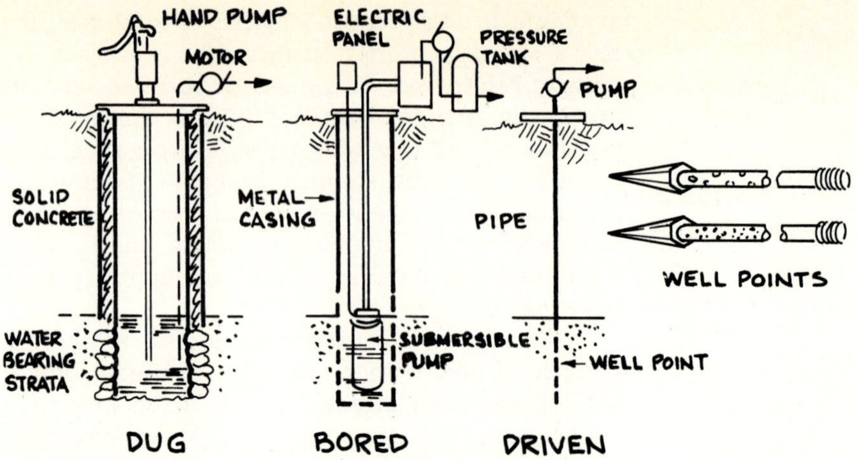


**Figure 16-8** Wells.

in shallow sandy or gravel formations and consist of a perforated wellpoint attached to the suction pipe and may be limited in depth by the strength of the pipe or presence of adequate water flow. *Bored wells* (Fig. 16-8) are excavated with auger-type rotating equipment, and the resulting loose material is removed periodically as the auger is pulled up or may be flushed out with water jets. Holes to about 36 in. in diameter are possible and linings of steel, tile, or concrete are fitted in place to prevent cave-ins. *Drilled wells* are either of the rotary drill type or made by percussion drilling. Rotary drilling is done with a multitooth bit attached to a hollow drill rod. As the drill is rotated, a slurry is pumped down the hollow drill rod, lubricates the bit, and returns to the surface with the loosened material around the side of the rod. In percussion drilling, a drill bit with a chisel edge is dropped by some reciprocating action again and again, breaking the rock into small bits, which may then be removed by bailing. Casings are usually required for either type of drilled wells.

Pumping equipment for wells of less than 25 ft in depth is generally mounted at ground level and operates by suction. For wells to about 100 to 125 ft deep, the water is lifted by centrifugal pumps assisted by recirculating a part of the water by jets or ejectors. For deep wells, the pump head is submerged in the well and the water is forced to the surface rather than lifted. In wells of depth greater than about 150 ft, submerged multistage pumps may be installed or a combination of deep-well pumps and surface suction pumps may be used. Deep wells provide a built-in storage capacity to partially overcome unexpected drawdown or diminishing of underground supply. Most well water requires some filtering through a sand filter and may require chemical dosing to eliminate remaining bacteria, color, or harmful mineral compounds.

## 16.9 EVAPORATION

Distillation of water to remove impurities is not a new process, and the ability to obtain usable water from seawater by evaporation and subsequent condensation has been a standard practice as long as steam engines have been used in ships. Evaporators are also used in connection with shore stations using water for steam boilers and to a limited extent to extract potable water from saline water. Several factors must be considered that govern evaporation: (1) space and heat requirements and availability, (2) ease of operation, (3) control of residual scale, (4) accessibility for cleaning, (5) purity of condensate, and (6) quantity of raw water available. Raw water may be pretreated or may be obtained directly from the source, and the amount and disposition of scale produced by the evaporation process is of vital concern.

The principles of the evaporative process are relatively simple: a distilling chamber, in which the water is heated to form a vapor; and a second chamber, in which the vapor is condensed back to a liquid (see Fig. 16-9). In this system most of the suspended solids in the raw water are deposited as the vapor is formed, and condensate with a solids count of as low as 2 ppm is possible. There are several types of evaporators (submerged-tube, film, flash, and multiple-effect), and most use steam as the heat source for evaporation. Single-effect evaporators will deliver about 0.9 lb of distillate for each pound of steam used, and when units are connected in multiple, they will produce in direct relation to the number of sections involved. In multiple-section use, the steam in the first unit is returned to the boiler for reuse, and the vapor at high temperature from the first unit is used as a heat source for the second unit and is drawn off as water. This may be continued until the heat of the vapor no longer reduces the water to vapor.

The removal of scale after evaporation is accomplished by several methods. When the scale is exposed to heat, it tends to be more hard and brittle than when submerged. Brittle scale may be removed from tube surfaces by alternating heat with cold in rapid sequence, which cracks off the scale, which then falls to the bottom of the tank and is easily removed. Soft scale requires mechanical or chemical methods for removal. A partial vacuum may also be used to cause moisture beneath the scale to flash into steam and blow off the scale. Corrosion of evaporator coils is also a major problem. Raw water may contain oxygen, carbonates, carbon dioxide, or sewage waste with mineral or organic acids or magnesium chlorides. Pretreatment with sodium zeolite will eliminate most of these scale-forming pollutants. Completely distilled water is not very palatable, so when distilled water is to be used for drinking, it is either mixed with other potable water or small quantities of minerals and chemicals are added.

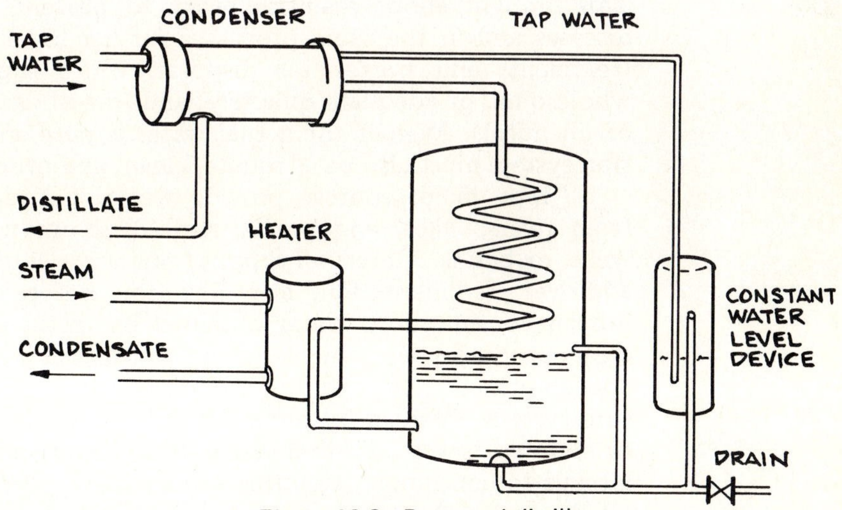


**Figure 16-9** Barnstead distiller.

## 16.10 RECLAIMED WATER

With the increasing demand for water and the decreasing amount of usable water available, another source is being considered in greater favor: water recycled from sewage. In earlier days the raw sewage was simply dumped into the nearest lake or river, thus fouling that source for potable water. Later the sewage was treated and dumped, but the water content was still wasted. Thoughts of drinking sewage water are repugnant to most people, but quite a few cities have been mixing reclaimed water with other potable water for a number of years. Only about 1% of sewage is solid materials, so sewage treatment engineers have classified water as *storm water*, which usually contains a minimum of solids and no human waste; *gray water*, which is waste from sinks, showers, lavatories, and similar fixtures without solid waste; and *black water* from water closets, urinals, and other fixtures that include solids. Storm water is best handled in separate systems and used to recharge underground water. Optimistically, gray water would require only minimum treatment if it was collected separately, but gray water and black water are normally collected together.

Water reclamation from sewage is only an extension of the sewage-treatment system reviewed in Chapter 17. After the sewage has passed through the full system and would normally be dumped into a river or stream, additional settling and treatment are applied and the water is eventually added to the potable-water supply. Treated water is oxidized in open ponds, usually through several stages, then is percolated through filter beds, treated with chemicals (usually chlorine), and is ready for use. A number of such reclamation systems are in daily use. The New York Institute of Technology, Old Westbury, New York, the municipal plant at Baileyville, Maine, and the extensive plant at Santee, California, are some of the most successful. The Santee (Padre Dam Municipal Water District) plant not only provides potable water but also includes seven lakes for public recreation.

## 16.11 SUMMARY

The fact that a considerable number of diseases are transmitted through water and sewage was not discovered until the end of the nineteenth century. This brought about scientific ways to prevent the transmission of these diseases and in the same process a further investigation into water supply, treatment, and wastewater disposal. Water supplies must be attractive, wholesome, of adequate quantity, and free of excessive amounts of minerals or chemicals. Even if the actual water is pure and acceptable, the distribution system must also be adequate, clean, and properly designed.

Three major sources provide water: (1) rainwater; (2) surface water from rivers, lakes, and ponds; and (3) groundwater from natural springs, wells, or basins. Almost all types of raw water must be treated with chemical additives, sedimentation, and filtration before it is potable, and fit for human consumption. Water obtained by distillation is usually required for steam-producing equipment but may also produce potable water for human consumption. Transportation and distribution of treated water is by a pressure system obtained by pumping and reservoir storage in ground-level tanks or elevated towers. Further distribution by street mains allows for individual laterals to consumers' premises and for fire-hydrant installations. In spite of all the water that is apparently available today, the increasing population explosion and resulting community expansion, together with higher daily per person usage and industrial requirements, have made collection, treatment, and distribution of water a matter of primary concern for current and future existence.

# 17

# Sewage Disposal

## 17.1 INTRODUCTION

The disposal of sewage is becoming more and more of a problem as a result of three major factors: the migration of rural population to the cities, expansion of city boundaries into rural areas to accommodate the influx, and the use of sewer systems to handle increasing volumes of kitchen and laundry waste and comparable commercial/industrial waste. The vast majority of this human waste is handled inadequately by municipal sewage treatment plants. The existing treatment plants are undersized for the load imposed, money and time do not keep pace with increased load, and partially treated residue is dumped into oceans, lakes, or rivers, thereby causing further environmental problems. In addition to municipal sewage treatment, it is estimated that at least 15 to 20 million individual sewage-disposal systems in the form of primitive privies or septic-tank installations serve more than 60 million people in the United States.

Sewage treatment in almost any form is dependent on the principle of decay and oxidation. Under normal conditions, limited amounts of nitrates and phosphates are produced which become nutrients for growing plants, which then recombine the nitrates and phosphates with carbon dioxide from the air to produce new living tissue. Municipally treated sewage is provided with accelerated conditions for the production of nitrates and phosphates, which in turn are washed out with the treated effluent. Limited or regulated amounts of these materials released into soil or water are relatively harmless, but when present in concentrated amounts in lakes or rivers, they tend to increase dramatically the growth of aquatic plants. These plants then use up the oxygen in the water, which eventually destroys fish and other water life. Individual sewage systems, such as cesspools and septic tanks, generate the same chemicals but in such small quantities that they are actually beneficial as fertilizers when introduced into the soil by absorption.

Proper sewage disposal is necessary for the protection of public health, regardless of the process used, and improper or incomplete treatment may result in such diseases as hepatitis, dysentery, diarrhea, or typhoid. The improper disposal of waste may contaminate rivers or lakes that also provide potable water for a community, or may pollute ocean beaches, killing fish, crustaceans, and reducing recreational possibilities. It is possible to treat sewage waste completely enough to eliminate all objections, even to the point that the final liquid may again be used for drinking purposes without ill effects. Initial treatment (*primary*) takes place by partial biological action in septic tanks and to some extent in privies and cesspools. More advanced treatment (*secondary*) occurs in municipal disposal plants, which encourage bacterial action further by oxidation and separation. Final treatment (*tertiary*) is accomplished by treating the resulting liquid with chlorine or similar chemicals and drying and pulverizing the sludge as a commerical fertilizer.

The disposal of human waste is generally accomplished by means of one of three principal methods. The most primitive of these is a simple hole in the ground, often provided with some sort of shelter, in which the excrement ultimately decays and the liquids percolate into the surrounding earth—in short, a privy. The second is by use of a septic tank and absorption field or cesspool, where the waste is decomposed by bacteria and the liquid absorbed in the earth. The third, and most universally used, is the municipal underground sewer system and treatment plant.

## 17.2 THE PRIVY

There still exist today thousands of individual privies, termed "outhouses," "two-holers," "latrines," as well as a number of colloquial names that are even more expressive. Circumstance as well as property location determine the existence of these units. Before cities had extensive municipal sewer systems, each household, limited by building lot size, had its own privy. Farm households, despite acreage available for possible disposal fields, also have used the privy extensively, because it was simple to build and maintain.

In basic form, the privy is a hole in the earth of convenient size and depth covered by a small building, which provides shelter and privacy as well as protection against accidental missteps or filling by rain or snow. Excrement is deposited in the hole through a rustic seating arrangement and is often aided in decay by the introduction of lime. Any liquid percolates into the surrounding soil and the remaining sludge is ultimately removed, or the hole filled and a new unit provided elsewhere. Comfort is obviously at a minimum since heat, lighting, and other personal values are usually not present, and offensive odors are generated by the decay process. Most cities now ban construction or use of a privy, but their presence is often indicated by stars or cresents cut in the doors of these unique structures in villages and farm lots.

## 17.3 SEPTIC-TANK SYSTEMS

The normal private sewage-disposal system, where it is allowed by a municipality, consists of a septic tank where the sewage is retained long enough to separate and decompose the solids, and an absorption field to allow the liquids to percolate into the surrounding earth (see Fig. 17-1). In this process the sewage is treated by *anaerobic bacteria*, which decompose the organic

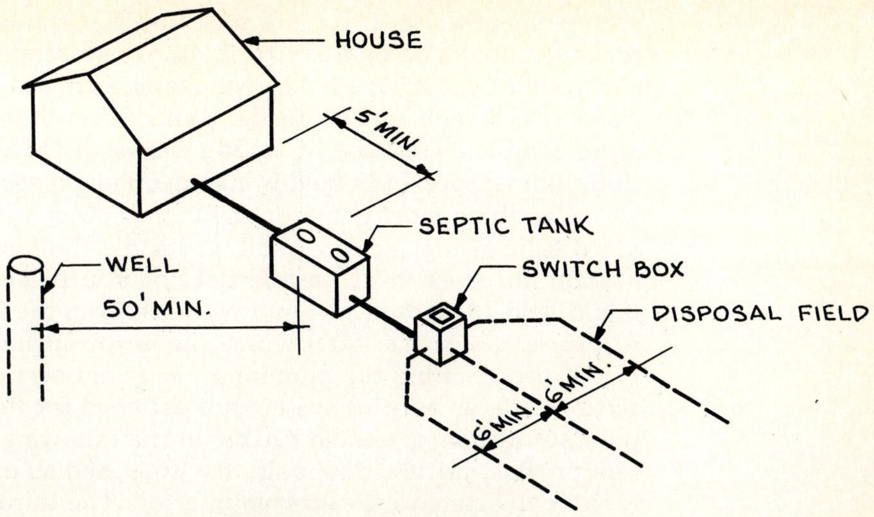


**Figure 17-1** Typical septic tank system.

solids into harmless compounds. Unfortunately, this decomposition is accompanied by the emission of considerable odor and the presence of disease-producing (*pathogenic*) organisms. In addition to the fairly large area required for the absorption field, a septic-tank system depends almost entirely on proper percolation of resulting effluent into the soil and the periodic removal of sludge from the tank.

Septic tanks may be constructed of cast-in-place reinforced concrete, masonry sealed with cement plaster, or they may be prefabricated. Prefabricated units are generally made of steel, coated to prevent rust and corrosive action, or of preformed plastic, which may be coated or not to add protection. Influent and effluent openings should be arranged so that the actual opening is above the bottom sludge level and below the top mat level. The invert elevation of the effluent opening through the tank wall must, of course, be lower than the influent or the liquid in the tank will not "flow." Baffles are provided to allow the tank contents to divide and settle and the liquid to continue through the tank. Some provision for periodic inspection and sludge removal must be made in the top of the tank. If properly sized and constructed for assumed use, the sludge will not normally require removal more often than every three to four years. Manholes, removable slab tops, or other methods should be provided to facilitate sludge removal.

When the sewage in the septic tank has separated the solids from the water (water is 70 to 85% of total sewage) and the water has been partially treated by the bacteria in the tank, the water flows out of the tank at the effluent end. Some municipalities allow a vertical cesspool to act as an absorption field, while others require an open tile field of some length. A *cesspool* is very similar to a water-supply cistern, except that septic-tank effluent is dumped into the cesspool to be absorbed by the surrounding earth, while a cistern is intended to collect or to hold potable water. Stone, open-joint masonry, or even wood may be used to line the walls of a cesspool to retain the earth in position and allow percolation of liquid into the earth. Cesspools are now prohibited in most parts of the United States.

Most septic-tank systems rely on an absorption field of clay tile or perforated plastic pipe installed within a covering of crushed rock or gravel, all buried in the earth. The length of the tile field as well as its arrangement are determined by the absorption of the earth. The disposal field must be kept at least 50 ft from the residence and 100 ft or more from a water supply, preferably downhill from any water wells. Clearance from waterlines under pressure should be at least 10 ft from the septic tank and 25 ft from the absorption field. If the field is on sloping ground, or if the field is to be divided into several smaller fields, control or switch boxes may be necessary. More complete coverage of testing the earth for absorption and the installation of the disposal tile field is included in later sections.

The three most common faults that occur with septic-tank systems are: (1) design of a tank too small to handle the load imposed on it, (2) poor absorption of earth at the tile field or insufficient length to dispose of the liquid, and (3) failure to remove sludge from the tank periodically. The lack of proper design size often comes about through miscalculation of the number of people using the plumbing, or by connection of additional plumbing fixtures into an existing system not designed for this load. The second factor, poor absorption, is treated further in the following section and holds true for septic fields, privies, cesspools, dry wells, and all other methods where liquid is to be dissipated into surrounding soil. The third factor, proper removal of septic-tank sludge, is only a matter of proper maintenance. Over a period of time the sludge in the bottom of the tank builds up, reduces the amount of raw sewage and liquid space, plugs the inlet or outlet, and the tank ceases to function. At that point plumbing pipes fill, fixtures fail to operate, and a real problem develops. Since septic tanks are normally buried under several inches to several feet of earth, lawn, or flowers, it is very easy to overlook cleaning them every few years. Stoppage may also be caused by planting trees or other plants with extensive root systems which grow toward the readily available liquid in the tile fields.

## 17.4 ABSORPTION TESTING

Privies, cesspools, and septic-tank fields all rely on the ability of the surrounding earth to absorb the liquid contained in the waste. This is really quite a requirement, given that most sewage is primarily liquid. Therefore, the first step in the design of a septic sewage-disposal system is to determine if the soil is suitable as an absorption field and the approximate capacity for such absorption. A number of basic requirements must be met: (1) the proposed trench should be at least 4 ft above any rock, clay strata, or groundwater level; (2) the soil should be tested for adequate percolation time (Table 17-1); (3) there should be sufficient area for the maximum tile field; and (4) the disposal area should be at a safe distance from any water source. Local health authorities, agricultural soils maps, experience from well drilling and other subsurface work, as well as on-site testing and surveys, will answer these requirements.

Visual inspection of the soil will often give reasonably reliable clues to the ability of a soil to absorb liquids. The size of soil particles also governs spaces between particles, which in turn influences the rate of absorption or water flow. Gravel soil with many spaces between particles is the best, with a sand–gravel mixture a close second. Clay and silt in solid beds will not absorb readily, but some clay strata are mixed with gravel to divide the bed into layers, which may allow a liquid flow along a plane. Soil that is reddish-brown to yellow or soil that has considerable amounts of vegetable humus indicates good oxidation and probable good absorption possibilities. Grading cuts, foundation excavations, and similar disturbances of the earth may reveal subsurface conditions, or test holes may be made. Test boring logs of foundation information are often valuable sources for this use.

**TABLE 17-1**
Absorption Trench Required per Bedroom (Lineal Feet)

| *Time required[a] for water to fall 1 inch (min)* | *Neither garbage disposal or dishwasher* | *Both garbage disposal unit and dishwasher* |
|---|---|---|
| 2 | 50 | 85 |
| 3 | 60 | 100 |
| 4 | 70 | 115 |
| 5 | 75 | 125 |
| 10 | 100 | 165 |
| 15 | 115 | 190 |
| 30 | 150 | 250 |
| 60 | 200 | 330 |

[a]Percolation time longer than 60 minutes indicates that the soil is not suitable for this type of disposal system.

The ability of the soil to absorb liquid is determined by a simple percolation test, which is usually required as a prerequisite by health authorities before a permit for a private sewage-disposal system will be issued. Test holes may be 4 to 12 in. in diameter or 12 in. square and should penetrate at least 12 in. deeper than the proposed tile-field bottom elevation. Several test holes should be dug to ensure that a reasonable analysis of the entire area has been made. The bottom of the hole should be covered with 2 to 3 in. of gravel to prevent scouring by the water, and sides should be free of tool marks or smeared soil. The prepared hole should then be filled with water and maintained full for at least 24 hours in order to simulate a soaked-soil condition when in actual use. At the time of conducting the test, a measuring stick is installed in the hole, as shown in Fig. 17-2, and the hole is filled with 12 in. of water and kept at that point for 6 to 24 hours. If at the end of that time water still remains, it should be adjusted to 6 in. above the top of gravel and

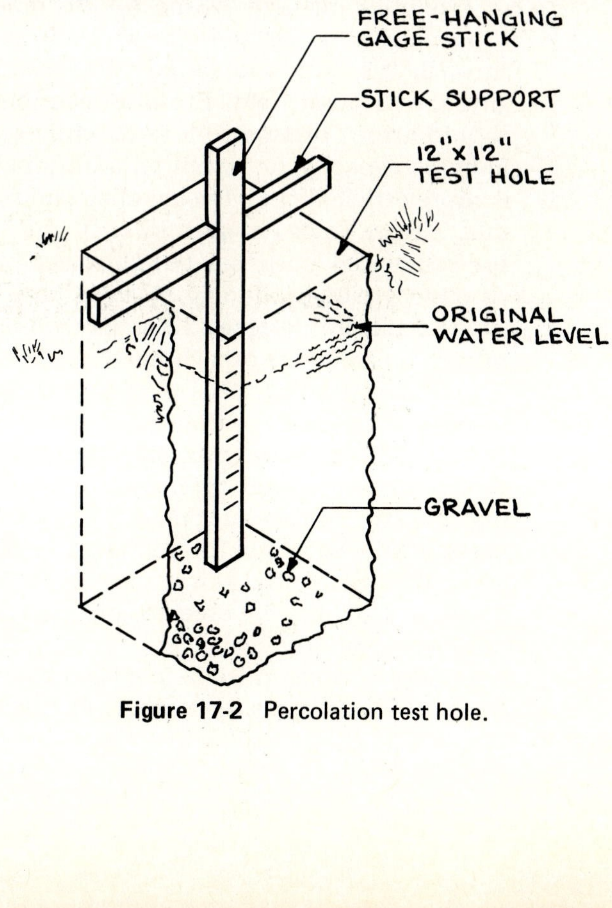


**Figure 17-2** Percolation test hole.

readings taken at 30-minute intervals for the next 4 hours. The drop in the final 30-minute period should be used to calculate the percolation rate as follows: a drop of 3 in. in 30 minutes equals 30/3 or 10 minutes per inch. If the water is absorbed in less than 30 minutes in porous soils, the same procedure should be used except that readings should be made at 10-minute intervals over a period of 1 hour, using the final 10-minute drop for calculations.

## 17.5 SEPTIC-TANK DESIGN

Once the capability of the earth to act as an adequate absorption field is determined, the next step is to design the various parts of the system. The septic tank is the first unit to receive the raw sewage from the residence and actually is not difficult to design properly as soon as the required capacity is determined. For the most part, the shape, size, and interior baffles are fairly uniform and determined by local practice, building codes, or health department requirements. The tank has three major functions to perform: (1) separation of solids, (2) retention of sludge and scum, and (3) adequate volume to allow bacterial action to take place.

As the raw sewage enters the septic tank, its rate of flow is decreased, thus allowing the heavier solids to sink to the bottom, where bacteria may reduce them to water-soluble compounds. While the heavier solids sink to the bottom, the lighter solids, grease, and fat rise to the top of the water level in a scum or mat, where they are decomposed by bacteria together with gas generated in the sludge. As the solids settle as sludge or float to the top as scum, the water is allowed to pass through the baffles and is eventually discharged from the tank. Although the anaerobic (oxygen-free) bacteria attack the contents of the tank, the process is not absolute and some pathogenic (disease-producing) organisms remain in the effluent water to be distributed into the soil by way of the absorption field. Septic-tank effluent should not be dumped into a pond, lake, or river.

Septic-tank size is one of the most important considerations. A tank too small to handle the load will allow inadequately treated sewage to pass through the tank too rapidly to be sufficiently decomposed, thereby clogging the disposal field. From an economic standpoint, it is also much easier to add a foot or two to a tank while under construction than to try later to increase capacity or install an additional tank. Furthermore, one larger tank is more efficient than two smaller tanks and will cost less to install and maintain. Grease traps between kitchen and septic tank are not normally required for residential purposes but should be installed where commercial kitchens or other grease-producing activities are involved. The grease trap is simply a type of fixture designed to separate grease from water, using baffles and flotation, but removal of accumulated grease by a regular maintenance schedule is often overlooked. This results in a clogged sewer line or passage of grease into the septic tank, ultimately killing the bacteria in the tank.

Most tanks are box-shaped, although several prefabricated tanks are vertical or horizontal cylinders, or other shapes. Figure 17-3 shows the normal construction requirements for a concrete tank, while Table 17-2 indicates gallon capacity for rectangular or cylindrical tanks. Most authorities agree that residential sewage averages about 75 to 80 gal per day per person, or Table 17-2 may be used to estimate the required tank size. No tank should hold less than 1000 gal.

Septic tanks must be constructed of decay-resistant materials, and steel tanks are usually coated with asphalt or coal tar to resist corrosion. The inlet

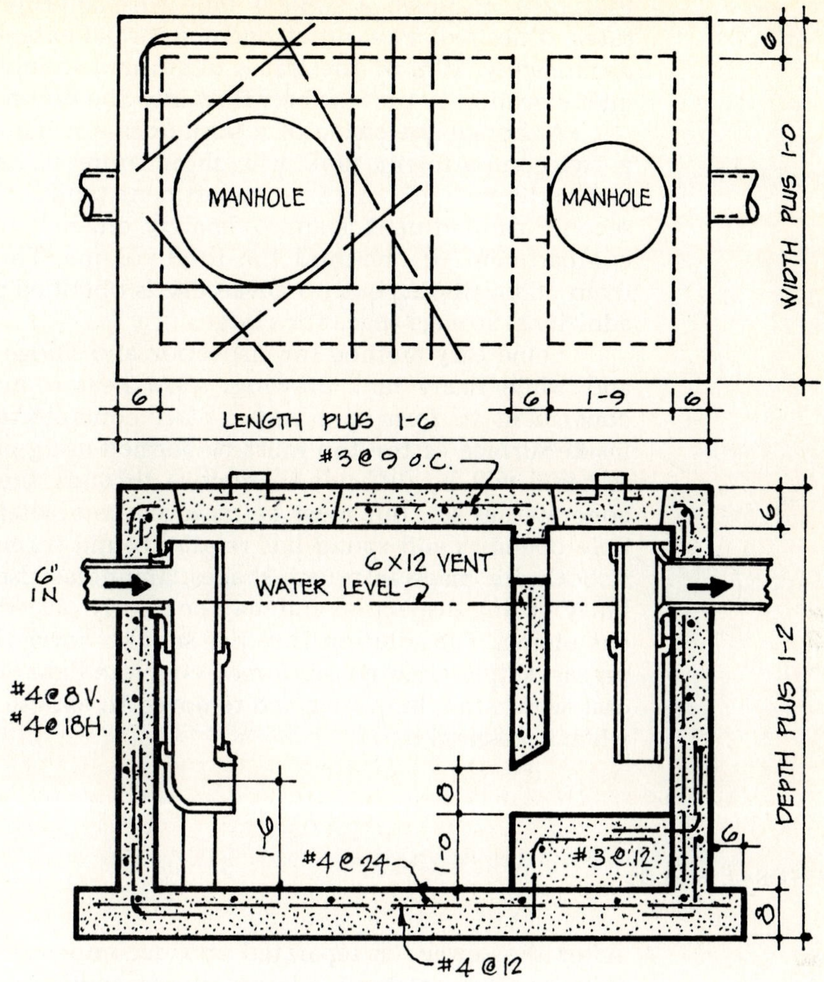


**Figure 17-3** Typical septic tank. (*From the Architectural Draftsman's Reference Handbook*.)

**TABLE 17-2**
Septic-Tank Capacity[a]

| Number of persons | Inside tank dimensions | | | | Capacity (gal) |
|---|---|---|---|---|---|
| | Length | Width | Depth | Liquid | |
| 4 | 6′-0″ | 3′-0″ | 5′-0″ | 4′-0″ | 500 |
| 6 | 7′-0″ | 3′-0″ | 5′-0″ | 4′-0″ | 600 |
| 8 | 7′-6″ | 3′-6″ | 5′-0″ | 4′-0″ | 750 |
| 10 | 8′-6″ | 3′-6″ | 5′-6″ | 4′-6″ | 900 |
| 12 | 8′-6″ | 4′-0″ | 5′-6″ | 4′-6″ | 1100 |
| 14 | 10′-0″ | 4′-0″ | 5′-6″ | 4′-6″ | 1300 |
| 16 | 10′-0″ | 4′-6″ | 5′-6″ | 4′-6″ | 1500 |

[a]Normal capacity is based on 4 $ft^3$/person/day. For schools and the like, use 2 $ft^3$/person/day.

pipe should enter the tank with an invert at least 4 in. higher than the effluent invert, and the discharge end should extend into the liquid level at least 6 in. Air space of 1 to 2 in. should be allowed between the influent pipe top or chamber to allow a free passage of gas throughout the tank. A similar arrangement should be used for the effluent pipe or outlet. One of the simplest methods used at both influent and effluent is a pipe tee, as shown in

Fig. 17-3, although a straight pipe with appropriate baffle may be substituted if desired. The outlet opening should extend below the liquid level approximately 40% of the liquid depth and should always be lower than the inlet elevation to prevent backflow into the house drain.

Although the shape of a tank does not materially affect its operation, a two-compartment tank will allow more bacterial action than a single-compartment tank. A two-compartment tank should be so designed that the second compartment is approximately one-half the total volume of the first compartment or one-third the total volume. The depth of liquid may vary from 30 to 60 in., but no advantage is obtained by greater depth except for additional storage space for sludge.

Some easy method for inspection and sludge removal must be provided, and while many tank drawings show neat round manholes, these are not convenient to form in a cast-in-place concrete tank top. To start with, the inside surface of the top must be formed using plywood or lumber, and this material will be difficult to remove through the small manholes after the concrete has set, but must be removed completely in any case. Round manhole openings and round lids require round forms, which may be simplified if both are made in square shapes. Manholes also limit ventilation and visibility during inspection and may be poorly placed for repair to inlet or effluent piping. The solution is rather simple: cover the tank top with a series of precast shiplap-jointed reinforced-concrete slabs or panels which may be easily cast with minimum forms and removed to expose any or all parts of the tank when necessary.

## 17.6 DISPOSAL FIELDS

After the sewage is separated and decomposed in the septic tank, usually over a period of about 6 hours, the remaining liquid passes to a disposal or absorption field. Depending almost entirely on the absorption value of the soil, the tile field may be anywhere from 100 to as much as 500 ft long. Also, depending on the absorption capacity of the soil, the tile field may be divided into several sections to allow one or more sections to "dry out" while another is in use. If only one section of tile field is required, there is no reason for a switch box as shown in Fig. 17-4, but this construction is necessary if more than one field is required. The switch box is nothing more than a box with the inlet from the septic tank higher than several outlets to separate fields. Plugs with adequate handles are provided to close outlets not in use. Periodic altering of plug locations will then allow a "wet" field to dry while a "dry" field is used.

Liquid from the septic tank or through the switch box is channeled to the disposal field. The tile field is installed in trenches of cross section, as shown in Figs. 17-5 and 17-6, and spaced not closer than 6 ft apart. The area at bottom, sides, and top of the tile is filled with gravel or crushed rock, and the tile line should not be deeper than 48 in. maximum to allow bacterial action (dependent on oxygen) to continue. While adequate gravity flow in sewer lines must be provided, the pitch of the tile field should be only a bit more than dead level in order to allow liquid to percolate into the surrounding rock between tile joints. Tile should be square-end (not bell-and-spigot), 4 or 6 in. in diameter, vitrified clay, with a space between tile ends of approximately ½ in. To prevent soil or other material from clogging the line, a

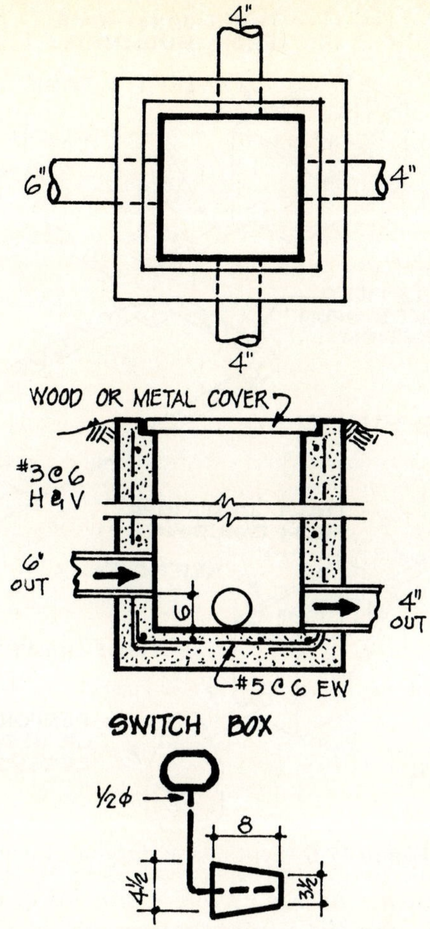


**Figure 17-4** Septic tank switch box. (*From the Architectural Draftsman's Reference Handbook.*)

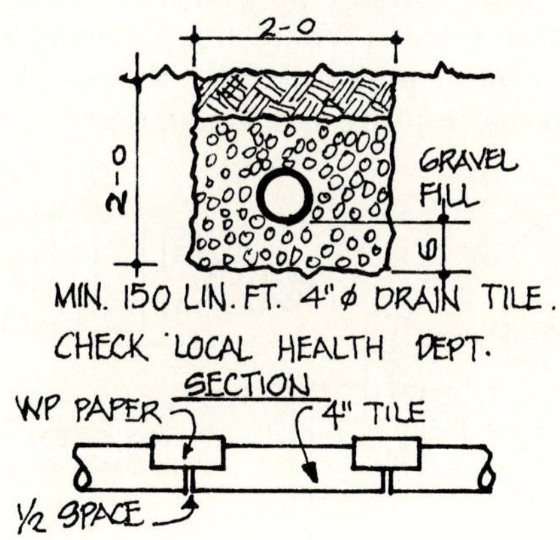


**Figure 17-5** Septic tank field tile. (*From the Architectural Draftsman's Reference Handbook.*)

piece of heavy waterproof felt, paper, or metal should cover the top and sides of each joint. Several shorter lines connected to a feeder line are preferable to one long line, because the end nearest the septic tank will usually absorb more liquid than the line farther along and a single long tile line may be partially dry and not operating a good part of the time.

Ideally, the absorption field should be reasonably level, but this may not always be possible. When it is not, the tile lines should be laid along the same contour lines to preclude excessive pitch in the line or should be constructed with a connecting system from one line to another. Several types of

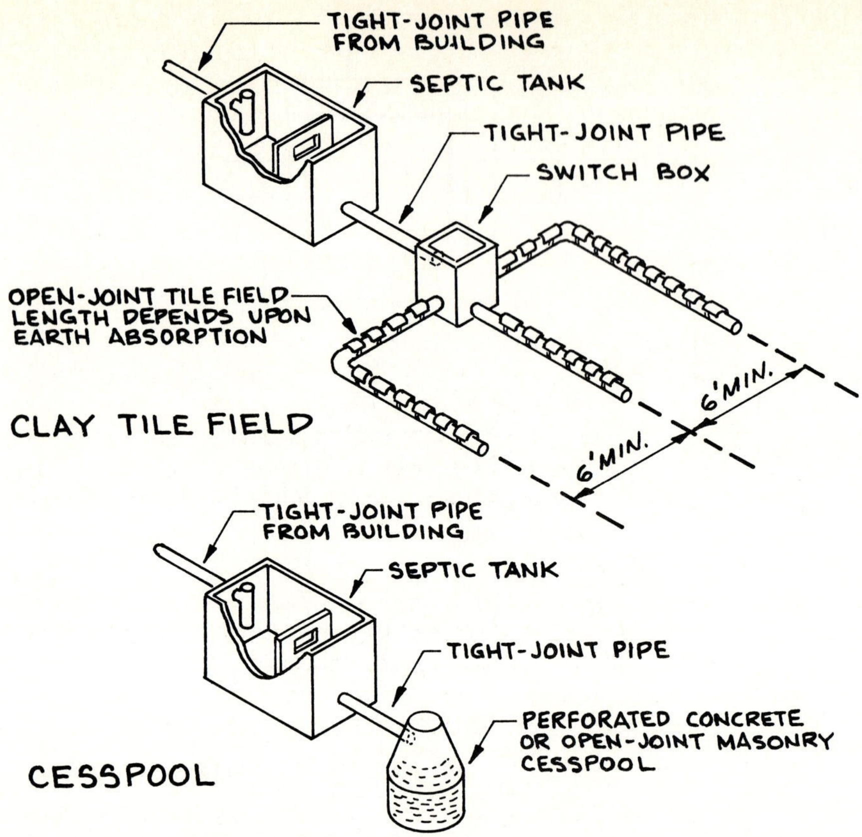


**Figure 17-6** Septic tank disposal systems.

connections are shown in Fig. 17-7. Connecting pipe between tile fields should be installed with tight joints rather than spaced joints as used in the tile fields. Surface water should be diverted from the absorption field by ditches, dikes, or other means. Where steep grading banks occur on the property, the tile lines should be maintained at least 6 ft from the bank top to prevent effluent percolating through the graded slope.

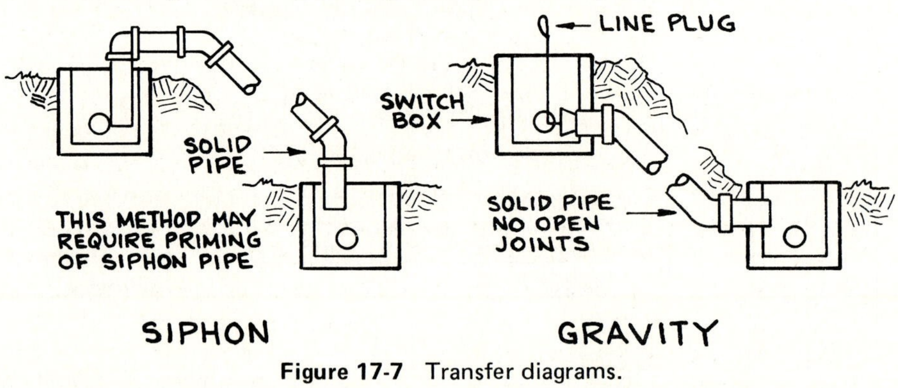


**Figure 17-7** Transfer diagrams.

## 17.7 MAINTENANCE

Septic-tank private disposal systems are not foolproof. When a system is properly installed, it should function efficiently for a great number of years with correct maintenance. Normal residential waste includes a limited amount of grease, fat, or various other materials, but these should not affect the operation of the system. The principal drawback is the accumulation of sludge in the tank, but this may be eliminated by periodic checking and pumping of sludge once every two or three years, depending on the load imposed. Failure of the absorption field may occur occasionally if the percolation is inadequate or the unattended tank allows partially digested raw sewage to flow into the tile field. For locations where municipal disposal systems are not available, and septic tanks are approved, the individual private construction in the only satisfactory method to dispose of human sewage waste.

## 17.8 MUNICIPAL SEWAGE SYSTEMS

Municipal sewage-treatment plants need to be designed with several very major items thoroughly researched: (1) character and volume of sewage to be treated, (2) the area served, (3) the present and future population, and (4) the degree of treatment desired. None of these items is fixed or stagnant, so ample consideration should be given to possible future requirements. Health departments, both national and local, also have regulatory rules that cover basic design, estimates of volume and treatment, disposal of effluent, and other "standards."

One of the most widely used of these standards is the *Standards for Sewage Works* used by the Great Lakes–Upper Mississippi River Board, as well as many other areas. The design of a treatment plant is also based on the time of utilization after completion and is usually anticipated as 15 to 20 years. Areas of potential growth may desire a plant that is operational, but not entirely complete, with the possibility of future additions or enlargements as population increases. In metropolitan areas, which may have a limited growth, the largest-size plant will be desired. Records of past and probable population growth may be obtained from the U.S. Census Bureau, school records, utility companies, building permits, and city planning departments. The nature of growth, or lack of it, may also be influenced by the degree of industrialization, transportation, local zoning, and similar factors.

The volume of estimated sewage is determined by flow measurements taken over weirs, 24-hour recording meters, or floats, in both wet and dry seasons. These, of course, must then be increased to try to arrive at a reasonable estimate of future volume. In many locations, rainwater from streets, roofs, and so on, normally considered storm drainage and handled separately, is channeled into "sanitary" sewers, thus drastically increasing the liquid volume in wet weather. Ground food waste from residences and a greater volume and variety of industrial waste must also be considered. Many sewer-plant-design sources list sewage flow as ranging from 50 to 500 gallons gal/day per capita, but a more reasonable rate may be about 100 gal/day capita, especially as waste from one person is normally considered to be about 75 to 80 gal/day. For additional storm water introduced into a treatment plant, a factor of three to four times the dry-weather flow is often used.

Sewage is from 80 to possibly 99% liquid. Included in the total volume are about 0.20 lb of suspended solids per day per capita and about 0.17 biochemical oxygen demand (BOD) per day per capita for sanitary sewers. Analysis of sewage usually includes the following: (1) suspended solids, which determine the sludge-handling requirements; (2) biochemical oxygen demand, which affects the amount of treatment required; (3) acidity or alkalinity, which indicates need for chemical adjustment; (4) grease or oil, which may affect grease- or sludge-handling ability; (5) freshness of sewage, which directly affects odor; and (6) need for chlorination or other chemical treatment. The presence of excessive grit, metal particles, industrial chemicals, and similar extraneous materials also needs to be analyzed as they affect piping, sludge handling, valves, and practically all parts of the system. Pretreatment or screening of raw sewage may reduce some of these items, but they must be planned for when preliminary design criteria are set up and provisions made for proper isolation.

Home garbage grinders and commercial or municipal garbage grinders may increase the suspended solids in sewage to 100%, with an average of about 60%. The effect on the volume of sewage is negligible but may increase the amount of grit by as much as 80%. An increase of up to 100% of sludge and scum in primary operations may be expected. Primary effluent may increase from 5 to 10%, thus requiring enlargement of secondary treatment facilities. Treatment requirements may vary as a result of seasonal changes of contributing population, seasonal food processing, low water or high water, varying temperatures, and the use or discard of final effluent or sludge.

In addition to the disposal of sewage, most municipal plants now in the planning stage are also being designed to permit retrieval of wastewater. Water from rain or storm drainage is usually channeled into "storm sewers" and is contaminated only by oil or other surface material and does not go into the sewage-treatment plant. Gray water and black water are processed as a part of sewage and discussed at greater depth in Chapter 16. Disposal of sludge from a municipal treatment plant is as much a problem as with private systems, and is discussed later in this chapter. A combustible gas, methane, is also produced as a by-product of sewage treatment and may be burned at the plant site or further processed as an energy source.

## 17.9 PLANT DESIGN CRITERIA

A number of important factors must be evaluated when construction of a municipal sewage-treatment plant is contemplated: quality and quantity of sewage, disposal opportunities for effluent and sludge, topography versus hydraulic requirements, location of the plant in relation to housing or industry, availability of operating personnel, and last but certainly not least, the ability of the municipality to afford the plant economically. Even if none of these items is problematic, the time required to satisfy all regulating agencies, the time required for architectural and engineering work, and the time needed for construction mean some very long-range activity.

The design of municipal sewage treatment plants is rather highly specialized and should not be attempted by untrained personnel or organizations which may not have the facilities or experience. Practically every plant is different, owing to the factors noted in the foregoing paragraph, but the various elements or units that may compose a plant are similar enough to be explained (see Fig. 17-8), as are the combinations described in the following section.

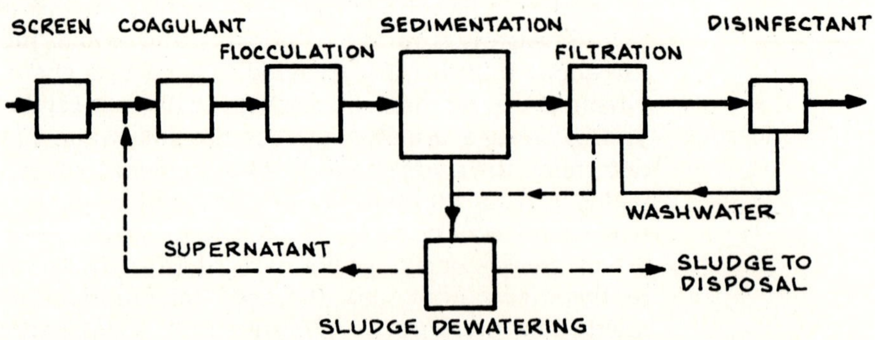


**Figure 17-8** Typical disposal diagram.

## 17.10 TREATMENT PROCESSES

Sewage-treatment processes may generally be considered as primary, intermediate, and secondary. *Primary treatment* is usually a process of sedimentation which may remove 50 to 60% of suspended solids and 25 to 40% of BOD. *Secondary treatment* involves biological processes and may remove approximately 90% of suspended solids and 75 to 90% of BOD. Chemical treatment, heat, air, or other modifications may be used in the *intermediate* stages and on the final effluent and sludge. The final disposal of treated liquid effluent and sludge determine to a very great extent the amount of secondary and intermediate or final treatment required. Sludge may be used for landfill or for fertilizer or may be incinerated with or without drying. Liquid effluent may be chemically treated, ponded, and used as a source of domestic water, may be dumped into a convenient sea or river, or may have to be evaporated or absorbed into a disposal field.

Most municipal sewage plants have some sort of screening device as a first unit when sewage enters the plant. Screens may be composed of equally spaced bars, slotted or perforated plates, woven wire, or similar material arranged in drums or racks installed parallel or perpendicular to the sewage flow. Screens with openings from $\frac{3}{32}$ to $\frac{3}{16}$ in. are termed "fine" screens and are used to remove industrial waste which may produce excessive scum or foam in digestion tanks. Bar screens with openings from $\frac{3}{4}$ to 2 in. are used ahead of sewage pumps, grit chambers, and primary sedimentation tanks, while bar screens with openings 3 to 6 in. (termed "trash racks") are used to prevent logs, boxes, or other large refuse from entering the plant. All screens should be provided with cleaning devices or rakes to prevent buildup of screenings, which may retard sewage flow or cause intermittent surges as screens are cleaned.

Comminuting devices are designed to cut up any retained material in the sewage flow without removal from the flow. Several major types are available, but all operate on the principle of a set of rotating teeth passing between stationary bars or grids, thereby cutting retained solids into a fairly uniform size for easier digestion. Comminutors have openings from $\frac{3}{8}$ to $\frac{3}{4}$ in. and are commonly located between a primary screen or grit chamber and the primary sedimentation tank to prevent any excessive damage to teeth or stationary plates by hard grit particles. Comminuting devices may also replace screens in locations where access may be difficult or screenings hard to remove.

The term *grit* includes small coarse particles of sand, gravel, metals, vegetable or fruit seeds, and similar nonputrescible material. Removal of grit protects moving mechanical equipment from abrasion, prevents pipe and equipment from clogging, and reduces cleaning of the system. For obvious reasons, grit chambers (Fig.17-9) are usually located behind preliminary screens and ahead of any pumps. Control of the velocity of sewage is of primary importance in grit chambers, and many designs provide a spiral flow which allows the heavier solids to drop to the bottom of the unit, while lighter-weight material passes through. Mechanical, airflow, or other grit-removal methods are necessary to keep the unit from eventual clogging and the amount of grit may vary from as little as 0.75 cubic feet per million gallons of sewage to as much as 600 cubic feet per million gallons depending on type of sewage, street conditions, and other factors. Final disposition of collected grit may be by burial or use in landfill, but all amounts of organic matter must be completely removed.

The next step after grit removal is the settlement process, where the

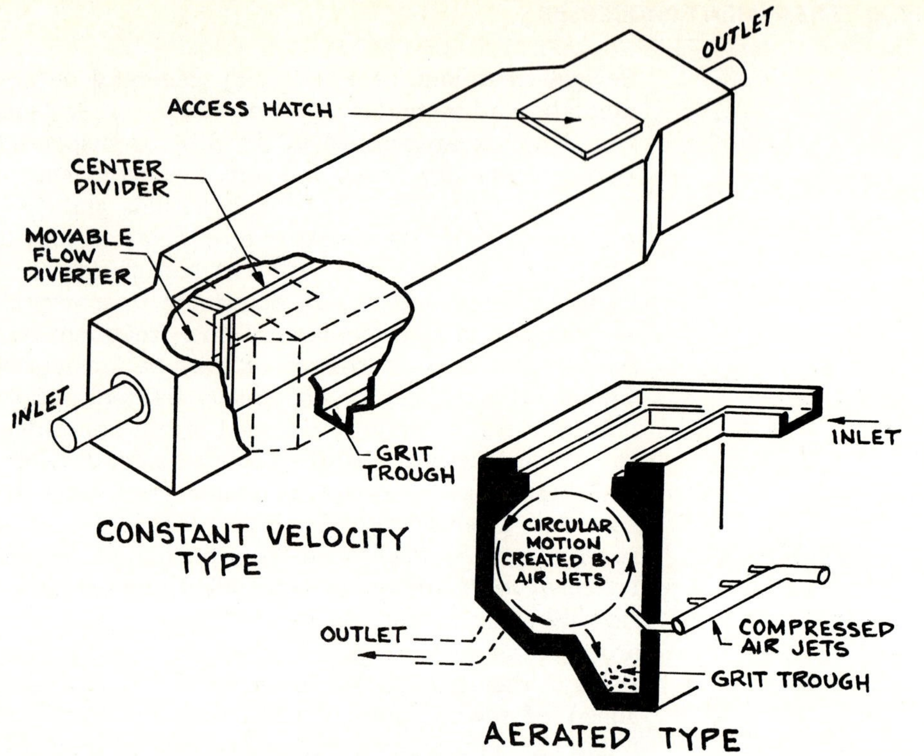


**Figure 17-9** Grit chambers.

remaining suspended matter in the sewage flow is settled out by gravity and slow flow of the liquid. Tanks (see Fig. 17-10) may use horizontal flow, vertical flow, or radial flow, and the longer the period of detention and the slower the velocity of flow, the greater the amount of suspended matter removed. About 50% of the suspended solids will precipitate in about 2 hours, and the design size of tanks should be based on the dry-weather flow (3 times normal = maximum rate), so most tanks have a designed flow of 9 to 15 hours' retention. Horizontal-flow tanks are simply rectangular troughs with length 3 to 10 times width, 6 to 10 ft deep, and with a bottom sloped at about 1 in 30. Vertical-flow tanks are deep tanks with vertical sides and with the bottom slope at 45° or more to facilitate removal of sludge by gravity. Several variations of vertical tanks are in successful use. The radial-flow tank is a shallow circular tank in which the sewage enters at the center, the heavier particles drop almost immediately, and the lighter particles are swept to the perimeter of the tank, where they settle, and the liquid is carried away in an exterior trough. Precipitated particles or sludge in horizontal and vertical tanks are usually pumped, while a rotating scraper is normal for circular tanks. Continuous-flow settlement tanks without chemical treatment will provide about 10 to 12 tons of wet sludge per million gallons of sewage and may be increased to 17 to 20 tons with the introduction of chemicals. Cost of chemical addition is about 25% more than for simple sedimentation.

Although settlement tanks or basins remove most of the suspended matter, the remaining solids, colloids, and dissolved matter must be removed or made harmless. This required semifinal treatment is made in percolation beds of broken stone, blast-furnace slag, clinkers, or other material that will allow contact with bacteria and oxygen. This contact causes a chemical-biological change in the organic matter still present to convert it to gases, nitrates and nitrites, and humuslike solids. The sewage is usually introduced

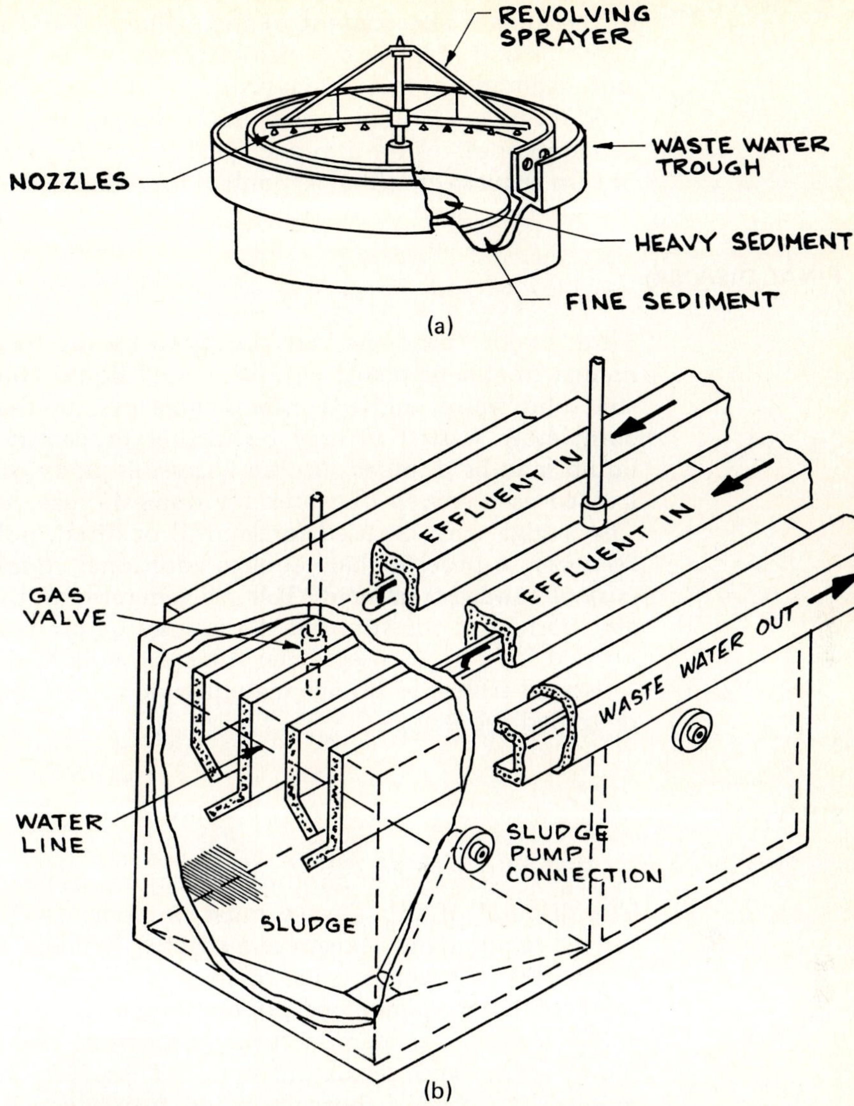


**Figure 17-10** Types of tanks. (a) Circular tank; (b) two-compartment Imhoff tank.

into a bacteria bed through a piping system, allowed to remain for about 2 hours, and drained. The remaining slime coating the ballast is then exposed to air for about twice the time the liquid is in the bed, thereby allowing aerobic bacterial action to further disintegrate any remaining solids. A newer and somewhat more efficient system is a percolating filter, where the liquid is sprayed over a bed of filter material similar to that used in bacteria beds, percolates through the filter material, and is carried away in a system of closely spaced pipe or tile at the bottom. Filter material used in either type varies from ½ to 1½ in. in diameter and from 3 to 6 ft deep. The average time of percolation, using ¾- to 1-in. gravel filter material, is about 18 to 20 minutes when sprinkled at a rate of 250 gallons per square yard of surface per day.

Chlorine is used principally to disinfect treated sewage effluent prior to its discharge into a stream or other body of water. A contact time of not less than 15 minutes is usually necessary to provide a residual of not less than 0.2 to 1.0 mg/liter. Effective chlorination generally results in a 99% reduction in the undesirable content of the effluent. BOD also may be reduced by as much as 35%. Chlorine is also effective in the oxidation of hydrogen sulfide and decomposition, which produces odor. In addition to these two end-product uses, chlorine is also used throughout a system to reduce the hydrogen sulfide effect on iron and steel parts; to control surface growths, which in turn reduces clogging; to control foaming; and other uses.

## 17.11 FINAL DISPOSAL

Three major "products" are produced by the treatment of sewage in a municipal treatment plant: (1) disinfected liquid (water), (2) sludge or humus/soil solid waste, and (3) inflammable gas (methane). All of these may be completely wasted or may be further processed into usable products. The liquid may be dumped into an accessible body of water as waste or further treated as a source of water for domestic use, as described in Chapter 16. The sludge may be used for landfill or dried, pulverized, and packaged for fertilizer, a process that requires additional equipment but may reduce the cost of sewage treatment. The gas generated in the treatment process is usually burned at the site, but it too may be processed and stored in tanks or bottled for various uses. Even the trash collected on the screens and the grit separated from the sewage may provide some economic return when cleaned or graded properly.

## 17.12 SUMMARY

Sewage of one sort or another is a major problem for any human habitation. The disposal of this sewage must be considered before any other human-service requirement, except in a strictly nomadic society. In our present situation, this disposal is a factor in any decision to allow additional building required by movement of population, increased commercial and industrial waste created by manufacturing processes, and by more stringent health requirements and regulating codes. These factors have forced many communities to place a moritorium on future construction, often even before questions of water supply, available land, and other services. As a parting admonition to those who plan for construction, be certain that adequate facilities for proper sewage disposal by septic tanks or municipal system are available before starting work and that these are allowed by regulating authorities.

# 18

# Plumbing Systems

## 18.1 INTRODUCTION

The design of plumbing systems may be very simple when only a few fixtures in a one-story, one-bathroom residence are considered but may be more or less complicated when more fixtures or floors are added. Regardless of the number of fixtures, the requirements for proper and economical design of sewer, venting, and water are essentially the same. The physical arrangement of the fixtures may determine their location, size, and venting. The type and use of the fixtures will also determine the water piping. A great deal of this planning is required by the plumbing code, but as no two systems are identical, differences generally occur with each new building.

In designing plumbing for multistory commercial buildings, a major factor is the number of people using the fixtures and the proportion of men and women to the total. Some types of installations must also consider frequency or certain time periods when plumbing is used. Larger installations often have back-to-back fixtures or multiple installations side by side. These types provide the possibility of different connections and economy but also involve some major considerations in regard to space. The type of custodial operations also have an effect on plumbing and may determine whether fixtures are floor-mounted or wall-hung, and the water supply, or lack of it, may determine if flush valves may be used on water closets and urinals.

Special fixtures and installation requirements are also necessary for some types of plumbing work. Schools with elementary, particularly kindergarten, children need special small fixtures installed lower than they normally are so the children may use them easily. Plumbing for the handicapped generally does not require special fixtures, but installation space and height are almost always special.

## 18.2 FIXTURE ARRANGEMENT

The arrangement of fixtures, especially in a bathroom, determines the method used to install the required plumbing, and as a result, the cost of the installation. The most simple arrangement is that shown in Fig. 18-1, where the sewer and water all occur on the same wall. In this arrangement, the vents are all tied together, the sewer is installed in a line beneath all the fixtures, and the water is supplied to each fixture by a tee connection from the main water supply. The principal objection to this arrangement is that the window is usually placed over the bathtub, which makes it rather inconvenient to adjust the window. Another arrangement, shown in Fig. 18-1, still has all the fixtures on one wall, but a short wall at the end of the square tub or shower unit divides the space and provides access to the window. Although a side entrance affords more privacy, as shown, the door also swings badly, forcing a person against the lavatory. Each of these arrangements require minimum floor space as noted.

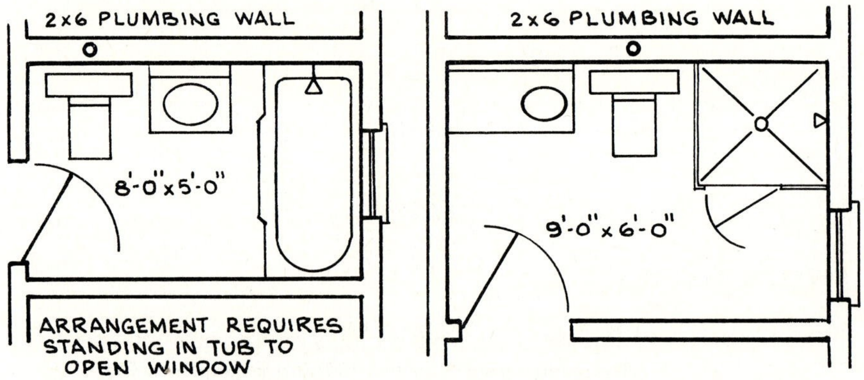


**Figure 18-1** Minimum "full" bathroom sizes.

When fixtures are not all on one wall, as shown in Fig. 18-2, additional sewer and water piping are required, as well as at least two vent stacks. The actual floor space required is more than shown in Fig. 18-1, and care in designing closets or other acoustically "dead" or absorbing spaces should be considered.

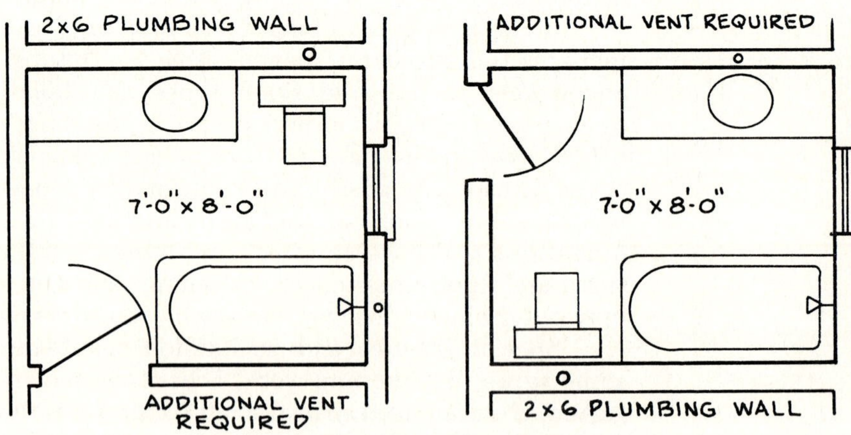


**Figure 18-2** Multiple-vent bathrooms.

Today more and more residential construction includes a private bathroom accessed from a master bedroom and another bathroom opening on a hall. In addition, many homes have washer–dryer space in the hall or immediately adjacent to bathrooms or kitchens. Arrangements that require minimum piping and number of vents are shown in Fig. 18-3. Quite often in this type of arrangement, one bathroom is equipped with a tub and shower, while the other has a shower unit only. In either case this does not change the plumbing requirements, except that the shower may require a bit more pipe.

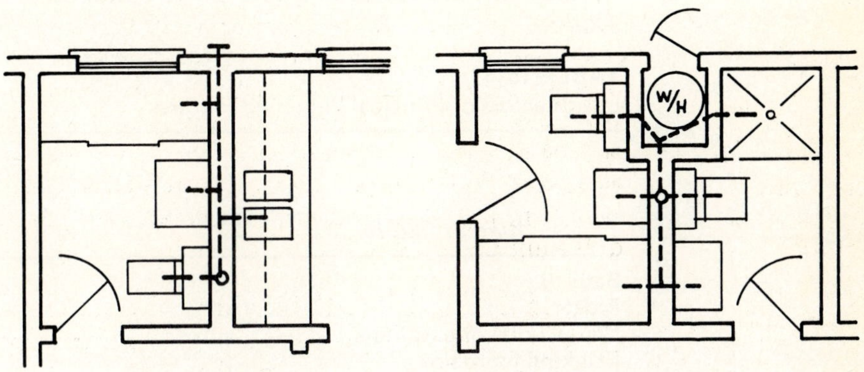


**Figure 18-3** "Back-to-back" vents.

Why all this concern about the location of fixtures? The proper location generally determines the location and amount of sewer and vent piping, and this work is the most difficult to install and the most expensive. The water closet is probably the most important, requiring the largest pipe, and therefore should be placed to require the least amount of pipe. As the other fixtures require smaller pipe, they should be considered after the water closet. In cases where the water closet is on one side of the room and another fixture is located on the other side, the drain from the second fixture should be made to the water closet drain, not vice versa. The larger vent is also off the water closet, so any other vent is of a minor size.

## 18.3 DRAIN AND WASTE

The primary concern in regard to the drain–waste–vent (DWV) or sewer–vent system is to remember that it functions by gravity, not pressure. Proper slope of pipe is essential to allow this system to operate, and a pitch of ⅛ to ½ in. per foot is generally considered adequate. This will provide for good drainage but at the same time will not be so steep that solids will be separated from the liquids. The size of pipe for each fixture is rather well established, and the Uniform Plumbing Code, Section 401, spells out the relevant information as follows:

> (a) Drainage pipe shall be cast-iron, galvanized steel, galvanized wrought iron, lead, copper, brass, ABS [acrylonitrile-butadiene-styrene], PVC [polyvinyl chloride], or other approved materials having a smooth and uniform bore, except: (1) That no galvanized wrought iron or galvanized steel pipe shall be used underground and shall be kept at least six (6) inches (152.4 mm) above ground. (2) ABS or PVC installations limited to residential construction, not more than two (2) stories in height.

In addition to the DWV material used, the size of the various pipes is important. A pipe that is too small to carry the load will have a tendency to clog easily. Too large a pipe wastes space and money, and the solids may settle out because of reduced scouring action of the liquids. The problem of proper pipe size has created some acceptable design standards based on measurements that determine that a normal lavatory discharges at a rate of approximately 1 $ft^3$ of water per minute. Thus a lavatory is assigned a load factor of 1 and various other fixtures assigned load factors in relation to the amount of discharge from each fixture. The principal fixtures and their load factors are indicated in Table 18-1, condensed from the Uniform Plumbing Code.

**TABLE 18-1**
Load Factors of Principal Plumbing Fixtures

| | *Minimum trap and trap arm size* | | |
|---|---|---|---|
| *Kind of fixture* | *inches* | *mm* | *Units* |
| Bathtubs | 1½ | 38.1 | 2 |
| Bidets | 1½ | 38.1 | 2 |
| Dental units or cuspidors | 1¼ | 31.8 | 1 |
| Drinking fountains | 1¼ | 31.8 | 1 |
| Floor drains | 2 | 50.8 | 2 |
| Interceptors for grease, oil, solids, etc. | 2 | 50.8 | 3 |
| Interceptors for sand, auto wash, etc. | 3 | 76.2 | 6 |
| Laundry tubs | 1½ | 38.1 | 2 |
| Clothes washers | 2 | 50.8 | 2 |
| Receptors (floor sinks), indirect waste receptors for refrigerators, coffee urn, water stations, etc. | 1½ | 38.1 | 1 |
| Receptors, indirect waste receptors for commercial sinks, dishwashers, air washers, etc. | 2 | 50.8 | 3 |
| Showers, single stalls | 2 | 50.8 | 2 |
| Showers, gang (one unit per head) | 2 | 50.8 | |
| Sinks, bar, private (1½ in. (38.1 mm) min. waste) | 1½ | 38.1 | 1 |
| Sinks, bar, commercial (2 in. (50.8 mm) min. waste) | 1½ | 38.1 | 2 |
| Sinks, commercial or industrial, schools, etc., including dishwashers, wash up sinks, and wash fountains (2 in. (50.8 mm) min. waste) | 1½ | 38.1 | 3 |
| Sinks, flushing rim, clinic | 3 | 76.2 | 6 |
| Sinks, and/or dishwashers (residential) 2 in. (50.8 mm) min. waste) | 1½ | 38.1 | 2 |
| Sinks, service | 2 | 50.8 | 3 |
| Mobile-home-park traps (one for each trailer) | 3 | 76.2 | 6 |
| Urinals, pedestal, trap arm only | 3 | 76.2 | 6 |
| Urinals, stall | 2 | 50.8 | 2 |
| Urinals, wall (2 in. (50.8 mm) min. waste) | 1½ | 38.1 | 2 |
| Wash basins (lavatories) single | 1¼ | 31.8 | 1 |
| Wash basins, in sets | 1½ | 38.1 | 2 |
| Water-closet flush-tank type, trap arm only | 3 | 76.2 | 4 |
| Water-closet flushometer valve trap, trap arm only | 3 | 76.2 | 6 |

Uniform Plumbing Code.

Fittings for the drain lines may be of several different types, but all should have the capability to maintain tight connections. Changes in direction of drainage piping must be made with standard fittings in angles presented by one-sixteenth (1/16), one-eighth (1/8), or one-sixth (1/6) bends. Where horizontal drains connect with vertical stacks, they must enter through 45° or 60° Y branches, one-eighth (1/8) bends, or similar approved fittings. No fitting may have more than one branch at the same level unless the fitting is designed to prevent the flow of one branch from entering the other branch. Each horizontal drain pipe must be provided with a cleanout at its upper terminal and at each 100 ft or major change of direction. The simple one-bathroom diagram in Fig. 18-4 indicates the major drain connections and probable fittings that may be used.

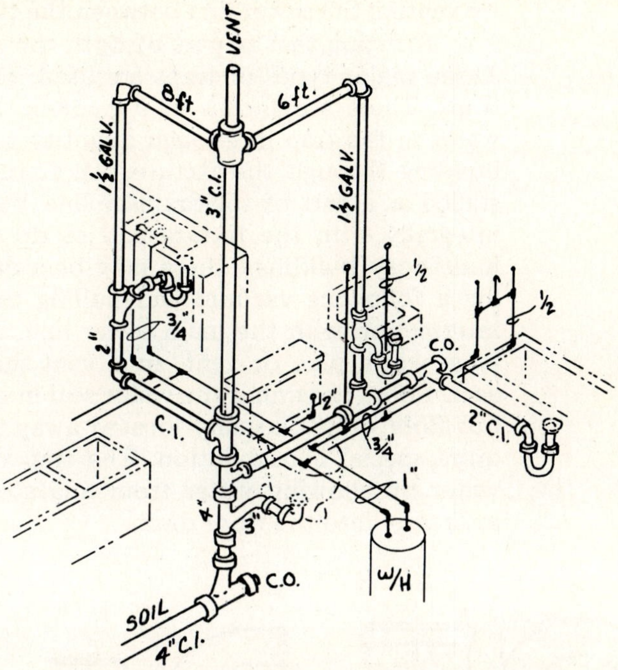


Figure 18-4 Typical one-bath/kitchen plumbing.

## 18.4 VENTING

Proper venting of plumbing fixtures is required to prevent sewer gas from entering the building and is installed as a part of the DWV system. Theoretically, every fixture has a separate vent, but in actual practice most vents are connected together above the fixtures served and only one main vent is carried through the roof (see Fig. 18-5). Circuit vents are allowed when

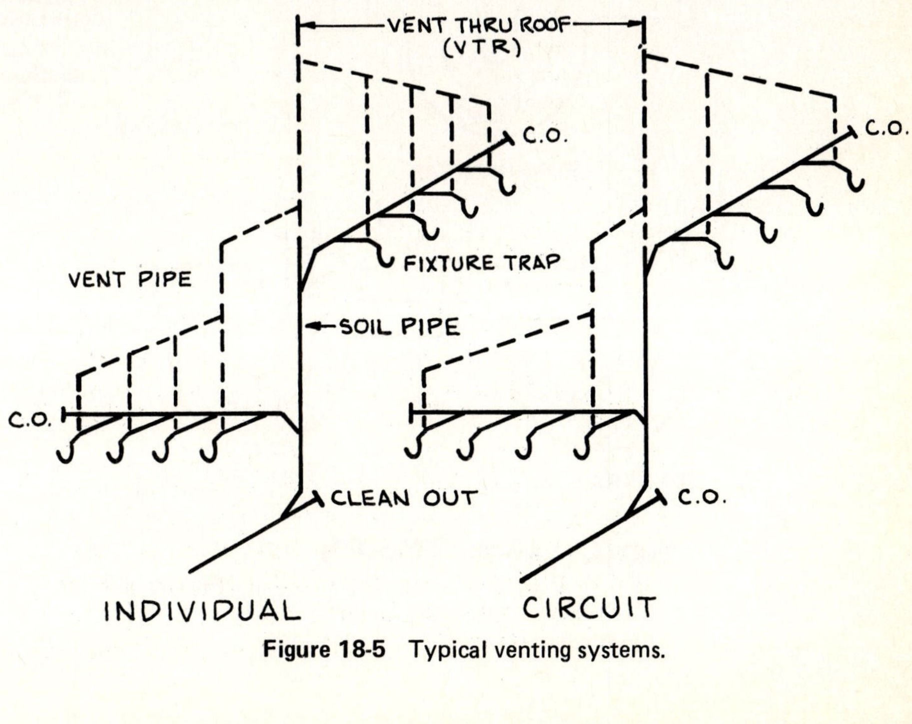


Figure 18-5 Typical venting systems.

banks of similar fixtures are installed together. This arrangement provides for a vent at the beginning and another at the end of the string of fixtures, thereby venting the sewer line between the two vents.

A fitting that is part of both the sewer and the vent is the fixture trap. Three major types of traps are used: (1) P traps, (2) S traps, and (3) drum traps. These are shown in Fig. 18-6. The function of the trap is to retain water in the trap in enough quantity to prevent sewer gas from entering the building through the fixture. All fixtures require traps. Most traps are installed as a part of the drainage line, but all water closets have a trap formed integrally with the fixture and so do not require an additional fitting. In multistory buildings, there may be a danger of rapidly falling sewage creating a following vacuum and pulling the water from traps as it passes their intersection with the main sewer line. This can be prevented by sufficiently large waste pipe or vent, or a vent may be connected to the main vent at each floor. Retaining the water seal in most fixtures is normally no problem, but floor drains, usually located away from a water source in a dry area, require special consideration. The trap in a floor drain may be kept full of water by allowing water from a lavatory or drinking fountain to discharge indirectly into the floor drain.

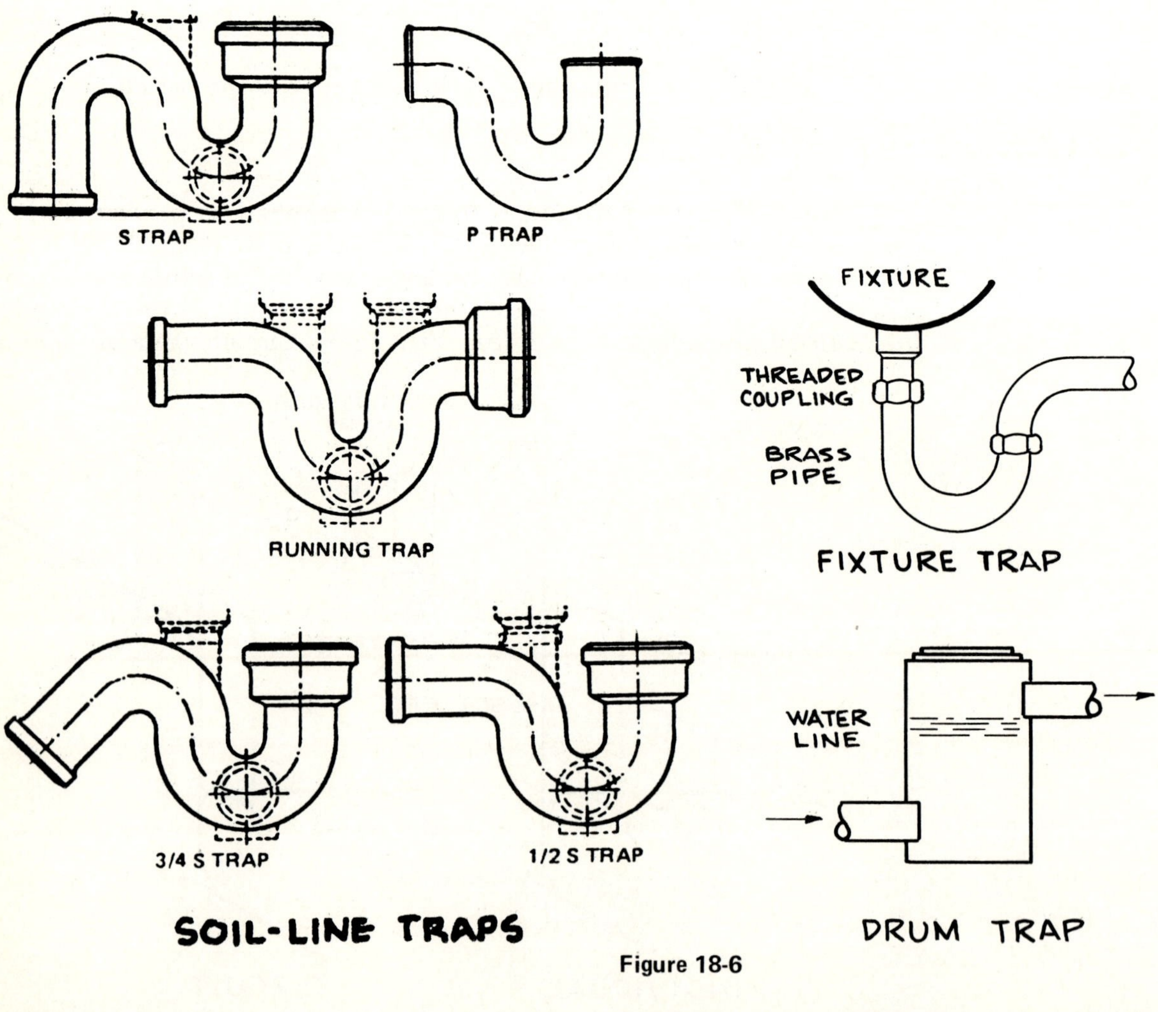


Figure 18-6

## 18.5 DWV DESIGN

The first requirement for DWV design is to determine the number of people to be served and the type and number of fixtures required. Next, determine the number of fixture units from Table 18-1 and the minimum trap size for each fixture. The unit number also indicates the gallons of water discharged by the fixture per minute. Then consider the minimum trap and drain size for each fixture (Table 18-1). Follow with size of the required drain and vent piping for Fig. 18-7, using Table 18-2.

1. The number and type of fixtures with their unit value and drain size are shown in Table 18-1. Add the unit numbers for bathtub plus lavatory (2 + 2 = 4). The drain from tub to water closet then requires 1½-in. pipe and the vent from each fixture is 1½ in.
2. Now add the water-closet unit of 4 (2 + 2 + 4 = 8). This will require a 4-in. soil pipe from water closet to main stack since no drain may be smaller than the fixture drains combined.
3. The kitchen sink has a unit value of 2 and the trap size is 1½ in. A 1½-in. pipe will accommodate three units and the vent should again be 1½ in.
4. The total fixture load on the main vertical and horizontal drain is bathroom (8) plus kitchen (2), or 10. Theoretically, the main drain could be 2 in. since a pipe of that size will accept up to 16 units. However, the code requires that no drain pipe be smaller than the size of the pipe immediately upflow, so the main drain will have to be 4 in.
5. The vent from the main drain must be "not less than the largest required building sewer," and so must be 4 in.

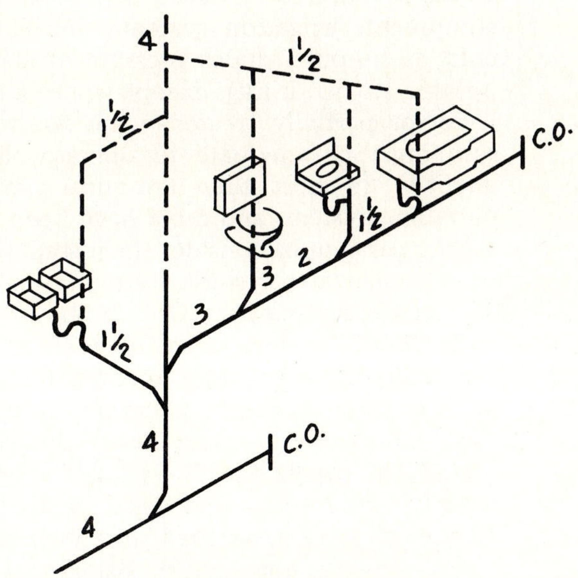


**Figure 18-7** Typical 4-fixture DWV.

**TABLE 18-2**
Maximum Unit Loading and Maximum Length of Drainage and Vent Piping

| | *Size of pipe, in., (mm)* | | | | | | | | | | |
|---|---|---|---|---|---|---|---|---|---|---|---|
| | 1¼ (31.8) | 1½ (38.1) | 2 (50.8) | 2½ (63.5) | 3 (76.2) | 4 (101.6) | 5 (127) | 6 (152.4) | 8 (203.2) | 10 (254) | 12 (304.8) |
| Max. units, drainage piping[a] | | | | | | | | | | | |
| Vertical | 1 | 2[b] | 16[c] | 32[c] | 48[d] | 256 | 600 | 1380 | 3600 | 5600 | 8400 |
| Horizontal[e] | 1 | 1 | 8[c] | 14[c] | 35[d] | 216 | 428 | 720 | 2640 | 4680 | 8200 |
| Max. length, drainage piping | | | | | | | | | | | |
| Vertical ft (m) | 45 (13.7) | 65 (19.8) | 85 (25.8) | 148 (45) | 212 (64.5) | 300 (91.2) | 390 (118.6) | 510 (155) | 750 (228) | | |
| Horizontal | | | | | (unlimited) | | | | | | |
| Vent piping,[f] horizontal and vertical | | | | | | | | | | | |
| Max. units | 1 | 8[c] | 24 | 48 | 84 | 256 | 600 | 1380 | 3600 | | |
| Max. length ft (m) | 45 (13.7) | 60 (18.2) | 120 (36.5) | 180 (54.7) | 212 (64.5) | 300 (91.2) | 390 (118.6) | 510 (155) | 750 (228) | | |

Uniform Plumbing Code.

[a] Excluding trap arm.

[b] Except sinks and urinals.

[c] Except six-unit traps or water closets.

[d] Only four water closets or six-unit traps allowed on any vertical pipe or stack and not to exceed three water closets or six-unit traps on any horizontal branch or drain provided any one of the three discharges is separated from the other two by a minimum horizontal developed length of 15 ft (4.6 m) whether connecting to vertical or horizontal piping.

[e] Based upon ¼ in. per foot (20.9 mm/m) slope. For ⅛ in. per foot (10.4 mm/m) slope, multiply horizontal fixture units by a factor of 0.8.

[f] The diameter of an individual vent shall not be less than 1¼ in. (31.8 mm) nor less than one-half the diameter of the drain to which it is connected. Fixture unit load values for drainage and vent piping shall be computed from Tables Nos. 4-1 and 4-2. Not to exceed one-third of the total permitted length of any vent may be installed in a horizontal position. When vents are increased one pipe size for their entire length, the maximum length limitations specified in this table do not apply.

## 18.6 WATER SUPPLY

Water is required at all fixtures, hose bibbs, water heaters, air-conditioning equipment, irrigation systems, and similar locations. Water, both hot and cold, is supplied under pressure of about 50 to 60 psi. Water must also be potable (fit to drink), except where a special system may be supplying seawater or partially treated water for water-closet flushing only. Water pressure must be adequate to operate all fixtures properly and in sufficient quantity. In short, there are quite a few requirements for usable water in a plumbing system, and most have been reviewed in earlier chapters. The following paragraphs consider the installation of water systems in buildings.

Older water systems were installed with galvanized steel pipe and malleable galvanized fittings. One of the problems with this material, when used with "hard" water, is the deposit of calcium inside the pipe which may eventually close the pipe opening. Copper pipe (type K) and tubing (type L) are excellent water-pipe materials, and plastic pipe (PVC and PVDC) are now used extensively in water service. Plastic pipe for hot water is limited to the use of ABS (acrylotnitrile-butadiene-styrene) only, but this type may also be used for cold water and drain systems. Copper and plastic pipe have some advantages over galvanized steel in that fittings require less actual space and make a neater appearance. Black-iron or lead pipe should not be used in water-supply systems.

The two basic types of water-supply systems are (1) upfeed, and (2) downfeed (see Fig. 18-8). The upfeed system is more dependent on

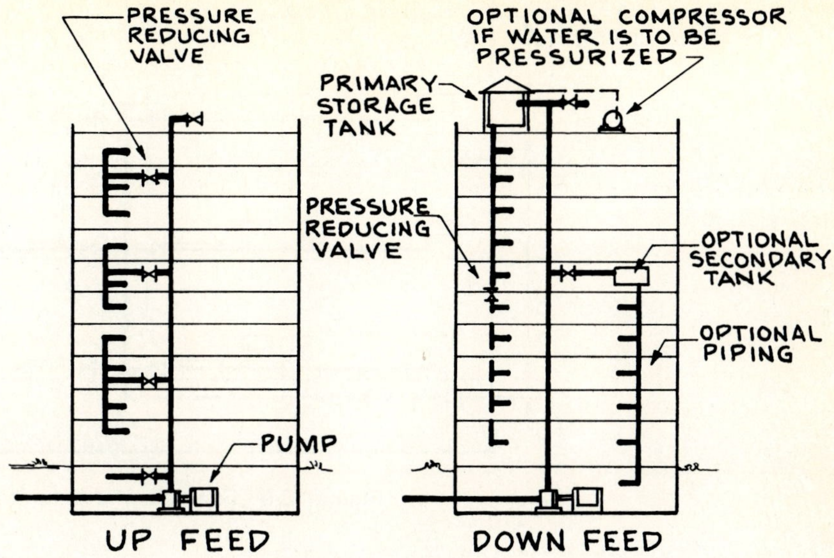


**Figure 18-8** Building water systems.

water pressure, as the water must overcome the static (downward) pressure of 0.434 psi per vertical foot, and the pipe and fittings friction generated in the height of the water riser. Also, there must be sufficient pressure at the topmost fixture to operate it properly, usually at least 12 to 15 psi. When upfeed cannot supply water in adequate quantity and pressure, the downfeed system is used. This is a combination of both pressure, supplied by pumping the water to tanks at the top of the building, and gravity from the tank down to the fixtures. With high-rise buildings the downfeed may be zoned to allow the pumped supply to be stored in tanks at intervals in the building height. Additional pumped supply may be taken from a tank at a lower floor and pumped to a tank at a higher floor. Downfeed also furnishes a storage tank for downfeed fire protection systems in many cities.

## 18.7 UPFEED WATER DESIGN

The upfeed water design must be able to supply water to the topmost fixture at sufficient pressure to operate the fixture. For all practical purposes, this height above the supply main is about 40 to 50 ft in order to provide a minimum pressure of 12 to 15 psi at the topmost fixture. Three factors determine the available pressure: (1) pressure in the street water main, (2) loss of pressure due to static head (height), and (3) friction loss of pipe and fittings.

### EXAMPLE

Assume a three-story building constructed 10′-0″ floor to floor, each floor having two apartments with four fixtures in each apartment (tub, lavatory, water closet, sink), as shown in Fig. 18-9. Water pressure in the street main is 65 psi. The supply main is 2 ft below the first floor.

Fixture units per floor:

$$2 \text{ tubs} \times 2 \text{ units} = 4$$

$$2 \text{ lavatories} \times 2 \text{ units} = 4$$

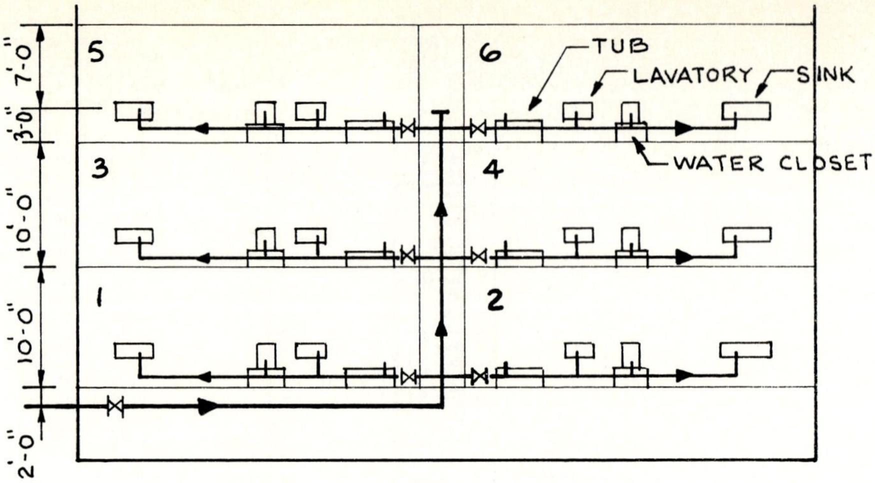


**Figure 18-9** Example of a six unit apartment.

2 water closets × 4 units = 8

2 sinks × 2 units = 4

20 per floor

Height from supply to top fixture:

| | |
|---|---|
| Supply to first floor | 2 |
| First to third floor | 20 |
| Third to sink height | 3 |
| | 25 ft |

*Static head loss:* 0.434 psi × 25 = 10.75 lb

Friction loss in pipe and fittings could be figured individually with each length of pipe and each fitting, but as this is a laborious process, a number of shortcuts may be used. Where the total length of supply pipe is easy to compute, add 50% for the equivalent length of fixtures and then adjust the friction loss to the largest pipe size. In another method, add the loss of static head to fixture pressure at the topmost fixture and subtract from the main supply pressure.

| | |
|---|---|
| Street water pressure | 65.00 psi |
| Static loss 10.75 + 15.0 | 25.75 psi |
| Pressure to overcome friction loss | 39.25 psi |

Gallons per minute of water flow is the total of all fixture units.

20 units per floor × 3 = 60 gal/min

If lawn sprinklers, air-conditioners, and other units that require water are connected, their requirement must also be included in the total gal/min value. The minimum pressure at the topmost fixture should be not less than 15 psi for flushometers or 8 psi for tank water closets. Water meters are not always required by municipalities, but a 1-in. meter will have a friction loss of about 20 psi when the flow is 60 gal/min.

To determine friction loss in the pipe system, estimate the total length of pipe from the meter to the farthest fixture, and then multiply by the loss per 100 ft, as shown in Fig. 18-10. In addition to the loss due to straight runs of pipe, a substantial amount of friction loss is added by the various fittings, as shown in Table 18-3. Determination of the actual number and type of fittings is variable, depending on the manner in which the piping is installed, so often only the straight runs are considered, *plus* an additional amount of approximately 25% for fittings. When actual layout of all the fittings and valves is possible, they should of course be included rather than taking a percentage.

**EXAMPLE**

| | |
|---|---|
| Assume 60 ft from street to building | 60 lin. ft |
| Assume 50 ft horizontally in building | 50 lin. ft |
| Floor to floor: 10 + 10 + 5 | 25 lin. ft |
| | 135 lin. ft |

$$\frac{\text{pressure to overcome friction}}{\text{total length of pipe}} \times 100$$

$$= \text{average permissible friction loss per 100 ft}$$

$$\frac{39.25}{135} \times 100 = 29.07 \text{ psi per 100 lin. ft}$$

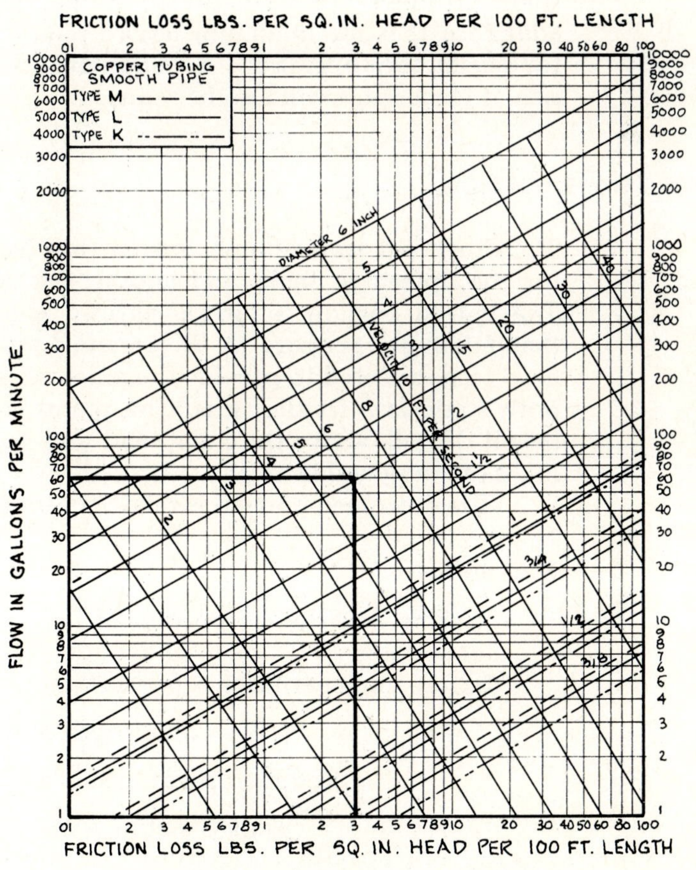


UNIFORM PLUMBING CODE

Figure 18-10

**TABLE 18-3**
Allowance in Equivalent Length of Pipe for Friction Loss in Valves and Threaded Fittings

| *Diameter of fitting (inches)* | *90° standard elbow (ft)* | *45° standard elbow (ft)* | *Standard T 90° (ft)* | *Coupling or straight run of T (ft)* | *Gate valve (ft)* | *Globe valve (ft)* | *Angle valve (ft)* |
|---|---|---|---|---|---|---|---|
| 3/8 | 1 | 0.6 | 1.5 | 0.3 | 0.2 | 8 | 4 |
| 1/2 | 2 | 1.2 | 3 | 0.6 | 0.4 | 15 | 8 |
| 3/4 | 2.5 | 1.5 | 4 | 0.8 | 0.5 | 20 | 12 |
| 1 | 3 | 1.8 | 5 | 0.9 | 0.6 | 25 | 15 |
| 1 1/4 | 4 | 2.4 | 6 | 1.2 | 0.8 | 35 | 18 |
| 1 1/2 | 5 | 3 | 7 | 1.5 | 1 | 45 | 22 |
| 2 | 7 | 4 | 10 | 2 | 1.3 | 55 | 28 |
| 2 1/2 | 8 | 5 | 12 | 2.5 | 1.6 | 65 | 34 |
| 3 | 10 | 6 | 15 | 3 | 2 | 80 | 40 |
| 4 | 14 | 8 | 21 | 4 | 2.7 | 125 | 55 |
| 5 | 17 | 10 | 25 | 5 | 3.3 | 140 | 70 |
| 6 | 20 | 12 | 30 | 6 | 4 | 165 | 80 |

Uniform Plumbing Code.

Find 29.07 (use 30.0) at the bottom of Fig. 18-10. Then read upward to the intersection of 60 gal/min, and find the diagonal line indicating a 2-in. pipe size. This will be the minimum size for pipe from the street to the building and to the fixtures on the first floor. The same process may be used to determine the requirements for the upper floors, remembering that the vertical distance is less and the number of fixtures is reduced for each additional floor. To these pipe sizes, the additional requirement for friction should be added, so that the minimum service pipe will probably require 2½ or 3 in. for satisfactory operation. As requirements are reduced for upper floors, the size of feeder pipe will also be reduced.

## 18.8 DOWNFEED WATER DESIGN

The downfeed water-supply system involves two separate calculations, one for the upfeed to tanks and another for downfeed from tanks to fixtures. The upfeed portion is figured exactly as indicated in Section 18-7, except that minimum pressure at the terminal tank may be zero and a pump will have to be selected to provide the pressure required to raise the water against both static and friction loss. The pump capacity to do this may limit the height of tank above the pump and may require a step system if the building is of considerable height.

The downfeed portion is relatively simple to calculate.

### EXAMPLE

Assume a downfeed portion of five stories, 10′-0″ floor to floor with tank water 5′-0″ deep and outlet 3′-0″ above the floor. Assume 20 fixture units per floor × 5 = 100 units.

| | |
|---|---|
| Total length of riser | 50 linear feet |
| Required psi at lowest fixture | 15 psi |
| Tank pressure = 5 × 0.43 | 2.15 psi |
| Static pressure = 50 × 0.43 | 21.50 psi |
| Lowest floor total pressure | 38.65 psi |

From Fig. 18-10 find 77 psi at the bottom, read upward to 100 gal/min and find a pipe size of 2¼ in. Remember that static pressure is additive in downfeed. In drops of many floors, a pressure-reducing valve or other controls may be necessary to prevent high-pressure buildup detrimental to fixtures at lower floors.

## 18.9 HOT-WATER SUPPLY

Hot-water heaters are of two principal types: (1) unit heaters, and (2) storage heaters. *Unit heaters* combine the heating unit with the storage tank and are primarily used in residential construction. *Storage heaters* have two separate units, a heater and a tank, and are essentially for commercial installations. Fuel is generally gas, oil, or electricity for unit heaters but in addition, commercial heaters may use hot water or steam through a heat exchanger. Direct heaters are designed to provide hot water only. Indirect heaters use a separate heat source and may be part of a heating or power system.

Unit heaters (Fig. 18-11) are generally cylindrical in shape, thoroughly insulated, with the water surrounding the heat and rated in gallons of capacity and rate of recovery. The general principle of the unit heater is that colder water at the bottom of the tank expands as it is heated, rises to the top of the heater, and forces cooler, more dense water toward the bottom. In commercial installations, the heating unit is a heat-transfer device, using heat to warm the water, which then passes into a storage tank for further distribution. In either system, a pipe loop from the water heater to the farthest fixture and return, sometimes including a pump, provides hot water at each fixture outlet without undue flow of cold water until hot water finally arrives. Most hot-water supply is controlled by thermostats connected into the hot-water portion and the fuel supply. Sizing of hot-water equipment includes an assessment of the amount of water per fixture (Fig. 18-9), hourly or daily requirement, peak loads, and capacity of heating unit.

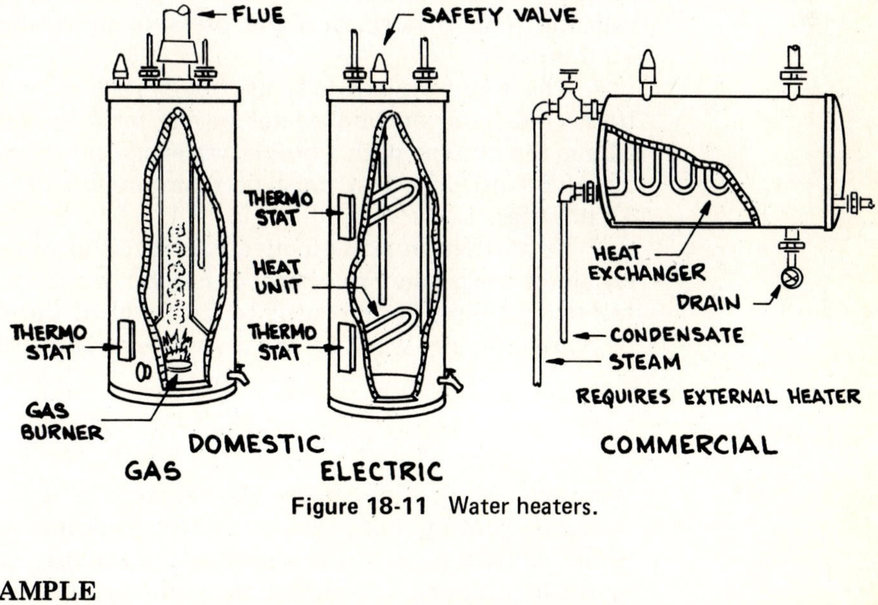


**Figure 18-11** Water heaters.

**EXAMPLE**

Assume a building with six apartments, with 2.5 persons per apartment.

Daily hot water: 6 apts. × 80 gal/apt. 480 gal/day

Maximum hourly demand: 480 × 1/7 68.57 gal/hr

Storage tank size: 480 × 1/5 96.0 gal

Heating capacity: 480 × 1/7 68.57 gal/hr

The amount of hot water available is usually about 75% of tank capacity. If the tank capacity is assumed at 100 gal, the required heating capacity may be found as follows:

$$100 \text{ gal} \times 0.75 = 75 \text{ gal available}$$

$$\text{peak load} = 6 \text{ apt.} \times 12.0 \text{ per apt.} \times 4 \text{ hr} = 288 \text{ gal}$$

$$288 - 75 = 213 \text{ gal to be heated}$$

$$\frac{213}{4 \text{ hr}} = 53.2 \text{ gal/hr minimum heater capacity}$$

## 18.10 TESTING

The Uniform Plumbing Code requires that "the piping of the plumbing, drainage, and venting systems shall be tested with water or air." It should be obvious that such testing should be done before all the rough or concealed piping is covered up with surfacing material, but unfortunately this is not always the situation. Testing is done, however, before fixtures are installed and is the responsibility of the installer.

The drainage and vent systems are tested by sections, or in their entirety, by using a water test. All openings must be tightly closed except the highest opening, and the system filled with water until it overflows. If the system is tested in sections, the head of water shall be at least 10 ft. The water shall be retained without evidence of leaks for at least 15 minutes before inspection. If the systems are tested with air, all openings except one must be sealed airtight, an air pressure of at least 5 psi shall be introduced through the one opening with a valve, and the pressure maintained for a period of at least 15 minutes.

The water system may also be tested in its entirety or by sections, and both cold-water and hot-water piping must be tested. Pipe shall be tested by filling the system with potable water at the normal working pressure or by air at 50 psi. The test must be maintained at required pressure for at least 15 minutes.

Tests that show defective construction require drainage of the piping, repair of leaks, and retesting either by water or air. Shower receptors are tested by filling the receptor to the level of the rough threshold. Tests may be waived on any work by the inspecting authorities.

## 18.11 SUMMARY

Plumbing, as considered in this chapter, includes piping systems for sewer drainage, venting, cold and hot water, and appropriate fittings and fixtures. Some additional work installed as "plumbing" may be storm drainage, gas or oil fuel piping, and irrigation work, but these are covered in other chapters. Installation of plumbing is regulated by codes or local ordinances, the most notable of which are the National Plumbing Code and the Uniform Plumbing Code, which are pratically identical.

Two of the most important things to keep in mind in regard to plumbing are that sewer, drain, and vent systems operate by gravity, while water piping is a pressure system. Friction loss due to rough interior pipe surface and multiple fittings plus static head of water tend to reduce operating pressure in water piping. Different types of pipe have different roughness characteristics, so the friction values change with type of pipe. Static head of water is the pressure developed at the base of a water column in relation to its height, 0.43 psi for a column 12 in. high.

Water systems are upfeed where the water flows up from the main supply line or downfeed where the flow is downward from elevated tanks. Friction and static head are deductive losses from supply pressure for upfeed systems and for the upfeed portion of the downfeed system. The supply to fixtures of downfeed are essentially by gravity. All piping, both drainage and water, is required to be tested either by a full head of water or comparable air pressure.

# 19

# Special Systems

## 19.1 INTRODUCTION

The word *plumbing* generally means the visible fixtures and connecting piping, but there are other systems of piping that may be installed by plumbers, steamfitters, heating installers, and a number of different craftsmen, depending on the material used and the location of installation. As this book and chapter deal primarily with support systems installed in buildings or on the immediate site and connected to the building systems, pipe work incorporated in the construction of oil wells, pipelines, special industrial installation, and highway construction is not considered here.

In this chapter several common systems and several special installations are considered. Some of these installations use material already reviewed, but for others special materials may be required. In addition to plumbing work in buildings, piping work for swimming pools and requirements for mobile-home parks are briefly covered.

## 19.2 FUEL-GAS PIPING

Gas piping is usually included as a part of plumbing work when installed to provide fuel for hot-water heaters, but it may be installed by others in connection with furnaces for heating or air-conditioning. The pipe used may be wrought iron, steel (galvanized or black), yellow brass, or PVC plastic.

Pipe must be new or used previously only for conveying gas. All valves or other controls must be designed for fuel gas. In actual practice, black-steel (iron) pipe and standard threaded fittings are normally used. Gas meters are required to be accessible and not installed in boiler rooms, electric meter rooms, or beneath stairs.

Gas piping must be installed above the earth when beneath a building, with at least 6 in. of clearance. Where gas piping is buried outside the building, the pipe must be covered with at least 12 in. of earth, and ferrous pipe must be protected from corrosion by approved coatings or wrappings. Risers must extend at least 6 in. above grade and be similarly protected. Shutoff valves are required at all meters and entrance supply locations. If plastic pipe is used, a No. 18 copper wire must be attached to the pipe and terminated as a ground wire.

To determine gas pipe sizes, the first factor to find is the load or demand from Table 19-1. Second, after the gas demand is established, the cubic feet per hour is found by dividing the demand by the average heating value (Btu) of gas in the location of the building. Next, the approximate

**TABLE 19–1**
Minimum Demand of Typical Gas Appliances

| *Appliance* | *Demand* | |
|---|---|---|
| | *Btu/hr* | *W* |
| Domestic gas range | 65,000 | 19,045 |
| Domestic recessed top burner section | 40,000 | 11,720 |
| Domestic recessed oven section | 25,000 | 7,325 |
| Storage water heater — up to 30-gal. (113.6L) tank | 30,000 | 8,790 |
| Storage water heater 40- (151.4L) to 50-gal. (189.3L) tank | 50,000 | 14,650 |
| Domestic clothes dryer | 35,000 | 10,255 |
| Fireplace log lighter (residential) | 25,000 | 7,325 |
| Fireplace log lighter (commercial) | 50,000 | 14,650 |
| Barbecue (residential) | 50,000 | 14,650 |
| Gas refrigerator | 3,000 | 879 |
| Bunsen burner | 3,000 | 879 |
| Mobile homes — each (See Appendix E) | * | * |
| Gas engines (per horsepower) | 10,000 | 2,930 |
| Steam boilers (per horsepower) | 50,000 | 14,650 |

Uniform Plumbing Code.

**EXAMPLE**

Assume a one-story house with a gas range, a 30-gal water heater, and a gas-fired clothes dryer. The distance from the meter to the house is 30 ft and in-house piping is an additional 35 ft to the fixture group.

| | |
|---|---|
| Gas range | 65,000 Btu/hr |
| Water heater | 30,000 Btu/hr |
| Clothes dryer | 35,000 Btu/hr |
| | 130,000 Btu/hr |

Natural gas contains about 1000 Btu per cubic foot. Therefore,

$$\frac{130{,}000}{1000} = 130 \text{ ft}^3/\text{hr}$$

Refering to Table 19-2, find 65 ft (use 70 ft), read down the column to 123. This might be adequate, but practice would generally dictate the next figure of 235. Read left to find a pipe size of 1 in. If the fixture outlets are in separate locations, each branch is calculated in a similar manner from a main and the main is calculated from the totals.

**TABLE 19-2**
Size of Gas Piping[a]

| Pipe size (in.) | Length (ft.) 10 | 20 | 30 | 40 | 50 | 60 | 70 | 80 | 90 | 100 | 125 |
|---|---|---|---|---|---|---|---|---|---|---|---|
| ½ | 170 | 118 | 95 | 80 | 71 | 64 | 60 | 55 | 52 | 49 | 44 |
| ¾ | 360 | 245 | 198 | 169 | 150 | 135 | 123 | 115 | 108 | 102 | 92 |
| 1 | 670 | 430 | 370 | 318 | 282 | 255 | 235 | 220 | 205 | 192 | 172 |
| 1¼ | 1,320 | 930 | 740 | 640 | 565 | 510 | 470 | 440 | 410 | 390 | 345 |
| 1½ | 1,990 | 1,370 | 1,100 | 950 | 830 | 760 | 700 | 650 | 610 | 570 | 510 |
| 2 | 3,880 | 2,680 | 2,150 | 1,840 | 1,610 | 1,480 | 1,350 | 1,250 | 1,180 | 1,100 | 1,000 |
| 2½ | 6,200 | 4,120 | 3,420 | 2,950 | 2,600 | 2,360 | 2,180 | 2,000 | 1,900 | 1,800 | 1,600 |
| 3 | 10,900 | 7,500 | 6,000 | 5,150 | 4,600 | 4,150 | 3,820 | 3,550 | 3,300 | 3,120 | 2,810 |
| 3½ | 16,000 | 11,000 | 8,900 | 7,600 | 6,750 | 6,200 | 5,650 | 5,250 | 4,950 | 4,650 | 4,150 |
| 4 | 22,500 | 15,500 | 12,400 | 10,600 | 9,300 | 8,500 | 7,900 | 7,300 | 6,800 | 6,400 | 5,700 |

| Pipe size (in.) | 150 | 200 | 250 | 300 | 350 | 400 | 450 | 500 | 550 | 600 |
|---|---|---|---|---|---|---|---|---|---|---|
| ½ | 40 | 34 | 30 | 27 | 25 | 23 | 22 | 21 | 20 | 19 |
| ¾ | 83 | 71 | 63 | 57 | 52 | 48 | 45 | 43 | 41 | 39 |
| 1 | 158 | 132 | 118 | 108 | 100 | 92 | 86 | 81 | 77 | 74 |
| 1¼ | 315 | 270 | 238 | 215 | 200 | 185 | 172 | 162 | 155 | 150 |
| 1½ | 460 | 400 | 350 | 320 | 295 | 275 | 255 | 240 | 230 | 220 |
| 2 | 910 | 780 | 690 | 625 | 570 | 535 | 500 | 470 | 450 | 430 |
| 2½ | 1,450 | 1,230 | 1,100 | 1,000 | 920 | 850 | 800 | 760 | 720 | 690 |
| 3 | 2,550 | 2,180 | 1,930 | 1,750 | 1,600 | 1,500 | 1,400 | 1,320 | 1,250 | 1,200 |
| 3½ | 3,800 | 3,200 | 2,860 | 2,600 | 2,400 | 2,200 | 2,100 | 2,000 | 1,900 | 1,800 |
| 4 | 5,200 | 4,400 | 3,950 | 3,600 | 3,250 | 3,050 | 2,850 | 2,700 | 2,570 | 2,450 |

Uniform Plumbing Code.
[a]Maximum delivery capacity in cubic feet of gas per hour of I.P.S. pipe carrying natural gas of 0.65 specific gravity, based on pressure drop 0.5 in. water column.

length of pipe is found, and from Table 19-2, under the proper length, the cubic feet per hour is obtained from the second step. Read to the left to find the pipe size.

## 19.3 LIQUEFIED PETROLEUM GAS

Most of the requirements noted for fuel gas also apply to liquefied petroleum gas (LPG). This type of gas is heavier than natural or manufactured fuel gas, so installations of heating units located below the tank are prohibited. In addition, all service piping and main supply must be outside the building, readily accessible, and easily recognized and individually controlled. The method of pipe sizing is the same as for fuel gas, except that different tables for propane or butane are found in plumbing codes.

## 19.4 SITE DRAINAGE

Site water drainage should be a completely separate system from the sewer drain system. This is desirable, as an excess flood of storm water could flush out a septic-tank system before bacterial action is completed or could overload a municipal system with water that does not require treatment. The storm drainage system designed to collect and transfer rainwater from roofs is an integral part of the building, and is discussed in this chapter. Disposal of surface water and ground-level collection and disposal of roof drainage are necessary and may be as simple as shaping the earth to allow surface flow or may be handled by an underground pipe system.

In almost every building situation, except where buildings abut concrete walks or other paved areas, the earth is required to be sloped away from the building for a distance of at least 5 ft. This slope may be continued to the property line, but most state laws prohibit dumping surface water from one property onto an adjacent property. This means that some form of open ditches, culverts beneath roads or walks, or underground pipe systems with adequate catch basins be installed. Proper grading of the earth's surface will usually provide unnoticeable swales or flat ditches for drainage to the street or other disposal.

The cross section of greatest efficiency of any drainage ditch is one that has the shortest wetted perimeter. Theoretically, the semicircle has the shortest perimeter and therefore the highest hydraulic efficiency. A half-hexagon is the most efficient of trapezoidal sections. If a rectangular section is used, the water depth should not be more than one-half the channel width. Loss of head is minimal in open channels owing to curves or bends. Culverts (see Fig. 19-1) may be necessary to carry surface water under roads and may be as large as 12 ft by 12 ft. Most state highways have standard designs which may be used with confidence.

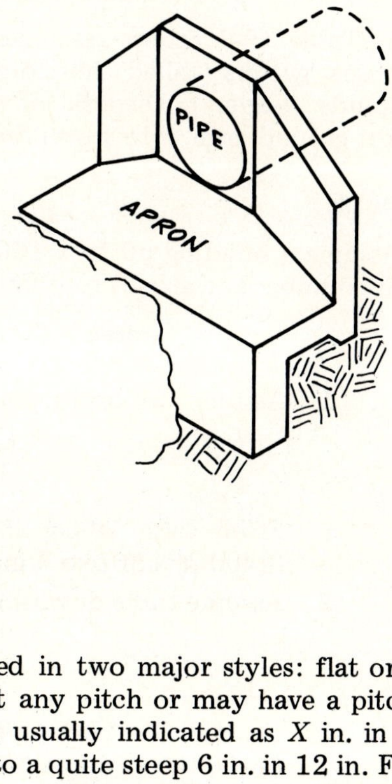


**Figure 19-1** Drainage culvert.

## 19.5 ROOF DRAINAGE

Roofs are designed in two major styles: flat or pitched. Flat roofs may be dead flat without any pitch or may have a pitch to perhaps 1 in. in 12 in. Pitched roofs are usually indicated as $X$ in. in 12 in. and vary from about $2\frac{1}{2}$ in. in 12 in. to a quite steep 6 in. in 12 in. Flat roofs often are depressed in midspan by the weight of the beam if only slightly overstressed. This creates a rain puddle, which in turns creates more weight and more depression, which holds more water and more weight and creates more depression. Roof drainage then becomes a problem. If the roof is pitched, the rainwater may run into gutters, where downspouts will carry it to the ground, or into storm drains (see Fig. 19-2).

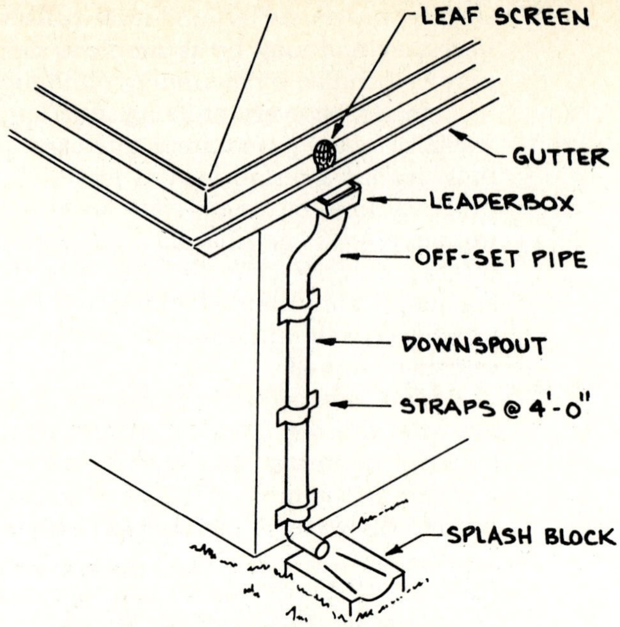


**Figure 19-2** Roof gutter and downspout.

As Table 19-3 shows, reasonably standard rules exist for the sizing of roof drains, gutters (called "horizontal rainwater piping" in Table 19-3), and downspouts (leaders). Disposal of rainwater may be onto concrete splash blocks at ground level, into storm drain systems, or into a cistern for storage.

**EXAMPLE**

Assume a building 50 ft × 100 ft. Rainfall is 2 in. per hour. Determine the number and size of roof drains.

$$\text{area} = 50 \times 100 = 5000 \text{ sq. ft.}$$

1. Assume two downspouts:

$$\frac{5000}{2} = 2500 \text{ ft}^2$$

From Table 19-3a, 2500 $ft^2$ will require one 3-in. downspout. (4400 $ft^2$) or two 3 in. for the building.

2. Assume three downspouts:

$$\frac{5000}{3} = 1666 \text{ ft}^2$$

From Table 19-3a, 1666 $ft^2$ will require one 3-in. downspout. (4400 $ft^2$) or three 3 in. for the building.

3. Assume four downspouts:

$$\frac{5000}{4} = 1250 \text{ ft}^2$$

From Table 19-3a, 1250 $ft^2$ will require one 2-in. downspout. (1440 $ft^2$) or four 2 in. for the building.

Therefore, the best solution is four downspouts.

**TABLE 19-3**
Sizing of Roof Drains and Rainwater Piping for Horizontal Projected Roof Areas ($ft^2$)

a. Sizing of roof drains and rainwater piping for varying rainfall

| *Rain fall in inches* | *Size of drain or leader (in.)* 2 | 3 | 4 | 5 | 6 | 8 |
|---|---|---|---|---|---|---|
| 1 | 2,800 | 8,800 | 18,400 | 34,600 | 54,000 | 116,000 |
| 2 | 1,440 | 4,400 | 9,200 | 17,300 | 27,000 | 58,000 |
| 3 | 960 | 2,930 | 6,130 | 11,530 | 17,995 | 38,660 |
| 4 | 720 | 2,200 | 4,600 | 8,650 | 13,500 | 29,000 |
| 5 | 575 | 1,760 | 3,680 | 6,920 | 10,800 | 23,200 |
| 6 | 480 | 1,470 | 3,070 | 5,765 | 9,000 | 19,315 |
| 7 | 410 | 1,260 | 2,630 | 4,945 | 7,715 | 16,570 |
| 8 | 360 | 1,100 | 2,300 | 4,325 | 6,750 | 14,500 |
| 9 | 320 | 980 | 2,045 | 3,845 | 6,000 | 12,890 |
| 10 | 290 | 880 | 1,840 | 3,460 | 5,400 | 11,600 |
| 11 | 260 | 800 | 1,675 | 3,145 | 4,910 | 10,545 |
| 12 | 240 | 730 | 1,530 | 2,880 | 4,500 | 9,660 |

b. Size of horizontal rainwater piping

| *Size of pipe in inches 1/8-in. slope* | *Maximum rainfall in in./hr.* 2 | 3 | 4 | 5 | 6 |
|---|---|---|---|---|---|
| 3 | 1,644 | 1,096 | 822 | 657 | 548 |
| 4 | 3,760 | 2,506 | 1,880 | 1,504 | 1,253 |
| 5 | 6,680 | 4,453 | 3,340 | 2,672 | 2,227 |
| 6 | 10,700 | 7,133 | 5,350 | 4,280 | 3,566 |
| 8 | 23,000 | 15,330 | 11,500 | 9,200 | 7,600 |
| 10 | 41,400 | 27,600 | 20,700 | 16,580 | 13,800 |
| 11 | 66,600 | 44,420 | 33,300 | 26,650 | 22,200 |
| 15 | 109,000 | 72,800 | 59,500 | 47,600 | 39,650 |

| *Size of pipe in inches 1/4-in. slope* | *Maximum rainfall in in./hr.* 2 | 3 | 4 | 5 | 6 |
|---|---|---|---|---|---|
| 3 | 2,320 | 1,546 | 1,160 | 928 | 773 |
| 4 | 5,300 | 3,533 | 2,650 | 2,120 | 1,766 |
| 5 | 9,440 | 6,293 | 4,720 | 3,776 | 3,146 |
| 6 | 15,100 | 10,066 | 7,550 | 6,040 | 5,033 |
| 8 | 32,600 | 21,733 | 16,300 | 13,040 | 10,866 |
| 10 | 58,400 | 38,950 | 29,200 | 23,350 | 19,450 |
| 12 | 94,000 | 62,600 | 47,000 | 37,600 | 31,350 |
| 15 | 168,000 | 112,000 | 84,000 | 67,250 | 56,000 |

Uniform Plumbing Code.

**EXAMPLE**

Assume the same size of building with a pitched roof. The ridge is the long dimension of the building. Rainfall 2 in./hr.

$$\text{area} = 50 \times 100 = \frac{5000}{2} = 2500 \text{ ft}^2 \qquad \text{each side}$$

1. Assume two 2-in. downspouts on each side, each with a capacity of 1440 $ft^2$ each.

2. Each gutter drains 2500 ft$^2$ with two downspouts and a slope of ¼ in. From Table 19-3b, find the roof area in 2 in. rainfall column nearest to 2500 ft$^2$.
3. Table 19-3b shows 2320 ft$^2$ in 2 in., rainfall column for 3″ gutter which may be used but common sense would indicate 4 in. to allow for heavier rainfall.

If the four downspouts are connected into an underground drain system the two downspouts on each side will require a horizontal connecting pipe (Table 19-3b) of 4 in. at a ¼-in. slope, and the pipe to the storm drain system must accommodate water from 5000 ft$^2$ using a 6-in. pipe.

## 19.6 SPECIAL PIPING

Various types of piping, and their corresponding uses, are covered in a number of locations in this book and in other works. For the most part, they are the more common types: galvanized and black steel, cast-iron bell and spigot or hubless, plastic pipe, nonferrous copper, brass, and lead, and their fittings. Most of these are regularly incorporated in plumbing, and air-conditioning work, but occasionally other types of pipe are used in building construction or as an adjunct to supplying utilities to the building.

Water piping installed by municipalities may be of several types: flanged heavy-duty cast iron, asbestos cement, steel-lined concrete, or combinations. Sewer lines are customarily of concrete precast pipe, clay tile, or cast iron. Acid and chemical lines may be of glass, or tin-lined pipe may be used for distilled water or similar use where more complete protection against contamination or leakage is necessary. In many cases, sewer or waterlines that have been deteriorated by long use or other damage are relined by installing a heavy concrete slurry on the interior surfaces by machines operated from one manhole to another. One of the major reasons for the use of nonmetallic pipe is the opportunity to eliminate or reduce corrosion by electrolysis between metals or by earth-contained chemicals. This protection may also be provided by specially installed anticorrosion systems incorporating grounding or waste-metal devices.

Cast-iron pipe for water service ranges from 3 to 48 in. in diameter and is designed for pressures from 50 to 350 psi. Cast-iron pipe for gas service is available in diameters of 4 to 48 in. and in pressure ranges of 10 to 250 psi. Required wall thickness varies with the installation method indicated by the American Water Works Association (AWWA) Standard H1-67 as "A: pipe laid on flat-bottom trenches, backfill not tamped; B: pipe laid on flat-bottom trenches, backfill tamped; or F: pipe bedded in gravel or sand, backfill tamped." This AWWA standard also covers pipe installed on piers or beneath traffic conditions. A number of additional AWWA standards include cement-mortar-lined pipe, rubber-gasket joints for pressure pipe, ductile iron pipe, and their proper fittings. Cast-iron pipe has the advantage of long life, often as much as 100 years, and some salvage value at the end of its working days. In some sections of the United States, water supply also includes minute vegetable matter (tubercles), which attach to the interior of metal pipe, creating more friction and reducing effective diameter of the bore. Where this may occur, cement-mortar-lined pipe is used to reduce or prevent the growth.

Asbestos-cement pipe is also extensively used for water distribution and is available in 3- to 18-in. diameter. Joints are made up by using a sleeve coupling. Interior surfaces are quite smooth, and the problem of metal corrosion does not exist. This same pipe is used for on-site drainage work. Clay tile, either in 12-in. square end or 4-ft or 5-ft bell-and-spigot type, in diameters of 3 to 36 in., is used for drainage work. When used in septic systems or other dispersal areas, the 12-in. length is laid with a slight space between units to allow liquid to percolate into the adjacent soil. When clay tile is used as a drain pipe, the longer lengths of bell-and-spigot pipe are used, and joints are made up with cement mortar. Pipe made of asphalt-saturated fiber (Orangeburg) is also used for drainage and may be perforated to collect underground liquid or distribute it, or may be unperforated for use as a conducting pipe.

Glass and tin-lined pipe are quite special and normally used only in some chemical or industrial situations. Small-diameter glass pipe may be bent or shaped by heating, but sizes larger than about 1 in. require proper glass fittings, usually connected with solvent or clamps. Tin-lined pipe is often used for distribution of distilled products (water, liquor, wine) and may be a standard steel or copper pipe with the interior lined with pure tin deposited as a heavy coating. Complete coverage of any base metal at joints and valves is a major problem, so much of this type of installation is now being done with stainless steel pipe. Wood as a "pipe" material has no place in modern building except perhaps in the construction of irrigation flumes, where slight leakage may be expected and repair with simple portable tools is possible.

## 19.7 SUMMARY

The dividing line between "plumbing" and "piping" when used in construction, particularly high-rise apartment and industrial construction, is often very thin and what the designer feels to be the plumber's work may not always be so. Installation of on-site work from municipal systems (water, sewer, gas, steam) may require trades other than plumbing and nearly always when installed in public property these systems have little or nothing to do with the "plumber" but are installed by the forces of the municipality or the utility involved. Even in development of large sites such as residential tracts, industrial areas, schools, and similar construction, most of the piping work outside the buildings may be done by specialized trades.

Even within the building there may be a conflict of interest regarding "who does what." Rainwater on small construction is usually collected with sheet-metal gutters and downspouts, a sheet-metal contractor's responsibility. However, in most commercial and high-rise building, this work is installed in steel pipe with cast-iron roof drains and thus may be plumbing work. In residential construction, practically all sewer, water, and gas piping is installed by the plumber, but in other construction where central heating or boiler plants are included, a great deal of this work may be done by related trades. Unfortunately, there is no uniform method to determine the extent or overlap of work as viewed by various trade unions, so the construction designer can only use his best judgment in designing a workable system.

# 20

# Fire Protection

## 20.1 INTRODUCTION

Fire in a building, any building, can be a disaster! In addition to the possible loss of lives, the building may be totally destroyed or damaged to the extent of being of little future value, the contents of the building may be destroyed by fire or suffer irreparable water or smoke damage, and the neighborhood or adjacent buildings may also be damaged or deteriorated. Building protection requirements have been established by the various building codes, the National Fire Protection Association, and local fire departments. In the past few years, a number of fires in major high-rise buildings have brought the matter of fire protection and resulting smoke and water damage to the notice of the general public, as well as to others more closely concerned with such catastrophes.

A normal residence does not, unfortunately, have the same opportunity for better fire protection as do most commercial, industrial, or retail buildings. The current popularity of smoke alarms provides a greater possibility for the occupants of a home to escape death by fire or smoke inhalation but in most cases does little to save the structure. Sprinkler systems or other fire-quenching apparatus is rarely installed in a residence or apartment building. Various requirements for stairwell protection, automatic alarms, fire-hose cabinets, sprinkler systems, and siamese connections for fire departments have favored the commercial buildings, but there are still disastrous fires occurring in old as well as new construction.

## 20.2 CODE REQUIREMENTS

Fire-extinguishing systems required by codes vary a bit from one code to another but are generally reasonably similar. The Uniform Building Code (UBC) requires "standard fire-extinguishing systems" for every basement or cellar "when floor area exceeds 1500 $ft^2$ and there is not provided at least 20 $ft^2$ of opening entirely above adjoining ground level . . . , in the tie and fly galleries and in all places behind the proscenium walls of stages, over enclosed platforms in excess of 500 $ft^2$ in area." In hospital operating rooms, delivery rooms, x-ray rooms, and patient rooms 600 $ft^2$ or smaller, the "automatic fire-extinguishing system" may be omitted "when each room is provided with detectors of products of combustion other than heat. . . . For Group H, Div. 1 and 2 occupancies (storage and handling of highly flammable or explosive materials . . .) protection [is required] for areas of 1500 $ft^2$ or each 3000 $ft^2$ for Div. 3 (woodworking plants)."

Zoning ordinances, of course, have a great influence on code requirements for fire-protection equipment. The type of construction—wood masonry, concrete, or steel—is mandatory, so the type and amount of fire protection is predictable. Requirements for minimum pipe size, location of outlets, length and size of fire hose, and a number of other items are also spelled out in codes. Three major types of fire protection systems are in general use: (1) dry standpipes, (2) wet standpipes, and (3) combination standpipes. These are permanent systems and do not include portable extinguishers of various types which may be required where other systems are not required.

## 20.3 DRY STANDPIPES

A dry standpipe system is installed for the exclusive use of the fire department, does not have a constant water supply, and has inlet and outlet fittings to match those used by the fire department. The Uniform Building Code requires that "all buildings, four or more stories in height, shall be equipped with one or more dry standpipes." Dry standpipes are installed with galvanized-steel or wrought-iron pipe and fittings and must be able to withstand a water pressure of 300 psi. Minimum standpipe size is 4 in. for buildings 75 ft or less in height and 6 in. above 75 ft in height. A drainage pitch of not less than ¼ in. in 10 lineal feet must be provided in all horizontal lines.

The *dry standpipe* remains dry until the fire department connects the system either through a department pumper or directly to a hydrant. This is accomplished through a fitting termed a *siamese connection* (Fig. 20-1), which is installed at the lower end of the standpipe. Standpipes 4 in. in diameter require a two-way connection and those 6 in. or larger require a four-way connection. Inlet connections must be located in plain sight on the street side of the building and adequately marked. Outlets at floors must be 2½ in. and threaded for fire-department hose connections, which are not the same as standard pipe threads. Each outlet is equipped with a valve and pipe

**Figure 20-1** Siamese connection.

cap. Connections to the dry standpipe are usually located at or near stairwells on each floor. Permanent hose cabinets are *not* provided for dry standpipes.

## 20.4 WET STANDPIPES

*Wet standpipes* contain water at all times, and are installed from the basement to the uppermost story of a building. Minimum pipe size is usually $2\frac{1}{2}$ in. with $1\frac{1}{2}$-in. outlets with valves at each floor. Outlets must be located so that all portions of a room or other space may be reached with 100 ft of hose not less than $1\frac{1}{2}$ in. in diameter. The hose must be capable of resisting 500 psi and be equipped with a hose reel or rack and proper nozzle. The wet standpipe shall be capable of providing not less than 35 gal/min of water at not less than 25 psi.

In taller buildings, the wet standpipe usually has a wet storage tank, or tanks, to keep the system "wet" and maintain the quantity and pressure required. In this system, a reserve tank connected to the main water supply is also connected to pumps that raise the water to the top tank. The top tank thus provides downfeed arrangement to hose outlets. The arrangement shown in Fig. 20-2 is typical of a normal downfeed system, except that connections to the toilet room or other fixtures are not shown on every floor. The top reserve tank is sized to provide adequate water for normal plumbing uses, but if a fire occurs, the reserve may be drained into the wet standpipe and hose system. At the street level, a siamese connection is provided for the fire department. In tall buildings, the system may be zoned with several tanks or

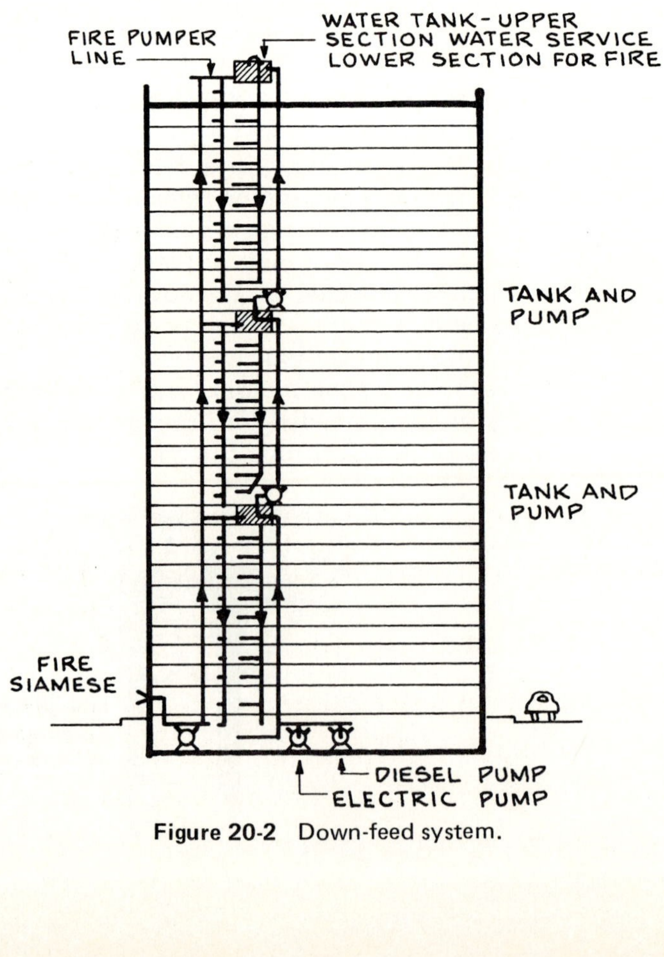


**Figure 20-2** Down-feed system.

the down-pipe may have control valves or orifice devices to control the pressure buildup.

Where buildings are not tall enough to make the downfeed system operate in a satisfactory manner to provide volume and/or pressure required, an upfeed system may be used (Fig. 20-3). In this system, a reserve tank, underground cistern, or deep wells are connected into the system through a series of pumps. In case of fire, the pumps automatically start and provide reserve or backup water to the wet standpipe. Two disadvantages should be considered if this system is to be used. Of primary concern should be the adequate water source to be used as a supplement to any municipal water supply. In case of a major fire one of the first things that firefighters shut down, if possible, is the electrical system. If this occurs, the pumps will not operate and the water cannot be pumped.

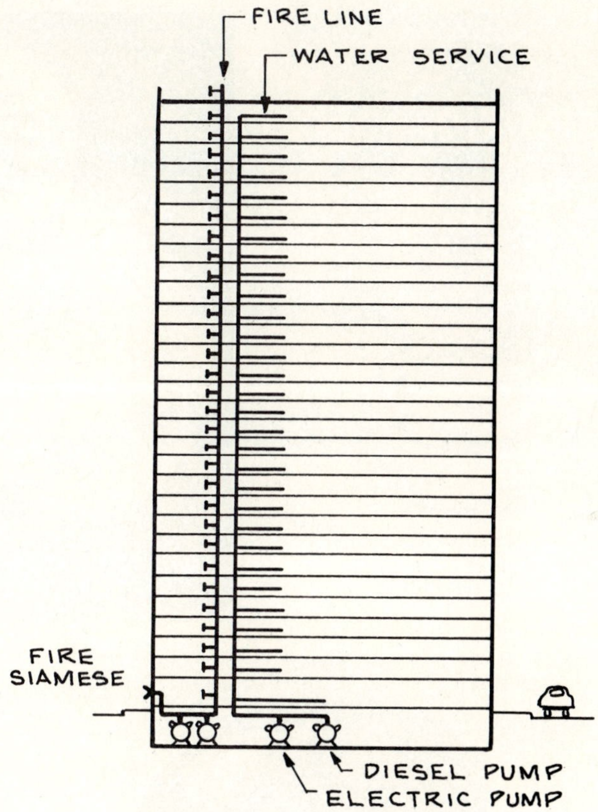

**Figure 20-3** Up-feed system.

## 20.5 COMBINATION SYSTEMS

Combination systems include a number of dry and wet standpipe arrangements, usually installed in stairways and smoke-proof enclosures that extend from ground level to the roof of buildings exceeding 150 ft in height. Connections for fire departments are the standard siamese connections and are the four-way variety, installed at street front and at required elevations. Interior connections at each floor are $2\frac{1}{2}$ in. and include a roof outlet. In general, all the requirements for either wet standpipes or dry standpipes apply and the systems are usually cross-connected through a valve arrangement so that the fire department may have adequate outlets to fight a fire properly at various floor levels.

## 20.6 SPRINKLER SYSTEMS

Automatic sprinkler systems are one of the most common fire-extinguishing methods. This equipment is usually installed in a grid pattern in the ceiling of a room or space and is connected to the wet-standpipe system. Pipe systems may be concealed above the ceiling structure with pipe extension to outlets located beneath the ceiling level, or may be exposed below the ceiling and supported with appropriate hangers. Sprinkler heads are shown in Fig. 20-4 and consist of a fusible link that will melt at temperatures from about 130 to 160°F, thus releasing water from the wet system onto the fire.

Spacing of sprinkler outlets is governed by several factors: (1) fire rating of the building, (2) type of occupancy, (3) type of construction, and (4) total area. Most of these are considered by insurance companies and under-

**Figure 20-4** Types of sprinkler heads. (Courtesy of "Automatic" Sprinkler Corp. of America)

writers who may actually establish requirements together with the National Fire Protection Association. Coverage of a sprinkler outlet is from about 200 $ft^2$ for light hazard to perhaps 100 $ft^2$ for high hazard. Supply pipes are spaced at 10 to 15 ft apart with nozzle outlets spaced at 8 to 12 ft apart on each supply line. Outlets along parallel pipes are usually staggered so that maximum coverage is achieved by overlapping of circles of distribution. A good rule of thumb is to provide one nozzle outlet for each 100 to 125 $ft^2$ of floor area. Additional piping or outlets may be needed when deep beams obstruct the plane of the ceiling.

Sprinkler systems should always have a fire department connection in addition to the connection to the wet standpipe. Gravity tanks designed to supply sprinkler heads should be capable of operating at least 25% of the sprinkler heads for at least 30 minutes. In connection with an automatic sprinkler system, there should be an alarm installed outside the building and activated by the water system. Such an alarm will alert building inhabitants as well as the general public that a fire, or at least excessive heat, has started the sprinkler system. The principal objection to sprinkler systems is the lack of quick shutoff. Once a sprinkler head is opened, either by heat or accidentally, a considerable amount of water is introduced into the immediate area, and water damage may be greater than fire damage. Controls are not possible for each sprinkler unit, so even if a sprinkler section is shut off, most of the water in the pipe will still be drained.

## 20.7 SPECIAL EQUIPMENT

Water has been, and still is, the major fire-extinguishing material used. There are several inherent disadvantages to the use of water, however. As a liquid it soaks into fabrics, plaster or gypsum drywall, and furniture. It is difficult to remove as it has no real form and tends to flatten into a sheet or film over any horizontal surface. Even when removed, it leaves a stain and is particularly objectionable when it functions as a carrier of soot or ashes. Unless fire-extinguishing systems are protected, freezing weather may make them inoperative. To eliminate or minimize these objections, several other fire-extinguishing systems using chemicals have been devised, often for special installations.

Where fires occur in plants that handle or manufacture flammable materials such as paint, oil products, or some plastics, any water used may simply spread the fire as these materials will float on the water and will not readily mix or be smothered by it. Carbon dioxide ($CO_2$) puts out a fire by diluting the oxygen in the air, thus depriving the fire of oxygen below the level required for combustion. This gas is generally stored in cylinders under about 300 psi of pressure and may be piped to a convenient location. A temperature-regulated release valve or manual control will allow the gas to escape, and as it is an inert gas, it will smother the fire. No residue is left, but spaces should be thoroughly ventilated after using carbon dioxide.

Foam, created by a special detergent and water mist composed of millions of bubbles, has been successfully used in fighting fires in large areas. A ratio of about 1000 gal of bubble foam produced from 1 gallon of water makes the system interesting and economical. The basic operation includes a fine-mesh net which is coated automatically with detergent while water mist is blown against the net to create the foam. When a fire occurs, the equipment is started and the moisture in the bubbles turns to steam by the heat of the fire. As the steam rises more foam is drawn toward the fire, and, as bubbles burst because of inside steam pressure, the water contained tends to cool the fire and reduce the oxygen in the air to starve the fire. Advocates of the system claim that the foam is 99% air, so personnel are able to wade through the foam if necessary without endangering their lives. Damage by a high-pressure stream of water is practically eliminated, but the foam and detergent-laden residue must be carefully cleaned out, as it penetrates every crack and crevice.

Nitrogen and several other nonflammable gases are used under pressure to extinguish fires in tanks or other closed containers, and are especially useful for fighting oil fires. The gas is injected into the tank above the liquid on fire and replaces the oxygen above the liquid, eventually smothering the fire or forcing the oxygen-laden air out of the tank.

## 20.8 PORTABLE EQUIPMENT

Probably the best-known piece of fire-extinguishing equipment is the portable unit used in almost every conceivable situation. These units are available in sizes from 2 gal to 1 quart using foam, gas, powder, and various chemicals. Many extinguishers are combined with hose cabinets of a wet-standpipe system, as shown in Fig. 20-5. Portable units should be light in weight, require no extensive maintenance, and should be easily operated by unskilled personnel. Extinghishers should also be universal in use: they should be suitable for fighting any type of fire, but this is usually not possible.

Extinguishers using water as a major ingredient are most common and are of three general types: (1) soda-acid type, (2) gas-cartridge type, or (3) pressure type. *Soda-acid* units (usually 2-gallon size) are one of the oldest and consist of a solution of bicarbonate of soda and water contained in the metal container. A small container of sulfuric acid is retained upright at the top of the unit when in normal position. If the unit is turned bottom side up, the acid and soda water solution mix, pressure and foam are generated, and the foam is automatically ejected through the short hose and nozzle. The *gas-cartridge* type is a container (usually 1½ to 2 gal) filled with water. A gas cartridge, usually carbon dioxide ($CO_2$), is enclosed and when the cartridge is pierced or broken, a pressure of about 125 psi is produced to force

**Figure 20-5** Portable fire fighting equipment.

the water through the attached hose. The third general type, the *stored pressure* type, is simply that. The container holds water permanently pressurized at about 80 to 100 psi by use of air or nitrogen introduced through a valve in the container. A pressure gauge is an integral part of the extinguisher, so a visual check of the condition is always possible.

*Gas extinguishers* usually use carbon dioxide or nitrogen under pressure. Either gas is a nonconductor of electricity, so this type may be used on all types of fires. The gas is contained at rather high pressure (up to 700 psi) and is released through a trigger-and-funnel arrangement. The high pressure requires a heavy container, but this disadvantage may be offset by the fact that no residue is generated by the gas. *Dry chemical* (powder) *extinguishers* are among the most efficient units manufactured today, but have some inherent disadvantages. Powder extinguishers may use $CO_2$ gas to propel the powder onto the fire, or the entire container may be under pressure. Unfortunately, the time of operation may be in the order of 7 to 30 seconds, so some skill is required to direct the powder onto the fire, and ambient temperatures may cause caking or compaction within the container.

*Vaporizing-liquid extinguishers* use chemicals that travel, under pressure, from the container to the fire as droplets, which are then vaporized and retard the combustion. The two most common chemicals used are bromochlorodifluoromethane ($CBrClF_2$) and bromotrifluoromethane ($CBrF_3$). These vaporized chemicals are toxic and so may be dangerous, but they do not leave a residue after use, and the containers are light in weight.

## 20.9 MECHANICAL FIRE PROTECTION

The earlier portion of this chapter has dealt with the suppression and extinguishing of fire in a building, but attention should also be given to prevention of the spread of a fire. Most fires are preventable or may be contained in a small area if a few simple devices are installed at openings, ducts, plenum chambers, and similar areas that may be subjected to fire. Occasional inspection and maintenance is necessary to keep protective devices in operating condition, but this should be only a very minor factor compared with the damage that may result from a serious fire.

Most building construction, other than single-family residences, is subject to a variety of code requirements related to fire safety. These requirements usually deal with width of halls or corridors, elimination of dead ends. out swinging doors, location of fire escapes, exit signs, and similar items—primarily subjects considered in early planning. Spread of a fire may be reduced or controlled if certain "mechanical" things are done. Probably the most useful item in this field is the *fusible link.* This link is a formed piece of antimony alloy rated to melt at given temperatures. In most installations, it is used as a link in a tension line that holds a device open until the link melts because of exceeded temperature, at which time the link parts and the device closes by its own weight or by a spring arrangement.

Fire doors use a fusible link in their "hold-open" closer arm. If the temperature becomes excessive, the link parts and the door is automatically shut by the closer action. Similar arrangements are installed with sliding doors and skylight and roof vents. In heating ducts, a fire damper is hinged at the top. Inside, a fusible link is arranged to hold it open, and if excessive temperature occurs, the link parts and the damper falls on the pressure side to close off the duct. Another type operates by sliding vertically. In this form, the damper is installed outside and above the duct and is held up with a fusible link. If and when the link parts, the damper falls and closes the duct.

Attics and under-floor spaces used as plenum spaces need special consideration. Most of these areas are too wide or high to be closed successfully with a damper. Any surfaces should be nonflammable and covered with gypsum board or metal. This may reduce the possibility of fire spread while it reduces turbulence of air passage. Fire dampers are appropriate at entrances or exits of connecting ducts. A vast number of pressure, temperature, and velocity devices and valves control fuel, water, gas, air, and almost every other phase of plumbing and air-conditioning and may be manually activated or automatic. Many of these units are described in other chapters and are probably familiar to the reader.

## 20.10 DETECTION DEVICES

Detectors may be divided into three different types: (1) flame detectors, (2) heat detectors, and (3) smoke detectors. Each of these must be capable of proper operation 24 hours per day 365 days per year.

Photoelectric cells are used in *flame detectors* to "see" a fire directly or when radiation is reflected onto them. *Infrared detectors* employ a masking tube or a monitor that reflects rays into the infrared photocell. A built-in circuit and scanner view a radiation flicker for a few seconds and move on without giving an alarm if the radiation is not consistent. If radiation is constant, the equipment will sound an alarm. *Ultraviolet detectors* detect the ultraviolet radiation emitted by flames but may be actuated by lightning, ultraviolet lamps, or similar devices. Integrated circuiting may be used to eliminate these problems.

*Heat detectors* respond either when the temperature is greater than a predetermined level, or when the rate of temperature increase is greater than desired. Several systems are used in heat detection. The fusible solid is a metal link that melts at a given temperature and turns on the alarm. Metals expand at a known rate, so a device using the movement of expanding metal is used in a detector where a thin strip of metal is secured between two fixed posts. As the metal strip expands with temperature increases, the strip arches upward and makes contact with another strip, completing the circuit and sounding an alarm. A device using gas expansion operates somewhat like a barometer. A closed metal chamber having a flexible diaphragm attached to an electrical contact may be flexed by a change in enclosed air pressure as temperature changes.

Smoke is a suspension of minute solid particles in air resulting from combustion. *Smoke detectors* fall into two groups: (1) optical detectors, and (2) ionization detectors. *Optical detectors* use a light beam focused on a photocell. When smoke or other particles pass through the light beam, the photocell activates an electrical circuit and an alarm. *Ionization detectors* contain two electrodes across which a potential difference is maintained, and a radioactive source. The radioactive source ionizes the air into positive and negative gas ions, and as these ions are moving between poles they vary in speed. The amount of smoke particles causes a measurable reduction of speed and adjustment is possible to a given level. Any increase in the predetermined amount of smoke then sets off an electrical circuit and an alarm.

## 20.11 SUMMARY

Fires are destructive to people, buildings, inventories, and the general environment of a community. Prevention methods and construction requirements have been able to reduce destructive fires, but at the same time more high-rise buildings, more flammable products, and greater concentrations of people create more fire problems. Insurance companies also contribute to fire prevention by their requirements for fire-resistant materials. All these factors together tend to make buildings safer from fire before a problem arises, but what happens when a fire actually starts?

Truthfully, fire science has not really solved the problem. Fires in high-rise buildings still occur, lives are still lost in fires, and methods and equipment used are inadequate. Building designers need to design better fire-alarm systems, better fire-extinguishing methods, and better escape systems for human beings trapped in buildings. The methods for fire suppression outlined in this chapter are essentially obsolete and generally ineffective but are the best that are available at this time.

# 21

# HVAC Principles

## 21.1 INTRODUCTION

The study of heating, ventilation, and air-conditioning (HVAC) and its proper application in building design involves not only the heating or cooling of air, but also insulation of the building, movement of the air, humidity, building use and location, interior and exterior temperature, cost of equipment and its installation, control of the process, and cost of operation. Most new construction in the United States now includes air-conditioning either as immediate installation or through provisions for later remodeling. Some buildings in some climatic areas do not really need air-conditioning but need good heating or good ventilation instead. Industrial plants may need neither heating or cooling but may require extensive ventilation. Retail establishments almost invariably need heating, cooling, and ventilation, and most new living spaces (homes, apartments, condominiums, and hotels/motels) provide all three in some sort of unit package.

Heat may be separated into three different types: (1) latent heat, (2) specific heat, and (3) sensible heat. *Latent heat* is heat that cannot be felt but is hidden in a substance and may be used to change the state of the substance. This change may be from liquid to solid, liquid to gas, or the reverse of any of these. *Specific heat* is the amount of heat, measured in British thermal units (Btu), required to raise the temperature of one pound of water one degree Fahrenheit. *Sensible heat* is the change in temperature of a material caused by heat. Cold is the absence of heat or the slower movement of a material's molecules. Heat is never removed but is transferred from one substance to another.

Most of the earth's compounds and elements can exist as solids, liquids, or gas. Water, normally a liquid, can absorb heat and turn into a gas (steam) or may lose heat and turn into a solid (ice). Heat always moves from higher-temperature substances to those with a lower temperature. By the same token, heated air and heated water tend to rise above air or water at lower temperatures. In addition to these properties, there are three major natural processes by which heat is transferred: (1) convection, (2) radiation, and (3) conductance. The human body may provide all of these, whereas other heat sources may be affected by only one or more. All these factors play an important part in heating, cooling, or ventilation and are discussed later in further detail.

## 21.2 ENERGY SOURCES

Energy sources for heating, ventilation, or air conditioning may be quite different, and each type has its own characteristics, advantages, and disadvantages. At one time or other, almost anything that was combustible has been used as a heat source. Wood, coal, oil, and gas have all been used for producing heat. By-products of some of these have also been used. Gas (methane) generated at sewage plants is combustible, and natural gas or gas manufactured from coal is used in great quantities for heating. Distillate, diesel, kerosene, gasoline, and other products obtained from crude oil are available in a number of grades. Heat generated by the self-incineration of garbage and other solid waste may produce steam, hot water, or hot air for municipal distribution. Steam or hot water from geysers or hot springs is a limited source of energy but one that is being developed commercially. Last but not least, the natural phenomenon of solar energy provides heating or cooling when properly designed and installed. Electricity may be considered a source of energy for heating or cooling but is in reality a method of transferring heat potential rather than an actual heat source, as some other energy source must first produce the electricity.

## 21.3 HEATING FUELS

Fuels for generating heat include a wider range of substances available as solids, liquids, or gases. Examples of solid fuels are wood, coal, and coke. Liquid fuels are generally confined to crude oil products such as fuel oil, distillate, diesel, and perhaps kerosene or gasoline in special cases. Gaseous fuels are natural gas, manufactured gas, and liquefied petroleum gas. Each of these major types requires a different method of firing; these are covered in Chapter 24. Electric heating does not depend directly on a fuel as such and in Chapter 13 is considered as a heat source.

One of the oldest heating fuels is *wood*. Any wood or other cellulose plant will deliver heat if dry and available in adequate quantity, and wood is a material that is reproducible. For many years, wood was used for fuel in direct-radiation fireplaces, in locomotives and other boilers, and for household furnaces as they came into use. Waste sawdust, log trimmings, and cabinet mill scrap are still used in many cases for fuel plant boilers. Soft woods such as pine, fir, and cottonwood are low in heat value, whereas hardwoods such as oak, hickory, or maple are relatively high in heat value. Wood is easy to ignite, burns with minimum smoke, and leaves little ash, but it requires more labor to produce in usable form and requires large storage space. A standard cord of wood measures 4 ft $\times$ 4 ft $\times$ 8 ft, or a total of 128 ft$^3$. Since this is wood cut and split irregularly, there are numerous air spaces, so only about 70 to 75% of the space is actually usable fuel. At the present time, wood as heating fuel is limited to use in fireplaces and small wood stoves and thus is not very important commercially as a major heat source.

Another older fuel is *coal* and its by-products. Coal is a solid created by deposits of vegetable matter eons ago which were then exposed to pressure, moisture, heat, and other natural changes. Several stages of "production" of coal determine its value. *Peat* is produced at an early stage under proper conditions of heat and moisture and a minimum of pressure, and is used as fuel in many households throughout the world. This fuel is found in "peat bogs" or surface areas, has a low fuel value, and is not a commercial fuel. With increased pressure, heat, moisture, and age, the peat may begin to form lignite, then bituminous coal, and finally anthracite. *Lignite* (brown coal) is considered a low-grade fuel compared to other coals. In structure, lignite is between peat and bituminous coal, ignites slowly, burns with little smoke, contains considerable moisture, and will not coke. *Bituminous* (soft coal) is the most plentiful of all coals and the least expensive. It ignites slowly and burns with a medium-length flame, producing little smoke. Semibituminous coal, midway between lignite and regular bituminous coal, is often termed "smokeless coal." A clean hard coal that burns with a uniform low flame is *anthracite*. It is more difficult to ignite but burns with little or no luminous flame and no smoke if fired properly.

*Coke* is the residue remaining after distillation of certain bituminous coals. Coke may also be obtained from petroleum residue, pitch, and similar materials, and burns rapidly with even heat and little required draft. Ignition is slower than for bituminous coal, but burning produces little smoke. Solid-fuel *briquettes* made from coal dust, coal fines, and a pitch or coal tar as a binder are available in a number of sizes and calorific values. A rapidly vanishing coal by-product is *coal oil*, derived by distillation from coal tar residue and often termed *paraffin oil*. Today this product has generally been replaced by kerosene, a distillation product of crude oil.

Various mixtures of hydrocarbons obtained from crude petroleum are termed *fuel oils*, depending on their viscosity, flash point, pour point, ash content, carbon residue, and water or sediment content. These fuel oils have standard established grade designations from No. 1 (low viscosity, easily vaporized) through No. 6 (heavy, require preheating, "bunker" type) and with PS 300 and PS 400 replacing No. 5 and No. 6 on the west coast of the United States. No. 2 (general purpose) is used for most residential heating and requires no preheat treatment.

Probably the most easily used and generally available fuel is gas in its several forms. *Natural gas* is a generic term for gas found in or near deposits of crude petroleum but is not connected with the production of oil. Natural gas is the richest in methane and is classified in one of the three following categories: (1) associated gas is a free undissolved gas found in close contact with crude petroleum, (2) nonassociated gas is found in areas other than those near crude petroleum deposits, and (3) dissolved gas is extracted from crude petroleum solution. Any gas made from coal, oil, coke, or any other gas is termed *manufactured gas*. Most have low calorific value and are generally used in industry rather than residential heating. Propane, butane, "bottled gas," and similar products are *liquefied petroleum gas* (LPG) extracted from "wet" natural gas. The terms "wet" and "dry" gas refer to the amount of gasoline in natural gas and "wet" gas contains more than 0.1 gal per 1000 $ft^3$ of gas. Propane gas is heavier than air, so any leakage or spillage creates a danger of explosion. Liquefied petroleum gas is delivered and stored under pressure to force it to remain liquid, and tanks may vary from recreation-vehicle size of 5 gal to 1500 gal for homes in cold-winter climates.

Table 21-1 lists the heat values of the various types of fuel discussed.

**TABLE 21-1**
Fuel Values

| *Fuel* | *Heat value* | |
|---|---|---|
| | *Btu* | *$ft^3$* |
| Wood (cord) | | |
| Ash | 28,300 | 2,176 |
| Cottonwood | 19,400 | 1,492 |
| Maple | 30,400 | 2,338 |
| Oak (white) | 32,500 | 2,500 |
| Pine | 18,100 | 1,392 |
| Coal (lb) | | |
| Peat | 7,300 | 262,847 |
| Lignite | 12,400 | 446,481 |
| Bituminous | 13,400 | 468,085 |
| Anthracite | 14,500 | 522,094 |
| Oil (gal) | | |
| No. 1 | 137,400 | 18,370 |
| No. 2 | 139,600 | 18,664 |
| No. 4 | 145,100 | 19,399 |
| No. 5 | 148,800 | 19,894 |
| No. 6 | 152,400 | 20,375 |
| Gas ($ft^3$) | | |
| Natural | 1,000 | 1,000 |
| Propane | 2,550 | 2,550 |
| Butane | 3,200 | 3,200 |

## 21.4 FUEL RESIDUE

As any fuel is burned, products of combustion are formed, usually as remaining residue such as ashes, clinkers, slag, soot, or fly ash. Most of these have no further use and require adequate disposal; however, slag in some forms and adequate quantity and fly ash may have further commercial value. *Ash* is the noncombustible mineral residue after the fuel has been consumed. Wood and coal and their derivatives produce ashes. In most cases, ashes are fine, light, powdery material which may include rocks or other substances that had been introduced into the fire with the original fuel. As ashes continue to be exposed to additional heat, they customarily shrink, resulting in the fusion of semiliquid globules into *slag*. Slag is a major by-product of some manufacturing processes, is crushed for roofing cover, and is used in the manufacture of concrete and concrete masonry units in place of stone aggregate. Under proper temperature conditions, the semiliquid slag globules may solidify and form *clinkers*, especially in coal-fired boilers. Clinkers obstruct the proper draft flow and are therefore undesirable. The inside surfaces of chimneys may become coated with *soot*, which is carried along with the smoke, leaving the combustion chamber, or may be deposited on the inside of fire-tube boilers if not properly controlled. Soot is of no great value, although it is used in some paint and plastic manufacturing processes. *Fly ash*, which is also deposited on the inside of stacks of coal-burning boilers, is essentially minute particles of carbon and iron compounds and is classed as an air pollutant, but about 95% of it can be removed with scrubbers in the smoke path.

## 21.5 HEATING FUNDAMENTALS

Heat energy is energy moving from one point to another as a result of temperature differences. The energy of heat is measured in British thermal units (Btu), the amount of heat required to raise one pound of water one degree Fahrenheit. One Btu is also equivalent to 251.99 International Steam Table (IT) calories, 778.26 foot-pounds of mechanical energy and ⅓ watt-hour. The latter values are more readily used, as the amount of heat required to raise one pound of water one degree Fahrenheit is rather difficult to determine in most cases. *Work* is measured in the unit *foot-pounds*, the amount of energy expended to lift one pound of weight a distance of one foot.

*Convection* is the transfer of heat from one warm surface to a cooler surface. Motion of the heated substance is required in convection, so only liquids and gases have this ability. When the human body is considered, the heat of the body passes to cooler surrounding air (except when the air temperature is higher than body temperature), which in turn rises and is replaced by cooler air and the process is repeated.

*Radiation* is the transfer of heat by rays traveling in a straight line from a warmer surface to a cooler surface and does not require movement of the source or receptor. The intensity of heat radiation lessens directly as the distance from the source increases. If the human body should be at a higher temperature, it is possible that the body might radiate heat, but radiation usually occurs from fixed objects.

*Conduction* is the transfer of heat through a material to another material or surface. Insulation is a poor conductor as it transfers heat poorly, but metals are good conductors as they transfer heat readily. Again, the human body is rather a poor conductor unless a cooler body is actually contacted, as occurs when a person touches a cold surface.

*Evaporation* may be an additional process of heat transfer by which moisture becomes vapor. When moisture evaporates from a warm surface, heat is removed and the surface cools. Moisture as perspiration evaporates from the human body, removes heat, and the body is cooled.

To illustrate these heating fundamentals, assume a condition in which a teakettle or boiler is being heated by fire (Fig. 21-1). The heat of the burn-

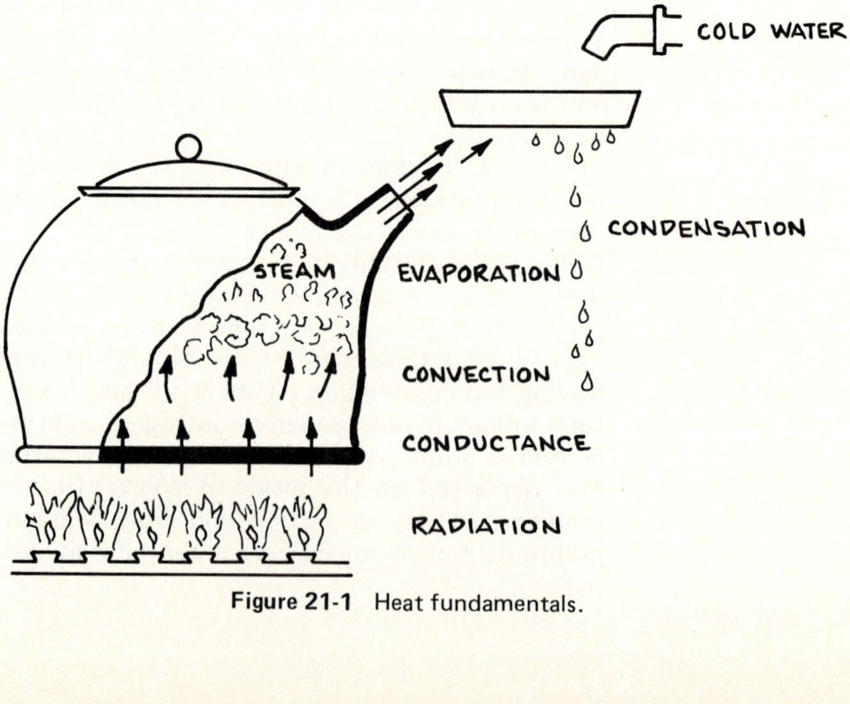


**Figure 21-1** Heat fundamentals.

ing material passes to the metal by radiation, then passes through the metal by conductance to the enclosed water, and then into the water by convection. The water circulates by variation of weight, following the principle that warmer water rises and cooler water replaces it nearer the heated metal. As more heat is applied, the water turns to steam by evaporation. If this high-temperature steam then passes over a cooler surface, it will give off heat by condensation back to liquid water. All heating and cooling is based on these fundamentals.

## 21.6 PRESSURE

"Atmospheric pressure is a unit of pressure equal to 14.69 pounds per square inch or $1.01325 \times 10^5$ newtons per square meter." This is a dictionary definition, but what actually is "pressure" and what is its place or function in heating? The pressure of the earth's atmosphere is measured by inches of mercury as in a mercurial barometer and is "standard" at 29.921 in. when balanced against the atmospheric pressure of 14.69 psi. For every increase in elevation on the earth, the atmospheric pressure will decrease approximately ½ lb for every 1000-ft increase in elevation above sea level. This means that 1 mile above sea level the pressure is 12.3 psi, but at Salton Sea, California (236 ft below sea level), it is approximately 14.92 psi.

In connection with heating and cooling, there are several types of pressure that must be considered: (1) absolute pressure, (2) atmospheric pressure, (3) gas pressure, (4) gauge pressure, and (5) static pressure. *Absolute pressure* (psia) is a pressure measured from a point of true zero in a perfect vacuum. This condition cannot exist in nature but is created in the laboratory as a point of measurement. *Atmospheric pressure* is the pressure created by a column of air 1 $\text{in.}^2$ in area at sea level. *Gas pressure* is measured either as absolute (psia) or gauge (psig) pressure in pounds per square inch (psi). *Gauge pressure* uses normal atmospheric pressure as zero and 0.00 psig is equal to 14.69 psia at sea level. *Static pressure* is the measure of pressure exerted by a cube of water 12 in. × 12 in. × 12 in. (weighing 62.4 lb) on the bottom area of 144 $\text{in.}^2$ or 0.433 psi.

*Boyle's law* for pressure states that the volume of a gas varies inversely with the pressure, provided that the temperature remains constant. If the volume is changed, the pressure also changes, so if a gas of 10 $\text{ft}^3$ in volume at a pressure of 10 psi is compressed to 5 $\text{ft}^3$, the pressure will increase to 20 psi. If the volume changes, so does the pressure, or vice versa. *Charles's law* states that at a constant pressure, the volume of a gas varies directly with the absolute temperature; and at a constant volume, the pressure varies directly with the absolute temperature. Absolute pressure (psi − 14.69) must be used in any computations involving these laws and absolute temperature (F + 460°) must be used for temperatures.

**EXAMPLE**

$$\text{Boyle's law:} \quad V_1 \times P_1 = V_2 \times P_2$$

$$\text{Given: } V_1 = 12\ \text{ft}^3,\ P_1 = 10\ \text{psig},\ V_2 = 4\ \text{ft}^3.\ \text{Find } P_2.$$

$$P_1 = 10 + 14.69 = 24.69\ \text{psia}$$

$$12 \times 24.69 = 4 \times P_2$$

$$296.28 = 4P_2$$

$$P_2 = \frac{296.28}{4} = 74.07 - 14.69$$

$$= 59.38 \text{ psi}$$

**EXAMPLE**

$$\text{Charles's law:} \quad P_1 \times T_2 = P_2 \times T_1$$

Given: $P_1 = 10$ psig, $T_1 = 30°\text{F}$, $T_2 = 60°\text{F}$. Find $P_2$.

$$(10 + 14.69) \times (60 + 460) = P_2 \times (30 + 460)$$

$$24.69 \times 520 = 490P_2$$

$$P_2 = \frac{12{,}838.8}{490} = 26.20 - 14.69$$

$$= 11.51 \text{ psig}$$

Boyle's law and Charles's law may be combined into an equation more useful in solving most gas problems:

$$\frac{P_1 \times V_1}{T_1} = \frac{P_2 \times V_2}{T_2} \qquad \text{or} \qquad PV = WRT$$

where $P$ is the pressure per square foot absolute, $V$ the volume in cubic feet, $W$ the weight of gas in pounds, $R$ the gas constant, and $T$ the absolute temperature. To use this formula, it is also necessary to know the figure for $R$, as given in Table 21-2.

**TABLE 21-2**
*R*-Values of Common Gases

| *Gas* | *R-value* |
|---|---|
| Air | 53.34 |
| Alcohol | 41.55 |
| Ether | 23.11 |
| Oxygen | 48.55 |
| Refrigerant 744 | 38.82 |
| Steam | 82.23 |

**EXAMPLE**

Find the pressure of 20 ft$^3$ of oxygen weighing 0.5 lb/ft$^3$ at a temperature of 30°F.

$$PV = WRT$$

$$P \times 20 = (20 \times 0.5) \times 48.55 \times (30 + 460)$$

$$20P = 10 \times 48.55 \times 490$$

$$P = \frac{237{,}895}{20} = 11{,}894$$

$$P = \frac{11{,}894}{144 \text{ in.}^2} = 82.60 \text{ psia} - 14.69 = 67.91 \text{ psi}$$

## 21.7 WATER

Water may exist as a liquid, heated to become steam, or frozen as ice. The normal weight of liquid water at 32°F is 62.41 lb/ft$^3$ or 8.3311 lb/U.S. gallon. The weight of water varies inversely with temperature. The higher the temperature, the less water weighs. At 212°F water generally turns to steam and weighs only 59.82 lb/ft$^3$. Interestingly, the density of water increases as it gains temperature or loses temperature. This phenomenon causes the circulation of water in heating systems, and the maximum density is obtained at 39.1°F.

Steam is a colorless, invisible gas with expansive properties created by exposure to adequate heat. Most people visualize steam as a white cloud, but this is really a partially condensed vapor. *Saturated steam* is normally the first vapor from a boiler; it is invisible and under high pressure. When steam is saturated with moisture, it is termed *wet steam*; *dry steam* has had the moisture removed. *Superheated steam* is steam that has a higher temperature than the corresponding pressure. The process of producing steam is known as *vaporization* of the water. The amount of heat necessary to create steam is the sum of sensible heat, internal latent heat, and external latent heat. These types of heat have been noted in Section 21-1. The boiling point of water depends on heat, pressure, and the type of liquid. When there is a temperature rise in a closed vessel containing water, the pressure will also rise until there is equilibrium between the two.

When steam is exposed to a colder surface, such as cooled air or a system of other cooled surfaces, the steam loses its heat and condenses back to water. This occurs when the temperature of the steam falls below that corresponding to its pressure. In this process, about 5% by volume of the steam is liberated as air, which is removed in most heating systems by proper air valves. This requirement of air release is discussed further in Chapter 24.

## 21.8 PSYCHROMETRIC PROCESS

The psychrometric process, employing an instrument called a psychrometer, measures atmospheric humidity by comparing the temperature of a wet-bulb thermometer with that of a dry-bulb thermometer. A *wet-bulb thermometer* is an ordinary thermometer whose bulb is covered with a cloth saturated with water during the time that temperature is being registered. A *dry-bulb thermometer* is an ordinary thermometer without modification. In use, the two thermometers are mounted on a sling psychrometer (Fig. 21-2),

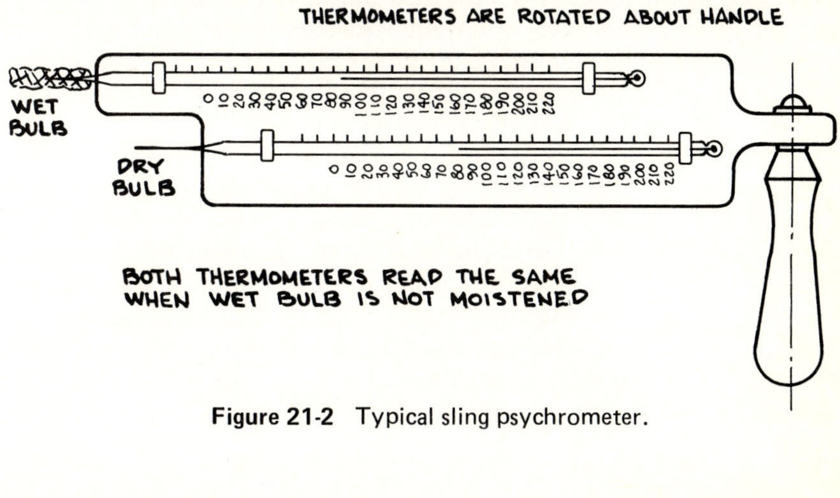


**Figure 21-2** Typical sling psychrometer.

the wet bulb is saturated, and the sling is rapidly passed through the air until the water of the wet bulb is evaporated. The dry-bulb thermometer registers the actual air temperature, while the wet-bulb thermometer registers a lower temperature by evaporation, or cooling. When both bulbs are dry, they should indicate the same temperature.

The difference in the reading of wet-bulb and dry-bulb thermometers depends on the amount of moisture in the air. If the moisture content of the air is high, the evaporation from the wet bulb will be slow and less heat removed. If evaporation is rapid, the wet-bulb thermometer indicates a dry atmosphere, which means that the air is dry and may absorb more moisture. Dry air or air with a low moisture content has a low wet-bulb temperature, while high wet-bulb temperatures indicate humid air. At 100% relative humidity, evaporation stops, the air cannot absorb more moisture, and both wet-bulb and dry-bulb thermometers show the same temperature. Three definitions are important in psychrometric studies: (1) relative humidity, (2) dew-point temperature, and (3) grains of moisture. *Relative humidity* (RH) is the actual amount of moisture in the air compared to the total or maximum moisture that the air can hold. *Dew-point temperature* is the temperature at which moisture condenses on a surface. *Grains of moisture* is the measurement unit used to determine the amount of moisture in the air (7000 grains = 1 lb).

Psychrometric charts simplify the calculations relating to study of moisture content in the air and eliminate long algebraic processes. Such charts are available from major air-conditioning equipment suppliers as well as being included in most books on air-conditioning. These charts can show the number of Btu that must be removed from proper cooling, the amount of moisture removed from an air-conditioning system, grains per pound converted to gallons per hour, and many other factors essential to design of a good air-conditioning system. Typical psychrometric charts and their use is discussed in further detail in Chapter 27.

## 21.9 ENTHALPY

All substances are made up of molecules that are constantly in motion. The higher the temperature, the more rapid is this rate of movement. This movement is called *internal kinetic energy* and is contained in solids, liquids, and gases. Because of a material's position or arrangement of parts, both liquids and gases contain *potential energy*. Solids do not have this potential energy, as they have no possibility for movement. Both kinetic and potential energy are affected by temperature and the state of the substance. The total energy (kinetic plus potential) is termed *enthalpy*.

Enthalpy is measured in Btu/lb and is calculated from reference points established for certain materials. True zero enthalpy or heat content of all materials is absolute zero (−460°F). The zero reference point for water is 32°F, and the zero reference point for refrigerants is −40°F. Between 32°F, the freezing point of water, and 705°F, the critical temperature of water, pressure has an effect on the boiling temperature. Above 705°F, water exists only as a gas, regardless of pressure. Below 32°F, water may exist as either a liquid or a solid but not as a gas. If the pressure is reduced to 0.0886 psia (latent heat of melting or fusion), the boiling temperature is lowered to 32°F and water may exist as a liquid, solid, or gas. The enthalpy of water at its boiling point of 14.7 psia (200 Btu) and the enthalpy of saturated water vapor (1170 Btu) can be calculated to show that 1170 Btu − 200 Btu = 970 Btu, the latent heat of vaporization or condensation of water at sea level.

The determination of enthalpy depends almost entirely on psychrometric wet-bulb temperature. The enthalpy lines on a wet-bulb psychrometric chart are the same as the wet-bulb temperature lines. Use of enthalpy values may also be another method for determining the sensible heat factor by dividing the sensible heat removed by the total heat removed.

## 21.10 INSULATION

Any pipe or duct containing heat at a temperature above or below the ambient temperature surrounding that pipe or duct will either lose heat or gain heat from its surroundings. Heat is usually not intended to be transferred until it reaches a grille or louver into a room, so any heat lost through the pipe or duct material is a waste. To reduce this waste as much as possible, thermal insulation is applied to the exterior of the carrier. Insulation, however efficient, will not stop the flow of heat but only retard it. Therefore, to gain the greatest efficiency from a heating system, insulation of various types is usually needed at several different locations in a system.

The degree of economy that may be achieved by insulation depends on two factors: (1) the temperature difference between the heated material and the surrounding air, and (2) the type, efficiency, and installation of the insulation. Water piping installed outside the building in areas where freezing occurs should either be buried beneath the frostline or be well insulated to prevent the water from freezing. Hot-water piping within a building may be insulated with preformed sections held in place with wire or metal straps or may be wound with batt-type material. Cold-water piping may also be insulated if water is to remain cold as in refrigerated systems or if condensation on the pipe surface is to be avoided. Warm-air ducts are usually insulated with flexible batt material wrapped around the pipe or slabs of insulation installed with wire, metal bands, adhesive, or metal clips. Boilers and steam piping are often covered with a layer of cementitous "lagging" and finished with a cloth covering. Complete enclosure of joints and control units (except operating handles) is necessary for utmost insulation efficiency.

Insulation materials suitable for thermal insulation are numerous and may be applied as batts, slabs, preformed shapes, or in plastic form. Among the best known are fiberglass, cork, asbestos, and magnesia. Magnesia and asbestos are normally applied in plastic form, a messy job but one that presents a nice appearance when properly covered with cloth. These materials are most often used on boiler installations, as the irregular shapes are more easily covered than with more solid materials. Round piping may be insulated with preformed sections or wrapped with insulation batts or blankets. Rectangular-shaped ducts may be obtained with insulation factory-installed on the *inside surface*, or blanket or slab insulation may be installed on the *outside surface*. Plastic duct tape about 2½ to 4 in. wide is used to cover joints and other openings and is self-adhesive. Some typical values for insulation are shown in Table 21-3.

**TABLE 21-3**
Duct Insulation Values

| *Material* | *Conductivity* |
|---|---|
| Asbestos paper | 0.40-0.60 |
| Cellular glass | 0.25-0.35 |
| Corkboard | 0.30 |
| Expanded paper fiber | 0.039-0.108 |
| Gypsum lightweight plaster | 1.50 |
| Hair or kapok between paper | 0.24 |
| Mineral fiberboard | 0.22-0.29 |
| Polystyrene block | 0.029-0.037 |
| Urethane plank or block | 0.022-0.026 |
| Urethane spray | 0.019-0.026 |

## 21.11 SUMMARY

The application of heating, ventilation, and air-conditioning for buildings has come a long way in the past 50 years and will undoubtedly continue to advance as new requirements and their solutions are found. Human beings are now accustomed to pleasurable living, which means adequate heat in cold weather and ventilation and cooling in warm weather. Location seems to have little to do with demand, as HVAC applies in equatorial regions, near arctic conditions, and in all areas between.

Fuel supply and its economical use is a major concern, even though various types of fuel are in adequate supply throughout the world. Of all the fuels, wood is the only one reproducible on a long-range basis, but waste in manufacturing may provide more opportunity in the future to supply some exotic combinations that have fuel value. This reuse of waste material may also help to solve such fuel residue problems as ashes, smoke, fumes, and fly ash. More and better insulation use, both for building construction and local insulation of pipes, and other HVAC units, should provide the possibility of greater economy of heating and cooling, and this area will undoubtedly progress, together with advances in "man-made" weather.

# 22

# HVAC Materials

## 22.1 INTRODUCTION

The range of raw materials, fabricated parts, and complete units of equipment for use in heating, ventilating, and air-conditioning is probably greater than anyone actually realizes. Raw materials include the various fuels used to create heat (coal, oil, and gas), water used as a heat-transfer material or as a component of steam, and air as a coolant or as a heat-transfer substance. The number of fabricated parts includes pipe of various materials and hundreds of sizes; ductwork of sheet metal, clay, plastic, or fiber for air-conditioning; and grilles and registers that may be fixed or adjustable and manufactured in many sizes and shapes. Major equipment varies from fairly simple heating units that depend primarily on radiation, to very complex combinations of heating, cooling, and ventilation units. Hundreds of types of control devices must also be considered which may be used to control temperature, humidity, air-flow, pressure, and the numerous other physical properties of the different systems.

Some materials used in "mechanical systems" for buildings are common and are used in other kinds of work. Pipe is one of these materials and is used for plumbing work, miscellaneous steel construction, and dozens of fabrications in other fields. Sheet metal, either aluminum or galvanized steel, is used in ductwork, as well as being fabricated into furnace casings, fan housings, tanks, and many other objects. Even some of the control devices are fairly well known. Thermometers, barometers, thermostats, and all sorts of valves are used every day. The more complicated or less familiar items such as compressors, evaporators, heat exchangers, and heat pumps may require some explanation to understand their function. Most of these materials are discussed in this chapter.

## 22.2 PIPES AND DUCTS

In most respects, pipe that is used in HVAC work is the same type of pipe used for plumbing installations. Schedule 40 galvanized-steel pipe is used for water lines, condensate piping, and other locations, while black-iron pipe is used for gas and fuel lines. For hot-water or steam mains, galvanized-steel pipe is generally used, as long, straight pieces are easier to install, there are adequate fittings readily available, valves usually have threaded connections, and proper suspension of runs is not a major problem. Rigid copper pipe (type K) is used in some special conditions where the thicker pipe walls are an advantage. Type L copper tubing in coils is economical and easy to install with minimum fittings but does require more support to eliminate sags and low spots. Most frequent applications for type L copper tubing are in sizes of $1/4$ to $3/4$ in. while type K hard-drawn copper pipe or tubing is used where larger sizes are required. Tubing is available in coils of 25, 50, and 100 lineal feet, while pipe is standard in 20-ft lengths. All the various fittings for galvanized-steel, black-iron, or copper pipe and tubing are also used in HVAC work as appropriate. Some clay tile, precast concrete, or asbestos-cement pipe is also used for construction in contact with the earth, but their qualities are no different than when they are used in plumbing installations. Cast-iron pipe and fittings are seldom used in connection with HVAC, as any design or construction of soil or waste systems is considered plumbing work. Plastic pipe is also seldom used, as it has minimal desirable properties needed for installation of hot-water, steam, refrigeration, or pressure systems.

Ducts are fabricated from aluminum, sheet steel, fiberglass, reinforced fiber, and nonmetallic pipe. Their primary purpose is to transport air at low pressure from a heat source to the required room, or to return "used" air to the heat source. Pressurized air in a duct tends to take a helical movement, so ducts with a round cross section use space more economically but are more difficult to connect to with branch piping or other fittings. The standard types of fittings, regardless of materials, are indicated in Figure 27-1. All ductwork must be carefully fitted to prevent loss of air, and joints are usually further sealed with joint tape applied over all joints on the outside of the duct. Adequate support is usually not a problem, as ducts are directly supported by the ceiling joist or are enclosed in walls. Where ducts are beneath floor construction, metal straps approximately $2\frac{1}{2}$ in. wide are used to support the ducts from the floor joist. Dampers or other air-control devices installed in ductwork should be located for easy adjustment, and any fire damper should have extra attention with regard to tight fit and the possibility of replacing the fusible link.

## 22.3 FURNACES AND BOILERS

As noted in other chapters, there are many types of fuels and many different arrangements for using the combustion heat of those fuels. The *type* of heat-producing device is based on these factors and on the kind of heat required. The kind of heat may be hot-water, steam, warm-air, or any combination of these, but not all types of heat-producing units are convenient for all heating requirements. The physical size of a heat-producing unit is a primary determinant. Only a few steam-producing boilers are small enough for residential work, yet may be more desirable for apartments, hospitals, office buildings, and similar construction where demand is greater, an operating engineer is employed, or steam is used for other purposes in addition to heating. Compact oil- or gas-fired units with heat exchangers may be ideal for residential heating or air-conditioning installations but may not have adequate capacity for larger installations.

There are essentially two types of boilers (Fig. 22-1): (1) water-tube, and (2) fire-tube. Size is not a determining factor but may influence selection in large installations. As boilers are used to generate hot water or steam, water is used in both types. The boiler consists of a steel or cast-iron container in which is arranged a firebox to accommodate the fire, a series of tubes or pipes between the firebox and the chimney stack, a water or steam storage compartment, and proper controls for maintaining the fire, water, and steam pressure. An ash pit is also required if the boiler is coal fired. If the fire and hot combustion gases pass through tubes which are surrounded by water, the boiler is termed a *fire-tube boiler.* If the tubes are filled with water and the fire and combustion gases circulate around the tubes between the firebox and chimney, the unit is a *water-tube boiler.* Small "boilers" for residential use are generally of the water-tube type, operating a hot-water system. Exterior surfaces of a boiler may be covered with insulation "lagging" (noncombustible asbestos cement or magnesite), or the boiler may be enclosed in a sheet-metal covering or casing.

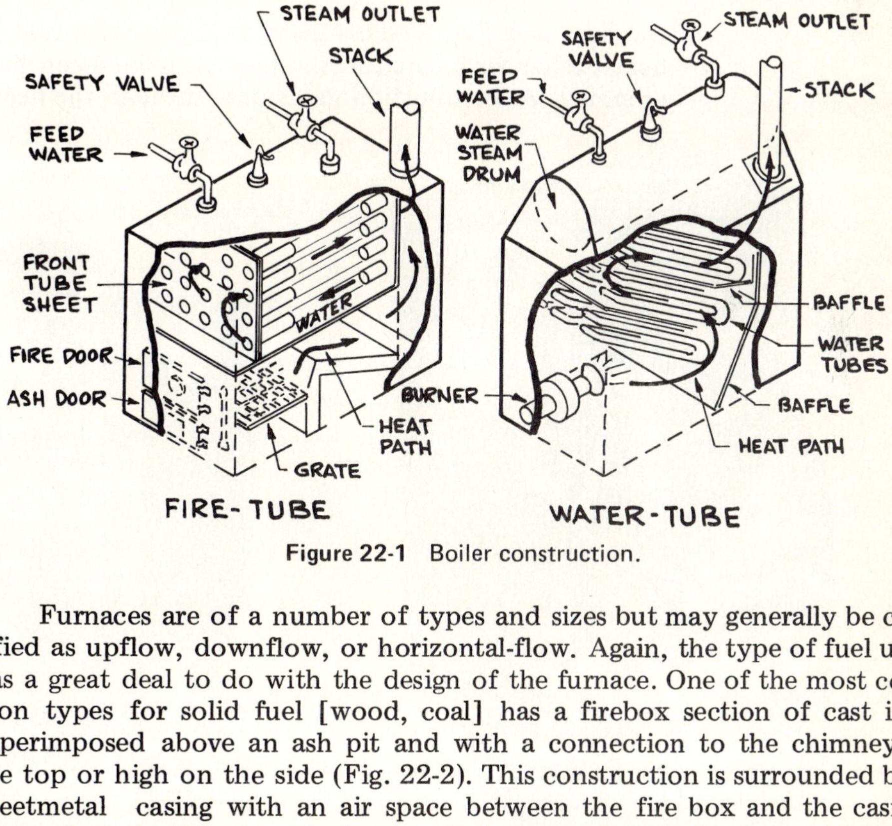


**Figure 22-1** Boiler construction.

Furnaces are of a number of types and sizes but may generally be classified as upflow, downflow, or horizontal-flow. Again, the type of fuel used has a great deal to do with the design of the furnace. One of the most common types for solid fuel [wood, coal] has a firebox section of cast iron superimposed above an ash pit and with a connection to the chimney at the top or high on the side (Fig. 22-2). This construction is surrounded by a sheetmetal casing with an air space between the fire box and the casing. Warmed air in this space is drawn off through pipes at the top, and cooler return air is introduced near the floor. Draft control is obtained by dampers in the firebox and ash-pit doors, and in the smoke pipe to the chimney. Ashes and clinkers produced by the solid-fuel combustion are dumped into the ash pit through a system of movable grates and are removed through the ash-pit door. Fuel is hand-fired or automatically stoker-fired, but some original installations or conversions may use oil or gas as fuel.

Most air-conditioning systems use relatively compact oil- or gas-fired

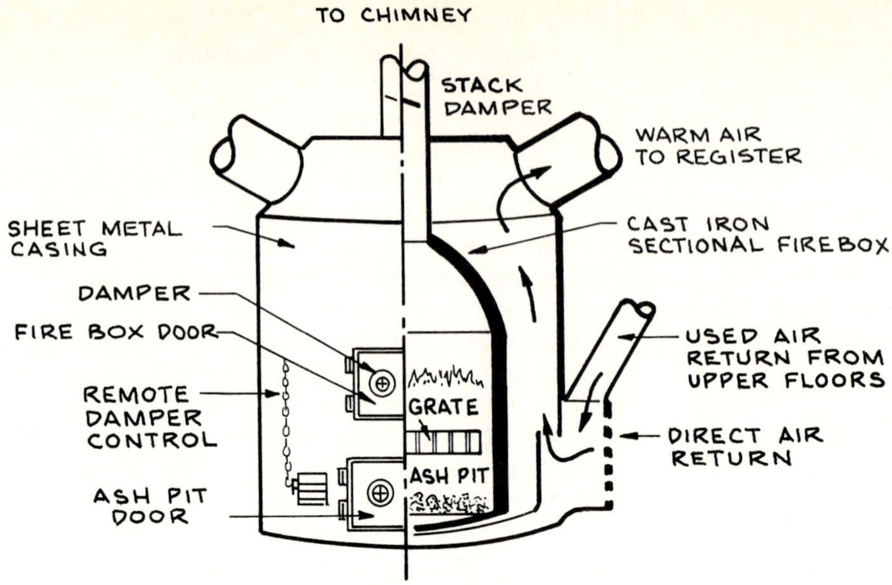


**Figure 22-2** Coal-fired furnace.

furnaces (see Fig. 22-3). Almost without exception, these furnaces include a heat exchanger located in the heat flow between the burner and the vent or chimney. As combustion gases pass through the heat-exchanger portion, the

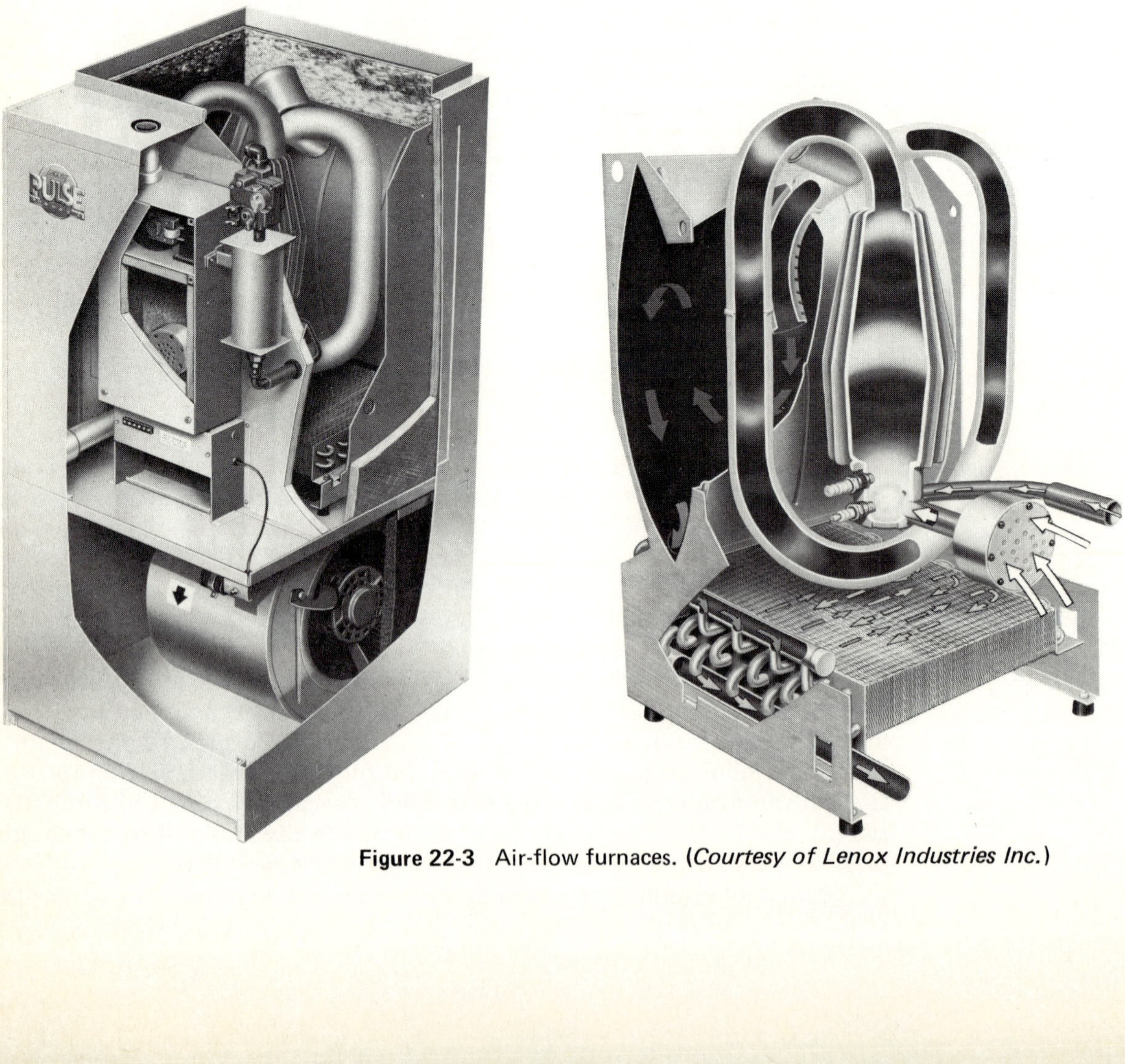

**Figure 22-3** Air-flow furnaces. (*Courtesy of Lenox Industries Inc.*)

air is heated and the spent gases go up the chimney and the warmed air goes into the heating system. In an *upflow* unit, the burner is at the bottom near the floor level but above the return air inlet. A fan forces air through the heat-exchanger portion and the warmed air is taken off through the top of the unit. In the *downflow* type, the supply air is taken in at the top of the unit, passes through the heat exchanger, and is forced out near the bottom. The horizontal-flow system is similar to both types but is designed to function horizontally in attics or similar spaces where vertical clearance is limited. The upflow design is best suited to distribution systems installed above the ceiling, while the downflow unit is better when distribution ducts are beneath the floor. Where floor space may need to be conserved, the *horizontal-flow* unit may be successfully installed in the attic or in the crawl space beneath the floor but "out of sight, out of mind" may also affect filter changes and other maintenance. Most furnaces of these types are controlled automatically but do require venting of used combustion gases.

Another type of "heater" is not truly either a boiler or a furnace. This is a wall or floor unit, generally gas-fired, with the combustion gases heating a series of pipe heat exchangers as the gases pass on their way to the vent. These are not connected to piping or ducts but function by direct radiation, and are used for one space or several small connected spaces. The wall unit fits into the wall construction between the studs and may be flush with the wall surface or extend 3 to 6 in. into the room. Supply air is taken off the floor, heated, and returned to the room at a higher level, generally about 5 to 6 ft above the floor. Some units are constructed to supply heat on each side of a wall by using an additional face panel. Floor "furnaces" are simply units recessed in the floor, covered with a lattice or grid, and transmitting heat through the steel shell or casing surrounding the burner. Each type of furnace requires a vent for used gases and may be manually controlled or operated with a thermostat.

## 22.4 COMPRESSORS

Early compressors for refrigeration were normally single-action reciprocating units built entirely of steel and operating at low speeds. This was necessitated by the one primary refrigerant used (ammonia), which caused corrosion and required high pressure. The introduction of new refrigerants and self-contained hermetically sealed units allowed use of nonferrous metals and eliminated belt-driven compressors, and by 1940 most residential refrigerators, air-conditioning units, and other small compressors were of the hermetically sealed type. Compressors are of many types and range in capacity from fractional horsepower size to a capacity of several hundred tons.

A *reciprocating compressor* (Fig. 22-4) is one in which a piston, connected to a rotating shaft, moves up and down in a cylinder, alternatingly compressing the cylinder contents and drawing in more refrigerant. This *positive-displacement* unit operates to provide capacity as a product of the speed or strokes per minute times the volume of the cylinder. The *rotary compressor* operates with a circular motion around a shaft, compressing the refrigerant against the sides of the enclosing cylinder by an eccentric roller. Inlet and outlet valves are located at proper position in the cylinder to allow the refrigerant to be sucked into the cresent-shaped space, compressed as the roller rotates, and ejected under higher pressure at the outlet valve. Another

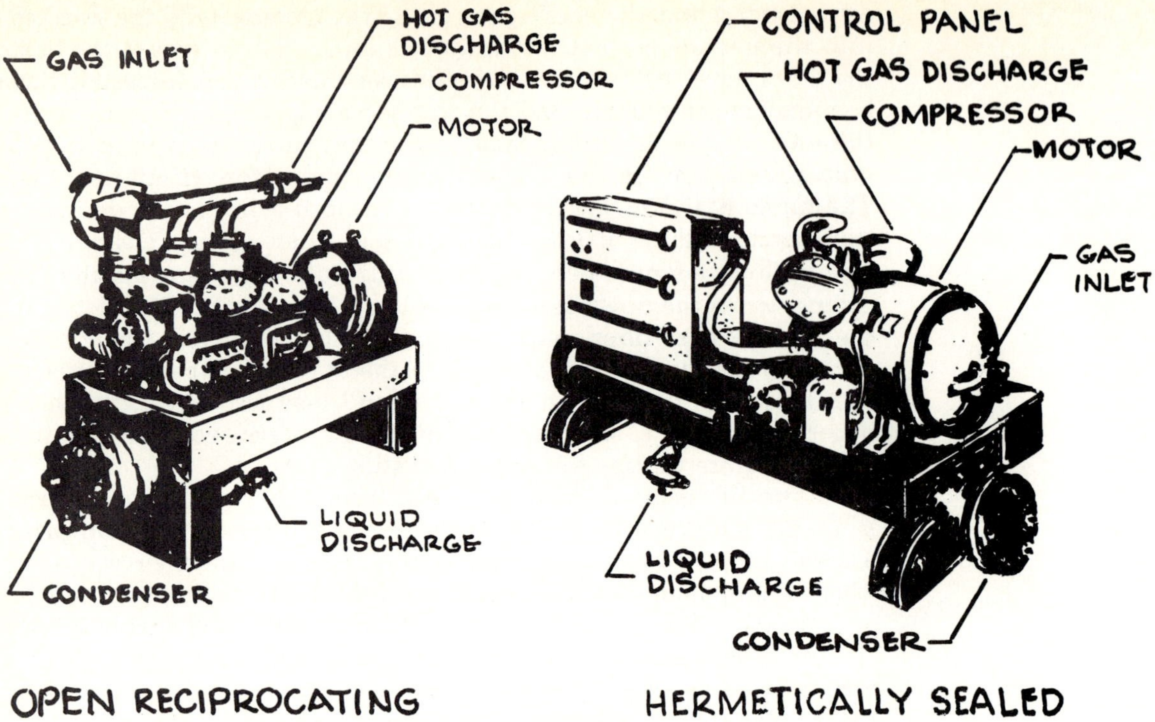


**Figure 22-4** Compressors. (*Courtesy of Carrier.*)

type of positive-displacement compressor is the *helical screw* type. In this unit, two intermeshed helical units, turning in opposite directions, compress the refrigerant as it passes the length of the screw. The inlet is between the two shafts at one end, and the outlet is in a similar position at the opposite end. One end of each screw is open to match with the inlet or outlet locations.

*Open-drive compressors* (not hermetically sealed) are driven by belt connections to the drive motor or directly coupled to the motor shaft. This has some advantages in the fact that different sizes or types of motors may be used even if the electrical supply is direct current or alternating current from the 60-Hz service common in the United States. In many cases a difference in motor size simply means a change in the belt-and-pulley arrangement. Open-connected compressors are also usually field-serviced, a condition that is not possible with hermetically sealed units, which require complete replacement of integrated motor and compressor. *Hermetically sealed compressors* (see Fig. 22-4) are smaller, lighter in weight, freer of vibration, are cooled as a function of the evaporator system, and have no belts or other connections to be maintained. The open-drive compressor is rapidly losing favor to the hermetically sealed unit, mostly because of the increased use of refrigeration in air-conditioning and food preservation. This increase has created a shortage of knowledgeable, experienced service personnel capable of field-servicing refrigeration units, which in turn has caused manufacturers to use hermetically sealed units that require factory servicing.

## 22.5 CONDENSERS

The condenser of a refrigerating system or air-conditioning system transfers heat from the refrigerant to another medium (usually air or water), where the heat may be dissipated. In this process, the refrigerant, (gas) at a high temperature, condenses to a lower-temperature liquid. Two primary systems are used to obtain the required condition: (1) water-cooled units, and (2) air-cooled units. A number of different types of construction are used in either case, but the process is the same.

The refrigerant coming from the compressor is at high temperature as well as high pressure, and is in a gaseous state. The main body of the condenser consists of pipe coils containing the refrigerant (see Fig. 22-5). With a *water-cooled* unit, horizontal or vertical, the pipe is enclosed in a casing where cooling water is circulated around the refrigerant piping. Baffles in the casing, fins on the refrigerant piping, and similar attachments tend to increase the contact surface of the piping and give greater condensing area. The condenser for an *air-cooled* system is also a series of pipes or coils but not enclosed in a casing. Most coils are composed of layers of tubing supported in a framework open to the air. With a *natural-draft* condenser the air circulates over the tubing by convection, entering at the bottom as cooler air, absorbing heat from the tubing, and exhausting at the top as the air heats. This type of condenser is not very effective, however, so most air condensers use a fan to blow cooler air through the coil system. This cooler air picks up heat from the exposed tubing and is wasted into the atmosphere. The heat gain or loss is equal in any case. The amount of heat lost by the refrigerant is equal to the amount of heat gained by the coolant.

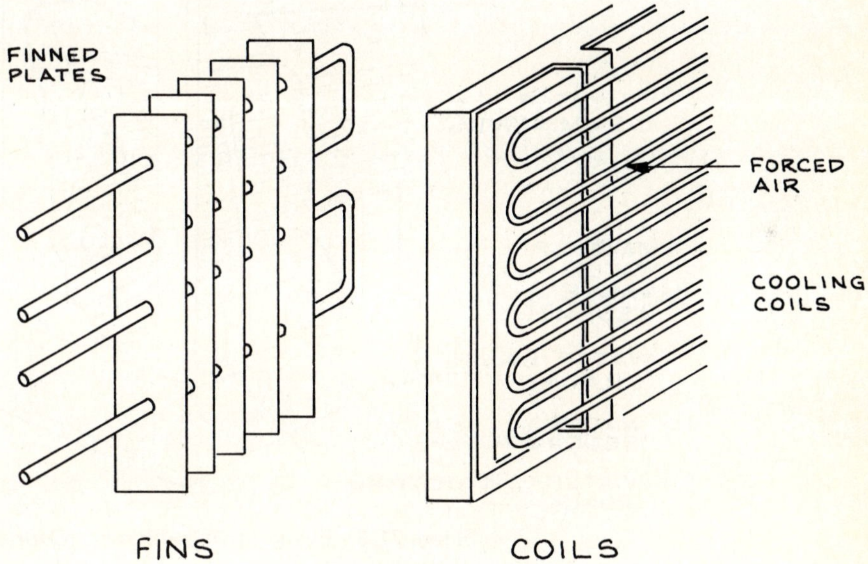


**Figure 22-5** Coolers.

The amount of cool air or cool water required to operate a condenser properly is considerable and in large refrigeration of air-conditioning plants may have to have some backup source to provide precooling. One of the most common is a lake or pond, where the surface evaporation will reduce the temperature of the cooling water. Another source is the cooling tower, where the water is circulated vertically over a series of baffles exposed to a natural or mechanical air draft and evaporation again cools the water. A third method is a forced-draft system, which uses large fans to blow outside air through a cooler unit or even directly through the condenser. In absorption-expansion systems a portion of the chilled water may be reused as a condenser fluid.

## 22.6 EVAPORATORS

An *evaporator* is a unit, usually fabricated with a series of copper tubes, that absorbs heat from the surrounding air and transfers it to the refrigerant (Fig. 22-6). In this process, the room air is cooled and the liquid refrigerant is transformed into a gas or vapor. When the high-pressure liquid passes through the control valve and enters the evaporator, it loses some of the pressure but allows for absorption of heat. The more surface provided by the tubing, or fins around the tubing, the more exposure to the heat. Air passing over the exposed surfaces are therefore cooled and used to reduce the temperature.

The evaporator provides the end result of the refrigeration cycle. For air-conditioning systems, the air is cooled by passing over or through the evaporator. In many commercial installations, such as refrigerators, cold-storage boxes, meat cases, frozen-food cabinets, and even food-handling spaces, the evaporator is exposed to the air or food being processed. In some systems, such as beverage manufacturing, the evaporator coils are immersed in the liquid to cool it. Although the physical shape and arrangement of piping may vary, the evaporator section always absorbs heat from the warmer surroundings and replaces the heat with cooler air or liquid.

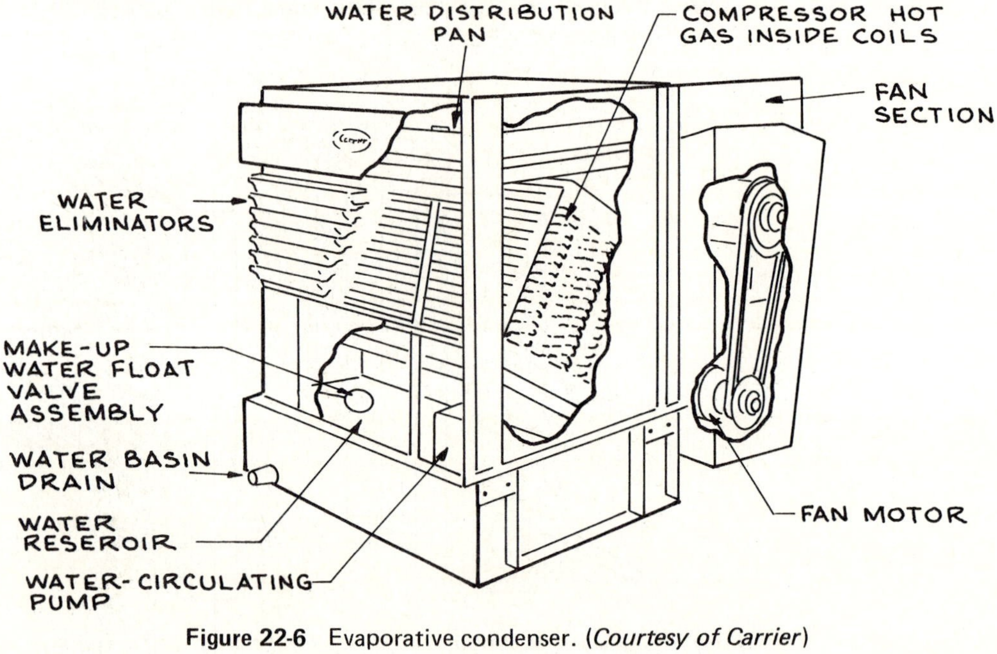


**Figure 22-6** Evaporative condenser. (*Courtesy of Carrier*)

## 22.7 HUMIDIFIERS

Humidity is the measure of the amount of moisture in the air. A measure of 50% *relative humidity* means that the air contains one-half the amount of moisture that it is capable of holding. Measurement of humidity is a unit called a *grain of water vapor*, which is equal to 7000 grains in 1 lb of water, or 8 grains of vapor per cubic foot at 70°F. Humidity may be too low to provide comfortable living conditions, in which case more moisture vapor needs to be added; or humidity may be too high, in which case moisture can be removed by dehumidification. High temperature of the air tends to increase the amount of moisture the air will hold, while a decrease in temperature reduces the capacity accordingly.

Several types of humidifiers are available: (1) open-pan type, (2) atomizing type, and (3) wetted-element type. The *open-pan* type is the oldest form and generally consists of an open pan arranged to provide water vapor into the warm-air stream. To increase vaporization, a series of vertical metal plates may be arranged within the pan of water, or an immersion heater may be used. Improvements on the simple pan type have resulted in several methods used to vaporize or spray water into the air. One method employs a rotating brush which dips into water at the bottom of its rotation and throws off water into the air as it turns. The rotating-disk type is similar but uses a rotating disk which is partially submerged in the water. A high-pressure nozzle may also be used to spray water against a splash plate. The *wetted-element* device comes in several forms but usually consists of a pad or screen holding moisture, with a fan blowing vapor through the pad. All types are capable of automatic control of the water in relation to the humidity of the air.

The removal of excess moisture is termed *dehumidification*. This process is based on condensation of the excess moisture when temperature of the air falls below the dew point. When this happens, the water obtained by condensation falls into a pan and is wasted. Other systems surround the refrigerant piping with a heating element. A reverse of the wetted-element humidifier pulls air through an absorbing filter in the fresh-air supply, which reduces the moisture by retaining a part on the absorbing pads.

## 22.8 CONTROLS

There are a great many controls that may be used or required in connection with heating, refrigeration, and air-conditioning. For the most part, these control devices regulate the flow of liquids and gas, fluctuate with changes in temperature or pressure, and may operate as safety devices. These units may also be physical controls, such as manually operated valves and switches; automatically operated valves, switches, and dampers; sensing devices which operate other devices; gauges of various types that indicate the physical condition of a substance but do not "control" it; and many types of safety devices. Many controls are multifunctioning and may be electrically, electronically, or mechanically operated. In large commercial installations, recording information may also be a part of some control systems, and with more reliance on computers and similar electronic equipment, many of the control functions are regulated automatically.

One of the most commonly known *regulating* or *primary* devices is the *thermostat*, which automatically turns heating or air-conditioning equipment on or off, depending on the temperature setting. A *pressurestat* operates on similar principles but is actuated by pressure, usually through a small bellows, which fills or deflates as pressure varies and thereby opens or closes electrical circuits controlling various equipment. The *bourbon-tube* pressurestat (Fig. 22-7) is composed of a "shepherd's crook" tube of mercury. As pressure on the mercury increases, the tube tends to straighten, closing an electrical circuit. A third type of primary control is the *humidistat*. Most types are made of a human hair stretched between a stationary point and a movable point. As the hair becomes moist because of higher humidity, the hair expands and allows an electrical contact to be made. Another type uses a nylon material bonded to a spring that expands or contracts as moisture is absorbed.

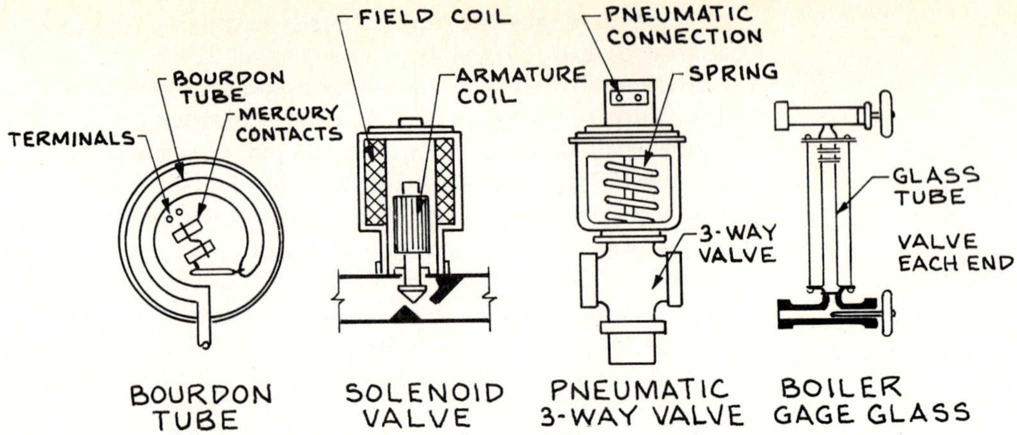


Figure 22-7

*Secondary* controls regulate a cycle during normal operation or act as safety devices to prevent accidents due to fuel failure, malfunction of equipment, excess fire, water, temperature, or pressure. A *contactor*, which consists of a coil with free-moving armature, is used when heavy electrical current loads pass through a control device. When the coil is energized, the armature moves into the coil and a protruding prong makes the electrical contact. A *water valve* is used to control the flow of water through a condenser and operates by fluctuating pressure of the condenser to automatically open or close the valve port, which controls the water flow. One of the more widely used control devices is a *solenoid valve*. This valve consists of a coil around a plunger which operates to open or close a valve port as current is applied to the coil, in much the same way that a contactor operates. Several *backpressure valves* are used, particularly in refrigeration systems. In general, these are operated by the action of pressure on a bellows, which in turn operates a valve stem against a valve opening.

Two simple types of electrical protection devices are also widely used. Several cartridge or screw types of fuses use a small *strip of metal* which will part when excessive load is placed on it. Another type is a *bimetallic disk*, which will flex on application of heat to the point where electrical contact is broken. A *current control relay* is a device used when reset is desirable without delay or inconvenience. This relay consists of a fluid-filled tube in which an iron core moves in a surrounding coil. When an overload occurs, the fluid slows the travel of the core, providing a short time delay until the core reaches the contact piece and disconnects the circuit.

*Gauges* are of many different types, sizes, and shapes but the most common type is a circular case with a movable needle that indicates information against a dial. The needle is moved by a simple gear train, by spring arrangement, or by pressure on the metal of the sealed case. Boilers and other equipment that contain large quantities of water or other liquids use *gauge glasses* to indicate the level of the liquid inside the unit. The normal gauge glass is a length of glass or plastic tubing, ½ to 1 in. in diameter, held between two stopcocks. One stopcock is installed at or near the bottom of the tank, while the other is installed near the top or above the normal liquid line. When the valves are opened, the liquid level between the two stopcocks is shown in the glass tubing.

Manually operated *valves* are the same type as those used in plumbing work. Gate valves, globe valves, angle valves, and check valves are used to control the flow of liquids, such as water, oil, condensate, and most any other liquid. Dampers are installed in ducts and smoke pipes to control the flow of air or other gases. Many dampers are installed so that they can be temporarily or permanently set after a system is balanced. Other dampers are manually or automatically operated to allow more draft or greater volume of air. Fire dampers are installed with a fusible link that melts at a high predetermined temperature, thereby allowing the damper or firedoor to drop into place and help segregate a possible fire.

## 22.9 INSULATION

Although not really a material of heating or air-conditioning, insulation is a necessity in such installations. Insulation of various types, *R*-values, materials, and methods of installation are a part of building construction, but the insulation of ducts, piping, and other pieces of equipment *is* a part of the installation of these items. Hot-water lines, warm-air ducts, water heaters, boilers and furnaces, and other equipment require insulation to retain the heat or cold.

Piping is generally covered with preformed semicircular sections of insulation composed of cork, fiberglass, magnesite, or similar materials. The sections are placed around the pipe and secured with wire or metal straps. In some types of insulation, the material is further covered with canvas and painted with appropriate colors to indicate the contents. Ducts are covered with fiberglass batts or blankets or perhaps with rigid material. Semirigid insulation is usually secured with wire or metal straps, while rigid board is often fastened with metal prongs secured to the duct and penetrating the rigid board. Equipment such as furnaces may have rigid board attached to the flat surfaces wherever possible, but curved surfaces may be covered with a number of different types of semiplastic material. Joints in ducts and around valve connections, as well as other open spaces, are usually sealed with plastic duct tape. Cold-water pipes in some installations may be covered in the same manner as for hot water to prevent the escape of cold. Buried gas pipe and other underground work is usually not insulated, except under special soil conditions, and may be wrapped with weatherproof tape or coated with asphaltic paint to reduce corrosion.

## 22.10 REMOTE (PIPED) SYSTEMS

Although most air-conditioning is installed in spaces closely enough associated to allow ductwork to carry the conditioned air, some spaces or situations must be supplied by using hot-water piping and chilled water to local units. These conditions require a central machine space and a heat source as well as a water-chilling possibility. Control of the system may be manual but is usually accomplished with three-way motorized valves, as shown in Fig. 22-8. In both the two-pipe and the three-pipe systems, the heat source, chiller, and ultimate coil use a common piping system. In the four-pipe system, there are really two complete and separate systems, one for heating and one for cooling, with individual coils for each system where required. Pipe systems are generally more adaptable to multifloor buildings, or where required, heating/cooling units may be remote from the actual machine space, as in factories and more-spread-out buildings. Calculations for air-conditioning of individual spaces are made in the same manner as for spaces using duct-

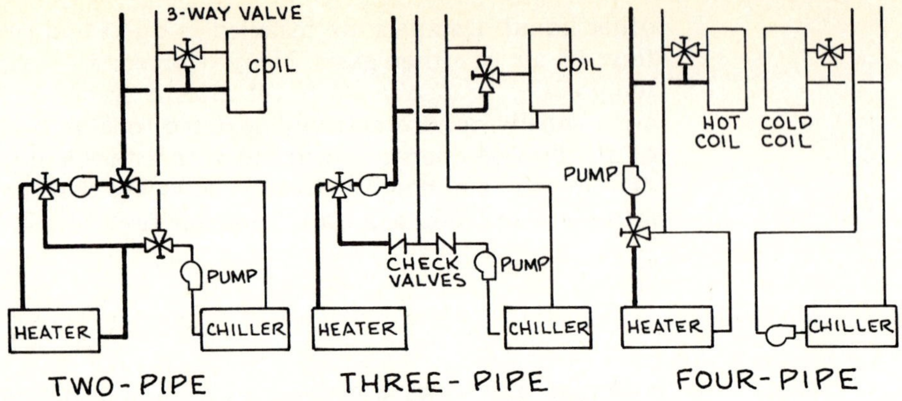


**Figure 22-8** Piped air-conditioning systems.

work, but the pipe size, insulation of both pipes, and the size of heater and chiller are calculated based on the entire system connected to them.

## 22.11 MISCELLANEOUS ITEMS

In addition to the items indicated so far, it should be obvious that many minor items of equipment or installation may be required. Electric motors and their controls in a variety of sizes and arrangements are necessary for proper operation of fans, pumps, compressors, and other units that require a "driver." Hanger brackets and other supports are necessary to hold pipe and ductwork in place and may be as simple as a length of No. 10 wire or may be an adjustable swivel-jointed combination. All furnaces require a smokestack, chimney, or vent to carry smoke or combustion fumes out of the building. A great variety of registers and grilles are available for supply openings as well as return-air openings, and may be fixed, partially adjustable, or fully adjustable in ceilings, walls, or floors.

## 22.12 DRAWINGS AND SYMBOLS

Drawings and symbols used in HVAC work are a combination of others used in electrical, heating, and plumbing, with some additional symbols peculiar to this work, as shown in Fig. 22-9.

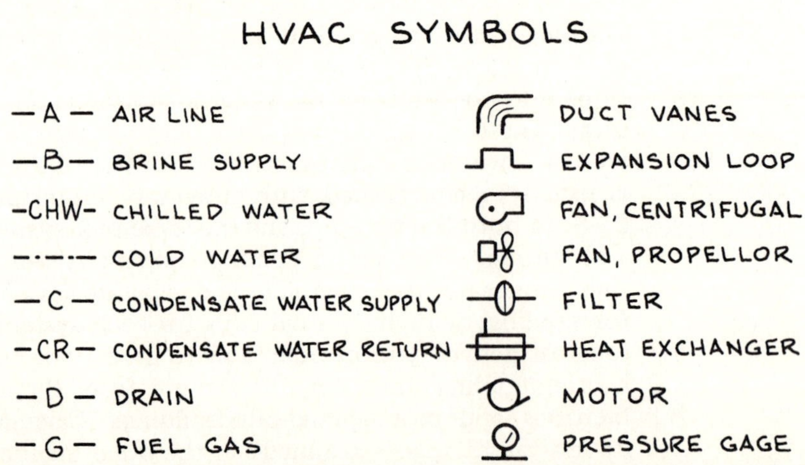

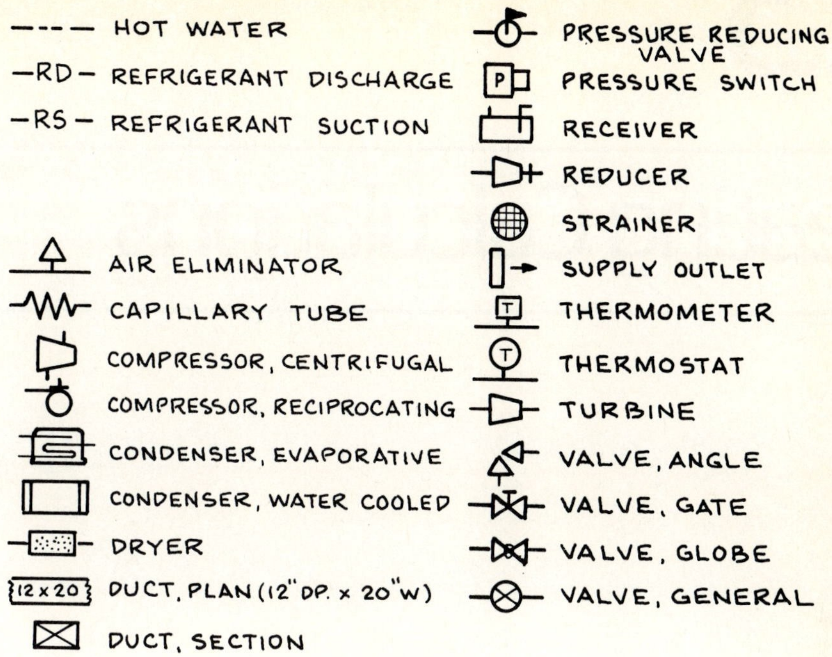


Figure 22-9

## 22.13 SUMMARY

The materials and equipment required for heating, ventilating, and air-conditioning are much like those used in other installations, with a few exceptions in the areas of refrigeration and air-conditioning. Pipe of different materials, sizes, and types is used in plumbing work, some structural work, "ornamental iron" items, and a host of minor uses. Sheet metal used in ductwork is the same material as that used for gutters, flashings, downspouts, and other fabricated parts that are not a part of heating or air-conditioning. Insulation for pipes and ducts is the same material as that used for other thermal insulation, except in a different form.

The major items that are special to HVAC work are boilers and furnaces, which are heat producers, and compressors, condensers, evaporators, and special valves relating to refrigeration and air-conditioning. Although the order and operation of these units is similar, the difference between installations may be considerable because of location or required function. A system of air-conditioning equipment for a residence may be totally improper for a commercial situation, and vice versa. Unit refrigeration in a home has little in common with an installation in a food-processing establishment except for the general principles of refrigeration. The type and cost of fuels may also be a major factor in heating and air-conditioning, and certainly the actual construction of the building will determine the amount of heat loss or heat gain for which HVAC equipment must be designed.

# 23

# Heating Calculations

## 23.1 INTRODUCTION

Heat-loss calculations are necessary for the proper design of heating or air-conditioning systems. The loss of heat from the building to outside air will determine the size of heating or cooling equipment, duct sizes, radiator or convector sizes, and all related parts. To determine the amount of heat loss from a building, it is necessary, and relatively easy, to calculate the loss through each wall, floor, ceiling or roof, and through each door, window, or similar opening. Insulation is very important. The proportion of glass to opaque walls and roofs is also important. To provide indoor climate comfort, the interior temperature must be constantly maintained at a higher level than the outside temperature in winter (heat), and cooler than the outside temperature in summer (air-conditioning).

Several criteria must be considered in the comparison or evaluation of building construction. For *transmission* of heat through a structural material, a $U$-coefficient is used. Conductivity is designated as $k$ or $C$, and the thermal resistance (reciprocal of conductivity) is shown as $R$. The flow of heat from the interior of a building may be calculated by multiplying the heat-flow characteristics ($U$) by the area of the material ($A$) times the difference between interior and exterior temperature in degrees Fahrenheit ($\Delta T$).

$$U \times A \times \Delta T = \text{Btu/hr}$$

To obtain the total heat loss and required heating or cooling, each room or space must be figured separately and then totaled.

## 23.2 TRANSMISSION

Transmission of heat through a wall or other surface is influenced by conductance, radiation, and convection. This applies to piping and ducts as well as to building materials. The amount of heat transmitted by conductance through a wall is dependent on a number of factors: (1) resistance to heat flow of wall material, (2) wall thickness, (3) temperature difference between each side of a wall, (4) quantity of heat, (5) radiation and absorption of wall surfaces, and (6) time. For convenience, conduction factors are expressed in constants such as Btu/hr, square foot, degrees Fahrenheit, or minutes or hours. The coefficient of conductivity of any material varies as the density of the material increases and as the level of operating temperature increases. The rate of heat conducted through a wall decreases as the thickness increases, or the rate of heat conductance increases as the wall thickness decreases. The coefficient of conductivity ($k$) is based on a standard thickness of any material, so the amount of heat conducted through a wall of any thickness equals the coefficient of the material divided by the thickness of the wall in inches. Conductance is the amount of heat (Btu) transmitted through a wall of any thickness by conduction per hour per square foot per 1°F temperature difference. Resistance to heat flow through a wall is equal to the sum of separate resistance values for each material in the wall or by each inch of thickness of each material in a wall.

$$R = r_1 + r_2 + r_3 + \cdots$$

where $R$ is the total resistance and $r_1$, $r_2$, $r_3$, . . . , are resistance values of separate materials in the wall. Also,

$$R = \frac{1}{U}$$

where $R$ is the total resistance and $U$ is the reciprocal of transmission. If $k$ represents conductivity, $L$ represents the total thickness of a wall in inches, $f_1$ or $a$ represents the value for conductance of air space, and $C$ designates coefficiency of conductance.

$$U = \frac{1}{L/k + 1/a} \quad \text{or} \quad C = \frac{k}{L}$$

**EXAMPLE**

Determine the coefficient of transmission for an 8-in.-thick common brick wall, both sides smooth, with still air on each side. Value for $k$ is given in Table 23-1. Value for $a$ in still air is 1.65.

$$U = \frac{1}{L/k + 1/a} = \frac{1}{8/5.0 + 1/1.65} = \frac{1}{1.6 + 0.06} = 0.602$$

$$= 0.602 \text{ Btu/hr-ft}^2\text{-}^\circ\text{F}$$

**TABLE 23-1**
Table of Building Materials and Combinations

| *Material or combination* | *Conductance, C* | *Conductivity, k* | *Transmission, U* |
|---|---|---|---|
| Materials | | | |
| Built-up roof, 3/8 in. thick | 3.00 | | |
| Brick, common clay 8 in. thick | | 5.00 | |
| Brick, face on 8-in. brick backing | | 9.00 | |
| Concrete wall, 8 in. thick | | 9.00 | |
| Concrete block, 8-in. hollow core | | 5.00 | |
| Glass, single thickness, 1/8 in. | 1.13 | | |
| Gypsum board, 1/2 in. thick | 2.22 | | |
| Metal lath and plaster, 3/4 in. | 7.70 | | |
| Floors | | | |
| Concrete, 4-in. deck, AC tile, air space, plaster ceiling | | | 0.25-0.34 |
| Frame, 5/8-in. plywood, 25/32-in. wood subfloor, plaster ceiling | | | 0.19-0.23 |
| Insulation | | | |
| Cane fiberboard | 0.53-0.80 | 0.33 | |
| Corkboard with asphaltic binder | 0.32 | 0.25-0.32 | |
| Fiberglass blanket, 3 in. | 0.27 | 0.32 | |
| Glass wool batts | 0.27 | 0.24-0.30 | |
| Mineral-wool blanket, 1 in. | | 0.26-0.28 | |
| Mineral-wool board, 1 in. | | 0.25-0.32 | |
| Vermiculite board, 1 in. | | 0.92 | |
| Masonry | | | |
| Brick, 8 in. thick, no finish | | 5.00 | |
| Cinder block, 4 in. thick, no finish | 0.58 | | 0.40 |
| Concrete, 8 in thick, no finish | | 9.00 | 0.37-0.57 |
| Concrete block, 8 in. thick, no finish | 0.86-0.90 | | 0.49-0.51 |
| Hollow tile, 6 in. clay, exterior stucco | 0.47-0.66 | | 0.27-0.37 |
| Partitions | | | |
| Wood frame, 2 X 4 stud, 1/2-in. gypsum board each side, no insulation | | | 0.31 |
| Wood frame, 2 X 4 stud, 1/2-in. gypsum board each side, 3-in. insulation (R-11) | | | 0.08 |
| Roofs | | | |
| Asphalt shingles | 2.27 | | |
| Wood shingles on 1-in. deck | 1.06 | | 0.25-0.26 |
| Concrete deck, 4 in. lightweight, no insulation | | | 0.14-0.15 |
| Steel deck, no insulation | 1.30 | | 0.64-0.86 |
| Steel deck, 2-in. rigid insulation | | | 0.15-0.16 |
| Wood deck, 1 in thick, no insulation | 1.06 | 0.80 | 0.40-0.48 |
| Wood deck, 1 in. thick, 1-in. insulation | | | 0.19-0.21 |
| Wood deck, 2½ in. thick, 1-in. insulation | | | 0.15-0.16 |
| Walls, exterior | | | |
| Concrete, 8 in. thick, furred gypsum board, no insulation | | | 0.21-0.34 |
| Concrete block, 8 in. thick, furred, 1/2-in. gypsum board interior, no insulation | | | 0.29-0.30 |
| Hollow clay, 12 in. thick, furred 1/2-in. gypsum board interior, no insulation | 0.52 | | 0.16-0.25 |
| Wood frame, 2 X 4 studs, 1/2-in. plywood sheathing, wood siding, no insulation | 0.50 | | 0.22-0.23 |
| Wood frame, 2 X 4 studs, plywood sheathing, 1/2-in. gypsum board, 3-in. insulation (R-11) | | | 0.07 |
| Wood frame, 2 X 4 studs, 1/2-in. gypsum board interior, sheathing, 1-in. stucco | | 5.60 | |

**EXAMPLE**

Determine the coefficient of transmission for a wall constructed of softwood studs, ½-in. wallboard inside, ¾ in. × 10 in. bevel siding outside, and a 3½-in. space between inner and outer wall coverings, filled with fiberglass insulation.

½-in. wallboard: $k_1 = 0.80$
¾-in. × 10 in. siding: $C = 0.95$
3½-in. insulation: $k_2 = 0.25$
Air movement: $f_1 = 1.65$

Therefore,

$$U = \frac{1}{1/C + L/k_1 + L/k_2 + 1/f_1}$$

$$= \frac{1}{1/0.95 + 0.5/0.8 + 3.5/0.25 + 1/1.65}$$

$$= \frac{1}{1.05 + 0.63 + 14 + 0.6}$$

$$= \frac{1}{16.28} = 0.0614 \text{ Btu/hr-ft}^2\text{-}^\circ\text{F}$$

Floors and ceilings, or roofs, are calculated in exactly the same manner but using values for the appropriate materials involved. Where concrete floors are in direct contact with the earth, heat transmission is usually neglected, except where refrigeration may be required in commercial installations. A temperature difference of 5°F may be used beneath refrigerated spaces if desired.

## 23.3 INSIDE/OUTSIDE TEMPERATURES

In heating or air-conditioning design, the *outside design temperature* is the normal temperature expected, *not* the coldest temperature on record. A list of these temperatures for selected localities in the United States is shown in Table 23-2. For locations not shown, or where temperatures may vary widely, the average local temperature over a 5-year period should be used. The *inside design temperature* is directly related to the use of the space. Normal residential space may be considered at 70°F, but other types of occupancy may dictate different inside design temperatures, either higher or lower. Commercial kitchens may want to consider a temperature of 62 to 65°F, while adjacent refrigerated rooms may require 0°F. In this case the inside temperature of the refrigerated space is 0°F, while the outside design temperature is 65°F and has nothing to do with the location. The temperature difference is the number of degrees Fahrenheit between the inside and outside design temperatures. Whereas the inside design temperature, except in refrigeration, is usually not lower than about 65°F, the outside design temperature during winter may be below 0°F (in Fairbanks, Alaska, −50°F). If the inside design temperature is considered at 70°F and the outside design temperature at San Diego, California is 40°F, the difference is 70 − 40, or 30°F. However, if the same inside temperature of 70°F is used at Fairbanks, Alaska, and the outside design temperature is taken as −50°F, the difference is 70 − 0 = 70 + 50 = 120°F. Attics, basements, garages, and other nonheated spaces will generally have a "room" temperature about midway between the outside design temperature and the inside design temperature.

**TABLE 23-2**
Average Temperatures of U.S. Cities[a]

| | | | Summer (cooling) | | | | | Summer (cooling) | |
|---|---|---|---|---|---|---|---|---|---|
| State | City | Winter (heating) dry-bulb (°F) | Dry-bulb (°F) | Wet-bulb (°F) | State | City | Winter (heating) dry-bulb (°F) | Dry-bulb (°F) | Wet-bulb (°F) |
| Alabama | Birmingham | 10 | 95 | 78 | Mississippi | Vicksburg | 10 | 95 | 78 |
| | Mobile | 15 | 95 | 80 | Missouri | Kansas City | –10 | 100 | 76 |
| Arizona | Flagstaff | –10 | 90 | 65 | | St. Louis | 0 | 95 | 78 |
| | Phoenix | 25 | 105 | 76 | Montana | Butte | –20 | | |
| | Yuma | 30 | 110 | 78 | | Miles City | –35 | | |
| Arkansas | Little Rock | 5 | 95 | 78 | Nebraska | Omaha | –10 | 95 | 78 |
| California | Bakersfield | 25 | 105 | 70 | Nevada | Reno | –5 | 95 | 65 |
| | El Centro | 25 | 110 | 78 | New Hampshire | Concord | –15 | 90 | 73 |
| | Fresno | 25 | 105 | 74 | New Jersey | Newark | 0 | 95 | 75 |
| | Los Angeles | 35 | 90 | 70 | New Mexico | Albuquerque | 0 | 95 | 70 |
| | San Diego | 35 | 95 | 68 | New York | Buffalo | –5 | 93 | 73 |
| | San Francisco | 35 | 85 | 65 | | New York | 0 | 95 | 75 |
| Colorado | Denver | –10 | 95 | 64 | | Syracuse | –10 | 93 | 75 |
| | Pueblo | –20 | 95 | 65 | North Carolina | Asheville | 0 | 93 | 75 |
| Connecticut | Bridgeport | 0 | 95 | 75 | | Raleigh | 10 | 95 | 78 |
| Delaware | Wilmington | 0 | 95 | 78 | North Dakota | Bismarck | –30 | 95 | 73 |
| Dist. of Columbia | Washington | 0 | 95 | 78 | Ohio | Akron | –5 | 95 | 75 |
| Florida | Jacksonville | 25 | 95 | 78 | | Dayton | 0 | 95 | 78 |
| | Miami | 35 | 91 | 79 | | Toledo | –10 | 95 | 75 |
| Georgia | Atlanta | 10 | 95 | 76 | Oklahoma | Tulsa | 0 | 101 | 77 |
| | Savannah | 20 | 95 | 78 | Oregon | Portland | 10 | 90 | 68 |
| Idaho | Boise | –10 | 95 | 65 | Pennsylvania | Philadelphia | 0 | 95 | 78 |
| Illinois | Chicago | –10 | 95 | 75 | | Pittsburgh | 0 | 95 | 75 |
| | Springfield | –10 | 98 | 77 | Rhode Island | Providence | 0 | 93 | 75 |
| Indiana | Indianapolis | –10 | 95 | 76 | South Carolina | Charleston | 15 | 95 | 78 |
| | Terre Haute | 0 | 95 | 78 | South Dakota | Sioux Falls | –20 | 95 | 75 |
| Iowa | Des Moines | –15 | 95 | 78 | Tennessee | Nashville | 0 | 95 | 78 |
| | Sioux City | –20 | 95 | 78 | Texas | Austin | 20 | 100 | 78 |
| Kansas | Topeka | –10 | 100 | 78 | | Dallas | 0 | 100 | 78 |
| Kentucky | Lousiville | 0 | 95 | 78 | | El Paso | 10 | 100 | 69 |
| Louisiana | New Orleans | 20 | 95 | 80 | | Houston | 20 | 95 | 78 |
| Maine | Augusta | –15 | 90 | 73 | Utah | Salt Lake City | –10 | 95 | 65 |
| Maryland | Baltimore | 0 | 95 | 78 | Vermont | Burlington | –10 | 90 | 73 |
| Massachusetts | Boston | 0 | 92 | 75 | Virginia | Richmond | 15 | 95 | 78 |
| | Worcester | –5 | 93 | 75 | Washington | Seattle | 15 | 85 | 65 |
| Michigan | Detroit | –10 | 95 | 75 | | Spokane | –15 | 93 | 65 |
| | Lansing | –10 | 95 | 75 | West Virginia | Charleston | 0 | 95 | 75 |
| Minnesota | Duluth | –25 | 93 | 73 | | Wheeling | –5 | 95 | 75 |
| | Minneapolis | –20 | 95 | 75 | Wisconsin | Milwaukee | –15 | 95 | 75 |
| | | | | | Wyoming | Cheyenne | –15 | 95 | 65 |

[a]A more complete list obtained from ASHRAE is included in Table A10 of the Appendix.

## 23.4 NET AREA

The easiest way to determine net area is to take dimensions from a drawing or blueprint of the building. This method generally requires a bit of calculation, however, as most plan dimensions are shown from the outside face of wood studs along all boundary walls of a frame building and to the centerline of interior partitions. To obtain net wall lengths (Fig. 23-1), therefore, approximately 4 in. (for 2 × 4 framing) must be subtracted from dimensions to outside walls and approximately 2 in. at partitions. Concrete and masonry buildings show dimensions to the face of walls and the thickness of walls, so net lengths are easier to calculate. Use of net wall lengths and height of walls and net areas of floors or ceilings may be necessary for exact calculations, but they also entail a considerable number of small fractions that may be cumbersome to include and are not really that necessary. If net lengths or heights are limited to increments of ¼ ft or multiples (3″–6″–9″–12″), they may be more easily inserted into formulas as decimals (3 in. = 0.25, 6 in. = 0.50, 9 in. = 0.75) and more easily handled in longhand multiplication or with hand calculators. The very small increments between fractional measurements will be indistinguishable when multiple calculation operations are carried out for most wall, floor, or ceiling areas.

In most areas the location of doors and windows in frame construction is shown by a dimension from a known wall line to the centerline of the door or window. This requires that the carpenter properly locate the rough frames around the opening. In concrete or masonry construction, the opening, including the door or window frame, is shown and the mason builds to the dimension shown. In those few cases where metal window fins are built into the masonry work, the exact size of the frame must be known for proper detailing.

All door and window openings should be calculated to the nearest square foot of area. Most doors are 2′-0″, 2′-4″, 2′-6″, 2′-8″, or 3′-0″ wide and for most purposes will be 6′-8″ or 7′-0″ high. A door 3′-0″ wide by 6′-8″ high ($3^0 \times 6^8$) will have an area of 19.98 ft$^2$ or 20.0 ft$^2$ for calculation purposes. Windows are of many sizes but are also figured by total square feet of each window. For easier calculation, any door or window may be given the same heat transmission as fixed glass of the same area at a temperature difference of 70 − 30 = 40°F or a value of 40 Btu/hr per square foot of area.

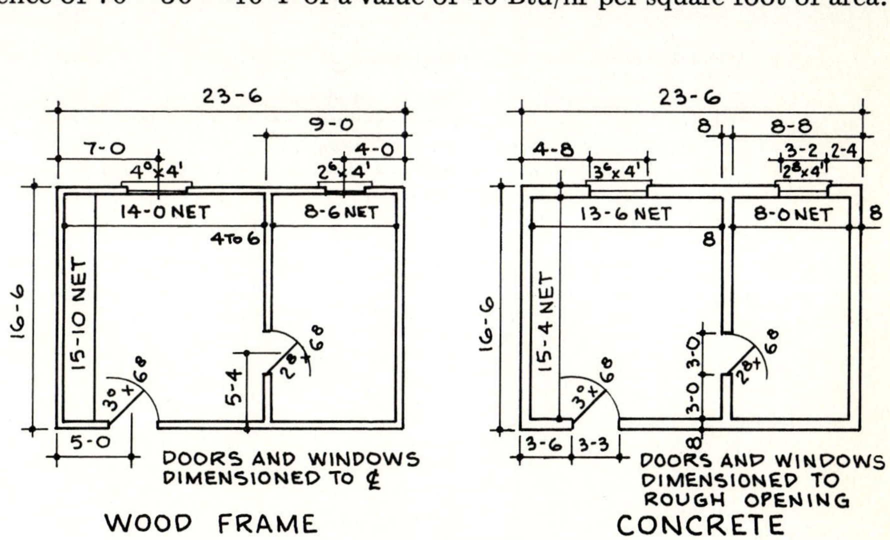


**Figure 23-1** Net dimensions.

**EXAMPLE**

Given a door size of 2′-6″ × 6′-8″. Then

$$2.5 \times 6.66 = 16.65 \qquad \text{say, } 16.75 \text{ ft}^2$$

$$16.75 \times 40 = 670 \text{ Btu/hr heat loss}$$

If more detailed calculation is needed, the type of door (solid core, hollow core, insulated) or type of window (casement, double-hung, sliding) will have to be known and Table 2 of *Manual J* of the National Environmental Systems Contractors Association (NESCA) used to determine heat transfer at various degree differences. The areas of doors or windows should be subtracted from the areas of the walls in which they occur and calculated separately, as the rate of heat transfer will usually be different. The total heat loss, or gain, will then be indicated as a separate item for doors or windows in any wall when total heat loss, or gain, is figured.

## 23.5 INFILTRATION

Wind, or a difference in air pressure, will cause some heat loss through cracks in door or window installations and other small openings. When the wind blows on the windward side of a building, outside air is forced into the building through cracks and a suction action is set up on the leeward (downwind) side, which pulls heated air from the building through similar cracks. The heating system must be designed to warm the infiltrated air to inside temperature. Two methods are available to calculate heat loss by infiltration: (1) the crack method, and (2) the air-change method.

The *crack method* is the more accurate because it considers actual loss at different pressures. The accepted infiltration rate for various types of windows is shown in Tables 23-3 and 27-3. Not all types of windows or all wind velocities are shown, so the listing most closely approaching the situation may have to be used. Table 23-3 expresses the heat loss in cubic feet per foot of crack. The design temperature difference must be calculated as described in Section 23-3. The following formula may then be used when $H$ is the heat loss in Btu/hr or amount of heat required to raise the temperature

**TABLE 23-3**
Infiltration of Air

| *Condition* | *ft³/hr/ft²* |
|---|---|
| Room | |
| No windows or exterior doors | 0.3–0.5 |
| Window/door: one side | 0.7–1.0 |
| Window/door: two sides | 1.0–1.5 |
| Corridors, entrance hall | 1.3–2.0 |
| Wood frame | |
| Wood siding | 0.008–0.025 |
| Exterior plaster | 0.09–0.29 |
| 9-in. brick wall | |
| Plain | 0.42–2.03 |
| Exterior plaster | 0.004–0.017 |
| Furred plaster | 0.003–0.008 |
| Wood window, average | 0.36–0.77 |
| Metal window, average | 0.32–0.77 |

of the air to that inside the building, $Q$ the volume of air in cubic feet per hour entering the structure, $\Delta t$ the design temperature difference, and $L$ the lineal feet of crack. Also required is the constant value ($O$) 0.018, which is the specific heat of air (0.24) multiplied by the density of outdoor air (0.075).

$$H = D \times Q(\Delta t) \times L$$

**EXAMPLE**

Assume a double-hung weather-stripped metal window 3′-0″ wide by 4′-0″ high, a wind velocity of 15 mph, and a temperature difference of 30°F.

$$D = 0.018 \qquad \Delta t = 30^\circ\text{F}$$

$$Q = 32 \qquad L = 3 + 3 + 4 + 4 = 14$$

$$\begin{aligned} H &= D \times Q\,(\Delta t) \times L \\ &= 0.018 \times 32(30) \times 14 \\ &= 0.018 \times 960 \times 14 \\ &= 241.92 \text{ Btu/hr} \end{aligned}$$

The *air-change method* assumes air leakage infiltration by air changes per hour per room. The number of air changes depends a great deal on the type of use of the room and the number of walls exposed to outside temperatures. In this method the formula

$$H = \text{CV} \times W \times D \times \Delta t$$

is used, where CV (the cubic volume) equals the room volume (width $\times$ length $\times$ height), $W$ is the number of exposed walls, $D$ is the constant 0.018, and $\Delta t$ is the design temperature difference.

**EXAMPLE**

Assume a room 10 ft $\times$ 12 ft $\times$ 8 ft high, two walls exposed to outside temperature, and a temperature difference of 30°F.

$$D = 0.018 \qquad W = 2$$

$$\text{CV} = 10 \times 12 \times 8 \qquad \Delta t = 30^\circ\text{F}$$

$$\begin{aligned} H &= CV \times W \times D \times \Delta t \\ &= 960 \times 2 \times 0.018 \times 30 \\ &= 1036.8 \text{ Btu/hr} \end{aligned}$$

## 23.6 INTERNAL HEAT LOSS

Adequate ventilation of residential spaces is usually supplied rather automatically by infiltration, but this is not true in larger buildings such as office buildings and similar spaces. Smoking, dust, fumes, and other airborne irritants must be removed by proper ventilation. The heat loss is equivalent to the heat required to warm the air introduced by ventilation. The recommended air changes for different space uses is shown in Table 23-4. The following formula, using some of the same factors as for infiltration (Section 23-5), is used to calculate ventilation heat loss.

$$H = D \times Q\ (\Delta t)$$

**TABLE 23-4**
Outdoor Air Requirements[a]

| *Application* | *Smoking* | *Cfm per person[b]* *Recommended* | *Cfm per person[b]* *Minimum[c]* | *Cfm per sq ft of floor[b]* *Minimum[c]* |
|---|---|---|---|---|
| Apartment | | | | |
| Average | Some | 20 | 10 | |
| Deluxe | Some | 20 | 10 | |
| Banking space | Occasional | 10 | 7½ | |
| Barbershops | Considerable | 15 | 10 | |
| Beauty parlors | Occasional | 10 | 7½ | |
| Brokers boardrooms | Very heavy | 50 | 20 | |
| Cocktail bars | | 40 | 25 | |
| Corridors (supply or exhaust) | | | | 0.25 |
| Department stores | None | 7½ | 5 | 0.05 |
| Directors' rooms | Extreme | 50 | 30 | |
| Drugstores[d] | Considerable | 10 | 7½ | |
| Factories[e,f] | None | 10 | 7½ | 0.10 |
| Five-and-tens | None | 7½ | 5 | |
| Funeral parlors | None | 10 | 7½ | |
| Garages[e] | | | | 1.0 |
| Hospitals | | | | |
| Operating rooms[f,g] | None | | | 2.0 |
| Private rooms | None | 30 | 25 | 0.33 |
| Wards | None | 20 | 10 | |
| Hotel rooms | Heavy | 30 | 25 | 0.33 |
| Kitchens | | | | |
| Restaurant | | | | 4.0 |
| Residence | | | | 2.0 |
| Laboratories[d] | Some | 20 | 15 | |
| Meeting rooms | Very heavy | 50 | 30 | 1.25 |
| Offices | | | | |
| General | Some | 15 | 10 | 0.25 |
| Private | None | 25 | 15 | 0.25 |
| Private | Considerable | 30 | 25 | 0.25 |
| Restaurants | | | | |
| Cafeteria[d] | Considerable | 12 | 10 | |
| Dining room[d] | Considerable | 15 | 12 | |
| School rooms[d] | None | | | |
| Shop, retail | None | 10 | 7½ | |
| Theater[e] | None | 7½ | 5 | |
| Theater | Some | 15 | 10 | |
| Toilets[e] (exhaust) | | | | 2.0 |

ASHRAE Handbook, 1981 Fundamentals.
[a]Taken from present-day practice.
[b]This is contaminant-free air.
[c]When minimum is used, take the larger of the two.
[d]See local codes which may govern.
[e]May be governed by exhaust.
[f]May be governed by special sources of contamination or local codes.
[g]All outside air recommended to overcome explosion hazard of anesthetics.

Human body heat is one of the major sources of internal heat gain in any building. In residences, body heat is not very important because there is normally adequate air in relation to the number of people. In more crowded spaces, the latent heat and the sensible heat need to be considered. People seated in a theater emit about 350 Btu/hr per person. Dancing causes each person to give off about 900 Btuh. Lighting and various appliances also produce internal heat. Table 23-5 lists some of the more common sources for such heat-producing units.

**TABLE 23-5**
Heat Gain (Btu/hr)

| *Unit* | *Electric* | *Gas* | *Other* |
|---|---|---|---|
| Coffee maker | 290 | 500 | |
| Coffee urn | 3,500 | 5,000 | |
| Fluorescent light | | | |
| Fry kettle, ft.$^2$ area | 5,000 | 9,000 | |
| Griddle, ft.$^2$ area | 4,600 | 7,000 | |
| Grille, ft.$^2$ area | 5,000 | 10,000 | |
| Hair dryer | 2,500 | | |
| Incandescent light | | | |
| Motors ¼ to 3 hp | 2,550 | | |
| People | | | |
| At rest | | | 330 |
| Office work | | | 450 |
| Heavy work | | | 1,450 |
| Recreation | | | 850-1,000 |
| Stove top, open | | 7,500 | |
| Toaster, residential | 50-100 | | |
| Waffle iron | 250-1,000 | | |

## 23.7 HEATING CALCULATIONS

Reading a how-to book may be interesting and provide knowledge, but working out an actual problem brings the entire process into better focus and with more understanding. Using the one-bedroom frame house shown in Fig. 23-2, follow the calculations required to determine the total heat loss, or the amount of heat required for winter comfort, to determine capacity of the primary heating unit. Calculation tabulation forms are similar to many used by different organizations.

### EXAMPLE

Using Fig. 23-2, calculate the heat loss if the outside temperature is 10°F and the inside temperature is assumed at 70°F (difference is 60°F). The construction is concrete slab on grade with carpet over pad in all rooms, stud walls with ⅝-in. exterior building board beneath 1 in. × 8 in. bevel wood siding, ⅝-in. interior gypsum board, and 3½-in. (R-11) mineral insulation batts. The ceiling is ⅝-in. gypsum board with 6-in. (R-19) insulation. Roof is ½-in. plywood, building paper, and wood shingles.

Several basic steps must be taken to determine the various factors in the calculation.

1. The difference in temperature is determined or given.

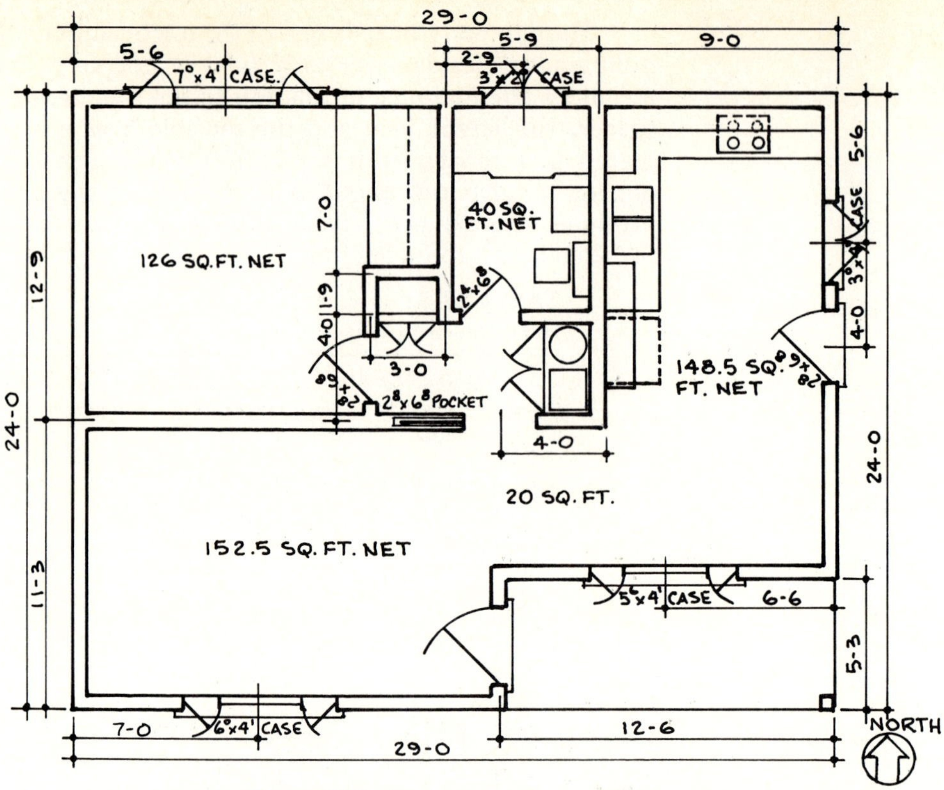


**Figure 23-2** One bedroom residence.

2. $U$-factors must be determined from tables and combinations of surface materials.
3. The net areas of walls, ceilings, floors, windows, and other openings must be determined.
4. The equivalent length of infiltration (crack factor) at windows and doors is determined by adding *all* perimeters of all openings (in lineal feet).

In this example, the difference between inside and outside temperatures at the walls is given. The ceiling difference may be the same but is usually several degrees less, so a value of 56 has been used. The concrete slab is directly on the earth, so a value of 25 is used, although a somewhat higher value (30 to 40) might be used if the floor was wood construction above grade. The first calculation process is to determine the $U$-values of the various surfaces as follows:

*Exterior walls*

| | |
|---|---|
| 1 in. × 8 in. bevel siding | 1.27 |
| ½-in. building board | 1.32 |
| 3½-in. insulation | 11.00 |
| ⅝-in. gypsum board | 0.56 |
| Inside air film | 0.68 |
| Outside air film | 0.14 |
| ¾-in. air space at insulation | 0.94 |
| | 15.91 |

$$U = \frac{1}{15.91} = 0.0628$$

*Ceiling*

| | |
|---|---|
| ⅝-in. gypsum board | 0.56 |
| 6-in. (R-19) insulation | 19.00 |
| Inside air film | 0.61 |
| Outside air film | 0.50 |
| | 20.67 |

$$U = \frac{1}{20.67} = 0.0483$$

**TABLE 23-6**
Heat Loss

| Room | Location | Size (ft) | Ht. (ft) | Gross area (ft) | Opening size | Opening area ($ft^2$) | Net area ($ft^2$) | U-factor | Temp. diff. (°F) | Heat loss (Btu/hr) | Room loss (Btu/hr) |
|---|---|---|---|---|---|---|---|---|---|---|---|
| Living room | W. wall | 10.75 | 8 | 86.0 | | | 86.0 | 0.062 | 60 | 320 | |
| | S. wall | 16.5 | 8 | 132.0 | $6^0 \times 4^0$ | 24.0 | 108.0 | 0.062 | 60 | 402 | |
| | E. wall | 4.5 | 8 | 36.0 | $3^0 \times 6^8$ | 20.4 | 15.6 | 0.062 | 60 | 58 | |
| | Ceiling | | | 152.5 | | | 152.5 | 0.048 | 56 | 409 | |
| | Dr./wd. | | | | | | | | | | |
| | S. window | | | | $6^0 \times 4^1$ | 24.0 | 24.0 | 0.61 | 60 | 878 | |
| | E. door | | | | $3^0 \times 6^8$ | 20.4 | 20.4 | 0.61 | 60 | 746 | |
| | Floor | | | 152.5 | | | 152.5 | 0.480 | 25 | 1,830 | |
| | | | | | | | | | | | 4,643 |
| Bedroom | W. wall | 12.0 | 8 | 96.0 | | | 96.0 | 0.062 | 60 | 357 | |
| | N. wall | 13.5 | 8 | 108.0 | $7^0 \times 4^1$ | 28.0 | 80.0 | 0.062 | 60 | 298 | |
| | Ceiling | | | 126.0 | | | 126.0 | 0.048 | 56 | 339 | |
| | Window | | | | $7^0 \times 4^1$ | 28.0 | 28.0 | 0.61 | 60 | 1,024 | |
| | Floor | | | 126.0 | | | 126.0 | 0.480 | 25 | 1,512 | |
| | | | | | | | | | | | 3,530 |
| Kitchen/ dining room | N. wall | 8.25 | 8 | 66.0 | | | 66.0 | 0.062 | 60 | 245 | |
| | E. wall | 18.75 | 8 | 150.0 | $3^0 \times 4^0$<br>$2^8 \times 6^8$ | 12.0<br>18.5 | 119.5 | 0.062 | 60 | 445 | |
| | S. wall | 11.75 | 8 | 94.0 | $5^6 \times 4^1$ | 22.0 | 72.0 | 0.062 | 60 | 268 | |
| | Ceiling | | | 148.5 | | | 148.5 | 0.048 | 56 | 400 | |
| | Dr./wd. | | | | | | | | | | |
| | E. wall | | | | $3^0 \times 4^1$<br>$2^8 \times 6^8$ | 12.0<br>18.5 | 30.5 | 0.61 | 60 | 1,116 | |
| | S. wall | | | | $5^6 \times 4^1$ | 22.0 | 22.0 | 0.61 | 60 | 805 | |
| | Floor | | | 148.5 | | | 148.5 | 0.480 | 25 | 1,782 | |
| | | | | | | | | | | | 5,061 |
| Bath | N. wall | 5.0 | 8 | 40.0 | $3^0 \times 2^4$ | 8.25 | 31.75 | 0.062 | 60 | 118 | |
| | Ceiling | | | 40.0 | | | 40.0 | 0.048 | 56 | 107 | |
| | Window | | | | $3^0 \times 2^4$ | 8.25 | 8.25 | 0.61 | 60 | 302 | |
| | Floor | | | 40.0 | | | 40.0 | 0.480 | 25 | 480 | |
| | | | | | | | | | | | 782 |
| Kitchen/ living room | S. wall | 4.0 | 8 | 32.0 | | | 32.0 | 0.062 | 60 | 119 | |
| | Ceiling | | | 20.0 | | | 20.0 | 0.048 | 56 | 54 | |
| | Floor | | | 20.0 | | | 20.0 | 0.480 | 25 | 240 | |
| | | | | | | | | | | | 413 |
| Crack loss | | | | 124.0 | | | 124.0 | 0.71 | 60 | 5,282 | 5,282 |
| | | | | | | | Total heat loss (Btu/hr) | | | | 19,711 |

| *Roof* | | *Floor* | |
|---|---|---|---|
| ½-in. plywood sheating | 0.62 | Carpet and pad | 2.08 |
| Building paper | 0.06 | | |
| Wood shingles | 0.94 | | |
| Inside air film | 0.17 | | |
| Outside air film | 0.61 | | |
| | 2.40 | | |

$$U = \frac{1}{2.08} = 0.4807$$

$$U = \frac{1}{2.40} = 0.4166$$

$$U\text{-factor for windows} = 0.61$$

$$U\text{-factor for crack loss} = 0.71$$

Now refer to the floor plan and calculate all the exterior wall areas, areas of windows and doors to be subtracted from the exterior walls, and the "crack factor" of all exterior windows or doors. When this has been completed, insert the various factors in Table 23-6, multiply the net area by the $U$-factor by the temperature difference for each wall or area. Finally, add these together for each room and then for the total heat loss that must be supplied to maintain 70°F when the exterior temperature is 10°F. Tables A9 and A10 in the Appendix give outside temperatures for other locations, but 70°F is more or less standard for interior temperature.

In the preceding example, the total heat loss to be supplied is approximately 20,000 Btu/hr *if* the weather cooperates and the outside temperature does not go below 10°F. This means that the *minimum* heating unit to be installed must supply 20,000 Btu/hr, but if the temperature goes lower than 10°F, the residence will be "cold" unless some arrangement is made to close off a part of the structure. The room arrangement in this example does not lend itself to such a solution. For the sake of comfort in such conditions, it is probably wise to consider a heating unit that will supply more than the required minimum, just in case.

The reader will probably notice that none of the interior walls between rooms has been considered. These are generally assumed to be similar in requirements on both sides, but some conditions may exist where one room is to be maintained at a lower, or higher, temperature than adjacent rooms. This would require an additional calculation to determine the $U$-factor of each wall between the rooms. A similar condition may exist between a kitchen and the adjacent attached garage wall where the garage is not heated. In general, the method above, although a bit more work than an estimate, will provide adequate information for most residential construction.

## 23.8 AVERAGE-VALUE METHOD

Accurate heat loss/heat gain computation is a rather long and exacting process that may not always be necessary. The average-value method is a little faster, as it uses some averages for transmission rather than calculating each material. Four major factors are assigned averages as follows: (1) wall factor: 0.32 Btu, (2) glass factor: 1.0 Btu, (3) contents: 0.02 Btu, and (4) radiation: 240 Btu. Each of these is based on a temperature of 1°F and must be multiplied by the design temperature difference ($\Delta t$). The averages shown for the four factors above indicate the amount of heat in Btu that will pass through that material in 1 hour.

Several simple steps are necessary, as follows:

1. Determine the design temperature difference $\Delta t$ = °F (highest temperature, $t_1$, minus lowest temperature, $t_2$).
2. Determine the cubic volume (CV) of the room and multiply by the content factor 0.02 = Btu/hr.
3. Subtract the total window area from the exterior wall area and multiply the net wall by the wall factor 0.32 = Btu/hr.
4. Multiply the glass area by the glass factor 1.0 = Btu/hr.
5. Add all Btu/hr figures for total loss.

**EXAMPLE**

Assume a room 10 ft × 12 ft × 8 ft high, two walls exposed to outside temperature, one 3′-0″ × 4′-0″ window in each wall, and a design temperature difference of 70°F inside and 30°F outside.

1. $\Delta t = 70 - 30 = 40°\text{F}$
2. $\text{CV} = 10 \times 12 \times 8 = 960 \times 0.02 \times 40(\Delta t) = 768$ Btu/hr
3. Walls $= A(10 \times 8 - 12 = 68) + B(12 \times 8 - 12 = 84) = 152 \times 0.32 \times 40 = 1945.6$ Btu/hr
4. Glass $= 3 \times 4 \times 2 = 24 \times 1.0 \times 40 = 960$ Btu/hr
5. Total Btu/hr $= 768 + 1945 + 960 = 3673$ Btu/hr

## 23.9 HEAT GAIN

For purposes of air-conditioning or refrigeration, some of the same approaches and factors are used to calculate the amount of heat that is gained in a room or space. Additional factors, such as latitude of the construction, orientation, shade or overhang of roof, wind velocity, and the type and density of materials all need to be included for an accurate figure. All of these are available from local sources or various handbooks, but a faster approximation which is usually adequate is also simpler. The following step-by-step method illustrates this approach.

1. Find the area of walls exposed to outside temperatures.
2. Find the area of windows.
3. Subtract the window area from the wall area.
4. Find the value factor in Table 23-1.
5. Multiply the net wall area by the wall factor = Btu/hr.

**EXAMPLE**

Assume a room 14 ft × 16 ft × 8 ft high, one 16-ft wall exposed to outside temperature, one window 6′-0″ wide × 4′-0″ high, with wall construction of wood studs covered with ⅜-in. gypsum plaster inside and 3-in. insulation and 5/16-in. plywood on the exterior.

1. Wall area $= 16 \times 8 = 124\ \text{ft}^2$

2. Window area = $6 \times 4 = 24$ ft$^2$
3. Difference = $124 - 24 = 100$ ft$^2$
4. Wall factor = 0.081
5. Btu/hr = $0.081 \times 100 = 8.1$

## 23.10 COOLING LOADS

Calculation of cooling loads for air-conditioning systems is an extension of those calculations for heating with additional consideration of shading of walls, orientation of the building, people, and other internal heat loads. An air-conditioning system must be capable of adding or reducing humidity, heating or cooling quite a bit of fresh air introduced deliberately for ventilation, and overcoming friction in ducts and room pressures. The heat load is calculated the same as for heating, but the low temperature will be inside, not outside, and the crack loss will be from inside to outside except where exterior wind pressure is high. The refrigeration load may be calculated as follows:

1. Calculate total heat gains.
2. Calculate the heat gain due to fresh air by the formula

$$\text{Btu} = \frac{V(\text{ft}^3 \text{ per hour fresh air})}{\text{ft}^3 \text{ per pound}} \times (H_{\text{fa}} - H_{\text{dp}})$$

   where $H_{\text{fa}}$ is the total heat in 1 lb of fresh air at the outside temperature and $H_{\text{dp}}$ is the total heat in conditioned air at the dew-point temperature.
3. Total heat gain (Btu/hr)/12,000 = tons of refrigeration.

## 23.11 COMMERCIAL CALCULATIONS

In calculating heating or air-conditioning requirements for commercial installations, a number of additional considerations enter into an accurate survey. For heating, it is usually adequate to use the normal material transmission values plus the internal heat generated by people, lighting, cooking, and similar items. Air-conditioning, however, involves other factors to arrive at an accurate load calculation.

Foremost among these additional factors is time of *peak load*. An office building will have a rather high demand for air-conditioning during operating hours from about 8:00 A.M. to 11:00 A.M., a lower demand from 11:00 A.M. to 1:00 P.M., about the same afternoon demand as in the morning, but a very low demand from 5:00 P.M. to 8:00 A.M. A restaurant or nightclub will have little demand during the day but a high demand from perhaps 5:00 P.M. to 2:00 A.M. Cafeterias may have greatest demand at noon, while "24-hour" coffee shops have approximately the same demand all day long but less after the evening dining hour. Factories and retail stores will have day-long loads but little at night unless night shifts are employed. Time of year will also have an effect, as loads generated at Christmas or other holidays will not be the same as "off-season" loads and should not be considered for normal calculation purposes.

Although all these factors may be figured separately and then coordinated for a final load analysis, it would be a waste of time to do so. Various tables and other aids are available by request from the National Environmental Systems Contractors Association (NESCA) or the American Society of Heating, Refrigerating and Air-Conditioning Engineers, Inc. (ASHRAE).

## 23.12 UTILITY COSTS

A survey of heating or air-conditioning requirements for new construction, or remodeling work, may be a guide for heating or cooling, but is the suggested system economical in regard to fuel? This brings up the question of which fuels are available locally and which may be least expensive if a choice is possible. In making any comparison of systems or fuels, the same formulas should be used, as different formulas may be somewhat biased toward the organization proposing the formula. To prevent such disparity, two formulas are suggested in this chapter, the heat-loss formula and the degree-day formula.

The *heat-loss formula* includes the following factors:

1. Difference between inside and outside design temperatures
2. Heat loss, expressed in Btu
3. Average winter temperature difference
4. Total hours of heating season
5. Btu per unit of fuel
6. Efficiency of utilization

The design difference is the same as for any heating calculation, and the total heat loss is arrived at in a similar manner. The average winter temperature is calculated by subtracting the average low temperature from the average high temperature. The total number of hours for operating the heating system depends on location of the building and is obtained by estimating total days multiplied by 24 hours. The Btu value of various fuels is obtained from tables, and the efficiency of gas- or oil-fired heaters is about 80% and of electricity, 100%.

**EXAMPLE**

Assume the difference between inside and outside temperature as 70°F − 40°F, the heat loss as 50,000 Btu, the average winter temperature as 50°F, total heating hours as 5025, and efficiency for oil and gas as 80%, and for electricity as 100%. Oil has 140,000 Btu/gal, gas 100,000 Btu/therm, and electricity 3143 Btu/Wh.

$$\frac{\text{Btu loss} \times \text{hr}}{\text{Btu fuel} \times E} \times \frac{t_1 - t_2}{t_3} = Q$$

Oil fuel:

$$\frac{50{,}000 \times 5025}{140{,}000 \times 0.80} \times \frac{70 - 40}{70} = 9601.3 \text{ gal}$$

Gas fuel:

$$\frac{50{,}000 \times 5025}{100{,}000 \times 0.80} \times \frac{70 - 40}{70} = 13{,}441.8 \text{ therms}$$

Electricity:

$$\frac{50{,}000 \times 5025}{3143 \times 1.0} \times \frac{70-40}{70} = \frac{343{,}124}{1000} = 34{,}312 \text{ kWh}$$

The *degree-day formula*, designed by the American Gas Association, is based on the theory that interior heat created by people, sunlight, and equipment will continue until the outside temperature drops to 65°F. The amount of fuel used will be directly proportional to the difference between 65°F and the mean outside temperature. The mean outdoor temperature is the sum of the highest and lowest outside temperatures of a 24-hour day divided by 2. Each degree in temperature below 65°F is considered as a *degree-day*. In the degree-day formula, the heat loss is divided by 1000, and the result is multiplied by the figure for the unit fuel consumption per degree-day per 1000 Btu design heat loss. The resulting figure is multiplied by the total number of degree-days in the heating season.

**EXAMPLE**

Assume the heat loss as 50,000 Btu, outside design temperature as zero, degree-days of 5025, oil at 140,000 Btu/gal, gas at 100,000 Btu/therm, and electricity at 3143 Btu/Wh. For gas efficiency, use 0.00429 and for oil efficiency use 0.00304 at 80%.

$$\frac{\text{Btu loss}}{1000} \times E \times D \text{ (degree-days)} = Q$$

Oil fuel:

$$\frac{50{,}000}{1000} \times 0.00304 \times 5025 = 763.8 \text{ gal}$$

Gas fuel:

$$\frac{50{,}000}{1000} \times 0.00429 \times 5025 = 1078 \text{ therms}$$

Electricity:

$$\text{Therms per year} \times \text{Btu/therm} \times E$$

$$1078 \times 100{,}000 \times 0.80 = \frac{86{,}240{,}000}{1000} = 86{,}240 \text{ kWh}$$

The cost of various fuels may be easily found by multiplying the cost of each fuel by the amount of fuel necessary for the time used.

## 23.13 SUMMARY

The calculation of heat loss and the ultimate estimate of fuel costs to maintain human comfort, either by heating or air-conditioning, is a modern necessity. Although the determination of heat loss in a structure may seem at first to be a long, complicated process, this is not true when a careful and accurate approach is made and good results are obtained. For the most part, estimates of heat loss or heating costs are made by specialized engineers, and this should always be the case for multistoried construction or special construction such as hospitals and similar, complicated structures.

Residential work is somewhat repetitive in that certain combinations of materials are used in building after building. This allows for reuse of many of the calculations and more rapid answers. But even if you do not intend to do the work yourself, some knowledge may be helpful in reviewing the work of others. The information in this chapter will also help toward an understanding of the systems discussed in later chapters, as well making that information more useful and easier to apply to actual construction.

# 24

# Heating

## 24.1 INTRODUCTION

Proper heating design is the first aspect of HVAC (heating, ventilation, and air-conditioning) that we cover in the following chapters. Many different types of heating systems have been used in buildings, but some former methods are now obsolete or are primarily ornamental or aesthetic, such as fireplaces. Each type of heating has its own characteristics and its own advantages or disadvantages. Several factors influence the selection and design of any heating method: (1) heat-conveying medium, (2) fuel to be used, (3) type of heat, (4) efficiency desired, and (5) assumed cost of operation.

The *heat-conveying medium* may be air, water, steam, or electricity. Air is the medium used in connection with air-conditioning and various types of warm-air furnaces. Hot water is used as a conveyor of heat either by means of the gravity flow of heated water or through forced circulation by pumping, and is the system most often used. Steam, formerly the primary medium in use, is somewhat similar to hot-water usage but is being rapidly replaced by warm-air and hot-water systems in residential as well as in some commercial construction. Electricity, as a system, is rather limited to various small units or as supplemental heat. Heat pumps and solar energy can also be used, but as they are combination systems, they are not usually treated as conveying media.

Various types of fuels are discussed in Chapter 21, but their selection, firing requirements, caloric heat value, efficiency, and cost are factors that must be considered for each fuel. Fuel, as well as the type of fire chamber or burner used, are also influential in determination of a heating system. Every fuel has an efficiency factor, depending on the heat value, type of combustion, comparative cost of fuel, and climatic conditions. Finally, the type of building construction and its occupancy type and load also influence heating.

## 24.2 HEATING DRAWINGS AND SYMBOLS

As noted earlier, every set of drawings for construction or any mechanical work must use symbols to convey part of the information to the manufacturer and/or installer. Heating drawings are no different. In addition, it has been standard practice for years to show "mechanical" work rather diagrammatically; that is, plumbing and heating pipes and ducts are usually shown by a single line, and often not in the exact location where they are ultimately installed. The symbols shown in Fig. 24-1 are those most commonly used in heating drawings:

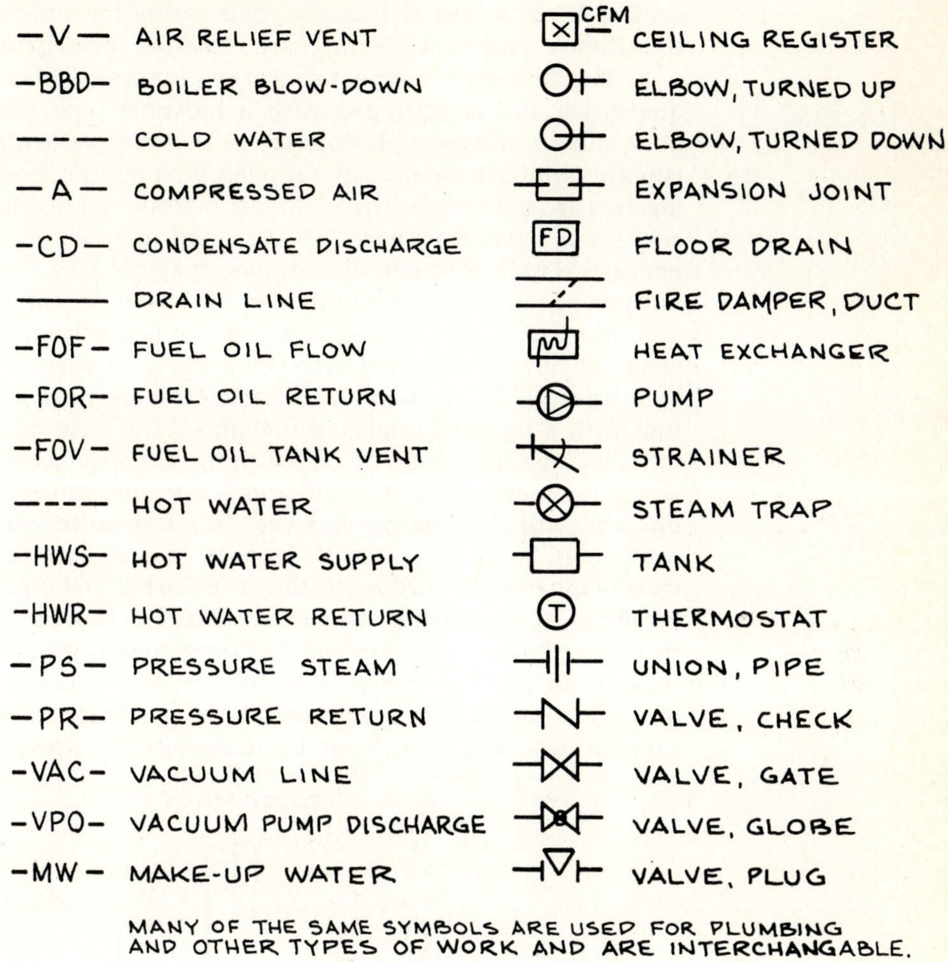


**Figure 24-1** Heating symbols.

## 24.3 COAL-FIRED SYSTEMS

Solid material such as coal or coke is one of the major fuels used in the eastern and midwestern United States and other locations, where coal is more available and less expensive than some other forms of fuel. Coal furnaces and boilers are either hand-fed or automatically fed, depending primarily on their use and installation. Gravity and forced warm-air systems, as well as hot-water systems, may have a coal-fired furnace as a heat source. As a matter of nomenclature, *furnaces* are used for heating air, while *boilers* are used for heating water or producing steam. Boilers are discussed in Section 24.7.

The coal-fired furnace is rather simple in its construction, with only a few basic parts (see Fig. 24-2). The firebox, grate, and perhaps the ash-pit base are usually manufactured of cast iron in several sections bolted together or in units that interlock. Some furnaces are fabricated with a sheet-metal shell lined with firebrick or similar refractory material. The grate, usually of cast iron, is composed of a number of sections, with edges that have alternating "fingers" and openings designed to hold the ash and burning fuel properly when in a horizontal position but so constructed that the various sections may be rotated 90° or shaken when the ash is dumped. The central firebox and ash pit are enclosed in a sheet metal casing with an air space between the central firebox and the casing. The casing terminates in a "bonnet" above the firebox, and duct openings are provided for distribution of the heated air.

The firebox is connected to the chimney with a horizontal flue of adequate size and is equipped with a movable pipe damper. A cast-iron feed door and a similar ash-pit door are both equipped with adjustable dampers to control combustion air. Air supplied through the feed door affects the burning fuel from the top, while the air introduced through the ash pit (primary air) keeps unburned particles afire and supplies oxygen for the fire from beneath. Draft through the furnace may thus be controlled by adjustment of openings in the ash-pit door, the feed door, or the flue.

Most simple coal-fired furnaces are hand-fired; that is, the coal from a bin is physically shoveled into the furnace at varying periods and in varying quantities. This "system" creates uneven burning by covering the burning fuel with a layer of unheated fuel and if the burning fire is low may actually put the fire out instead of replenishing it. Frequent opening of the feed door also allows excess air to chill the flame, and excess fuel creates a smoking interval until the fuel catches fire. In addition to hand-firing, coal-fired furnaces may be equipped with *stokers* of several types (Fig. 24-3) and automatic controls. According to the *ASHRAE 1979 Equipment Guide and Data Book*, most residential stokers are Class 1 (10 to 100 lb/hr) with Class 2 through Class 4 for industrial or commercial installation. Stokers may feed from a bin or hopper that is filled at intervals from a storage bin, or may be directly connected to a storage bin. In operation, a worm feed continuously supplies raw fuel to the fire by underfeeding (from beneath the fire rather

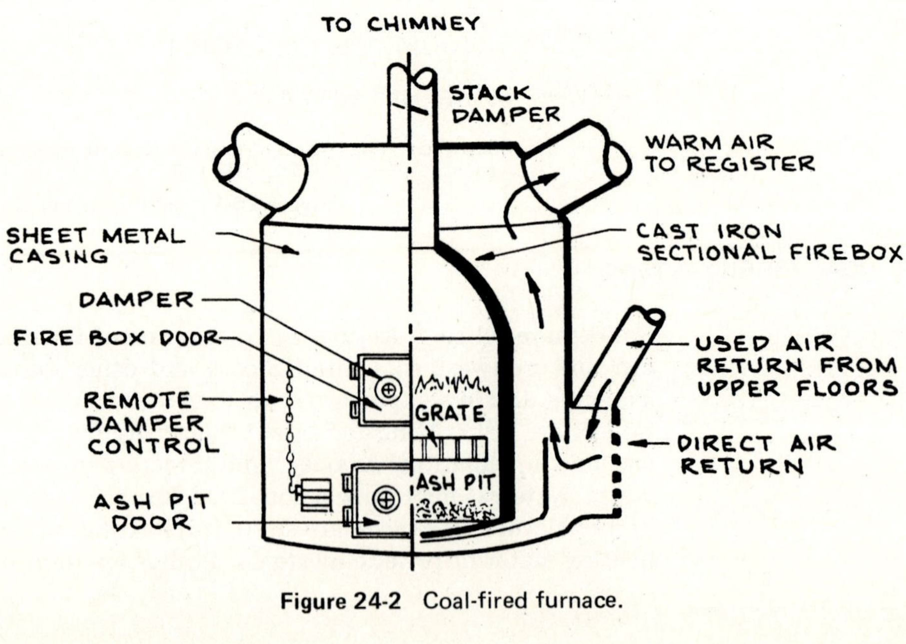


**Figure 24-2** Coal-fired furnace.

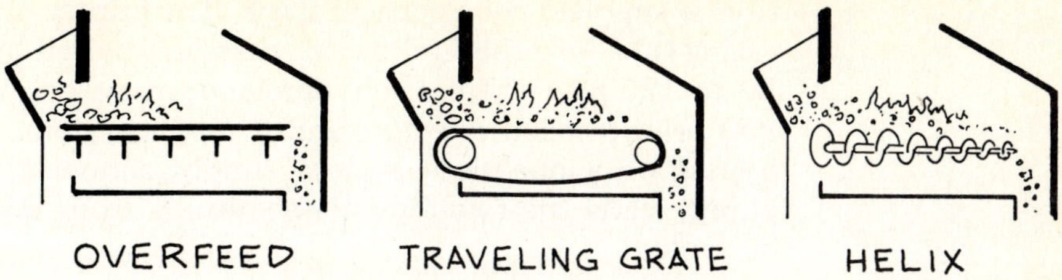


**Figure 24-3** Types of coal stokers.

than on top of it), and a blower supplies air to the retort or fire chamber. Ash is discharged around the edges of the retort and simply falls into an ash pit. Operation is controlled by thermostats, which in turn control the feed and blower motors.

## 24.4 OIL-FIRED FURNACES

Several types of oil-fired furnaces are available: (1) upflow, (2) downflow, and (3) horizontal-flow models. All furnaces should be listed by Underwriters' Laboratory (UL) and are usually supplied completely assembled except for external wiring, external controls, vents, fuel tanks, and similar parts. As furnace oil has a relatively heavy viscosity, the "oil burner" (see Fig. 24-4) atomizes or vaporizes the oil so that it will burn in a proper atmosphere. The system may be either low-pressure or high-pressure. Oil burners may also be of three major types: (1) vertical rotary burners, (2) horizontal rotary burners, and (3) wall-flame rotary burners. Vaporizing burners are usually used for residential buildings, while the rotary types are more common for industrial/commericial installations. The various grades of fuel oil are discussed in Chapter 21.

The construction of an oil-burning furnace includes a cast-iron or refractory-lined firebox, surrounded by a sheet-metal casing similar to that used for coal-fired furnaces. Since there is no ash, or at least a minimum of residue, there is no need for a grate or ash pit. Every furnace, however, requires a proper vent to a chimney. *High-pressure burners* are of the spraying type, not atomizers, and are also known as *gun, sprayer,* or *pressure burners.* In spraying burners, the oil is broken up into minute liquid particles by passing it through a series of strainers, and then into the nozzle, where air pressure breaks it up into a fine mist. Air is supplied by means of a *turbulator*, which gives the air a twisting motion and combines it with the oil mist. Igni-

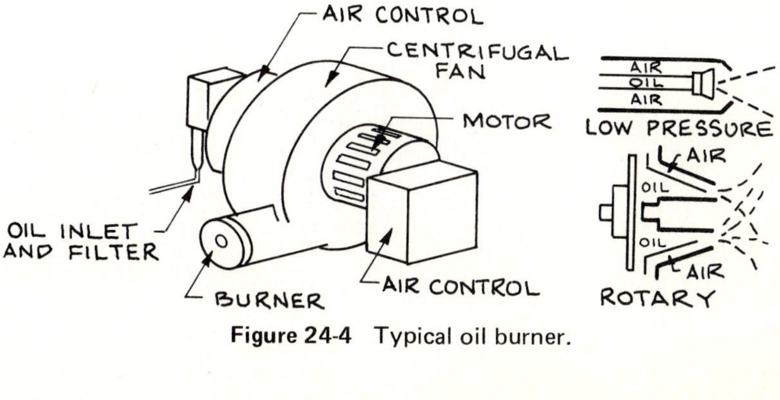


**Figure 24-4** Typical oil burner.

tion is supplied by electrical spark. The burner blows the oil spray through a casing tube into the firebox. Most residential oil burners require a pressure of 80 to 125 psi, whereas commercial or industrial burners require from 100 to 300 psi. So-called *one-pipe* and *two-pipe* systems are similar, except that the bypass of a one-piece system simply connects around the oil line at the pump, while the two-pipe connection is from the pressure relief back to the oil tank.

*Rotary burners* throw out oil by low pressure. In this type of nozzle, oil is supplied through a central tube and is sprayed by air pressure against a vaporizing rim or collar. The process further separates the oil into a fine mist, which burns against the rim in a blue flame. Another type, called a *pot burner*, is one in which the oil is supplied below the flame area and thus preheated, and the vapor then rises in the pot to a flame spreader. All these burners may have necessary air supplied by adequate chimney draft but in fact are almost universally equipped with fans or other mechanical devices for the air supply. Heating units for oil burning are usually classed as *upflow*, which means that the passage of air is normally from near the floor (inlet) through the fan and upward to the motorized burner, and then into a heat-exchanger unit before being discharged into a duct system (outlet).

It is not possible to hand-fire an oil-burning furnace, so the fuel must be stored in a tank. The location of the tank is subject to local ordinances but may be outside the building above or below the burner level or may be inside the building. Filler piping to the tank should always be terminated outside the building, and a tank vent pipe should also be terminated outside the building. Figure 24-5 shows how tanks may be installed in relation to burners and other equipment. Filler pipes should be a minimum of 2 in. in diameter. Vent piping should be at least 1¼ in. in diameter. Both piping systems should be wrought iron. Connection from the tank to the burner unit should be made with copper tubing with at least a ½-in. outside dimension and without joints if possible. Tank size will vary with heating requirements but should not be smaller than 200-gal capacity and thoroughly protected against rusting or damage.

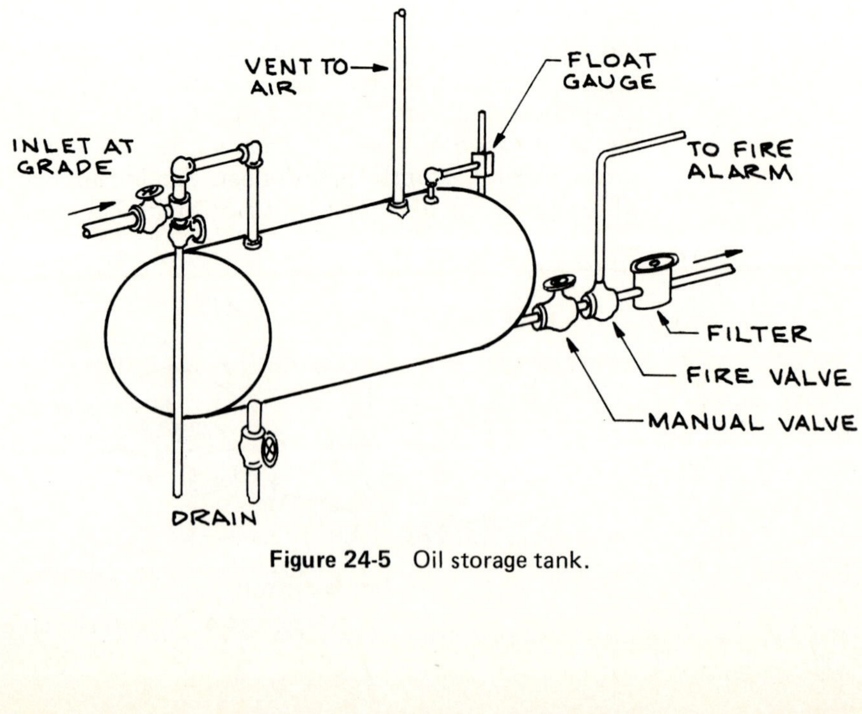


**Figure 24-5** Oil storage tank.

## 24.5 GAS-FIRED FURNACES

Gas-fired furnaces may be *downflow*, in which the air is forced down over the heating element; *upflow*, in which the air is taken in at floor level; forced upward over a heating element; or *horizontal*, in which the air moves from blower to burner to heat exchanger in a horizontal manner. Most gas burners function in the same manner as the burner of a gas-fired range: the gas is admitted into the burner at a pressure of approximately ¼ psi and is diluted or mixed with primary air at a ratio of about 1 part gas to 5 parts air for manufactured gas and about 1:10 for natural gas. Most burners operate at about 50% efficiency. Admission of primary air is accomplished by a circular adjustable shutter to provide an inner cone of flame, while secondary air surrounding the flame burns off any gas not fired in the inner cone. When the height of the inner cone is about 70 to 75% of the height of the visible flame, the burner is considered as properly adjusted.

Most modern gas-fired furnaces consist of several sections, each containing a burner and a heat-exchanger unit. To function properly all burners must be ignited simultaneously, or unburned gas may accumulate and explode. To accomplish this, most manufacturers provide a *crossover pipe*, which tends to feed all burners at the same time by distributing gas, and flame, to each at split-second intervals. The gas mixture is ignited by a *pilot light* or *pilot burner* activated by a thermocouple. The end of the thermocouple is held constantly in the pilot flame, which then generates an electrical current between the burner and the gas valve. If the pilot flame is extinguished, the gas valve closes.

Another type of burner uses a venturi-type mixing tube and an *inshot* or *upshot burner* (see Fig. 24-6). In this method the gas passes through a small jet at high velocity, which in turn pulls in air from the atmosphere by venturi-tube action. The gas mixture then passes into the feeder tube and either travels in a straight line (inshot) or the tube bends upward to a burner (upshot). The actual burner is usually a multiple-jet unit for upshot types, but a perforated disk at the end of the inshot straight tube. Ignition is most often accomplished by using direct spark rather than a pilot burner. The two types just described are generally used for industrial/commercial installations and the fan, blower, control functions, and burner tubes are supplied as a single unit.

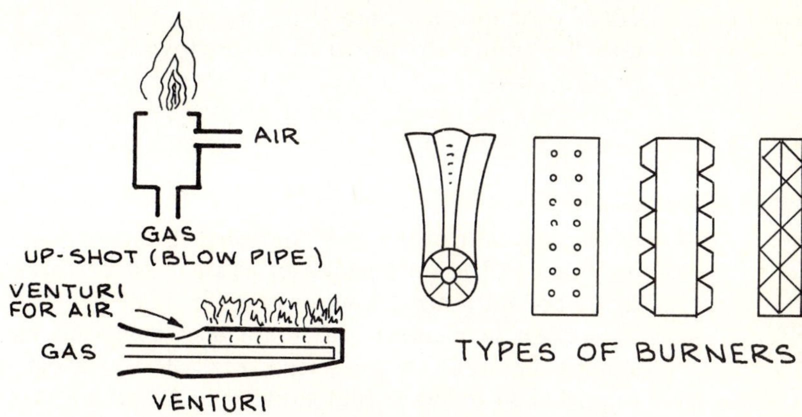


**Figure 24-6** Gas burners.

For a burner to operate satisfactorily, there must be adequate gas manifold pressure, proper orifice size, and proper air input to the burner. An orifice that is too small will cause delayed ignition and poor heating. If it is oversized, the primary air cannot be drawn in and incomplete combustion will result. Reduction of gas velocity will result in reduction of the amount of primary air. Most burners are manufactured from pressed steel or cast iron, with the drilled types favoring cast iron. Four major types are available: drilled ports, slotted ports, ribbon, and inshot. The type of burner will depend on the equipment used and each manufacturer will select the one they feel is the best for their units. The mixing tube is the portion of the burner which mixes the gas and the air, and it usually extends the full length of the burner.

## 24.6 ELECTRIC HEAT

Electric heat should not be generally considered as a "furnace" in the same manner that coal, oil, or gas furnaces are. In the first place, heating by electricity is expensive, not suited to heating large areas, and is most often used as supplemental or localized heat. Several forms of electric heating are available, and most operate on 60-Hz 110/120-V service. One of the most common electric heat sources is a coiled wire resistance unit that supplies radiant heat from about 1000 to 5000 W and is used as local heat in bathrooms and similar small spaces. Another heat source used in bathrooms or in areas where very concentrated heat is required is a 4½-in. heat lamp which is simply screwed into a standard socket or ceiling fixture and is rated at 100 to 300 W. These sources are often controlled with timer switches that automatically shut off the power at a preset time, thus preventing excessive waste of electricity and possible overheating or fire.

Baseboard electric units are used in homes as well as hotels, motels, and isolated offices or similar spaces. Units are available in a number of wattage values, are about 6 to 8 in. high by 6 in. wide and from 4 to 8 ft long. Resistance wiring is strung through their length, and some types have small fans to help distribute the heat generated. Although slower to provide heat than a central air system, the baseboard method will supply adequate heat to small rooms if operated for enough time. The principal advantage of baseboard heating lies in its inconspicuous location and its quiet operation.

*Electric furnaces* are available from 5 to 35 kW and are assumed to be 100% efficient, as there is no smoke, ash, or other residue. The basic design may be similar to the upflow or downflow gas-fired units, except that the heating element is a series of chrome-nickel-wire resistance coils. If all the coils were activated simultaneously, there would be a temporary excessive power demand on the electrical service, so *sequence relays* are often installed in the furnace to connect various heating units at predetermined times. Extensive internal fuse blocks are used to protect the unit from short circuits, overloads, and other malfunctions. A fan or blower is essential, and usually included in the unit package, as air must be forced over or through the resistance coil and into a duct system.

Electric heating grilles are also available for installation in ducts and similar enclosed spaces (see Fig. 24-7). Most grilles are resistance coils mounted in a frame that may be installed across the width of a duct system. The grille is controlled by a thermostat and does not include a filter or fan, as is typical of electric furnaces. The air in the duct is filtered at some other location and is moved through the electric heating grille by a fan or blower

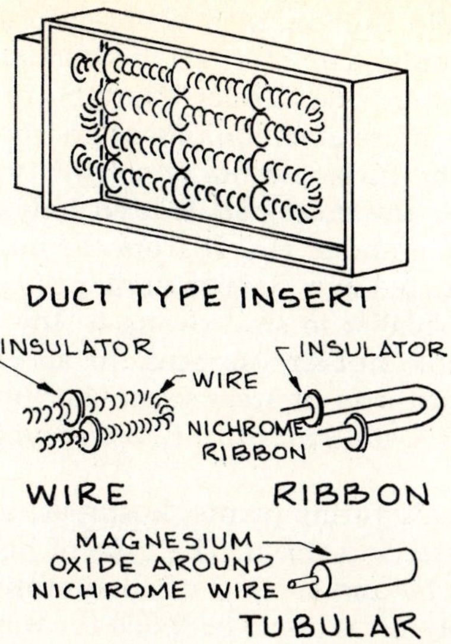


| WATTS - BTU/HOUR | | | |
|---|---|---|---|
| VOLTS | SIZE | WATTS | BTU/HR. |
| 120 | 28 IN. | 500 | 1700 |
| | 40 | 750 | 2400 |
| | 48 | 1000 | 3500 |
| | 60 | 1250 | 4300 |
| 208/240 | 28 | 375-500 | 1250-1700 |
| | 48 | 750-1000 | 2400-3500 |
| | 72 | 1150-1500 | 3800-5100 |
| | 120 | 1850-2500 | 6400-8500 |
| 240/277 | 48 | 750-1000 | 2400-3500 |
| | 72 | 1150-1500 | 3800-5100 |
| | 120 | 1850-2500 | 6400-8500 |

**Figure 24-7** Electric heat.

entirely separate from the grille. Like horizontal gas heaters, electric furnaces may also be used in duct systems and come complete with filters, fans, and heating elements.

## 24.7 BOILERS

The many types of boilers may be classified as water-tube or fire-tube, depending on where the water or fire is carried in the boiler, may be of sectional cast iron or steel construction, and may be low-pressure or high-pressure, depending on their use and whether hot water or steam is to be produced. Firing may be accomplished by the use of coal, coke, oil, or gas. Wood is used much less often than it once was, primarily because of its relative scarcity, and difficulty of storing and handling, although wood scrap and sawdust waste from woodworking plants is regularly used. Hot water or steam generated by boilers may be used directly in radiators or for hot-air systems through some type of heat exchanger.

*Hot-water boilers* are really specialized furnaces fabricated to provide hot water rather than hot air and are of various water-tube designs. Cast-iron sectional "boilers" (Fig. 24-8) are manufactured in units which, when secured

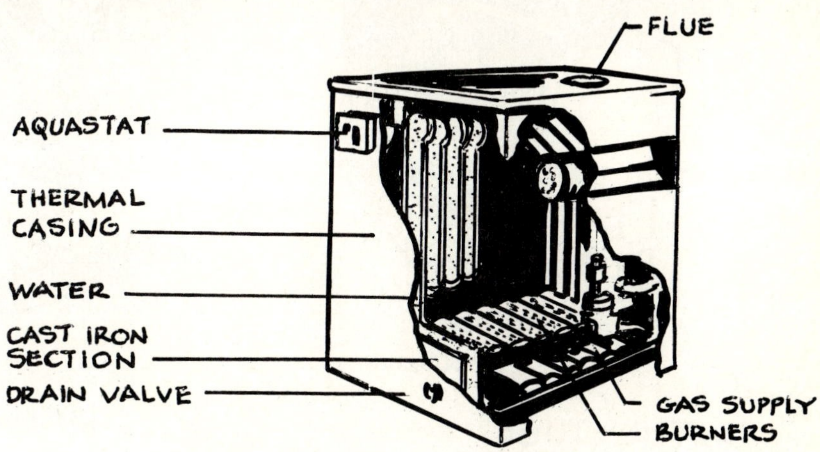


**Figure 24-8** Cast iron residential boiler.

together, form a large firebox with water-carrying capabilities at the sides and top of the assembly. The sections are connected together to allow free circulation of the water. The "boiler" may be fired with coal, oil, or gas, so grates, firing doors, and related equipment are the same as those indicated in the sections discussing fuels and burners. This type of "boiler" is essentially low-pressure and, for the most part is used only when the hot water is not transported more than about 100 ft from the heating source. Steel sectional "boilers" are constructed in a manner similar to cast-iron boilers, except that such units may be smaller in size, owing to the type of material used, and more closely resemble hot-air furnaces. In all cases these units should be heavily insulated. Cast-iron or steel sectional "boilers" are available to produce up to about 3½ million Btu/hr, but beyond that a different type is more economical.

Industrial manufacturing plants, hospitals, and similar structures often require both hot water and steam. The type of boiler used for this purpose is known as a *shell boiler* and consists of an appropriate firebox, a tube system for heating the water, and a dome or space for water or steam, all enclosed in an insulated metal shell. In terms of heating capacity, a boiler may be rated in thousands of Btu per hour or square feet of *equivalent direct radiation* (EDR). A *boiler horsepower* (bhp) is the evaporation of 34.5 lb of water into dry steam at 212°F. One bhp is considered equivalent to 140 $ft^2$ of steam radiation per hour. One square foot of direct radiation (EDR) is equal to 240 Btu per hour for steam boilers, and 1 $ft^2$ of EDR for water boilers equals 150 Btu/hr. Efficiency of an oil- or gas-fired boiler is usually considered at 70 to 80%, while a coal-fired boiler is about 60 to 75% efficient.

Industrial boilers (Fig. 24-9) are invariably *water-tube*, in which the tubes exposed to the heat are filled with water, or are *fire-tube*, in which the tubes are surrounded by water and the fire or hot gases pass through the tubes. Simple boilers have the tubes arranged for only one pass of the fired gases from the firebox to stack, but more economy is realized if double-pass or multiple-pass arrangements are used. Various other arrangements have been devised to add additional heat by passing the cooled gases over the steel shell

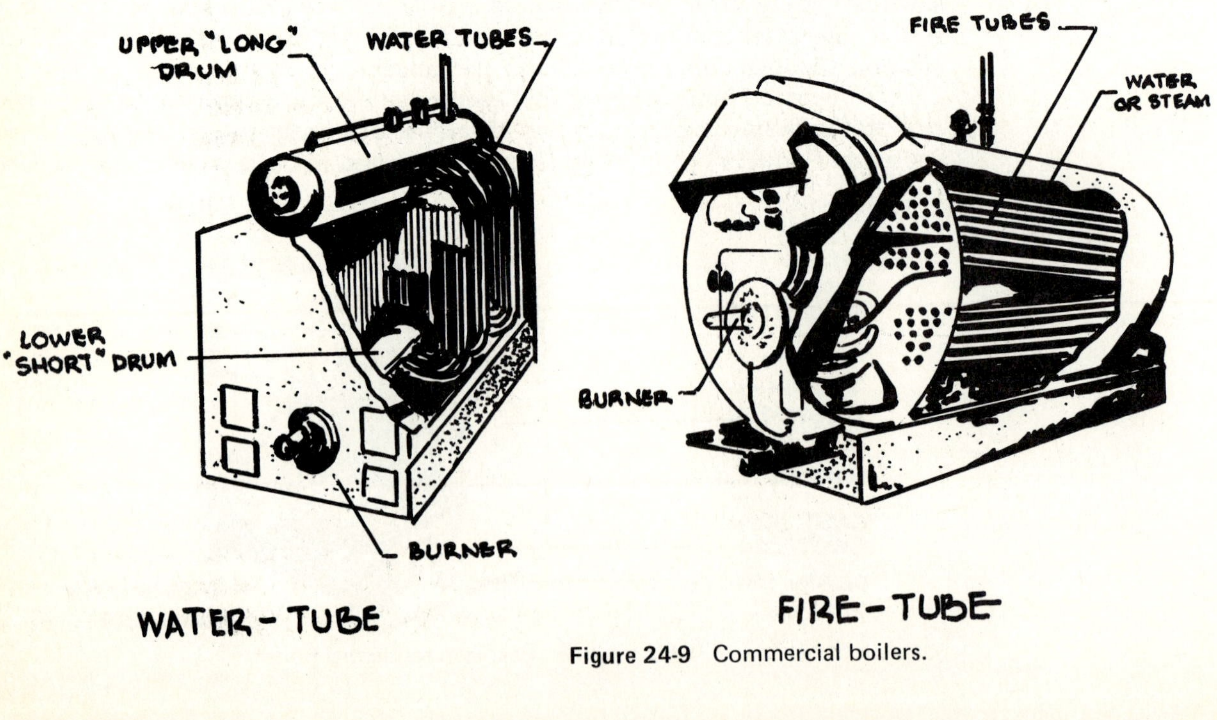


**Figure 24-9** Commercial boilers.

or by using this otherwise wasted heat to preheat boiler water. Draft for most industrial boilers is provided by fans or blowers, particularly if oil-fired burners are used.

A variety of fittings and accessories are necessary for all types of boilers (see Fig. 24-10). In addition to the controls required for coal stokers or oil- or gas-fired units, there must be a number of draft regulators at fire or ash doors and in the breeching between the boiler and stack. The water gauge, a vertical glass tube, indicates the water level in the boiler, and a steam gauge indicates the steam pressure in steam boilers. A most important device is the safety valve, which can be set to trip or "blow" if the internal pressure exceeds that desired. "Tying down" the safety valve to increase or maintain the boiler pressure is an old practice in steamboat races but often results in boiler explosions or other damage rather than an increase in quantity or quality of the steam. Low-water cutoff valves are also necessary to shut off automatic fuel-burning equipment when the water level drops below safe operation. In steam systems, a vacuum relief valve may be useful if the supply line is closed and a vacuum is created by condensation of steam. Other fittings include a number of types of valves, air separators, automatic regulators, aquastats, and similar devices. All boilers should have proper facilities and valves for complete drainage and control valves for both feedwater and hot water or steam.

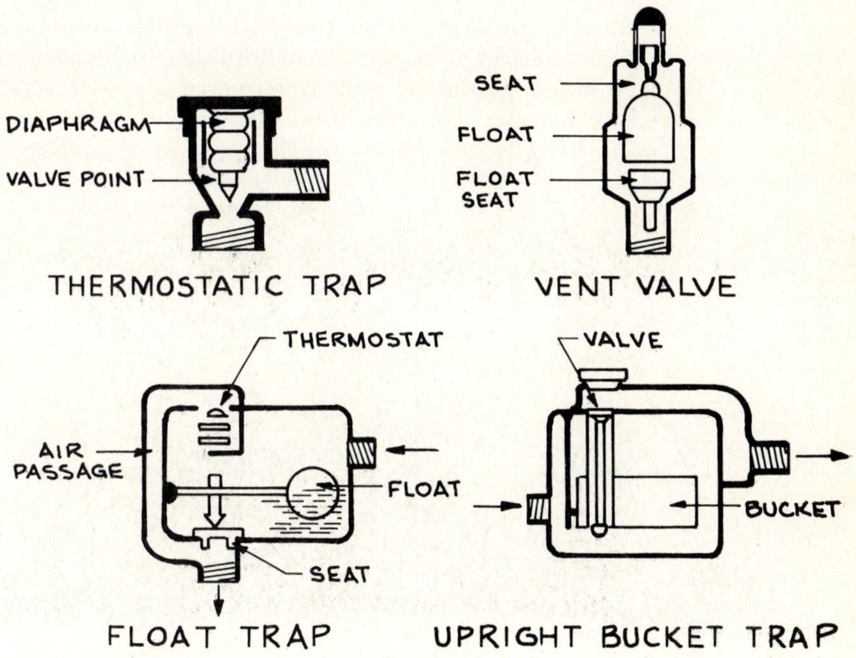


**Figure 24-10**

## 24.8 CHIMNEYS AND SMOKESTACKS

Chimneys and smokestacks serve the same purpose, conveying smoke and other combustion products from the fire to the free air. *Chimneys* are generally considered as a part of residential systems or small commercial installations, while *smokestacks* are a part of industrial or power-plant systems. Either type must be constructed of fire-resistant materials and of adequate cross section and height to provide draft enough to carry off the smoke. A smooth interior flue, accessibility for cleaning, proper separation from adjacent wood structures, and possible lighting of stacks as an airplane traffic hazard are additional requirements.

Residential chimneys originated when primitive peoples poked a hole in the roof of their dwellings to allow the smoke from their cooking fires to escape. Gradually, this simple hole became a vertical duct for the smoke and the chimney was born. In more modern times, the chimney was an addition to the fireplace and evolved through sticks or logs coated with clay, to stone, and to our present clay-lined masonry. In modern construction, two major types of chimneys are most common (see Fig. 24-11): (1) a brick or concrete masonry construction enclosing a preformed clay flue lining, and (2) an insulated double-wall sheet-metal pipe enclosed by, and separated from, wood framing. Cross-sectional area of the chimney is important to provide adequate draft. There are many books and tables that list the proper size of chimney flue required for different sizes of fireplace openings, and these should be used to construct a fireplace that does not direct smoke into the residence and that has adequate draft.

Masonry chimneys of brick, stone, or concrete units should have walls not less than 8 in. thick. This may mean careful arrangement of unit joints so that there are no open joints or cracks from the interior to the exterior of the chimney. To prevent this possibility, many locations require that a clay flue liner be installed to reduce the possibility of cracks and to provide a smooth interior surface that will decrease the amount of soot retained on the flue walls. In areas where earthquakes or high winds are anticipated, masonry chimney should be reinforced vertically with steel reinforcing rods buried in the masonry construction. Although fireplaces are no longer generally considered a major source for heating, they are desirable from an aesthetic view-

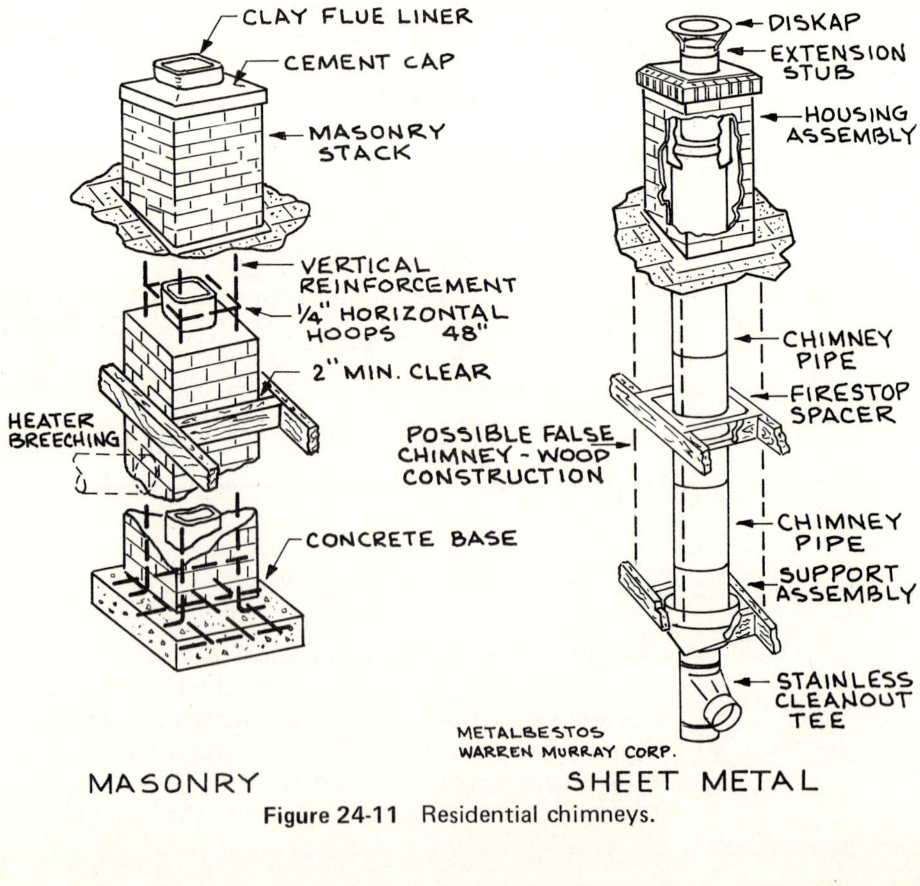


**Figure 24-11** Residential chimneys.

point. However, because masonry fireplaces and their chimneys are expensive to build and require time for proper construction, the preformed steel-shell fireplace and the sheet-metal chimney have come into existence. The sheet-metal chimney is simply a round pipe 8 to 24 in. in diameter, enclosed by, and separated from, another larger pipe, and installed a minimum of 2 in. from any combustible construction. This type of chimney is fastened to, and supported by, metal straps, which in turn are fastened to the building and maintain adequate separation. Wood framing, sometimes additionally lined with gypsum board is then built around the metal pipe to any exterior size desired and is covered with cement plaster, imitation brick, wood siding, or other material. The top of the framed chimney may be enclosed with a sheet-metal cover, plaster, or roofing. A similar type of sheet-metal chimney may be used for a furnace in residential work. All such chimney construction should be checked against the local building code for acceptance and compliance.

Industrial smokestacks are normally constructed with little concern for outward appearance. Their principal function is to provide proper draft for boilers and to carry the resulting smoke high enough into the air for adequate dispersion. Two major types of industrial smokestacks are common: (1) masonry or concrete stacks with refractory material linings, and (2) metal stacks without linings. Masonry or concrete stacks may be from 24 in. to 10 ft in outside diameter, enclosing a flue of 8 in. to 6 ft or more. Linings for these stacks are usually firebrick, especially manufactured for such use and laid up in fireclay with minimum exposed joints, and integrated structurally with the masonry or concrete stack walls. Steel stacks are constructed of a single layer of steel plate, usually $\frac{3}{8}$ to perhaps 1 in. thick, depending on height required, welded together and unlined. Most industrial stacks are designed freestanding and are connected to the smoke source by *breechings* or flues above the boiler level. Access is usually provided at the base of stacks for removal of deposited soot and other residue. Exterior, and sometimes interior, access to the stack is provided by integrated ladders of steel. Lighting of stacks at or near their top is required and red electric lights or laser-beam flashers are commonly used to warn aircraft. Industrial smoke is almost always cleaned with smoke washers, filters, or other devices before introduction into the smokestack.

Extension of chimneys and smokestacks above the roofs of buildings or the relation of the top of such chimneys to building walls and openings is most important. Most building ordinances specify the minimum distance or clearance of chimney openings from roofs, but Table 24-1 indicates some conditions that should always be considered. Chimneys that project more

**TABLE 24–1**
Chimney Clearance

| | *Minimum chimney wall thickness* | *Height above roof (ft)* | *Minimum clearance (in.)* |
|---|---|---|---|
| Residential masonry | 4 in. | 2 | 2 |
| Residential prefabricated metal | 18 gauge | 2 | 2 |
| Residential incinerator | 4 in. | 3 | 2 |
| Industrial | | | |
| 1400°F heat | 8 in. | 3 | 2 |
| 2000°F heat | 8 in. | 10 | 4 |
| 2000°F+ heat | 16 in. | 20 | 4 |
| Industrial incinerator | 8 in. | 10 | 4 |

than 4 ft above a roof should be braced in all directions to prevent damage by earthquake or high winds, and consideration should certainly be given to the direction of any steady wind and its effect on smoke dissipation.

## 24.9 WARM-AIR SYSTEMS

A warm-air heating system is one in which air is heated in a furnace and distributed through a series of ducts to spaces to be warmed. The heating unit may be any of those mentioned in this chapter, and the air may circulate by gravity or by power fans or blowers. In most situations, the air is heated with oil or gas and is distributed by a forced-warm-air system using a fan. Ducts may be fabricated of sheet metal, plastic, fiberglass, asbestos cement, or molded coated-paper compounds, and may be circular or rectangular in cross section. Prefabricated fittings such as tees, elbows, reducers, and wyes are available in most of the same materials. Finish grilles, registers, dampers, and other fittings are also furnished in a considerable variety of sizes, shapes, and materials.

Warm-air furnaces may be located in basements, attics, crawl spaces, closets or utility rooms, garages, or even on the roof. Several types of furnaces are available and are used in different locations. *Upflow (lowboy)* units are usually installed in crawl spaces or basements and the warm air is delivered upward into the building. When the furnace is installed at floor level in a closet or is freestanding, it is usually an *updraft (highboy)*, with return air admitted at floor level and warmed air distributed from the top of the furnace into attic ducts. Furnaces that admit air at the top and provide warm air at the bottom, *counterflow* or *downflow*, are usually used in crawl spaces or for duct systems beneath concrete slabs. Horizontal furnaces are used in crawl spaces, under floors, or other locations where space is minimal.

The three major systems of duct installation are generally installed beneath the floor or above the ceiling. With concrete floor slabs, the ductwork may be buried beneath the slab if the duct is properly insulated. Ducts buried in the earth are usually of asbestos cement or similar corrosion-resistant material. Ducts above the ceiling, in walls, or beneath suspended floors are usually of galvanized steel or aluminum, round or rectangular, and about 28 gauge. To reduce the transmission of sound or vibration, the furnace and ductwork are connected with rubber, cloth, or other flexible sleeves. Cooled air is returned to the furnace by a number of different methods: (1) central return at the furnace, (2) central return with a duct system, (3) individual room exhausts, or (4) by use of plenums integrated within the building construction. Supply grilles or registers usually have some type of air-control feature to regulate the amount of air (cubic feet per minute) introduced into a space (see Table 24-2). Return air grilles are sized to the system and normally do not have control features. Several types of splitters, dampers, firedampers, and access openings are used in duct systems to provide further control or fire protection.

Duct systems must be designed to provide adequate air in cubic feet per minute to all spaces to be heated or cooled, and are subject to two kinds of pressure, *static pressure*, present when the air is not moving, and *velocity pressure*, present when the air is moving. When air flows through a duct, some pressure is lost as a result of friction created by high velocity, small duct size, long duct length, direction changes of the duct, and sudden changes in the airstream. Heating estimating forms often give values for determining air quantities required, but this information can be easily calculated by using the following formula.

**TABLE 24-2**
Grilles and Registers

| *ft² /min* | *High-side wall (in.)* | *Perimeter floor (in.)* | *Ceiling (in. diameter)* | *Return (in.)* |
|---|---|---|---|---|
| 60 | 10 × 4 | 2¼ × 10 | | 10 × 6 |
| 80 | 10 × 6 | 2¼ × 12 | | |
| 100 | 14 × 6 | 2¼ × 14 | 6 | |
| 120 | 14 × 6 | 4 × 10 | 8 | 12 × 6 |
| 140 | 16 × 6 | 4 × 10 | 8 | |
| 160 | 16 × 6 | 4 × 12 | 8 | 10 × 8 |
| 180 | 20 × 6 | | 8 | |
| 200 | 20 × 6 | | 10 | 12 × 8 |
| 220 | 20 × 6 | | 10 | |
| 240 | 20 × 6 | | 10 | 18 × 6 |
| 260 | 24 × 6 | | | |
| 300 | 30 × 6 | | | 12 × 12 |

$$\text{ft}^2/\text{min} = \frac{\text{Btu/hr per room (sensible load)}}{1.08 \times \text{temperature difference between supply air and room air}}$$

For example, assume a sensible room load at 6000 Btu/hr and a temperature difference of 30°F.

$$\text{ft}^3/\text{min} = \frac{6000}{1.08 \times 30} = \frac{6000}{32.4} = 185\ \text{ft}^3/\text{min}$$

These same results may be obtained by using Fig. 24-12. Find the sensible load along the bottom of the chart (6000 Btu/hr) and follow upward until the degree-difference line (30°F) is reached. Then read the required cubic feet per minute at the left margin (185). For loads greater than those shown on the chart, simply multiply by 10 or 100 (i.e., 60,000 Btu/hr at 30°F equals 1850 ft³/min).

Sizing of supply- and return-air ducts may easily be obtained from tables provided by manufacturers. Three methods are used to determine sizes, an equal-friction method, a static-gain method, and a total-pressure method. The *equal-friction method* is mostly used for simple, symmetrical, low velocity systems and is based on maintaining the same air friction loss per 100 ft for the entire system. Duct velocity is reduced in the direction of the airflow but does not provide for equal static pressure at all openings, thus requiring balancing dampers in the system. The *static-gain method* is based on the theory that each section of duct is sized so that the increase in static pressure resulting from velocity reduction will balance the friction loss in the section. The initial duct size is determined by the required quantity of air and the velocity, but each succeeding duct section is sized to reduced velocity to offset the friction loss. The third method, the *total-pressure method*, is based on an initial duct velocity, with progressive reductions in velocity as each opening or terminal is reached. The system friction loss is then the total of the sum of the friction losses of the several duct sections and the terminal air pressures.

Complete charts and directions are given in the *ASHRAE Handbook, 1981 Fundamentals.* Equivalent lengths of straight duct for various fittings

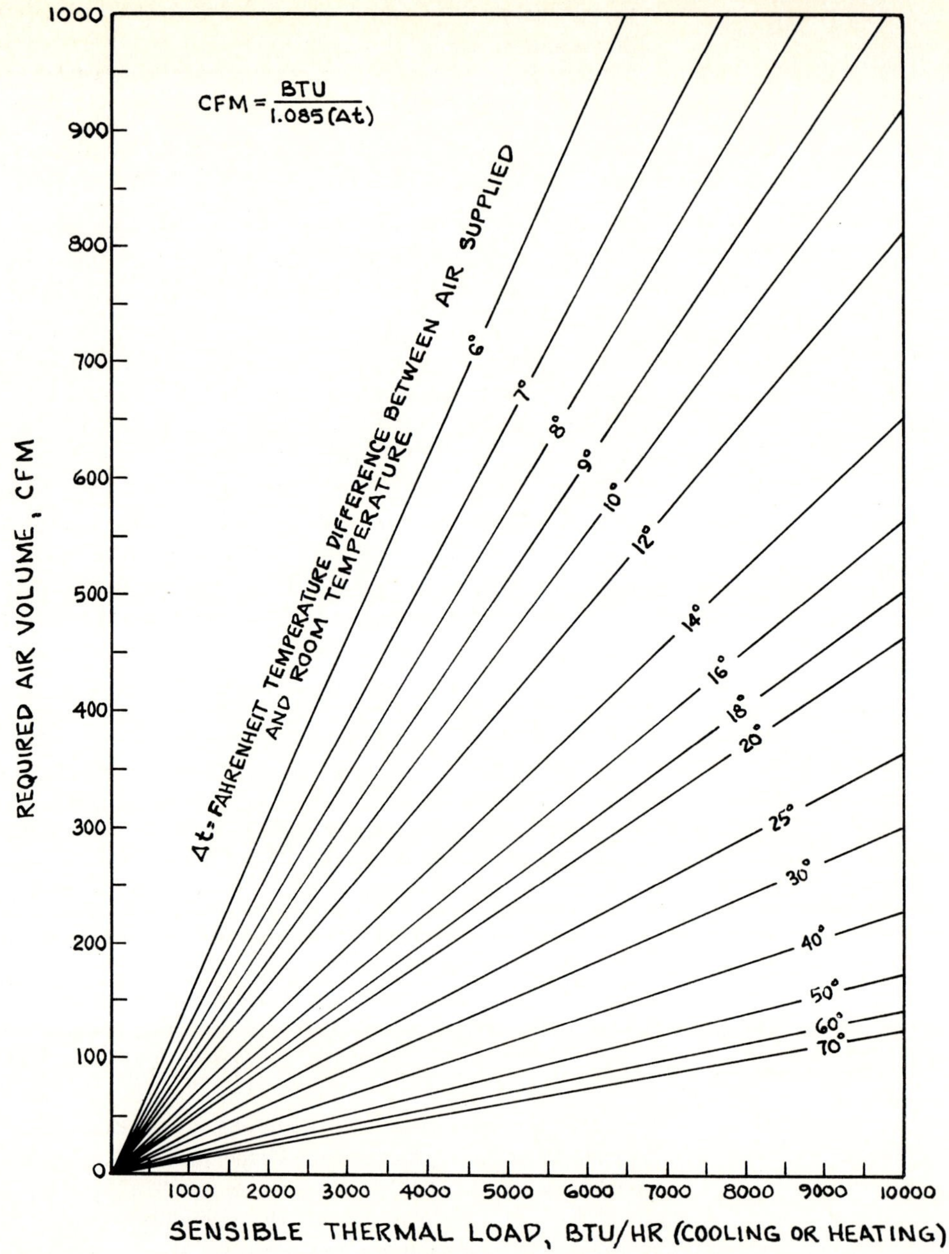


Figure 24-12

are shown in Fig. 27-4. As an example, a 90° elbow generates as much friction as 10 ft of straight duct, and a duct reducer has the same friction value as 5 ft of straight duct. In addition to the length of straight runs in a system, the total equivalent friction values for the various turns, connections, and other fittings must be added.

## 24.10 HOT-WATER SYSTEMS

Hot-water heating systems function by gravity flow of water or by pumping the hot water through the system. Four piping systems used are (1) the one-pipe system, (2) the two-pipe direct-return system, (3) the two-pipe reverse-return system, and (4) the series-loop system. Hot-water systems using water with a temperature below 250°F are used in residences and small buildings and are termed *low-temperature systems.* Those using temperatures greater than 250°F are termed *high-temperature systems* and are used primarily in commercial or industrial buildings.

In a *one-pipe system* (Fig. 24-13), one main pipe carries both the hot water to the various radiators or convectors and the cooler water back to the water-heating unit. Each radiator or convector is connected to the supply line by pipes from both ends of the radiator, and the hot water flows from one unit, through it to the second, through the second to the third, until it leaves the last unit and returns to the heater. The system may operate by gravity or with pumps. Where different spaces or areas require different temperatures, these may be *zoned* by having separate loop systems with proper controls at the water heater. In effect, this zoning provides several one-pipe systems operating from one water heater. Separate circulators or pumps and separate thermostats are necessary for each loop.

In a *two-pipe system* (Fig. 24-13), one pipe supplies hot water to each radiator and another pipe returns the cooler water back to the water heater. Each radiator still uses two connections, one to the hot-water line (supply) and the other to the cooled-water (return) line. The balancing of hot water among the various circuits is controlled with pipe sizing and valves. In the *two-pipe reserve-return system*, this balance is achieved by the length of piping used. The supply line serves each radiator as before, but the return

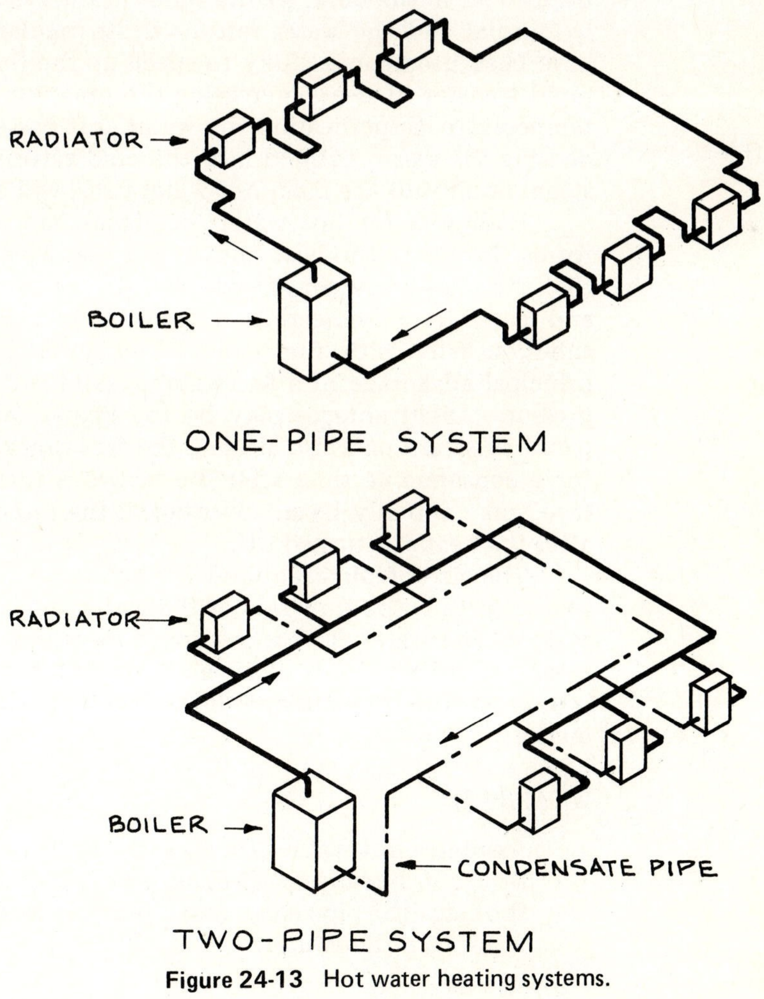


**Figure 24-13** Hot water heating systems.

line carries the cooled water parallel to the flow of hot water, or in reverse direction to the normal return. The first radiator has the shortest supply line but the longest return line, while the radiator farthest from the water heater has the longest supply and shortest return. This tends to balance the amount of total heat supplied to the space being warmed.

A *series-loop system* employs a length of pipe to connect the water-heating unit to the first radiator. The water passes through the first radiator and then travels to the second radiator and to succeeding radiators until it finally returns to the water heater. The only pipes required are the relatively short lengths between the radiators and the longer lengths for initial supply and return. The major difficulty with this system is that the hottest water goes to the first radiator and loses heat through every consecutive radiator until the coolest water finally returns to the water heater. Individual radiators cannot be shut off, as any valve in the line will stop circulation of the water through the entire system. Bypass piping around any radiator unit simply makes the system a one-pipe direct-flow system.

Water expands in volume whenever it is heated. When a hot-water system is completely full and the water is heated further, the volume increases. Unless some way is provided to hold the increased volume, the additional volume may rupture the piping or radiators. To prevent this, a *surge tank* is often installed between the water heater and the initial radiator. As the water volume increases, it flows into the surge tank and compresses the trapped air in the tank. As the water in the system cools, the pressure returns to normal and the water returns to its regular level. A number of methods have been used successfully to speed up the flow of hot water, and the more rapid transfer of heat. Increasing the pressure on the water will increase the temperature. Superheating the water as it passes through the heater and condensing the steam formed into the cold return water, or the introduction of steam or air into the main hot-water riser, will also increase the temperature.

Radiators for hot water, or steam, are manufactured in a number of forms. Probably the best known is a cast-iron sectional unit constructed so that the inlet is low on one end and the water circulates through the sections and leaves by a connection at the other end. Baseboard units consist of long tubes, usually with fins attached, enclosed in louvered metal covers. The principal advantage of a hot-water system is its flexibility and low operating pressure. Disadvantages may be the higher initial installation cost as compared to steam or warm air, and the fact that radiators and piping stay warm for a considerable time after the heater is turned off. The latter "disadvantage" may actually be an advantage if the retained heat carries into the night after the heater is turned off.

The size of pipe required for hot-water heating may be determined by two items: (1) the flow rate of the water, and (2) the friction loss in the pipe system. The water flow is measured in gallons per minute and the friction is measured in thousands of an inch for each foot of pipe. Water-pressure drop is expressed in milinches per lineal foot and 12 inches is equal to 12,000 milinches.

### EXAMPLE

Assume a hot-water heating system with a calculated heat loss of 75,000 Btu, 150 lineal feet of main pipe, 10 fittings using the equivalent of 10 ft of straight pipe each, and a temperature difference of 20°F from inlet water to outlet water.

$$\text{gal/min} = \frac{75{,}000 \text{ Btu}}{\text{temp. diff.} \times \text{minutes} \times \text{weight of water at } 215°\text{F}}$$

$$= \frac{75{,}000}{20 \times 60 \times 8}$$

$$\frac{75{,}000}{76{,}000} = 9.10 \text{ gal/min} \qquad \text{say } 9.5$$

$$150 + 100 = 250 \text{ equivalent feet}$$

Use 9.5 at the left of Table 24-3, then extend right to 285 under 400, or 228 under 500. Then use Table 24-4 under column 400 and find the value of 75,000 Btu (75 MBtu); the pipe size is between 1 and 1½ in. Doublecheck using column 500 and find 1 in. For adequate size use a 1½-in. pipe.

**TABLE 24-3**
Milinches per Foot of Pipe

| | *Equivalent length of pipe (ft)* | | | | | | | |
|---|---|---|---|---|---|---|---|---|
| *Head (ft)* | *50* | *100* | *200* | *300* | *400* | *500* | *600* | *700* |
| 1 | 240 | 120 | 60 | 40 | 30 | 25 | 20 | 17 |
| 2 | 480 | 240 | 120 | 80 | 60 | 48 | 40 | 34 |
| 2½ | 600 | 300 | 150 | 100 | 75 | 60 | 50 | 43 |
| 3 | 720 | 360 | 180 | 120 | 90 | 72 | 60 | 51 |
| 3½ | 840 | 420 | 210 | 140 | 105 | 84 | 70 | 60 |
| 4 | 960 | 480 | 240 | 160 | 120 | 96 | 80 | 68 |
| 4½ | 1080 | 540 | 270 | 180 | 135 | 108 | 90 | 77 |
| 5 | 1200 | 600 | 300 | 200 | 150 | 120 | 100 | 85 |
| 5½ | 1320 | 660 | 330 | 220 | 165 | 132 | 110 | 94 |
| 6 | 1440 | 720 | 360 | 240 | 180 | 144 | 120 | 102 |
| 6½ | 1560 | 780 | 390 | 260 | 195 | 156 | 130 | 111 |
| 7 | 1680 | 840 | 420 | 280 | 210 | 168 | 140 | 119 |
| 7½ | 1800 | 900 | 450 | 300 | 225 | 180 | 150 | 128 |
| 8 | 1920 | 960 | 480 | 320 | 240 | 192 | 160 | 136 |
| 8½ | 2040 | 1020 | 510 | 340 | 255 | 204 | 170 | 145 |
| 9 | 2160 | 1080 | 540 | 360 | 270 | 216 | 180 | 153 |
| 9½ | 2280 | 1140 | 570 | 380 | 285 | 228 | 190 | 162 |
| 10 | 2400 | 1200 | 600 | 400 | 300 | 240 | 200 | 170 |
| 12 | 2880 | 1440 | 720 | 480 | 360 | 288 | 240 | 204 |
| 14 | 3360 | 1680 | 840 | 560 | 420 | 336 | 280 | 240 |
| 16 | 3840 | 1920 | 960 | 640 | 480 | 384 | 320 | 275 |
| 18 | 4320 | 2160 | 1080 | 720 | 540 | 432 | 360 | 309 |
| 20 | 4800 | 2400 | 1200 | 800 | 600 | 489 | 400 | 340 |

**TABLE 24-4**
Heat-Carrying Capacity of Pipe (Thousand Btu)

| *Pipe size (in.)* | *50* | *100* | *200* | *300* | *400* | *500* | *600* | *700* |
|---|---|---|---|---|---|---|---|---|
| ½ | 5.2 | 7.2 | 10.0 | 13.0 | 16.0 | 17.5 | 19.5 | 22.0 |
| 1 | 21.0 | 30.0 | 44.0 | 55.0 | 66.0 | 72.0 | 80.0 | 89.0 |
| 1½ | 68.0 | 97.0 | 148 | 180 | 215 | 240 | 270 | 283 |
| 2 | 132 | 190 | 280 | 351 | 415 | 470 | 508 | 565 |
| 2½ | 215 | 312 | 462 | 570 | 655 | 723 | 800 | 860 |
| 3 | 390 | 565 | 805 | 1,010 | 1,250 | 1,350 | 1,505 | 1,710 |
| 4 | 800 | 1,200 | 1,700 | 2,100 | 2,600 | 2,950 | 3,150 | 3,475 |
| 6 | 2,500 | 3,700 | 5,400 | 6,700 | 8,000 | 9,000 | 10,000 | 11,000 |

## 24.11 STEAM HEATING

Steam heating systems are somewhat similar to hot-water systems, as some of the same equipment is used and the general principles are comparable. In this method, steam is created in a boiler, rises through pipes to the radiators or convectors, the steam condenses as it emits heat, and the resulting water returns to the boiler for reheating into steam (see Fig. 24-14). A *low-pressure system* operates at pressures from 0 to 15 psig (pounds per square inch gauge), while *high-pressure systems* operate on pressures greater than 15 psig. A *vapor system* operates on the condensation action to create a vacuum to move the steam, while a *vacuum heating system* requires a vacuum pump. Most condensate return pipes are designed to return the water to the boiler by gravity, but mechanical help may be used in the form of condensate pumps.

A *one-pipe system* employs a single riser from the boiler to the radiators, whereas a *divided system* uses two or more risers, similar to a hot-water zoned arrangement. The one-pipe system carries the steam to the radiators and in the same pipe returns condensate to the boiler. In the *two-pipe system*, one pipe carries the steam from the boiler, and another pipe carries the condensate from the radiator back to the boiler. A loop circuit and a pump are necessary when radiators are below boiler water level. Steam heating is also classified as *upfeed*, when the steam rises from the risers to the radiators, or a *downfeed*, when the riser is above the radiator and the steam must flow down to the radiator. If the condensate return is located above the boiler water level, it is termed a *dry return*, but it is called a *wet return* if the condensate return is below the boiler water level.

The *one-pipe gravity system* is the most economical to install but has several disadvantages. The pipe must be oversized to allow the condensate to return to the boiler against the steam flowing from the boiler. Water hammer may develop because the steam and condensate flow in opposite directions,

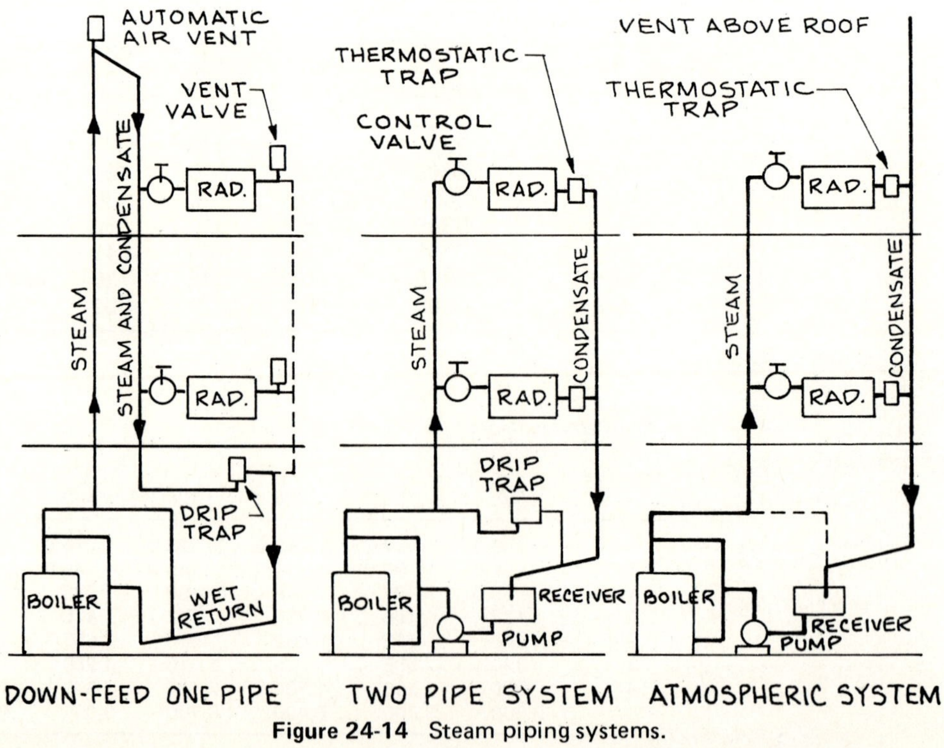


**Figure 24-14** Steam piping systems.

and air valves may fail to operate creating excess heat. In a *one-pipe reverse system*, the steam travels to the radiator and escapes through air valves at each radiator while the small amount of condensate returns to the boiler by gravity. Both upfeed and downfeed types offer a wide variety of installation possibilities but essentially are either one- or two-pipe systems.

When condensation of steam takes place, steam may flow in the system, which in turn creates a pressure change. When the water level in the several return drip lines is the same as in the boiler, the pressure will be the same as in the boiler. If the pressure in a return line decreases 1 lb, the water level in that return pipe will rise 1 lb $\times$ 2.3 (weight of water per each square inch) or 2.3 ft above the boiler level. If the water level rises 4.6 ft above the boiler level, the pressure will be the boiler pressure less 2 lb (2 $\times$ 2.3). Pressure drop in long lines may cause the water actually to back up into the end radiator and create a waterbound nonheating radiator.

A *vapor–steam heating system* is one in which steam is at approximately atmospheric pressure and operates with a slight vacuum. An *open* or *atmospheric system* is one in which the return line does not have a device to prevent return of air to the boiler and the condensate is wasted. In either case, the boiler pressure is usually less than 1 lb and only enough to overcome the frictional resistance in the piping system. In a vacuum system (Fig. 24-15), the low-pressure steam is admitted to the radiator at the top and "floats" on the trapped air, and any water present is forced out of the radiator at the bottom. As the steam drives out the air and water, heating the radiator, it also closes a trap valve to prevent steam from entering the return line. As the steam condenses, owing to the emission of heat through the radiator, the cooler water opens the trap and the condensate returns to the boiler. In a *closed vapor system*, no air is admitted to the system but a vacuum is created as the steam is condensed. A *mechanical vacuum system* uses a pump or *ejector* to create a vacuum by drawing off the air from the top of radiators and mechanically injecting the air into the main steam return line. Condensation of the steam creates the vacuum, and the ejector maintains it by removing air and condensate.

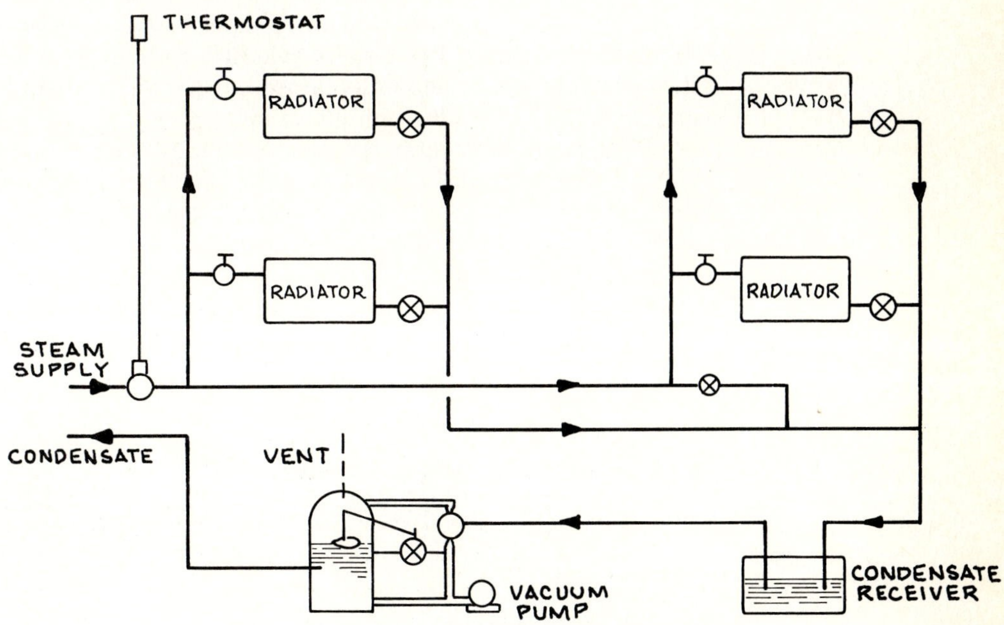


**Figure 24-15** Vacuum heating system.

High-pressure (25 to 150 psig) steam heating systems are usually found in large buildings where steam is also provided for manufacturing requirements. In this system the pipe sizes may be smaller, and the return lines may be installed above the radiators as the condensate can be lifted by the steam. Condensate is returned to the condensate pump or tank and is fed into the boiler together with other feedwater from manufacturing processes. Control valves, pressure-reducing valves, and similar devices may be necessary for proper control of steam from the boiler into the heating system, and in most installations an operating engineer is required to be present to supervise operation and control of high-pressure steam systems.

## 24.12 RADIANT HEATING

*Radiant heating* operates on the principle that a warmed surface creates waves of heat energy (thermal radiation) which will reduce body heat by radiation. Steam, hot water, or electricity are successfully used in radiant heating systems, although steam is more difficult to produce and control. Radiant heating is usually installed in floors or walls when hot water is used and in walls or ceilings when electricity is used. In most systems, an entire floor, wall, or ceiling surface is used as a heating panel. In operation, a radiant surface transfers heat to solid surfaces by radiation and to surrounding air by convection. Radiant heat rays do not appreciably heat the air through which they pass but move from warm to cooler surfaces, which then absorb heat.

When warm water is used as the heating source, the system may be installed in a ceiling and covered with plaster, in the upper portion of walls and again covered with plaster, or buried in a concrete floor slab. Installation in a wood floor may be very complicated and undesirable, since there is no mass of material to act as a radiator. Maximum temperature for ceiling panels should be not greater than 100°F and for floors not greater than 85°F. Wall panels should be installed in the upper half of a wall so that furniture will not block radiation, and should range from 100° to 110°F. Electric resistance cables may be buried in the plaster of a ceiling or wall but are not very efficient in concrete floor slabs. Electric radiant heat is more uniform than other systems but is generally more expensive to install and cannot be used as a part of an air-conditioning system.

Piping for hot-water systems is usually soft-copper tubing about ⅜ or ½ in. in diameter and should be installed with pipes not farther apart than 12 in. Steel or iron pipe should not be used. Figure 24-16 shows two methods for forming the radiant panel. Panel design A is the easiest to install, but the panel will provide the most heat at the water inlet, where the water is hottest, and the least heat at the outlet, as the water loses its warmth. Panel design B circulates the water so that a "hot" line is next to a "cold" line, and

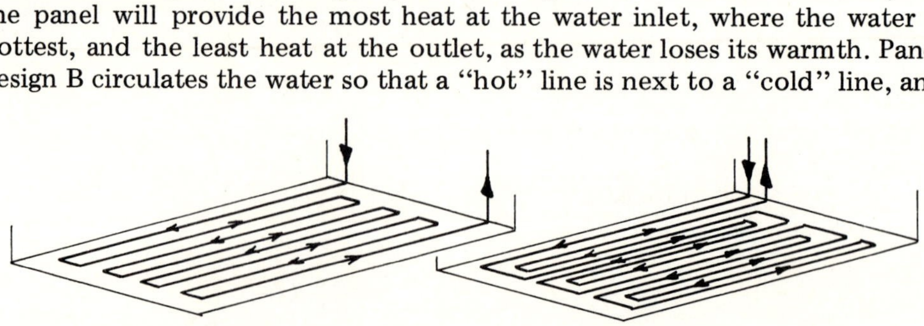

(a) (b)

**Figure 24-16** Diagrams of floor radiant heating.

a uniform heat is obtained over the entire system. Panels should be separately controlled from a main feeder line so that the amount of heated water may be adjusted to room warmth desired. The principal disadvantage of any radiant heating system is the fact that the wall, floor, or ceiling must be thoroughly heated before any radiation takes place and will continue to radiate heat after the system is turned off until the panel cools to surrounding temperature. Hot-water systems may develop pipe leaks, which require extensive remodeling of walls or ceilings and may require complete removal of a concrete floor. Such unfortunate occurrences may usually be avoided if the pipe is in one length, and a pressure test of at least 100 psig has been made before the piping is covered.

## 24.13 SUMMARY

Heating systems for buildings have many variables that should be considered before any one system is chosen. First, there is the matter of fuel type, as well as the type of heating unit using that fuel. Some areas have more fuel or more economical fuel than others, such that in some areas coal may almost be dictated over gas or oil, or vice versa. Space requirements for the heating unit are also of importance, as costly boiler rooms may not be necessary with other types of heating units. In addition to the fuel and heating unit to be used is the question of the most appropriate heat-distribution method. For many years, steam radiators were the most used, but increased availability of gas and oil and hot-water and forced-air distribution systems have altered that. Pipes take less physical space than ductwork, but if air-conditioning is also a consideration, a duct system may be designed for both uses. Radiant heat systems seem more ideal as there is no dust or odor, but the heat lag caused by the large radiator size may be a deterrent, and any accidental pipe leak or wire short circuit means a major remodeling job. The heating of living spaces is an absolute requirement in most areas, but all factors involved should be extensively and carefully considered before a final decision is made.

# 25

# Ventilation

## 25.1 INTRODUCTION

Ventilation in its normal context means moving air from one space to another different space. In this chapter two major uses for ventilation will be considered: residential ventilation and industrial ventilation. Residential ventilation is a rather limited subject, as there is usually a minimum of stale air or noxious gases to be considered and much of the ventilation required is supplied by air infiltration through cracks or opening of doors or windows. Those homes equipped with air-conditioning systems have a built-in ventilating system, which usually includes air filtering and humidity control, as well as heating or cooling.

Industrial ventilation is something else. Most industrial ventilation is pointed toward the abstraction of gases, dust particles, process odors, or similar airborne minute materials from machines or rooms where these health-affecting substances are produced. Industrial ventilation was brought on by increased crowding in the factories of the Industrial Revolution until someone realized that ventilation was needed to reduce fatigue and maintain production. Today, with the tremendous increase in use of many types of machinery, production of materials that involve chemical fumes, dust and airborne fibers, and the demand for air control in many processes, ventilation is a necessity. Both natural ventilation created by proper use of the wind and temperature and forced ventilation by fans, hoods, or suction equipment are now in general use, and various combinations are available for almost any type of ventilation required.

## 25.2 NATURAL VENTILATION

*Ventilation* is the movement of air from one space to a different space and is mostly a matter of *air volume*. *Circulation* of air is the movement of air in a confined space and is a matter of air *velocity*. Natural ventilation is created by a difference in temperature between the interior and exterior of a building, or by the velocity of the wind around a building. Ventilation possibilities by the wind must be based on three factors: (1) average wind velocity, (2) prevailing wind direction, and (3) interference of the wind around the building. In summer the wind is usually of less velocity than in winter, so a system designed to function properly by natural ventilation in the summer should function as well or better in the winter. Although this is generally true, the lower temperature in winter may also affect natural ventilation through temperature changes or a requirement to retain heat in a building at the sacrifice of ventilation.

When the wind blows steadily in any one direction without obstruction, the pressure may be considered constant at any point in the air stream. If the wind is obstructed by a building, however, the pressure is disturbed as it meets the building face, is pushed to each side, increasing the pressure by a proportion of the total; and meets again behind the obstruction to continue at constant pressure. As the wind passes along the sides of the building and starts again to pursue its normal path, it does not hug the rear of the building but forms an area of vacuum or suction along the building wall. This condition applies to vertical obstructions as it does to horizontal obstructions. To use this action effectively, a draft may be *induced* by providing an opening of the high-pressure side and a flue or chimney to the vacuum side.

Draft or ventilation may also be produced within a building by differences in temperature. A difference in the weight of warm air causes it to rise. If cooler air is admitted at a lower level, the warmer air will rise, and if there is an opening at the warm-air level, a draft will be induced. If the outdoor temperature is lower than the indoor temperature, the air will be sucked into the atmosphere through any opening at the warm-air level. This flow may be measured by using the formula

$$Q = 9.4A\sqrt{h(t_1 - t_2)}$$

where $Q$ is the airflow in cubic feet per minute (CFM), $A$ the square feet of inlets or outlets, $h$ the height of inlets to outlets in feet, $t_1$ the average indoor temperature in °F, and $t_2$ the outdoor temperature in °F. The factor 9.4 is a constant for 65% effectiveness of openings.

## 25.3 MECHANICAL VENTILATION

Mechanical ventilation may take many forms, all of which require some means to provide a change of air by blowing, suction, or other distribution using fans or similar equipment. Some of the required ventilation may be provided by an air-conditioning system for room temperature control, but much of the mechanical ventilation is localized to reduce fumes or dust. In a residence, the hood and fan at the cooking range location is probably the major consideration. Most building and health codes require an exhaust fan to be installed in bathrooms or toilet spaces when no natural window ventilation is provided, and some specialized-use areas such as workshops, photo darkrooms, or similar rooms may need mechanical ventilation.

Commercial ventilation, except for some roof ventilation, is almost entirely based on the use of some mechanical equipment. Modern industry uses many types of chemical compounds, some of them highly toxic, and the manufacturing process may develop a great many objectionable and unhealthy by-products. General ventilation of the workroom air may be required for body comfort and working efficiency, but more and more local ventilation is being required, and used, at individual machines or processing positions. This use of local systems results in lower heating or air-conditioning costs for general room comfort, allows better control of air pollution, and in most cases saves money in capital outlay for installation. In recent years, the federal Occupational Safety and Health Act (OSHA) has been in force, and certain standards, mostly already available from other national associations, have been included in the requirements for commercial ventilation.

Much of the commercial ventilation required falls into one of two major groups: (1) dilution, or (2) local exhaust. *Dilution* occurs when contaminants that are released into the workroom mix with the contained air. This method may be used when the quantity of contaminant is small, not toxic, nonflammable, and emitted at a sufficient distance from workers to allow dilution. *Local exhaust* systems typically consist of a hood or air cleaner connected to an exhaust fan and outlet to the atmosphere. Hoods must be carefully designed to draw the correct amount of air, and so shaped that fumes or dust from production are picked up at proper locations. The primary advantages of local ventilation are that contaminants are removed rather than diluted, and the required airflow is usually less than for dilution.

## 25.4 VENTILATION REQUIREMENTS

The ventilation of rooms or spaces is determined by room use, room size, and recommended *air changes per hour*. In Table 25-1, typical room uses are listed, together with recommended air changes per person. The cubic feet per

**TABLE 25-1**
Ventilation for Typical Areas

| *Room* | *Changes per hour* | *Room* | *Changes per hour* |
|---|---|---|---|
| Auditorium | 10-20 | Industry (cont.) | |
| Bakery | 8-12 | Machine shop | 8-12 |
| Bank | 8-10 | Mechanical room | 6-10 |
| Beauty shop | 8-10 | Painting | 15-25 |
| Bowling alley | 15-30 | Welding | 15-30 |
| Cocktail bar | 25-30 | Office | |
| Church | 15-25 | General | 10-15 |
| Conference room | 25-45 | Private | 15-25 |
| Dairy | 5-15 | Residence | 5-20 |
| Dry-cleaning establishment | 20-40 | Restaurant | |
| Hospital | | Dishwashing | 60-80 |
| Anesthesia | 8-12 | Eating | 10-20 |
| Autopsy | 8-10 | Kitchen | 10-30 |
| Delivery | 8-10 | Retail store | 5-20 |
| Laboratory | 6-10 | School | |
| Nursery | 10-15 | Classroom | 10-30 |
| Surgery | 10-20 | Corridor | 3-10 |
| Patient room | 15-25 | Laboratory | 6-10 |
| Industry | | Theater | 5-8 |
| Foundry | 6-20 | Washroom/toilet | 8-20 |

minute (CFM) per person is generally used as it gives a better result for total air requirements. This table is shown only as a guide, as some codes or city ordinances require different ventilating. In crowded rooms, the CFM should be at least 40 per person regardless of indications in the table.

Concentration buildup of any contaminant (gas, solvent fumes, chemical odors, heat) and the required air change to continue "pure" air are dependent on such factors as *room volume*, *ventilation rate*, rate of *contaminant generation*, and *concentration of the contaminant*. In industrial/commercial situations, where a reasonably accurate check of rate of generation and concentration of contaminant is obtained, a fairly accurate CFM factor may be calculated from the formula

$$\log_{10} \frac{G - Q_c}{G} = -\frac{1}{2.3} \times \frac{Q}{p}(t)$$

where $G$ represents the rate of generation, $Q$ the ventilation rate, $p$ the room volume, $t$ the time interval, 1/2.3 a constant, and $Q_c$ the concentration of the pollutant. Industrial ventilation is concerned primarily with the dilution or dissipation of many solvents used in production. From chemistry it is known that the volume of vapor formed by 1 pound-molecular weight at 70°F (room temperature) is approximately 400 ft$^3$. If it is then desired to find the amount of vapor generated by a solvent, the molecular weight is multiplied by 400.

**TABLE 25-2**
Dilution Rate for Some Common Industrial Solvents

| | *Molecular weight* | *Specific gravity* | *Vent design (ppm)* |
|---|---|---|---|
| Acetone | 58 | 0.79 | 150 |
| Amyl acetate | 130 | 0.87 | 75 |
| Benzene (benzol) | 78 | 0.88 | 25 |
| Butyl alcohol | 74 | 0.81 | 25 |
| Carbon disulfide | 76 | 1.26 | 7 |
| Carbon tetrachloride | 154 | 1.58 | 25 |
| Chloroform | 119 | 1.50 | 25 |
| Dichlorobenzene | 147 | 1.32 | 50 |
| Dichloromethane | 85 | 1.33 | 200 |
| Ether (ethyl) | 74 | 0.72 | 75 |
| Ethyl alcohol | 46 | 0.79 | 250 |
| Isopropyl acetate | 102 | 0.87 | 75 |
| Isopropyl alcohol | 60 | 0.79 | 150 |
| Methanol | 32 | 0.80 | 100 |
| Methyl chloride | 51 | 0.92 | 200 |
| Methyl isobutyl ketone | 100 | 0.80 | 150 |
| Naphtha | 110 | 0.75 | 200 |
| Nitroethane | 75 | 1.05 | 50 |
| Pentachloroethane | 202 | 1.67 | 2 |
| Propyl acetate | 102 | 0.89 | 75 |
| Tetrachoroethane | 168 | 1.58 | 5 |
| Toluol | 92 | 0.87 | 100 |
| Turpentine | 136 | 0.87 | 75 |
| Xylol | 106 | 0.88 | 75 |

**EXAMPLE**

What volume of vapor will be generated by the evaporation of 1 lb of acetone?

According to Table 25-2, acetone has a molecular weight of 58.

$$1 \text{ lb} = \frac{1}{58} \text{ pound-molecular weight}$$

$$\frac{1}{58} \times 400 = 6.89 \text{ ft}^3$$

The *ventilation design concentration* (VDC) in parts per million (ppm) is given in Table 25-2 for some of the most common solvents, and the maximum allowable concentration is given in Table 25-3. The task of industrial ventilation design is to determine the amount of a contaminant that may be present in a diluted form in a large workroom, as well as to determine the amount that may be present at the breathing level of a worker confined to one position during the work period.

**TABLE 25-3**
Threshold Limits (ppm) for Some Gases, Vapors, and Industrial Solvents

| *Concentration in parts per million* | | | |
|---|---|---|---|
| Acetaldehyde | 200 | Ethyl alcohol | 1000 |
| Acetic acid | 10 | Formaldehyde | 5 |
| Acetone | 500 | Gasoline | 500 |
| Ammonia | 100 | Hydrogen cyanide | 10 |
| Benzene | 35 | Hydrogen fluoride | 3 |
| Butadiene | 1000 | Naphtha | 500 |
| Carbon disulfide | 20 | Nitrogen oxide | 25 |
| Carbon monoxide | 100 | Nitroglycerin | 0.5 |
| Carbon tetrachloride | 50 | Ozone | 1 |
| Chlorine | 1 | Sulfur dioxide | 10 |
| Chloroform | 100 | Toluene | 200 |
| Dichloromethane | 500 | Turpentine | 100 |

## 25.5 RESIDENTIAL VENTILATION

Residential ventilation is similar to commercial ventilation in that it exhausts fouled air from a room or building. The major sources of fouled air are the kitchen cooking area and the bathroom. The air in the kitchen contains food odors, grease, and perhaps some smoke. To remove this fouled air, modern kitchens are equipped with range hoods and exhaust fans, items required by some codes. Air change is customarily at about one complete room change every 2 minutes. A second method is to size the hood and fan to provide an entrance velocity of 50 cubic feet per minute for normal cooking. A hood approximately 24 in. × 48 in. will have an area of 8 ft$^2$ and an average velocity of about 50 CFM. Quiet operation is also essential, so a fan with low outlet velocity should be selected.

Bathrooms without windows or other means of ventilation are usually required to have a ventilating fan connected with the light fixture so that positive ventilation is accomplished whenever the light is turned on. These fans may be connected with a short duct to the outside or may be installed in a wall for direct exhaust. Most fans of this sort provide about 50 to 100 CFM and draw about 1.0 A. Similar small exhaust fans may be used in other locations that may not be air-conditioned or otherwise adequately ventilated.

## 25.6 COMMERCIAL VENTILATION

Commercial ventilation uses both natural ventilation and power-actuated ventilation. Warehouses, storage sheds, and similar buildings are often equipped with roof ventilators that use the wind for ventilation. Grilles and louvers provide controllable access or exhaust possibilities. Manufacturing plants need a wide variety of mechanical exhaust systems. Most of these are directly related to the exhaust of fumes, but some are designed for other purposes. Woodworking plants generate a tremendous amount of wood chips, sawdust, and dust, but a minimum of fumes or odors. In this situation, the possibilities of fire are of major importance and exhaust systems to remove this waste were mandated very early by fire departments.

The development of standards for manufacturing-equipment ventilating have also led to further investigation of health needs for workers. The average person inhales about 30 to 40 lb of air per working day or from 300 to 500 $ft^3$ of air. The average of 350 $ft^3$ is usually used. Industrial processes are classified as types shown in Table 25-3, and most codes consider that small amounts of contaminant are not always detrimental to a human being, but dust in its various forms is inhaled, passes into the human bloodstream, and may reach vital organs. Most odors or irritants have not yet been classified for commercial or industrial consideration and control.

## 25.7 ROOF VENTILATORS

Roof ventilators provide a weatherproof outlet by *induction* for contaminated air and are manufactured in a variety of shapes and sizes. To be effective, roof ventilators must be installed to receive the full wind without interference. Air inlet openings should be equal to, or greater than, the throat areas of ventilators and should be located at a low level in the building structure. Generally considered the most efficient type of gravity ventilator, the *stationary-head* type has no moving parts, no upkeep is required, and it has high exhaust capability under any wind conditions. This type of ventilator is shown in Fig. 25-1a. Several forms of *revolving* roof ventilators are available, as represented by the sketch in Fig. 25-1b. In this type, the head swings on a pivot and is turned by a tail or vane so that the head opening is pointed downwind. Inductive action pulls air from below, as suction is created when the wind blows around the head. Major disadvantages of this type are that it

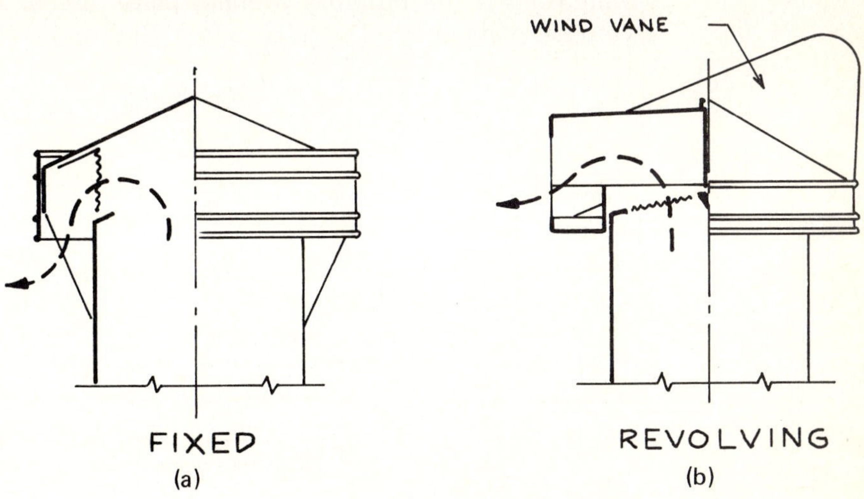


**Figure 25-1** Roof ventilators.

has a low capacity (down to zero) in still air and may be noisy or not revolve if the pivot is damaged.

Still another type of roof ventilator using the wind is the *turbine* (Fig. 25-2), of which several variations are available. The turbine vent consists of a series of blades shaped and secured to a revolving head in a manner to draw air up the vent and exhaust it through the blade openings. Wind from the windward side may enter the ventilator and must be expelled with the fouled air on the leeward side, thus decreasing its efficiency. As this type also revolves on a pivot, it has most of the disadvantages of any other revolving type. *Ridge*-type ventilators may be single units or a continuous length installed along the building ridge, and they operate by the flow of hotter air to the high point of the roof. Dampers automatically control the outward flow of air, but the wind must blow counter to the ridge direction to gain the most efficiency.

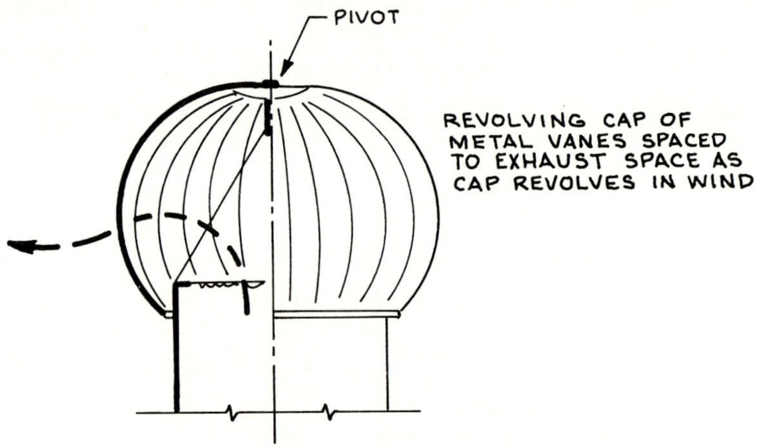


**Figure 25-2** Turbine ventilator.

The following formulas may be used to calculate the number and capacity of vents required for a building.

$$V_1 + V_2 = V$$

where $V_1$ is the building volume, plate line to the floor; $V_2$ the building volume, plate line to the ridge; and $V$ the total volume.

$$\frac{V \times \text{air change}}{60 \text{ (min.)}} = \text{CFM}_1$$

where the air change per hour is taken from tables.

$$\frac{\text{CFM}_1}{N} = \text{CFM}_2$$

where $N$ is the estimated number of ventilators.

$$\frac{\text{CFM}_2}{\text{vent capacity}} = \text{number of vents}$$

where the vent capacity is taken from manufacturers' catalog tables.

**EXAMPLE**

Assume a warehouse 100 ft × 60 ft, with a 15-ft wall height and the roof at a 4-in-12 pitch. The warehouse air change rate is two per hour.

$$V_1 = 100 \times 60 \times 15 = 90{,}000$$

$$V_2 = 100 \times 30 \times 12 = 36{,}000$$

$$V = 126{,}000 \text{ ft}^3$$

$$\frac{126{,}000 \times 2}{60} = 4200 \text{ CFM}_1$$

Assume four vents.

$$\frac{4200}{4} = 1050 \text{ CFM}_2$$

$$\frac{1050}{1165} = 3.6 \text{ vents} \qquad \text{24-in.-dia. vent} = 1165 \text{ ft}^3$$

Use four 24-in.-diameter vents spaced at 20 ft apart.

## 25.8 LOCAL EXHAUST SYSTEMS

General ventilation of a work space functions on the principle that contaminants are periodically flushed out as the concentration reaches a certain point. *Local exhaust systems* prevent dispersion of contaminants into the general air by withdrawing them at the source. The local exhaust *hood* is the point at which the contaminants are withdrawn from the work space and occur in many shapes and sizes. *Spray booths,* used in many manufacturing processes for the application of finish painting; *receiving hoods* at grinding wheels, used to catch the stream of dust from the process; *nozzles* of various designs either to catch chips and dust or to blow them into a hopper; and *suspended canopy hoods* are all "hoods" in its general meaning. In addition, there may be grilles or slots in benchtops, openings along tank rims, or simply pipes connected to tanks or bins. All serve the same purpose—inlets for local exhaust systems.

*Booths,* in their various applications, are one of the most efficient and positive of all local exhaust systems. A booth is a box arrangement, open on one side to accommodate the work and the worker, with an exhaust opening opposite the front opening. Since the contaminant-producing work is done inside the booth, and since the sides provide a barrier to outside air currents, almost all the contaminating material is generated in the booth, contained in the booth, and rather easily exhausted from the booth. The exhaust connection is normally at the back or top, except where heat is generated or used. In such cases, the exhaust should be from the top to take advantage of warm air rising. Air velocity at the front opening is usually low, from 50 ft/min to perhaps 200 ft/min. Adequate depth should be provided to ensure that air currents will not eddy and cause backflow around the sides or top of the booth.

*Receiving hoods* are of many sizes and shapes and are generally used where a stream of air and contaminants is characteristic of the process. Grinding is one of these processes. As the grinding wheel turns, it throws off a stream of air contaminated with wheel grit and dust from the material being ground. The velocity thus created carries the contaminated air into a receiving hood, where it is further removed by a suction system. Occasionally, an air jet is used to augment the airstream from the process, thus blowing the fouled air into the receiving hood.

*Exterior hoods* are those that function by suction but are not directly attached to machinery or booths. Types include longitudinal slots or narrow openings along tank edges, overhead canopies, openings in benchtops or floors, propeller fans in walls, and protected openings behind, over, or under work areas. This type of hood is used when other types of attached hoods would cause obstruction of the work area. In all "hoods" of this type, the contamination-producing area is not normally directly adjacent to the hood, so adequate suction must be provided to ensure that the contaminated air will be drawn into the hood. Dimensions, shape, and location are determined by the imaginary area to be ventilated, the contour area, the face area of the hood, and the air velocity required.

*Enclosure exhausts* are used primarily in tanks or other completely enclosed spaces in which manufacturing takes place or material is stored. Where crushed ore or similar materials are introduced into a bin, the passage of the material creates an air suction into the bin, thus forcing out air that is in the bin. Some of this air that is forced out also contains dust from the ore. For enclosures where a heated process is necessary, the heat will cause an increased air pressure within the enclosure. This must be reduced by an exhaust system to provide a negative pressure inside that is equal to or greater than the outside pressure.

## 25.9 EXHAUST FANS

Three types of fans may be considered in ventilation problems: (1) axial-flow fans, (2) centrifugal fans, and (3) special fans. Each of these groups also has several different types. *Propeller (axial-type) fans* are used to move large quantities of air against low static pressure and are reserved for general ventilation or dilution processes. The *narrow-blade fan* is commonly used in spray booths and for large areas and rapidly loses its efficiency when any resistance is placed in its airflow stream. *Vane axial fans* are usually installed in ducts and are economical of horsepower and space. They may develop high pressure and the airflow past the propeller unit forms a spiral or helix. Operation against 0 to 3 in. of static pressure is normal. A third form is the *vane duct fan*, which is similar to the vane axial fan but vanes are installed in the duct behind the propeller. These vanes straighten out the spiral movement of the air passing the propeller to a straight flow at higher static pressure. Vane fans may have adjustable blades, operate at about 2 to 10 in. static pressure, have a peak efficiency of about 70%, and are noisy, as are most duct fans.

*Centrifugal fans* include *radial-blade fans*, *forward-curve-blade fans*, *backward-inclined-blade fans*, and *airfoil fans*. Centrifugal fans are usually used in ventilating systems rather than axial fans, are quieter and less expensive, and cope better with fluctuating airflow conditions. *Radial-blade fans* are used for duct systems, as the flat radial blades tend to be self-cleaning. Heavy construction makes them heavy and more costly. All sizes are available. *Forward-curve-blade fans* (squirrel cage) have the leading edge of the blades curved toward the direction of rotation. This type of fan has low blade-tip speed, is used against low static pressure, (0 to 5 in.) but is relatively inefficient, even though the discharged air may be at a higher velocity than the movement of fan tip. Pulsation airflow may occur as a result of minor changes in system pressures. *Backward-inclined-blade fans* are used where system pressures fluctuate and air velocity leaving the wheel is low. This type of fan is used for handling large volumes of air, as the fan blades cause an increase in pressure directly as the wheel rotates. *Airfoil fans* are shaped like a cross section of an airplane wing or airfoil. This reduces noise and pulsation in the airflow.

Several standard formulas may be used when fans are considered. Velocity pressure is proportional to the square of the velocity and may be found by the formula

$$\mathrm{VP} = \left(\frac{V}{4005}\right)^2$$

where VP is the velocity pressure in inches of water, 4005 is a constant, and $V$ is the air velocity in ft/min. Brake horsepower varies directly with the cube of changes in fan speed. Doubling the fan speed increases brake horsepower by a factor of 8 (i.e., $2 \times 2 \times 2 = 8$).

**EXAMPLE**

A fan is rated by a manufacturer as delivering 1500 CFM of air at 4 in. of water static pressure at 300 rpm with 5.0 hp. If the fan speed is increased to 500 rpm, what is the volume, static pressure, and horsepower?

Volume:

$$V = 1500\left(\frac{500}{300}\right) = 1500(1.66) = 2499.99 \text{ CFM}$$

Static pressure:

$$\mathrm{SP} = 4\left(\frac{500}{300}\right)^2 = 4 \times 2.7776 = 11.1104 \text{ in.}$$

Horsepower:

$$\mathrm{hp} = 5.0\left(\frac{500}{300}\right)^3 = 5.0 \times 4.6259 = 23.129 \text{ hp}$$

**EXAMPLE**

Find the power cost difference in the preceding example. Assume an electric cost of \$0.03 per kWh. The fan operates 10 hours per day, 5 days per week, 50 weeks per year. One horsepower equals 745.7 W, or 0.75 kW.

$$\text{Time} = 10 \text{ hours} \times 5 \text{ days} \times 50 \text{ weeks} = 2500 \text{ hr/yr}$$

At 300 rpm:

Horsepower: 5.0 hp $\times$ 0.75 kW = 3.7 kWh
Yearly cost: 3.7 kWh $\times$ 2500 hr/yr $\times$ \$0.03 = \$277.50

At 500 rpm:

Horsepower: 23.2 hp $\times$ 0.75 kW = 17.4 kWh
Yearly cost: 17.4 kWh $\times$ 2500 hr/yr $\times$ \$0.03 = \$1305.00

The yearly cost difference is \$1027.50 if the fan speed is increased only by 66.66% (from 300 rpm to 500 rpm) while the volume is increased 66.66% (from 1500 $\mathrm{ft^3}$/min to 2500 $\mathrm{ft^3}$/min).

## 25.10 AIR CLEANING

Industrial manufacturing produces many different types of contaminants, some of which cannot be discharged into the community atmosphere because of legal regulations, salvage value, or other reasons. To provide a method for removal of such material, *air cleaners* are used before the air is finally exhausted. The use of air cleaners, or their nonuse, is also important because they usually represent the largest single source of resistance to airflow. Air cleaners may be required because a contaminant is toxic and damage to the environment would be sustained if such material were dumped into the air. The value of some solvents or metals may be sufficient that they should be salvaged from the airstream. The actual amount or volume of material carried in the exhaust system may dictate use of a cleaner to separate air from more solid materials, and of course governmental regulations may require cleaners.

As in most other equipment, several types of cleaners are available. One of the most common is a *filter*, either of the *fabric* type or of the "high-efficiency particulate air" (HEPA), or *absolute*, type. The fabric cleaner, or arrestor, is an effective, low-cost type of envelope or tubular shape that builds up a mat on the dirty-air side which filters out the contaminant particles contained in fouled air or gas. The usual fabric is cotton, but paper, glass cloth, and synthetic fibers are also used. Maximum temperature for these filters is about 180°F for cotton but may go to as much as 500°F for some filters of glass cloth.

*Wet collectors* may be used where moist conditions and high temperatures are present and are particularly adaptable when dust may be explosive or a fire hazard when dry. Another important consideration when considering wet cleaners is the matter of disposal of the polluted wastewater used to collect the contaminants extracted from the air. Since wet cleaners use water, the addition or subtraction of vapor affecting air humidity must also be realized. Two principal wet collectors are in common use: (1) a chamber or spray tower, and (2) a packed tower. With the *spray-tower method*, water is sprayed into the dirty air in a closed chamber, and the dust particles cling to the water droplets, which are then separated again by centrifugal force. Pressure drop is low. Water pressure may be from 10 to 400 psi but is generally rather low and may be required at about 5 to 10 gal/min per 1000 standard cubic feet per minute. For removal of gas, vapor, or mist, a packed tower may be used. In the *packed-tower method*, a tower or vertical cylinder is packed with a suitable medium, the contaminant is removed by filtration, and the air is extracted at the top. Another form is a continuous filter bed similar to that used for filtering water. The air is passed through the filter, which extracts solids, and the air is exhausted or further processed for removal of gas or fumes.

A number of other types of cleaning equipment are also produced which have special uses. The *cyclone collector* is often used at woodworking plants to handle the large amount of small wood scrap, chips, and sawdust developed by the various machines. This collector operates on a suction system with ducts to local exhaust hoods, and the collected waste is often used as boiler fuel. When grit or metal dust is present in the air, a *dry centrifugal collector* may be used. In this equipment, the dirty air enters on the centrifugal axis, is precipitated by centrifugal force onto specially shaped blades on an exhauster, and then exits to a collector or bin. A *wet precipitator* acts in much the same fashion, except that a water spray causes dust to collect on impeller blades to be removed later. When a gas or vapor mixture is passed through a bed of activated carbon, the airborne particles diffuse into the pores of the carbon and are 100% absorbed until all pores are filled, at which time the entire bed is replaced or regenerated. High temperature may be used to incinerate combustible high-toxic or nuisance gases, but special care is required with compounds containing chlorine or other halogens to avoid secondary pollution problems.

## 25.11 SUMMARY

Ventilation in its normal application does not usually create many problems for residential buildings and is usually accomplished by air-conditioning the entire structure or by local exhausting at kitchens or bathrooms. This is not true of most industrial manufacturing plants, although air-conditioning does do the job in most commercial buildings, such as office buildings, retail stores, and similar locations. In industrial applications, ventilation is approached in two ways: general air dilution of contaminants or direct local exhaust at the position of generation. Where the percentage of contaminants is light or relatively unimportant to normal health of workers, the workroom volume of air is simply changed or diluted by fans or ventilators on a basis of changes per minute. On the other hand, much of today's manufacturing includes use of chemicals, solvents, acids, metals, and heat processes to produce the final article. All of these produce pollutants in some form and in various quantities. A great many produce material that is toxic, irritating, or otherwise hazardous. Federal, state, and local governments are inaugurating more "standards" proposed to control undesirable emission of this by-product into the environment, and manufacturing companies are becoming more concerned as well. The subject of ventilation is relatively new when compared to other support systems for buildings but is one that will require more consideration in future planning.

# 26

# Refrigeration

## 26.1 INTRODUCTION

The subject of refrigeration and its use may be considered in two entirely different categories: (1) as a part of mechanical air-conditioning, or (2) as a method for preserving foods. In Chapter 25, refrigeration was considered as a feature of air-conditioning, and although the principles of refrigeration are the same in food preservation, their application is entirely different. In the home, food preservation is somewhat limited to a unit refrigerator, freezer, or combination refrigerator and freezer. In industry or commercial use, refrigeration includes flash-freezing of perishable vegetables and meat, storage of frozen foods, retail displays of frozen or chilled foods, chilled rooms or facilities to hold perishables awaiting further processing, truck and train transportation, production of ice, and a great number of related uses.

Most refrigerated units designed for industry also require extensive insulation of the container: walls, floors, ceilings, doors, and similar parts. Regardless of the use of a unit, the principles are the same and only the size or space available need be changed. The determination of desired temperature and the loss of heat are covered in Chapters 23 and 27. Information in this chapter is concerned primarily with portable residential-type refrigerators and freezers, and with built-in commercial refrigeration units, beverage coolers, and ice-producing equipment.

## 26.2 GENERAL PRINCIPLES

Unit or built-in refrigeration for food processing or preservation is based on the same principles as the cooling process used in air-conditioning systems. The primary difference is that while air-conditioning is used in a calculated relationship with heating to provide personal comfort, *refrigeration* is concerned with low cold levels at or below the freezing point of water. The basic method, called a *capillary-tube system*, consists of six components, each of which performs a different function (see Fig. 26-1).

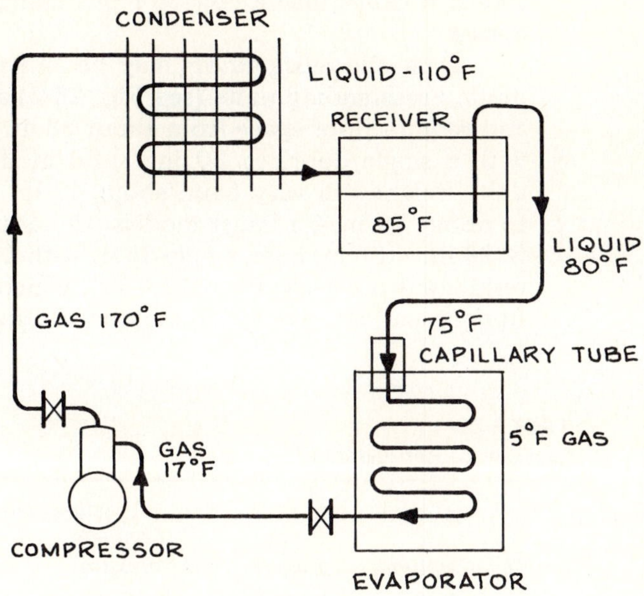


**Figure 26-1** Refrigeration diagram.

The principal parts of a refrigeration system function by the compression and evaporation of a number of refrigerants, which absorb latent heat when they change from a liquid to a gas at low temperature and pressure and then release the heat when they condense back to a liquid at high temperature and pressure. These refrigerants are listed by their refrigerant numbers, common names, and other properties in Table 26-1. In operation, the compressor reduces the pressure in the evaporator and at the same time compresses the vaporized refrigerant, forcing it into the condenser at high pressure and high temperature. The condenser removes the heat by exposing the coils to cooled air, thereby causing the refrigerant to return to a liquid. The liquid refrigerant is fed into a capillary tube, where it is restricted to a small opening, causing a reduction of pressure from one side to the other which vaporizes the refrigerant. The reduced pressure lowers the boiling point to the low-pressure side, the evaporator absorbs heat from the area being cooled, and the heat is absorbed by the refrigerant. Vaporization becomes more rapid and turns into a gas. The gas enters the compressor and the cycle is repeated. Cooling is provided in the evaporator section, and the release of heat is accomplished in the condenser section by exposure to cool air from a fan or blower.

## 26.3 RESIDENTIAL REFRIGERATION

Residential refrigerators and freezers are available in many different styles, shapes, and sizes. The food compartment is designed to maintain a temperature range of about 36°F, plus or minus about 2°F. The freezer compartment is usually held at 5 to 10°F for frozen foods and ice. Humidity is important for vegetable protection, and this is accomplished with pans or trays with covers to prevent water vapor escaping. So-called frost-free refrigerators allow the cold air to be primarily passed into the freezer section over the evaporator. Frost collects on the evaporator and is periodically removed automatically by heat from an electric coil. As the frost melts into water, it drops into a collector pan and is vaporized by heat from the condenser.

Domestic refrigerators may be either freestanding or under-the-counter units. Freestanding units (see Fig. 26-2) are the most common, are portable, and require floor space from about 28 in. × 26 in. for 9.0-ft$^3$ upright models with a single door, to 30 in. × 36 in. for a 15.0-ft$^3$ double-door frostless unit. Height will vary from about 48 in. for the smaller (apartment) model to about 70 in. for larger models. Under-the-counter models are about 26 in. × 26 in. × 36 in. high, single-door, with a 6.0-ft$^3$ capacity. Freezer chests for residential use vary from 15.0 ft$^3$, which requires about 36 in. × 48 in. of floor space and are approximately 30 in. high, to 66 in. × 48 in. for 23.1

**TABLE 26-1**
Qualities of Refrigerants

| *Refrig. no.* | *Common name* | *Evapor. pressure (psig at 5°F)* | *Condens. pressure (psig at 36°F)* | *Boiling temp. (°F) at 15 psia* | *Liquid volume (ft³/lb)* | *Vapor volume (ft³/lb)* | *Liquid density (ft³/lb)* |
|---|---|---|---|---|---|---|---|
| R-11 | Freon 11 | 24.0 | 3.6 | 74.7 | 0.010 | 12.27 | 91.4 |
| R-12 | Freon 12 | 11.8 | 93.2 | −21.6 | 0.011 | 1.49 | 80.7 |
| R-22 | Freon 22 | 28.3 | 159.8 | −41.4 | 0.012 | 1.25 | 73.4 |
| R-502 | R-22 + R-115 | 35.9 | 175.1 | −50.1 | | 0.82 | 87.2 |
| R-717 | Ammonia | 19.6 | 154.8 | −28.0 | 0.024 | 8.15 | 41.2 |
| R-718 | Water | 29.7 | 28.6 | 212.0 | 0.016 | 12444.4 | 62.4 |
| R-764 | Sulfur dioxide | | 66.5 | | | | |

| *Refrig. no.* | *Common name* | *Toxic* | *Odor* | *Molec. weight* | *Latent heat (Btu/lb at 5°F)* | *Liquid enthal. (Btu/lb at 5°F)* | *Vapor enthal. (Btu/lb at 5°F)* | *Leak tendency (1 to 10)* |
|---|---|---|---|---|---|---|---|---|
| R-11 | Freon 11 | No | Slight | 137.4 | 84.00 | 8.88 | 92.88 | 1 |
| R-12 | Freon 12 | No | Slight | 120.9 | 69.47 | 9.32 | 78.79 | 2 |
| R-22 | Freon 22 | No | Slight | 86.5 | 93.57 | 11.97 | 105.56 | 4 |
| R-502 | R-22 + R-115 | 0.6% | Slight | 111.6 | 68.86 | 11.89 | 80.75 | 3-4 |
| R-717 | Ammonia | Yes | Acrid | 17.0 | 565.00 | 48.30 | 613.30 | 10 |
| R-718 | Water | No | None | 18.0 | 1071.30 | 8.05 | 1079.35 | 9 |
| R-764 | Sulfur dioxide | Yes | Sharp | 64.0 | 172.3 | | | |

ft$^3$, which requires about the same height. Ice makers for home use will make about 2 lb of ice per hour, will store 35 to 40 lb of ice, and require about 18 in. × 24 in. of floor space and 35 in. in height.

**Figure 26-2** Residential refrigeration. (*Courtesy of Kelvinator.*)

The construction of refrigerators varies somewhat with each manufacturer, but the essential parts and their location in the case are similar (see Fig. 26-3). Most refrigerator cases are made of sheet metal inside and outside, with insulation sandwiched between. The interior is usually finished with white porcelain, while the exterior may be spray-enameled in white, color, or wood grains. Doors are finished with plastic liners that provide narrow shelves or pockets. Freezer chests usually have the evaporator piping installed behind the inner wall and the condenser piping behind the outer wall. Condensers on refrigerators are usually mounted on the back of the case exposed to the room air.

## 26.4 COMMERCIAL REFRIGERATION

Commercial refrigeration is the general term applied to all types of refrigerated equipment that is used in groceries, meat markets, hospitals, bars, florist shops, and similar retail establishments. The use of refrigeration in these

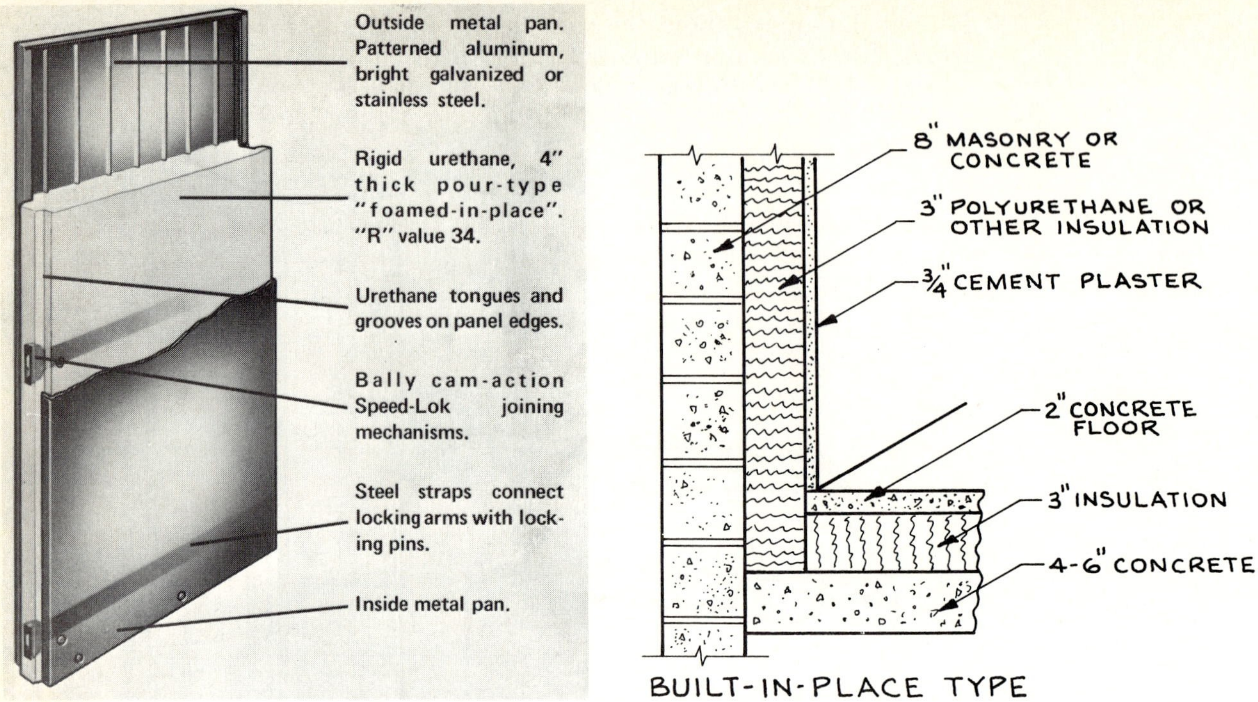


**Figure 26-3** (a) Typical Bally Prefab panel; (b) Refrigerator wall construction: sectional-demountable type. (*Courtesy of Bally Case and Cooler Inc.*

locations is commonly limited to display cases, storage cabinets, water and beverage coolers, and ice machines. More extensive refrigeration is required for "walk-in boxes" in some of these locations and manufacturing processes in connection with meat or other food production.

The major types of refrigeration used in retail stores relate very closely to those used in residential situations. Food stores have storage coolers for milk and similar items which are simply larger than those for residential use and have multiple doors for public access (see Fig. 26-4). These facilities are kept at about 35°F, while frozen food is maintained at about 5°F, or below. Open display cases for frozen pies, ice cream, and other frozen foods present a special problem, as considerable cold is lost to the store air. Restaurants and bars may use some refrigerators designed for residential use but of largest capacity, and for ice, cooled glasses, and other "products" they either use cooling systems on-site or may buy ice from commercial producers for storage and cooling. Hospitals, florist shops, and other users of commercial refrigeration also generally use refrigeration units that are portable or not completely built-in.

Built-in refrigeration may occupy small areas or may be large rooms with insulation of several layers of cork, Styrofoam, or fiberglass totaling 6 to 12 in., depending on the temperature required. Ceilings should also be insulated in the same manner, but concrete floors are often not insulated, resulting in considerable cooling loss into adjacent floors or into the earth. This loss may be reduced or possibly eliminated by pouring a subslab, installing rigid insulation, and finishing with another concrete slab. Insulation should be placed vertically at the edges of the slab. Interior surfaces should be finished with a coat of cement plaster painted with nontoxic washable

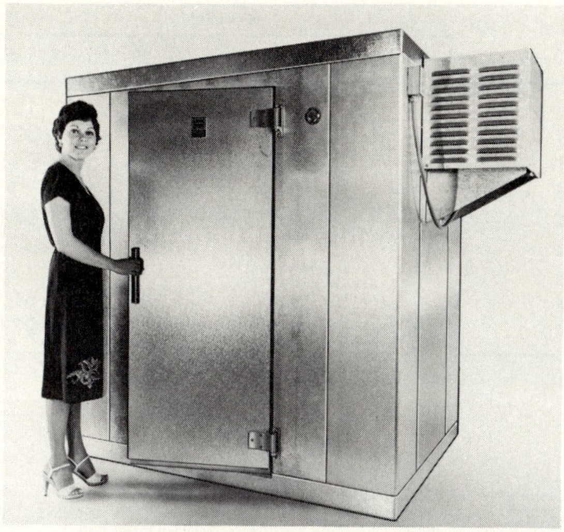

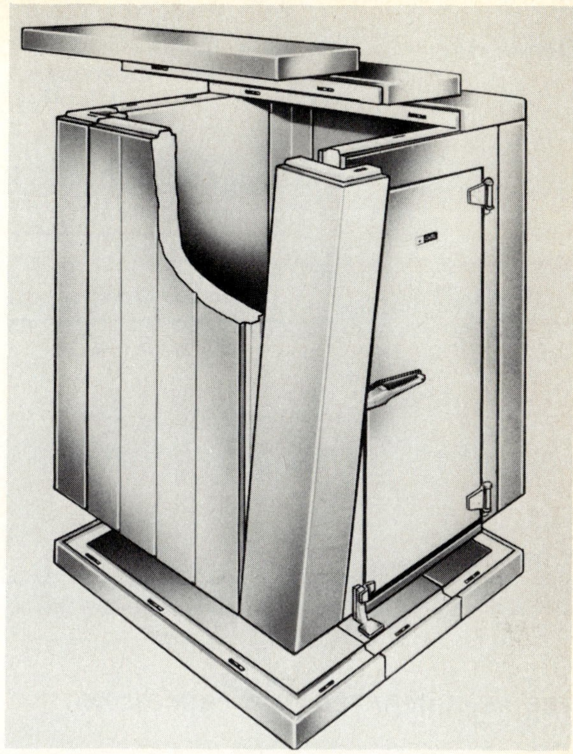

**Figure 26-4** Commercial retrigerators. (*Courtesy of Bally Case and Cooler Inc.*)

enamel. Doors, frames, and other accessories are generally purchased from specialty manufacturers. Also available are a great variety of meat hooks, overhead trolleys, and shelving that complete the installation. Exterior surfaces of built-in facilities are finished with wallboard or plaster in a normal manner.

## 26.5 PREFABRICATED BOXES

Prefabricated refrigerators of large size or as *walk-in boxes* are also available from a number of manufacturers. Most of these are fabricated with 22-gauge enameled steel, 0.042-in. aluminum, or 20-gauge stainless steel faces, using 4 in. of urethane as insulation. Panels for walls and ceiling construction are nominally 10 to 12 in. wide with standard heights from 7′-6″ to 12′-0″. Panels are connected with patented concealed fastenings to provide smooth interior and exterior surfaces. The range of refrigeration of these panels is 250 to −90°F, and thus they may be used either indoors or outdoors. Prefabricated floors for use with prefabricated walls and ceilings are also available and are usually installed in a depressed area in a concrete floor and covered with concrete or quarry tile. If installed in an existing building, the metal/urethane floor panels are simply placed over the existing floor and a slight ramp provided at the door.

Refrigeration systems are provided as "packages," ranging from 3500 to 9500 Btu/hr for −10°F. Smaller sizes use $\frac{3}{4}$-hp motors on compressors and $\frac{1}{15}$-hp motors on evaporators, while larger models use 2-hp motors on compressors and $\frac{1}{6}$-hp motors on evaporators. Almost all operate on 220 to

240 V and are rated from a 7.5- to a 14.0-A load. As might be expected, each manufacturer has slightly different "standard" sizes, but Table 26-2 gives those most used.

**TABLE 26-2**
Internal Cubic Feet for Prefabricated Refrigerators

| *Width and depth* | *Floor area ($ft^2$)* | *Height* 7′-6″ | 8′-6″ | 10′-6″ | 11′-6″ |
|---|---|---|---|---|---|
| 3′-11″ X 5′-10″ | 22.85 | 117 | 134 | 165 | 182 |
| 5′-10″ X 7′-9″ | 45.18 | 251 | 288 | 360 | 397 |
| 5′-10″ X 11′-7″ | 67.51 | 386 | 442 | 555 | 611 |
| 7′-9″ X 9′-8″ | 74.86 | 435 | 498 | 626 | 690 |
| 7′-9″ X 13′-6″ | 104.63 | 618 | 709 | 893 | 984 |
| 9′-8″ X 11′-7″ | 111.86 | 667 | 765 | 966 | 1064 |
| 9′-8″ X 15′-5″ | 148.96 | 906 | 1039 | 1305 | 1438 |
| 11′-7″ X 13′-6″ | 156.33 | 949 | 1088 | 1377 | 1517 |
| 11′-7″ X 17′-4″ | 200.68 | 1238 | 1420 | 1789 | 1971 |

Bally Case and Cooler Inc.

## 26.6 REFRIGERATED FOOD PROCESSING

Refrigeration plays a most important part in the production and preservation of food products of many kinds (see Table 26-3). Bacteria under certain conditions multiply rapidly and in turn will spoil food. On the other hand, some kinds of bacteria, when carefully controlled, are necessary in the production of one type of food from another type of food. Examples of this phenomenon are the processing of sour cream, cheese, butter, wine, and vinegar, where storage may prevent complete ripening. If fruit is picked when ripe additional storage and excess humidity may continue the ripening process until the fruit is unmarketable. Eggs lose moisture from the "white" in storage and so may be dehydrated by long refrigeration. Meat of most types does not show physical signs of deterioration at first but ultimately acquires an odor and a slimy film.

Many food products are processed by some refrigeration method before they are made available for sale. Receiving docks may be enclosed and cooled to about 60°F if desirable, but more often are simply open areas for temporary storage, weighing, and listing of raw products. Chilling or aging areas, where products are further sorted and cooled before being cut up or packaged, should be kept at about 36°F, as should areas where preparation, cutting, smoking, and curing of meat products are handled. Areas to be used for packaging or wrapping may be refrigerated or not, depending on the product. Food that is to be frozen is either quick-frozen by several different methods and stored at 0°F in storage rooms, or cooled and frozen on cold-plate shelving in the storage space. Frozen-food locker plants provide some of the same facilities, except that individual lockers, maintained at about 0°F, are used for personal food storage.

Two major systems for freezing food products are *sharp freezing* and *quick freezing*. The principal difference between the two is the size and number of ice crystals formed when the water in raw food solidifies. Ice crystals start forming in food at about 30°F, and continue until the major portion of water is solid at about 25°F. The zone between 25 and 30°F

**TABLE 26-3**
Storage Requirements for Perishable Products

| *Commodity* | *Storage temp. (°F)* | *Storage life* | *Commodity* | *Storage temp. (°F)* | *Storage life* |
|---|---|---|---|---|---|
| Apples | 30-32 | 2-6 months | Grapefruit | 50 | 4-8 weeks |
| Apricots | 31-32 | 1-2 weeks | Grapes | 31-32 | 3-8 weeks |
| Artichokes | 31-32 | 1-2 weeks | Honey | – | 1 year |
| Asparagus | 32 | 2-3 weeks | Horseradish | 32 | 10-12 weeks |
| Avocados | 45-55 | 4 weeks | Kohlrabi | 32 | 2-4 weeks |
| Bananas | – | – | Lard | 45 | 4-8 months |
| Beans, green and snap | 45 | 8-10 days | Lemons | 32-58 | 1-4 months |
| Beer, barreled | 35-40 | 3-10 weeks | Lettuce | 32 | 3-4 weeks |
| Beets | 32 | 10-14 days | Limies | 48-50 | 6-8 weeks |
| Blackberries | 31-32 | 7 days | Mangoes | 50 | 2-3 weeks |
| Blueberries | 31-32 | 3-6 weeks | Meat | | |
| Bread | zero | 2-3 weeks | Frozen bacon | −10-0 | 4-6 months |
| Broccoli | 32 | 7-10 days | Cured bacon | 60-65 | 4-6 months |
| Brussels sprouts | 32 | 3-4 weeks | Fresh beef | 32-34 | 1-6 weeks |
| Candy | 0-34 | – | Frozen beef | −10-0 | 9-12 months |
| Carrots | 32 | 3-4 weeks | Fresh lamb | 32-34 | 5-12 days |
| Cauliflower | 32 | 2-3 weeks | Fresh pork | 32-34 | 3-7 days |
| Celery | 31-32 | 2-4 months | Frozen pork | −10-0 | 4-6 months |
| Cherries | 31-32 | 10-14 days | Veal | 32-34 | 5-10 days |
| Coconuts | 32-35 | 1-2 months | Melons | 32-40 | 2-3 weeks |
| Coffee, green | 35-37 | 2-4 months | Milk | 34-36 | 2-4 months |
| Corn, sweet | 31-32 | 4-8 days | Mushrooms | 32-35 | 3-4 days |
| Cranberries | 36-40 | 1-4 months | Nut meats | 32-50 | 8-12 months |
| Cucumbers | 45-50 | 10-14 days | Oleomargarine | 34-36 | 1 year |
| Currants | 32 | 10-14 days | Onions | 32 | 3-4 weeks |
| Dairy products | | | Oranges | 32-48 | 3-12 weeks |
| Cheese | 30-45 | – | Parsnips | 32-34 | 4-6 months |
| Butter | 32-40 | 2 months | Peaches | 31-32 | 2-4 weeks |
| Ice cream | −15 | 2-3 months | Pears | 29-31 | 2-7 months |
| Milk | 33-55 | 7 days | Peas | 32 | 1-3 weeks |
| Dried milk | 45-55 | 2-3 months | Peppers | 32 | 2-3 weeks |
| Dried fruit | 32 | 9-12 months | Pineapple | 40-45 | 2-4 weeks |
| Eggplant | 45-50 | 10 days | Plums | 31-32 | 2-4 weeks |
| Eggs | | | Potatoes | 36-50 | 5-8 months |
| Shell | 29-31 | 6-9 months | Poultry | 28-30 | 10-14 days |
| Whole | zero | 1 year | Pumpkin | 50-55 | 2-3 months |
| Solid yolks | 35-40 | 6-12 months | Raspberries | 31-32 | 2-3 days |
| Figs, fresh | 28-32 | 5-7 days | Rhubarb | 32 | 2-4 weeks |
| Fish | | | Sausage | 31-36 | 1-7 days |
| Fresh | 33-35 | 5-15 days | Squash | 50-55 | 4-6 months |
| Frozen | −10-0 | 8-10 months | Spinach | 32 | 10-14 days |
| Frozen-pack fruit | −10-0 | 6-10 months | Strawberries | 31-32 | 5-7 days |
| Frozen-pack vegetables | −10-0 | 6-12 months | Tomatoes | 40-50 | 4-7 days |
| Furs and fabrics | 34-40 | 2-3 years | Turnips | 32 | 4-5 months |

ASHRAE Handbook, 1982 Applications.

should be passed through as rapidly as possible to eliminate concentrations of phosphates, chlorides, and other chemicals, or to keep them to a minimum. This is more important for fruits and vegetables than for meat, as the latter usually has a tougher cellular structure.

The latest method is called *snap freezing* and utilizes the cryogenic temperature supplied by liquid nitrogen at −320°F. In this process, meat is placed on a moving belt operating in a tunnel and is first cooled by circulating nitrogen gas. Continuing through the tunnel, the meat is sprayed with liquid nitrogen to reduce the surface temperature ±125°F in a few seconds and exposed in the next chamber to more nitrogen to equalize the interior and exterior temperatures. Meat is then completely frozen and ready for

packaging. This process greatly reduces the growth of water crystals during freezing, is very rapid, and the spent nitrogen is expelled into the atmosphere. Since nitrogen is inert, no danger is created by such a discharge.

## 26.7 ICE MANUFACTURING

Before the invention of mechanical refrigeration, people kept foods over an extended period by means of snow or ice cut from a river or lake and stored in sawdust. Today one of the major uses for refrigeration is the production of ice for bars, restaurants, and similar establishments. Ice is produced in two major forms: (1) large blocks, approximately 12 in. × 24 in. × 60 in.; or (2) small cubes, which vary from about ½ in. × ½ in. × ½ in. to 1½ in. × 1½ in. × 1½ in. Although cubes used in bars and restaurants are normally solid, those with a hole through the cube present more area, but the volume is decreased. The larger the contact surface, the quicker any surrounding liquid will be cooled. Large blocks are usually available from 50 to 400 lb and may be cut into smaller blocks for different uses.

Ice makers for commercial usage may be separated into cube machines and flake machines. Crushed or flake ice presents more surface for cooling surrounding liquid than cube ice. Flake-ice machines range in capacity from about 200 lb daily to about one ton. The ice is produced by freezing water on the outside of a cylinder until it is about 1/16 in. thick and then removing it with a cutting bar in chips or flakes. Cube machines (Fig. 26-5) freeze water in a series of pans, usually cubes but also cylinders, automatically controlled by the level of ice in the storage bin. Machines designed for restaurants, bars, and hotel and motel service will supply about 400 lb of cubes per day.

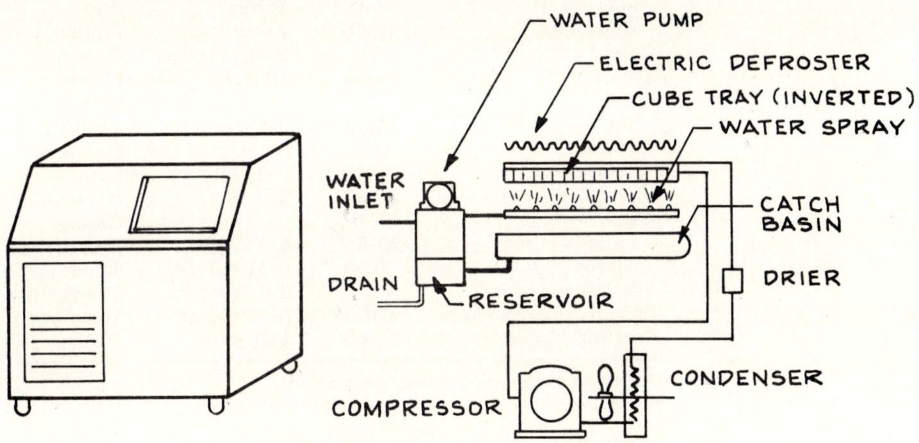


**Figure 26-5** Commercial ice-cube maker.

The amount of latent heat of fusion for ice is 144 Btu/lb, based on converting 32°F water to 32°F ice. The total heat (Btu) to be removed per pound of water for the manufacture of ice ($I$) may be determined as follows if the temperature of the water ($W$) is assumed to be 75°F and that of the brine refrigerant ($B$) 15°F.

$$
\begin{aligned}
W_1(75^\circ\text{F}) - W_2(32^\circ\text{F}) \times 1 &= 43 \text{ Btu} \\
W_2(32^\circ\text{F}) \text{ to } I(32^\circ\text{F}) &= 144 \text{ Btu} \\
I(32^\circ\text{F}) \text{ to } B(16^\circ\text{F}) = 16 \times 0.5 &= \underline{8 \text{ Btu}} \\
& \phantom{=} 195 \text{ Btu} \quad \text{required per pound}
\end{aligned}
$$

Operating conditions would generally increase this requirement about 15%

for various losses in production. If we assume a 15% loss, the refrigeration requirements per day to produce one ton of ice would be as follows:

$$\frac{\text{Weight (2000 lb)} \times \text{Btu/lb (195)} \times 1.15}{\text{Btu/lb (144)} \times \text{ton (2000 lb)}} = \text{tons refrigeration}$$

$$\frac{2000 \times 195 \times 1.15}{288{,}000} = 1.557 \quad \text{or } 1\tfrac{1}{2} \text{ tons per ton of ice}$$

The time required to freeze a block of commercial ice may be determined by the formula

$$\text{hours} = \frac{h^2}{t} \times K$$

where $h$ is the maximum thickness of block of ice, $t$ the temperature difference between 32°F water and brine temperature, and $K$ a constant for operation of the equipment (usually about 6 or 7). Using the formula with brine at 16°F and an average block thickness of $11\tfrac{1}{2}$ in., the time in hours may be found as follows:

$$\frac{11.5 \times 11.5}{32 - 16} \times 6 = \frac{132.25}{16} \times 6 = 49.59 \text{ hr}$$

## 26.8 COMMERCIAL EQUIPMENT

Refrigeration equipment for commercial installations operates in the same general manner as any other application of refrigeration, except that it is larger, occupies more space, requires more maintenance, and is more costly to buy and operate. Evaporators, which are the cooling portion of the system, may be *forced-air* or *blower* type for cooling rooms or refrigerated "boxes," or may depend on open piping supported from room walls or ceiling. For freezing, a system may be designed to use *cold plates* or nitrogen-tunnel snap freezing. Compressors of the *reciprocating* type may have many cylinders and occupy a great deal of space or may be of the smaller *rotary* or *centrifugal* type. Any of these requires a motor or other device to provide rotation.

Unless adequate cooling is provided by forced air or some other method, a *cooling tower* may have to be provided. In this method, the heat picked up by the refrigerant in the evaporator, as well as that added to the vapor in the compression phase, is dissipated by exposure to a cooling medium. This medium may be air or water. Water from wells is often used and may be quite a bit cooler than ambient temperature. By using a water-cooling tower, the temperature may be further reduced. Water towers (Fig. 26-6) are built

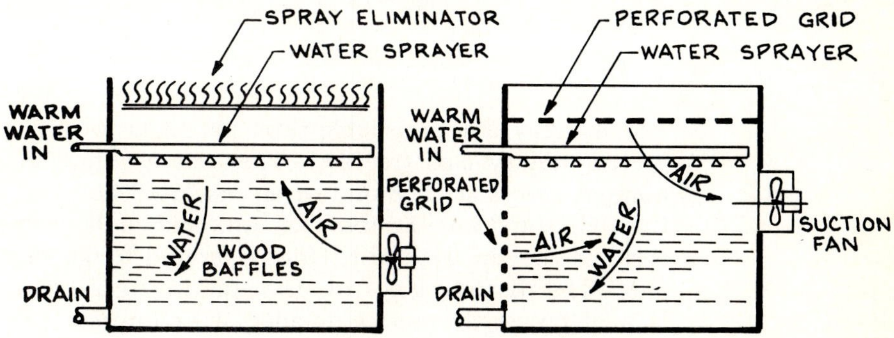


**Figure 26-6** Water towers.

around the heat-exchanger piping system and essentially consist of sprinklers above the piping, which allows water to fall over the pipe at the same time as the water is cooled by air passing through a louver system or by blowers. The water is collected beneath the tower and recirculated, so adequate water surface must be provided to allow warmed water to be cooled.

An entirely different type of refrigeration system may be used where steam is available either for heating or for manufacturing, or as a waste product from some other process. This method is termed an *absorption refrigeration system*. The capacity of commercial units ranges from 25 to 1000 tons. The operation of an absorption system (see Fig. 26-7) relies on water as a refrigerant and on a salt (usually lithium bromide) that has a great affinity for the water. The salt is dissolved in water until it constitutes about 65% of the total weight. The liquid chiller is manufactured in two parts, with a difference in vacuum from approximately 1.47 psia in the generator/condenser to 0.147 psia in the evaporator/absorber section. In operation, refrigerant is pumped to the spray head in the evaporator, where it falls over pipes containing the chilled liquid, to the bottom of the absorber. Another pump circulates the diluted salt solution from the absorber over heated pipes in the generator, and vaporizes a portion of the refrigerant. A pressure of 1.346 psia is maintained in the upper shell, which causes the refrigerant to condense at 112°F, after which it again flows over the evaporator to chill the enclosed liquid.

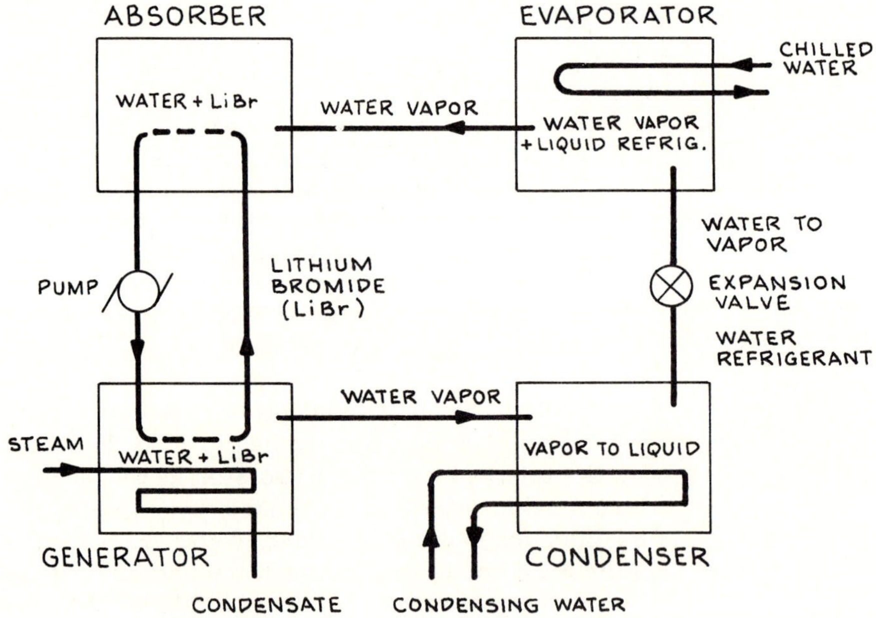


**Figure 26-7** Lithium bromide absorption system.

## 26.9 REFRIGERATED TRANSPORTATION

For many years, perishable farm products were carried to market in refrigerated railway cars (reefers). This was accomplished by using block ice in compartments at each end of an insulated box car. Ice could be replenished through trapdoors in the car top, and in later models air was circulated through the car by electric fans operated by generators turned from the car axle. The disadvantages of using railway cars, and later, trucks, was their lack of power to operate an adequate compressor and the weight and size of the equipment. Dry ice is a possible, but not entirely satisfactory, alternative.

Mechanical refrigeration for trucks and trains is similar to any other type of refrigeration. Compressors are driven by separate gasoline-powered engines, or through a connection to the truck engine. The condenser is mounted at front or side of the vehicle, and the evaporator is installed inside the truck body. Many trucks use a small fan to distribute the cooled air to all parts of the insulated truck compartment. Fiberglass is commonly used for compartment insulation because it is light in weight and has high insulating values. Vehicles may be insulated to temperatures as low as 0°F for frozen foods or to about 32°F for vegetables and similar products. To prevent the refrigerated truck from losing its cold during stops or overnight, most trucks use *eutectic plates*. These plates may be connected to garage takeoffs that continue operation of the compressor or to units mounted off the truck. The eutectic solution (calcium chloride hexahydrate, $CaCl_2 \cdot 6H_2O$) may also be frozen and will retain cold for quite some time if refrigeration breaks down.

## 26.10 SUMMARY

Refrigeration for the preservation of food has been in existence in many forms since very early in human history. Running water, snow, ice, cool caves, and chilled wind have all been used, but ice has lasted as a refrigerant longer than the others as it can be cut, flaked, or chipped, held in large chunks, blown into place, and used under a great many conditions. Mechanical refrigeration is a part of both normal air-conditioning and food preservation. The essential parts and their operation are very similar, and if properly designed and selected, may produce temperatures below 0°F regardless of outside temperature. For commercial installations, units are larger, more cumbersome, and costly but also produce more refrigeration than self-contained domestic refrigerators, freezers, or small portable commercial units.

# 27

# Air-Conditioning

## 27.1 INTRODUCTION

Air-conditioning has been called comfort control, environmental control, climate control, and perhaps a dozen other names. They all mean the same thing: the rearrangement of temperature, humidity, and air motion to provide more comfortable living conditions. A dictionary definition of air-conditioning is "a system for controlling the temperature, purity, and humidity of air in buildings." As early as A.D. 1500 Leonardo da Vinci built a fan arrangement to ventilate rooms, and the homes of upperclass inhabitants of India and southeast Asia were often equipped with fans operated by ropes or primitive waterwheels. Most air-conditioning depends on mechanical movement of some sort to provide comfort, but even the thick-walled small-opening structures of Egypt and southwestern "Indian country" of the United States had air-conditioning brought about by the massive construction and proper arrangement of openings to provide a draft. Today most new construction, as well as a number of older buildings, includes air-conditioning.

Some of the earliest air-conditioning was devised for factories, where candles, candy, and similar products affected by heat or humidity were manufactured. Refrigeration, supplied by tons of ice cut from frozen rivers or ponds and insulated with sawdust, was a natural means for the temporary preservation of meat, and on a hot summer day frequent trips to the "ice box" to cool off or obtain a cold drink were not uncommon. Before automobiles were equipped with air-conditioning units, it was common practice in California and Arizona to carry a block of ice in an uncovered pan to cool the interior temperature when crossing the desert areas in the high summer heat. Running water has always provided a source for removal of heat, and many farmers cooled their dairy products in "milk houses" provided with running water from a nearby stream or spring.

## 27.2 AIR-CONDITIONING DRAWINGS

As in the electrical, plumbing, and heating fields, conventional symbols are used in any drawings designed to represent air-conditioning work to be done. These drawings are similar in nature to those for heating: ducts are usually single lines, and symbols are extensively used to designate parts, controls, and other items. Figure 27-1 includes those symbols most used in depicting air-conditioning work.

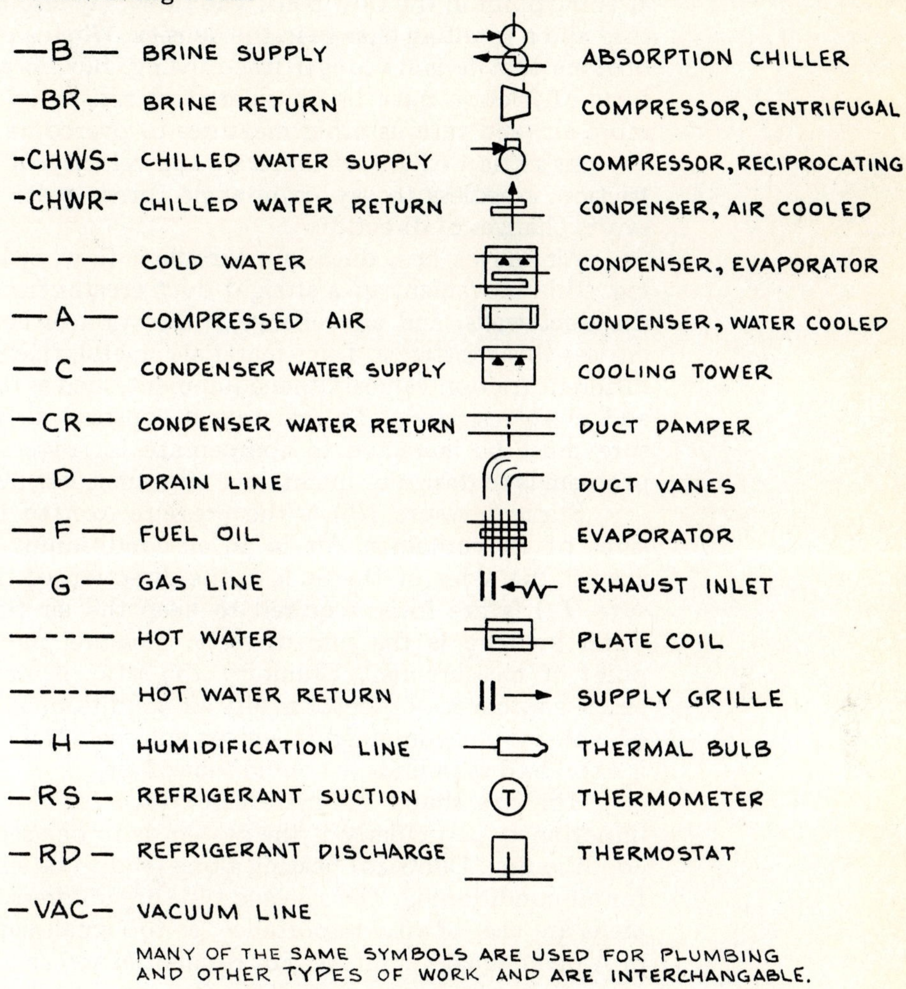


**Figure 27-1** Air conditioning symbols.

## 27.3 AIRFLOW PRINCIPLES

Before an intelligent approach can be taken to the subject of air-conditioning, it is necessary to understand the following principles of airflow.

1. Fluid flow
2. Resistance to airflow
3. Velocity of air
4. Humidity
5. Temperature differences
6. Type of equipment and ductwork

Air is a gas that consists of about 23.3% oxygen, 75.5% nitrogen, and 1.32% argon by weight, or 21% oxygen, 78.06% nitrogen, and 0.94% argon by volume. In addition to these elements, about 0.03 to 0.04% is carbonic acid and about 0.01% krypton and other gases. Air may also contain small quantities of industrial contaminants, such as smoke, ashes, fumes, and other gases.

The flow of air depends on pressure, either at the original air source or at some point in the flow path. The flow may also be influenced by temperature and a resulting change in the density. When air is enclosed in a duct, it is obvious that it is no longer free-flowing. But, in addition, resistance in the form of friction must be considered, as resistance requires more fan power, more air, and various other measures to overcome it. A change in direction or constriction of a duct increases the velocity of air but also increases the friction. Needless to say, right-angle turns create more problems than less severe changes of direction.

Air moves in a duct in a turbulent flow, twisting, churning, and mixing. This movement in a straight duct creates friction just by contact with the duct walls, and any change in direction or constriction of the duct increases this resistance. Duct materials—metal, fiberglass, or plastic—also have different friction values. Other equipment, such as filters, grilles, and dampers, add more resistance. To maintain a constant airflow, it is clear that pressure must be increased to compensate for resistance or friction. A major problem is to design or maintain a smoother, less obstructed duct interior.

Static pressure ($P_s$) is the pressure exerted by a material against the sides of its container. Air in an air-conditioning system exerts a pressure against the sides of the duct in the direction of the airflow. Velocity pressure ($P_v$) is the force required to keep the air flowing at the desired rate. Total pressure is the sum of static pressure plus velocity pressure at any point of measurement. Humidity, the ratio of moisture in the air to "dry" air, is another major factor in any air-conditioning system. The actual weight of water vapor contained in a given volume of air is the absolute weight, and is expressed as pounds per cubic foot of air.

The size, shape, and material of ducts for air-conditioning are of utmost importance, particularly if the system is to be used for both heating and air-conditioning. Ducts for heating alone tend to be smaller than may be needed for air-conditioning. The size and type of compressors and evaporation equipment are also of vital importance, as too small capacity may not be able to provide adequate comfort. For residential and small retail-store installations, air-cooled evaporators may be sufficient, but for many industrial or commercial systems, a water-cooled tower may be necessary. Fans, filters, dehumidifiers, and other equipment and controls add to or deduct from the efficiency of the design.

## 27.4 COMFORT FACTORS

The human body produces heat by changing ingested food into calories, and, in a healthy person, maintains body heat at 98.6°F. All body movement generates heat, which varies from about 350 to 1500 Btu/hr. This heat must be dissipated to obtain body comfort, and this is done by convection, radiation, and evaporation. These phenomena have been covered in other chapters, but a quick review here may not be out of place. Convection is the process in which heat rises and moves from a warm surface to a cooler surface. The body gives off heat to surrounding cooler surfaces, and as the warmer air rises, cooler air takes its place. Radiation is the process by which heat moves from a warming source to a cooler surface by direct heat rays. Evaporation is the process by which moisture becomes vapor, and as moisture evaporates, heat is removed and the surface becomes cooler. Perspiration is an indication of moisture that may be evaporated to cool the body surface.

The humidity of air is changed either by the actual moisture in the air or by the temperature of the air. A low relative humidity permits more moisture to be given off, so more evaporation is possible. At a temperature of 72 to 80°F a relative humidity of 40 to 50%, the average human being will normally be "comfortable." Air movement near the body increases the rate of evaporation. Heat radiates from the body at a faster rate because the heat from the surrounding air is speeded up. As air movement increases, all three factors, convection, radiation, and evaporation, also increase, or decrease if velocity decreases.

## 27.5 AIR-CONDITIONING CALCULATIONS

Heat flow in a building is the same whether the flow is from inside toward the outside, or vice versa. Many factors influence this flow, but several are easier to determine than others. The transfer of heat through walls of most material combinations—glass, roofing, and floors—has been adequately tabulated for easy reference. People, equipment, and other factors may also add, or subtract, heat that can be reasonably estimated. Orientation, shading, wind velocity, direction, and other external factors are more difficult to determine or may be considered so minimal that they need not be considered. The existence of insulation is a major factor, as 1 in. of insulation reduces the flow of heat by about 48 to 55%. An additional inch increases the resistance to about 62 to 68%, and a third inch increases the reduction to about 70 or 75%. Beyond 3 in., the value falls off rapidly and in most cases is not worth the expenditure. Roof insulation will be more effective than wall insulation, and in some locations as much as 6 in. of insulation may be used at roofs or ceilings.

Estimating the air-conditioning load requires that outside atmospheric conditions and the desired interior "comfort" conditions be considered. The information in Table 27-1 is used to obtain the outside dry-bulb temperatures of various locations, and is multiplied by 80% to obtain the outside humidity for heating calculations. For cooling, the same table gives both outside dry-bulb and wet-bulb temperatures in selected locations. (For more complete tables, use Table 2-1A of ASHRAE's *Cooling and Heating Load Calculation Manual.*) Inside design temperature is arbitrarily set by the designer at about 72 to 75°F dry-bulb (DB) for heating, and for summer cooling at about 75 to 80°F, with 50% relative humidity, dry-bulb.

Heat loss is the amount of heat lost through walls, ceilings and floors, windows, doors, and cracks. Heat gain is the amount of heat gained by solar action on the side of a building, by heat transmitted into a building through windows, and by the heat generated by people, machines, or equipment within a building. The thermal resistance of materials is the *R-value* and is the reciprocal of conductance ($C$) or the overall heat transfer ($U$). Typical $R$ and $U$ values are given in Table 27-2. The $U$-value is a measure of heat transfer in Btu/hr-ft$^2$-°F and is used when one material, or combination of materials, is known to determine heat loss.

**TABLE 27-1**
Design Temperatures for U.S. Cities

| State | City | Winter (heating) dry-bulb (°F) | Summer (cooling) Dry-bulb (°F) | Summer (cooling) Wet-bulb (°F) |
|---|---|---|---|---|
| Alabama | Birmingham | 10 | 95 | 78 |
| | Mobile | 15 | 95 | 80 |
| Arizona | Flagstaff | −10 | 90 | 65 |
| | Phoenix | 25 | 105 | 76 |
| | Yuma | 30 | 110 | 78 |
| Arkansas | Little Rock | 5 | 95 | 78 |
| California | Bakersfield | 25 | 105 | 70 |
| | El Centro | 25 | 110 | 78 |
| | Fresno | 25 | 105 | 74 |
| | Los Angeles | 35 | 90 | 70 |
| | San Diego | 35 | 95 | 68 |
| | San Francisco | 35 | 85 | 65 |
| Colorado | Denver | −10 | 95 | 64 |
| | Pueblo | −20 | 95 | 65 |
| Connecticut | Bridgeport | 0 | 95 | 75 |
| Delaware | Wilmington | 0 | 95 | 78 |
| Dist. of Columbia | Washington | 0 | 95 | 78 |
| Florida | Jacksonville | 25 | 95 | 78 |
| | Miami | 35 | 91 | 79 |
| Georgia | Atlanta | 10 | 95 | 76 |
| | Savannah | 20 | 95 | 78 |
| Idaho | Boise | −10 | 95 | 65 |
| Illinois | Chicago | −10 | 95 | 75 |
| | Springfield | −10 | 98 | 77 |
| Indiana | Indianapolis | −10 | 95 | 76 |
| | Terre Haute | 0 | 95 | 78 |
| Iowa | Des Moines | −15 | 95 | 78 |
| | Sioux City | −20 | 95 | 78 |
| Kansas | Topeka | −10 | 100 | 78 |
| Kentucky | Lousiville | 0 | 95 | 78 |
| Louisiana | New Orleans | 20 | 95 | 80 |
| Maine | Augusta | −15 | 90 | 73 |
| Maryland | Baltimore | 0 | 95 | 78 |
| Massachusetts | Boston | 0 | 92 | 75 |
| | Worcester | −5 | 93 | 75 |
| Michigan | Detroit | −10 | 95 | 75 |
| | Lansing | −10 | 95 | 75 |
| Minnesota | Duluth | −25 | 93 | 73 |
| | Minneapolis | −20 | 95 | 75 |

| State | City | Winter (heating) dry-bulb (°F) | Summer (cooling) Dry-bulb (°F) | Summer (cooling) Wet-bulb (°F) |
|---|---|---|---|---|
| Mississippi | Vicksburg | 10 | 95 | 78 |
| Missouri | Kansas City | −10 | 100 | 76 |
| | St. Louis | 0 | 95 | 78 |
| Montana | Butte | −20 | | |
| | Miles City | −35 | | |
| Nebraska | Omaha | −10 | 95 | 78 |
| Nevada | Reno | −5 | 95 | 65 |
| New Hampshire | Concord | −15 | 90 | 73 |
| New Jersey | Newark | 0 | 95 | 75 |
| New Mexico | Albuquerque | 0 | 95 | 70 |
| New York | Buffalo | −5 | 93 | 73 |
| | New York | 0 | 95 | 75 |
| | Syracuse | −10 | 93 | 75 |
| North Carolina | Asheville | 0 | 93 | 75 |
| | Raleigh | 10 | 95 | 78 |
| North Dakota | Bismarck | −30 | 95 | 73 |
| Ohio | Akron | −5 | 95 | 75 |
| | Dayton | 0 | 95 | 78 |
| | Toledo | −10 | 95 | 75 |
| Oklahoma | Tulsa | 0 | 101 | 77 |
| Oregon | Portland | 10 | 90 | 68 |
| Pennsylvania | Philadelphia | 0 | 95 | 78 |
| | Pittsburgh | 0 | 95 | 75 |
| Rhode Island | Providence | 0 | 93 | 75 |
| South Carolina | Charleston | 15 | 95 | 78 |
| South Dakota | Sioux Falls | −20 | 95 | 75 |
| Tennessee | Nashville | 0 | 95 | 78 |
| Texas | Austin | 20 | 100 | 78 |
| | Dallas | 0 | 100 | 78 |
| | El Paso | 10 | 100 | 69 |
| | Houston | 20 | 95 | 78 |
| Utah | Salt Lake City | −10 | 95 | 65 |
| Vermont | Burlington | −10 | 90 | 73 |
| Virginia | Richmond | 15 | 95 | 78 |
| Washington | Seattle | 15 | 85 | 65 |
| | Spokane | −15 | 93 | 65 |
| West Virginia | Charleston | 0 | 95 | 75 |
| | Wheeling | −5 | 95 | 75 |
| Wisconsin | Milwaukee | −15 | 95 | 75 |
| Wyoming | Cheyenne | −15 | 95 | 65 |

ASHRAE Handbook, 1981 Fundamentals.

**TABLE 27-2**
Heat-Transmission Coefficients

| *Construction* | *U* | *R* |
|---|---|---|
| Exterior wall | | |
| 2 X 4 stud, ½-in. fiberboard, wood siding, ½-in. gypsum board interior | 0.225 | 4.44 |
| 8-in. common brick, ½-in. gypsum board interior | 0.256 | 3.91 |
| 8-in. cinder block, 4-in. face brick, 1 X 3 furring, ½-in. gypsum board | 0.158 | 6.33 |
| 4-in. concrete block, 4-in. common brick, 1 X 3 furring, ½-in. gypsum board | 0.203 | 4.92 |
| 4-in. cinder block, plaster each side | 0.308 | 3.25 |
| Interior wall: 2 X 4 stud, ½-in. gypsum board each side | 0.306 | 3.27 |
| Floor: 2 X 8 wood joist, metal lath and plaster, ¾-in. wood subfloor, ⅝-in. plywood, resilient tile, *heat up* | 0.225 | 4.44 |
| Roof | | |
| Metal deck, 2-in. concrete, built-up roof, metal lath and plaster suspended | 0.211 | 4.73 |
| 2 X 8 wood joist, ⅝-in. plywood deck, rigid insulation, built-up roof, ½-in. gypsum board, ½-in. acoustic | 0.169 | 5.91 |
| Structural steel beam metal deck, rigid insulation, built-up roof, metal lath and plaster | 0.158 | 6.34 |

ASHRAE Handbook, 1981 Fundamentals.

**EXAMPLE**

A building has 500 ft$^2$ of wood-stud wall covered on the exterior with wood siding, sheathing board, mineral batt insulation, and lath and plaster interior walls. Assume an exterior temperature of 30°F and an interior temperature of 75°F.

$$\text{temperature difference} = 75 - 30 = 45^\circ$$

$$U\text{-value of wall (Table 27-2)} = 0.072$$

$$\text{heat transfer } (Q) = 500 \times 45 \times 0.072 = 1620 \text{ Btu/hr}$$

Doors and windows are a primary source of heat loss and in the case of windows a possible source of heat gain as well. The average window may be calculated for heat load by multiplying the glass area by the proper factor for the type of glazing used, or by using a predetermined tabular value, such as those given in Table 27-3.

Eave overhang contributes to the shading of windows, which in turn reduces the amount of heat gained from solar exposure through the window. The exposure direction and the latitude or declination of the sun are factors that may be used in connection with glass size to determine the actual area exposed. When the glass faces east or west, the average exposure factor is

**TABLE 27-3**
Approximate *U*-Values of Glass (Btu/hr/ft$^2$)

| | *Vertical* | | | *Horizontal* | | |
|---|---|---|---|---|---|---|
| *Number of sheets of glass* | *Number of air spaces* | *Indoors: still air* | *Outdoors: 15-mph wind* | *Number of air spaces* | *Indoors: still air* | *Outdoors: 15-mph wind* |
| 1 | 0 | 0.75 | 1.13 | 0 | 0.96 | 0.40 |
| 2 | 1 | 0.46 | 0.55 | 1 | 0.56 | 0.66 |
| 3 | 2 | 0.33 | 0.36 | | | |

0.60, south about 2.00 to 2.10, and southeast and southwest about 1.30. To determine the area of unshaded glass, multiply the exposure factor by the overhang in feet. Reduce this by the distance the window head is below the overhang, and multiply the result by the window width. Subtract this figure from the gross window area for the unshaded area.

**EXAMPLE**

Assume a window 4′-0″ high by 6′-0″ wide, with a 2′-0″ overhang and a window head 1′-0″ below the overhang. Exposure is toward the south.

Area of window: $4 \times 6 = 24 \text{ ft}^2$

Overhang: $2'\text{-}0 \times \text{factor } 2.00 = 4.00$

$4.0 - 1.0 = 3.0$

$3.0 \times \text{width } 6 \text{ ft} = 18$

$24 \text{ ft}^2 - 18 = 6 \text{ ft}^2$ exposed glass

Air infiltration is created by cracks around doors and windows, open doors, and other minor openings (see Table 27-4). The amount of infiltration may be determined by calculating the actual crack width and length times the air velocity, or by the simpler, but less precise, air-change method. The latter is based on the formula

$$\frac{W \times L \times H \times \text{AC}}{60} = \text{ft}^2/\text{min}$$

where $W$ is the room width, $H$ the room height, $L$ the room length, and AC the number of air changes per hour.

**TABLE 27-4**
Crack Infiltration at Windows[a]

| *Type* | *Condition* | *Wind velocity (mph)* 5 | 10 | 15 | 20 | 25 | 30 |
|---|---|---|---|---|---|---|---|
| Double-hung wood | Masonry wall | 1 | 2 | 3 | 4 | 5 | 6 |
| | Wood frame | 2 | 6 | 11 | 17 | 23 | 30 |
| | Average window 1/16-in. crack 3/64-in. clearance | 7 | 21 | 39 | 59 | 80 | 104 |
| Double-hung metal | No weather strip, locked | 20 | 45 | 70 | 96 | 125 | 154 |
| | Weather strip, unlocked | 6 | 19 | 32 | 46 | 60 | 76 |
| Metal | Industrial pivot | 52 | 108 | 176 | 244 | 304 | 372 |
| | Architect projected | 15 | 36 | 62 | 86 | 112 | 139 |
| | Resident casement | 6 | 18 | 33 | 47 | 60 | 74 |
| | Heavy casement | 3 | 10 | 18 | 26 | 36 | 48 |
| | Vertical pivot | 30 | 88 | 145 | 186 | 221 | 242 |

[a]Loss indicated for cubic feet per lineal foot of crack per hour. Crack opening not more than 1/16 in. Openings greater than 1/16 in. indicate a poor fit.

**EXAMPLE**

Assume a room 10 ft × 12 ft × 8 ft high, with two walls exposed to exterior summer conditions, with windows in two walls, and with one door.

$$\frac{W \times L \times H \times \text{AC}}{60} = \text{ft}^3/\text{min}$$

$$\frac{10 \times 12 \times 8 \times 0.90}{60} = \frac{864}{60} = 14.4 \text{ ft}^3/\text{min}$$

Add for average door openings:

$$50 + 14.4 = 64.4 \text{ ft}^3/\text{min}$$

Since most air-conditioning systems require more cooling than heating, it is expedient to consider the cooling cycle primarily. This may be a rather long and complicated procedure if exact requirements are needed, or a rather shorter and simpler process if a ballpark figure is acceptable. For most residential and small commercial work, the second method is adequate. Such items as heat generated by human bodies or machines are usually not large enough to influence residential work, although they may be a factor, together with lighting heat and continuous machinery operation, in factories and other larger projects.

To use the shorter method, follow these steps.

1. Determine the difference between the inside and outside temperatures.
2. Find the heat loss through glass areas.
3. Find the total wall area in square feet that is exposed to outside temperatures. Do the same for floor and ceiling areas.
4. Find the net wall areas by subtracting the glass area from the gross wall area.
5. Multiply wall, floor and ceiling, glass, and roof areas by the construction factor by the temperature difference to obtain total Btu gain.

If heat gain from lighting, human bodies, machinery, infiltration, and other sources is important, each should be calculated and added.

**EXAMPLE**

Assume a wood-stud building with exterior wood siding over sheathing, mineral wood insulation, and a lath and plaster interior finish ($U$ = 0.072). Floor construction is wood joist, with plywood sheathing and a hardwood finish floor, with lath and plaster ceiling ($U$ = 0.23). The roof has a pitch of 4 in 12, with wood rafters and sheathing, and composition shingles ($U$ = 0.48). The building is 20 ft × 30 ft with an 8-ft plate height and has two 3 ft × 4 ft windows on the southern 20-ft exposure, one 3 ft × 4 ft window and one 3 ft × 6′–8″ door on the eastern 30-ft exposure, one 6 ft × 4 ft window on the north, and no openings on the western wall. The temperature difference is 15°F.

1. Temperature difference: 15°F.
2. Heat loss through glass:

| | | |
|---|---|---|
| South: | $2 \times 3 \times 4 =$ | 24 ft$^2$ |
| East: | $1 \times 3 \times 4 =$ | 12 ft$^2$ |
| North: | $1 \times 6 \times 4 =$ | 24 ft$^2$ |
| West: | none | |
| | | 60 ft$^2$ |

3. Wall area:

| | | |
|---|---|---|
| South: | $20 \times 8 =$ | 160 ft$^2$ |
| East: | $30 \times 8 =$ | 240 ft$^2$ |
| North: | $20 \times 8 =$ | 160 ft$^2$ |
| West: | $30 \times 8 =$ | 240 ft$^2$ |
| | | 800 ft$^2$ |
| Ceiling: | $20 \times 30 =$ | 600 ft$^2$ |
| Floor: | $10 \times 30 =$ | 600 ft$^2$ |
| Roof: | $18 \times 30 =$ | 540 ft$^2$ |

4. Net areas:

| | | |
|---|---|---|
| South: | $160 - 24 =$ | 136 ft$^2$ |
| East: | $240 - 12 =$ | 228 ft$^2$ |
| North: | $160 - 24 =$ | 136 ft$^2$ |
| West: | $240 - 0 =$ | 240 ft$^2$ |
| | | 740 ft$^2$ |

5. Construction factors:

| | | |
|---|---|---|
| Walls: | $740 \times 0.072 \times 15 =$ | 799.2 Btu/hr |
| Ceiling: | $600 \times 0.23 \times 15 =$ | 2070.0 Btu/hr |
| Floor: | $600 \times 0.23 \times 15 =$ | 2070.0 Btu/hr |
| Roof: | $540 \times 0.48 \times 15 =$ | 3888.0 Btu/hr |
| Glass: | $60 \times 1.13 \times 15 =$ | 1017.0 Btu/hr |
| | | 9844.2 Btu/hr |

This shortcut method gives only the approximate heat loss and is adequate for some uses. For more complete calculations, it is recommended that the procedure used by ASHRAE be followed, which includes all the various heat loss or gain factors for each situation.

Basements and unheated spaces vary widely in heat loss or gain. Loss through a concrete basement floor or concrete slab-on-grade is usually negligible, but where floor construction is over a crawl space enclosed with concrete foundation walls, a factor of about 10 should be multiplied by the floor area. Basement temperature is often assumed to be 60°F, which indicates some loss through basement walls and floor. Loss through the floor is minimal, however, if more than 4 ft below grade. Leakage for slab-on-grade occurs primarily at the exposed slab edge and may be controlled with insulation at this location. This heat loss may be calculated by determining the perimeter length in feet and multiplying by 18 Btu/hr per foot. Unheated spaces such as closets are often figured at 50% of area or at 50% of adjacent heat loss, as they are usually completely surrounded by interior walls and acquire heat from these spaces by infiltration.

In addition to the Btu/hr required for a building, the air quantity in cubic feet per minute must also be calculated. This may be figured by using the formula

$$\text{ft}^3/\text{min} = \frac{\text{Btu/hr}}{\text{temperature difference between supply air and room air} \times 1.08}$$

**EXAMPLE**

Use the result of the preceding example, 9844 Btu/hr, and a 25°F temperature difference between supply air and room air.

$$\text{CFM} = \frac{9844}{25 \times 1.08} = \frac{9844}{27} = 364.59$$

The average residence requires about 12,000 Btu/hr (1 ton) for each 400 $\text{ft}^2$ of floor area.

## 27.6 AIR DISTRIBUTION

Air distribution includes the ductwork, grilles or registers, fans, and other apparatus that conducts the air from the air-conditioning equipment to the space to be air-conditioned and returns "used" air back to the original source. In a small installation, this may be a short, straight duct without bends or other constrictions, a single fan, and some sort of grille at the air-conditioning unit for returned air. This simple system is not that common, however. Because of the type of construction or room arrangement, most air-conditioning systems require ductwork with numerous horizontal and vertical turns, grilles at openings, reducers in duct size, and other devices that influence the airflow.

Ducts are available in square, rectangular, or circular cross-sectional shapes and may be fabricated from aluminum, galvanized sheet steel, fiberglass, plastics, or asbestos-cement pipe. Hundreds of standard elbows, reducers, and branch fittings are available in many different materials (see Fig. 27-4 for examples of these). Round ducts are more efficient, require less material for the same size, can carry more air in less space, and have less friction loss. Rectangular ductwork fits into the building construction more easily, is easier to install because of its flat surfaces at connections, looks better when exposed, and terminal grilles or registers do not require special fittings. In some systems, the "ductwork" may include a plenum, made by a space above a drop ceiling in a hallway or similar space that can be made airtight. Under-floor ducts beneath suspended or open floor construction are no different from those in other framed situations, but ducts beneath concrete floor slabs are usually of concrete, tile, or asbestos-cement pipe so as to eliminate corrosion and water damage.

Ductwork installed in wood-framed construction can usually be adequately supported with straps attached to the wood frame, or in the case of walls, by the plaster or drywall interior stud covering. In concrete or masonry construction, special arrangements must be made in the form of "chases" or other spaces formed to enclose the ducts. This is especially true when ducts are to be installed from one floor to another. One of the "gripes" most often heard on the job is that the architect and the engineers did not allow enough room for the ductwork, or piping, to be installed beneath concrete floor construction or in chases to other floors.

Three major duct arrangements are common, as shown in Fig. 27-2, and may be further divided into two groups: (1) heating duct systems, and (2) cooling duct systems. Generally, systems that supply processed air at or near the

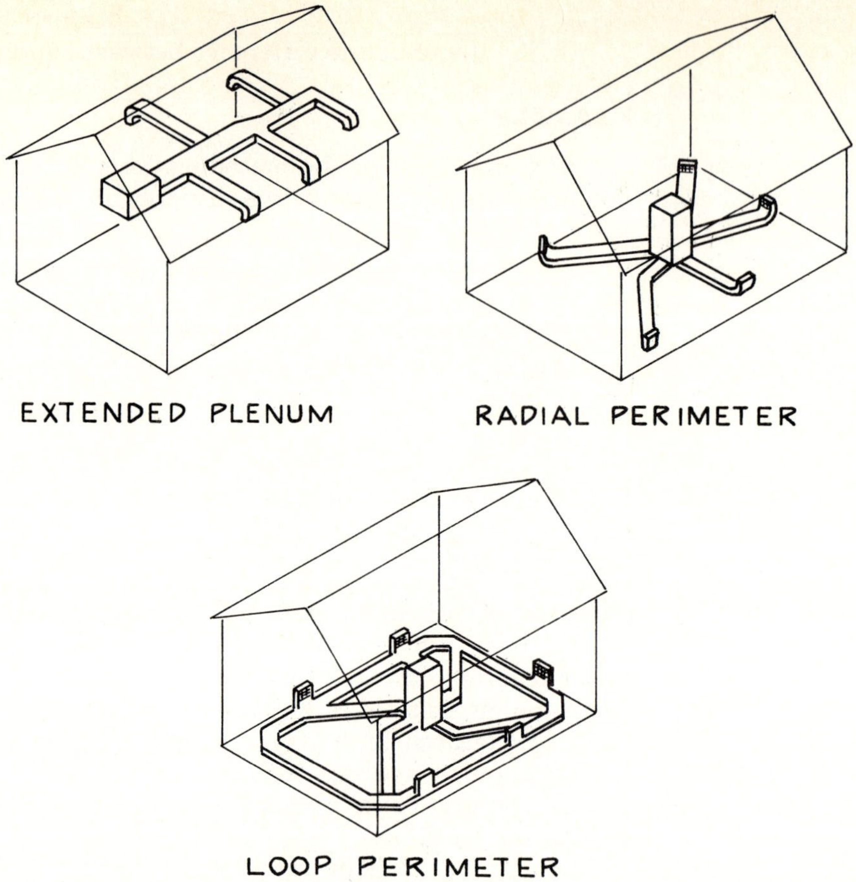


**Figure 27-2** Duct systems.

floor line and at outside walls are better for heating, whereas systems that supply air near the ceiling, or actually in the ceiling, are preferable for cooling. In the radial-perimeter system, a separate duct is run in a direct line from the air-conditioning unit to each outlet. As the duct is usually devoid of horizontal changes of direction and is installed beneath the floor construction, building designers should realize that additional space will be required if headroom is to be maintained. A loop-perimeter system is a continuous loop at the outside wall, with several feeders spaced at equal distances around the perimeter pipe. Registers are often spaced at points of greatest heating or cooling requirements. An extended plenum system is one in which the bonnet, or equipment plenum, on top of the unit is extended in one or more directions and branches are taken off as required. Each branch then supplies one or more outlets, depending on the physical arrangement.

The design of the duct system is directly related to the type of construction used. Both the *radial-perimeter* and *loop-perimeter* systems require some feeders that run diagonally to frame construction members, so these systems cannot be used in multistory buildings unless space for ductwork is provided between the lower ceiling and the bottom of the upper-floor joist. The extended plenum system requires a considerable amount of space for the plenum itself, which may be provided by a drop ceiling in a hall or by installation in the attic. The branch ducts must then be accommodated by proper framing, as the duct is often installed perpendicular to the length of floor or ceiling joist. Most multifloor installations are made with risers in the walls from one floor to another, but there are obvious limits to this system above the second floor. Where a second or third floor is to be served by a single unit, a vertical plenum or duct may be required to carry the air into an attic distribution system, and this requires a chase or vertical shaft space to house the duct.

Supply- and return-air openings must be placed to ensure that adequate air is supplied for uniform comfort in the room. The location of outlets is often determined by the major use of the system—heating or cooling. If both heating and cooling are required, the one used during the greater part of the year should be the primary consideration. For heating, the registers will provide the most comfort if installed at or near floor level beneath windows, where the rising heat will tend to overcome the cold downdraft off the glass. There are fewer problems of location for cooling units, as downdrafts are not common. However, large glass areas may pose a problem when solar heat is to be used: physical shading or specially placed outlet locations may be needed to compensate for this heat. High wall outlets are generally placed on inside walls of a building to force warm air toward a floor-installed return-air opening. Floor outlets are fine for both heating and cooling if they are adjusted to distribute the air along the wall, but they are also depositories for dust or small toys. All supply registers should be capable of adjustment, whereas most return-air grilles are fixed or stamped from flat sheet metal.

## 27.7 DUCT SIZING

Air ducts operate because of pressure, usually so low that a U-shaped manometer is used to measure the effect of air pressure on a column of water. One pound per square inch equals 27.7 in. of water column. Once the cubic feet per minute (CFM) value is known, information for sizing of supply- and return-air ducts can be acquired from most equipment manufacturers. Since return-air is warmer than delivery air, the ducts for return air need to be larger by about 20%. In general, it is better to use larger ducts, less air velocity (pressure), and therefore less power. For quick estimation, a reasonable figure may be obtained by using one CFM per square foot of floor area, with 6 to 10 air changes per hour.

To determine duct size, the CFM and the weight of air must be determined. At 60°F and 60% relative humidity, 13.4 $ft^3$ of air weighs 1 lb.

**EXAMPLE**

Assume 10,000 Btu/hr, a specific heat value of 0.24 Btu/hr, and a temperature difference of 10°F. Determine duct size for both supply and return of air.

$$\frac{10{,}000}{0.24 \times 10} = 4166.66 \text{ lb/hr}$$

$$\frac{4166.66}{60} = 69.44 \text{ lb/min}$$

$$69.44 \times 13.4 = 930.55 \text{ ft}^3\text{/min}$$

Using Fig. 27-3, locate the intersection of 930 $ft^3$/min and 400 ft/min (assumed velocity) and find a duct size of 21 in. for supply. Try again with 500 ft/min pressure and find 19 in. round.

Return air should be about 22 to 23 in. Some manufacturers'

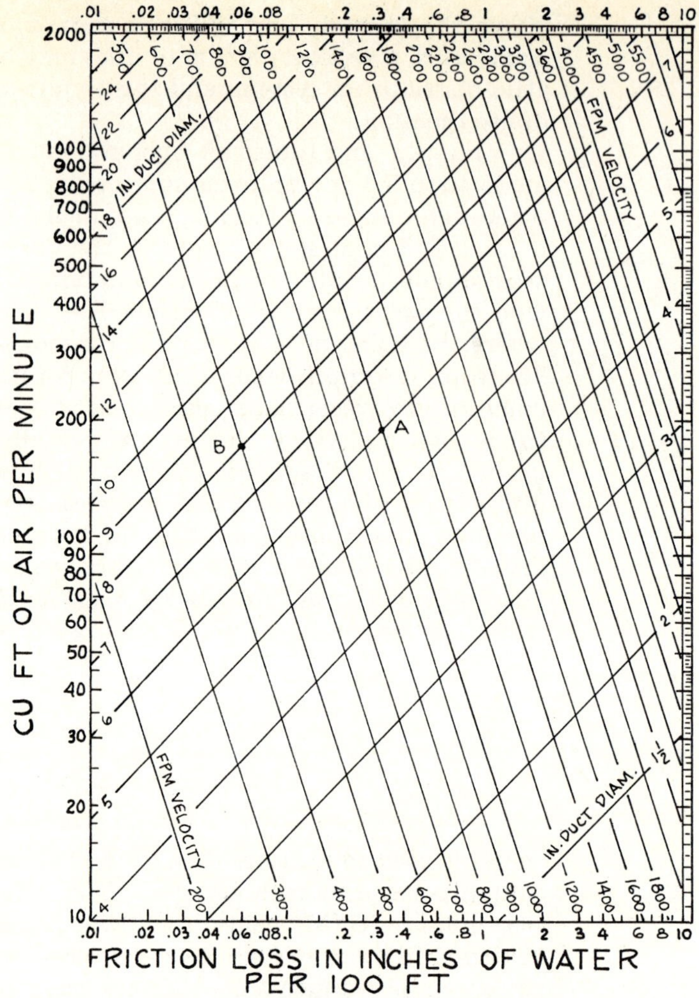


(BASED ON STANDARD AIR OF 0.075 LB PER CU FT DENSITY FLOWING THROUGH AVERAGE, CLEAN, ROUND, GALVANIZED METAL DUCTS HAVING APPROXIMATELY 40 JOINTS PER 100 FT.)
CAUTION: DO NOT EXTRAPOLATE BELOW CHART.

FRICTION OF AIR IN STRAIGHT DUCTS FOR VOLUMES OF 10 TO 2000 CFM. (REPRINTED BY PERMISSION FROM ASHRAE

**Figure 27-3**

equivalent charts might indicate 26 in. × 10 in. rectangular supply and 40 in. × 10 in. for return air.

Air velocity is a determining factor in obtaining the required cubic feet per minute. By most manufacturers' data, 500 ft$^2$/min is about minimum. For a residential main supply, 1000 CFM is more usual, with 500 to 800 CFM at supply outlets and about 600 to 800 CFM on air returns. Where longer runs occur and larger spaces are conditioned, such as banks, large general offices, and stores, the supply velocity should be between 1000 and 1500 CFM for outlets, 700 to 800 CFM for main supply, and about 1500 CFM for return-air ducts. Greater velocity than these may create noise control problems in ducts and outlet units.

Although a system using straight ductwork might represent the ideal installation, very few systems can be installed without using elbows and other fittings. These items are reasonably standardized in regard to shape, and friction loss is calculated in terms of equivalent number of feet of straight duct, as shown in Fig. 27-4. For registers and grilles, this value varies widely, depending on the size, shape, and operation. A flat, punched, sheet-metal return-air grille may have a value of 10 lineal feet, while an adjustable angled supply opening has a value of 60 lineal feet. Grille and register manufacturers are able to provide proper equivalents for their products.

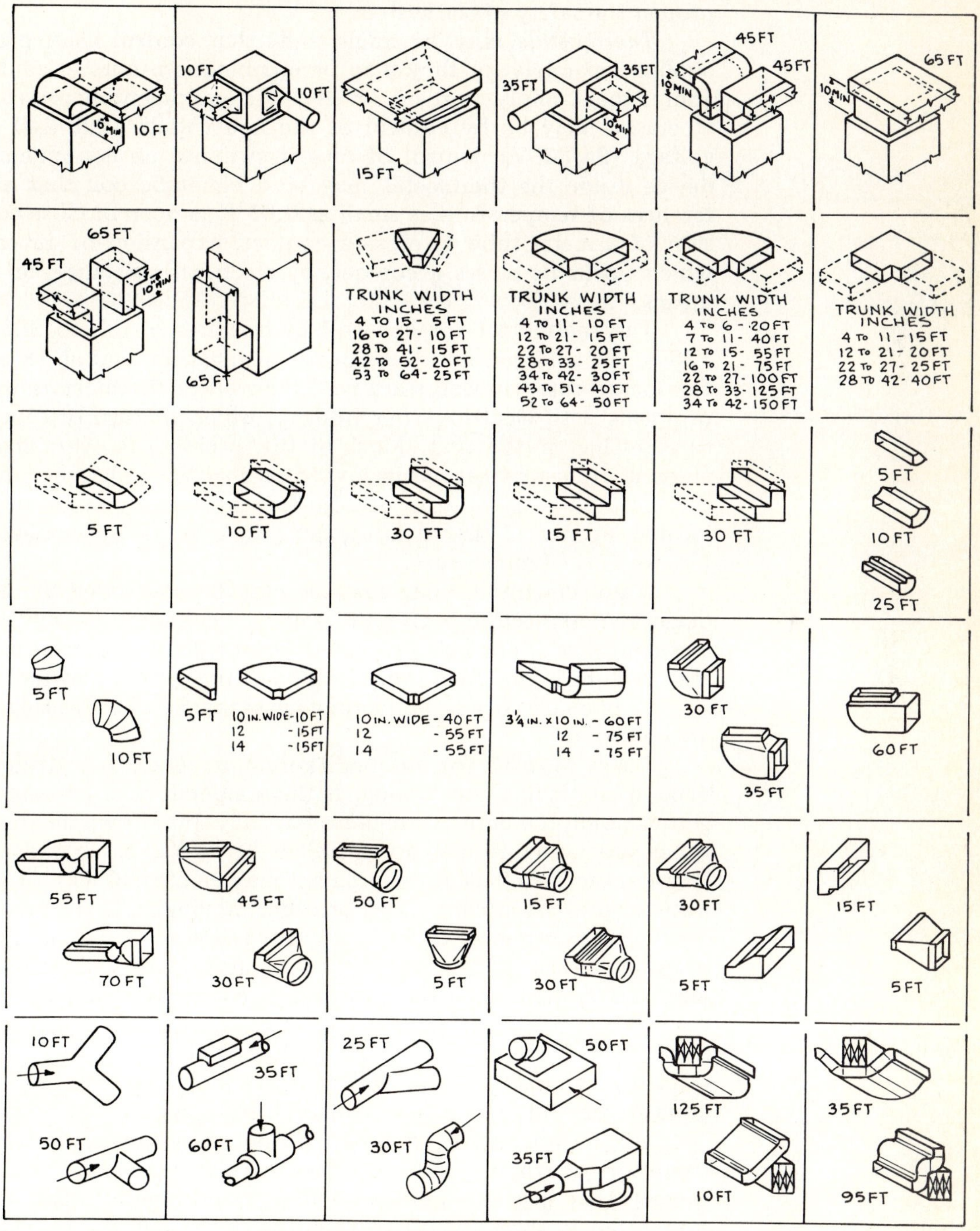


**Figure 27-4** Equivalent straight duct for various fittings.

## 27.8 CONTROLS, FANS, AND FILTERS

Most air-conditioning systems keep the equipment operating automatically by using an assortment of controls on heating, cooling, humidifying, dehumidifying, combustion, liquid flow, time, and a number of other items. Several types of controls may be used: electric, electronic, pneumatic or hydraulic, fluid-operated, and various combinations. Thermostats are operating controls that start or stop the heating or cooling systems. Limit controls protect the safety of the system.

*Thermostats* may be single units that control heating separately and cooling separately, or they may be combination units to control both. Generally, they simply turn on or shut off equipment at preset temperatures. Operation may be by control of the line voltage (120/240 V) or by low-voltage (24/36 V) control of on–off switches on equipment. The control device inside the thermostat may be a bimetalic coil that expands at differences of temperature as small as 0.05°F to open or close contacts, expansion of a metal tube to close a contact, expansion of air as it is warmed which opens or closes a contact, by electrical resistance, or by hydraulical means.

The thermostat for line voltage operates on 31 to 300 V at 24 A at 120 V for solenoid and transformer installation, and at 18 to 96 A 120 V for locked-rotor motor starting. Line-voltage thermostats may have some problems with electric-power heating, which in turn disrupts the temperature setting of the unit. Most of this problem is overcome by the small amount of current carried in low-voltage systems. Another major fault with thermostat use is the ease of adjustment. In many cases, the thermostat is located openly so that it is available to changes by whoever wants a difference in the temperature.

Many air-conditioning systems may be controlled by automatic *timer* devices in connection with temperature. These may be adjusted to turn off a system that is not being required (weekends) and turn it on again for normal working hours. In another form, a tilting mercury switch may be used to silently open or close several circuits in sequence when heating and/or cooling are controlled.

*Limit controls* for air-conditioning are used to provide safety disconnects if something goes wrong. If the temperature or pressure reaches a dangerous point, the limit switch automatically shuts down the operation. Several types are available and are installed in ducts, baseboards, and plenums. Physical limit controls for equipment may be pins, blocks, or similar obstructions to prevent moving slides or other moving parts from overrunning their maximum movement. Other types of regulation are sequence controls. When heavy electrical loads from motors or heating are suddenly imposed on an electrical system, they tend to create low voltage and light flickering. Sequence controls close circuits one after another to reduce this sudden load until normal operation is accomplished. Newer controls are equipped with electronic circuits, diodes, and transistors, and, of course, line-voltage circuits are protected with fuses or circuit breakers.

Air movement in an air-conditioning system is usually produced by some sort of fan, and may be either by positive pressure (above atmospheric pressure) or negative pressure (below atmospheric pressure). The air feed into a fan is always negative (induced draft) and the air exhausted from a fan is positive (forced draft). Two types of fans are commonly used in air-conditioning systems: the propeller fan or axial type, and a centrifugal or radial type (see Fig. 27-5). The *propeller fan* is mounted on a shaft with several blades that are slightly cupped or mounted at an angle on the hub and is usually used to draw air into a space or expel it through a wall. The *centrifugal* or *radial fan* is constructed with a revolving cage on which the blades are mounted at the circumference, and the entire unit is mounted in an enclosure. Either type of fan may be direct-drive or belt-driven.

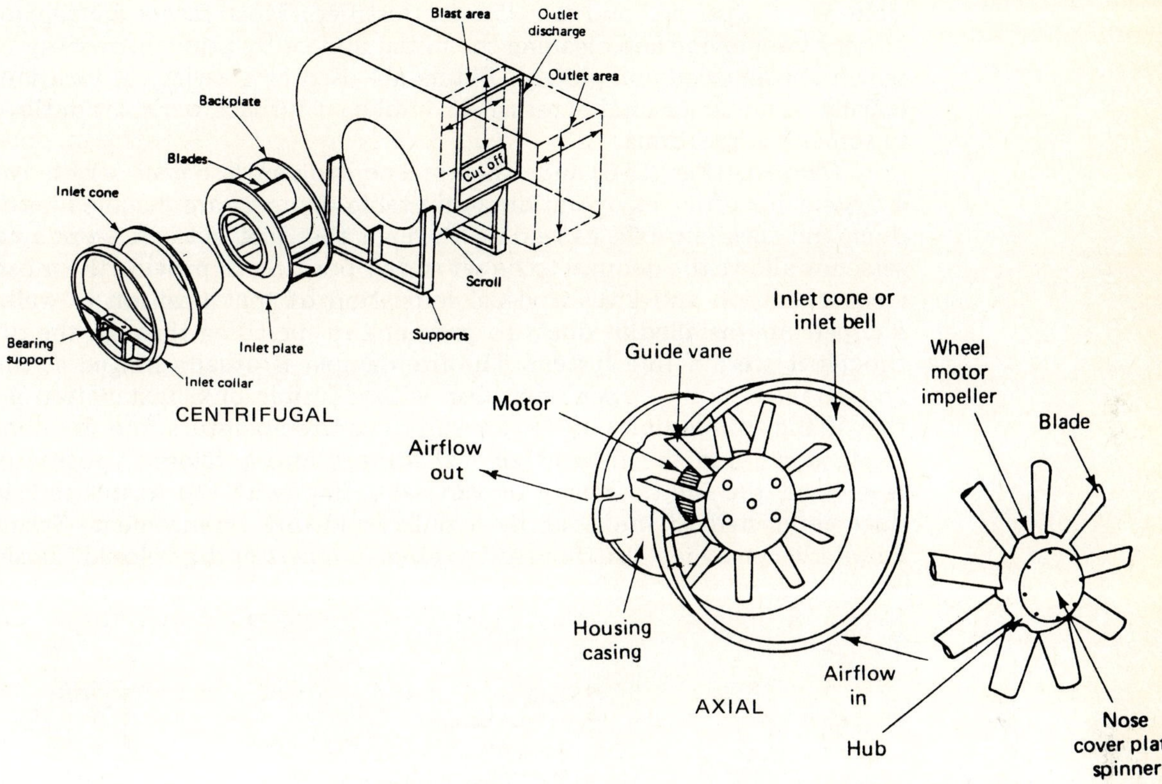


**Figure 27-5** Types of fans. (*Courtesy of AEROVENT INC.*)

Cleaning the air has become a major problem with air-conditioning owing to the increasing amount of pollution created by industry and the growing population. Pollution may be classified as solids, liquids, and vapor, and vary from about 0.004 in. in diameter (dust) to 0.01 micron (0.00006 in.). Dust is formed when solid matter is disturbed by nature or a manufacturing process reduces a solid to finer particles. Fumes are solids that have been turned into gas by some process, usually manufacturing. Smoke is the result of incomplete combustion of a material, usually a fuel. Liquids are impurities created by atmospheric changes (fog) or by evaporation or splashing of liquids in a manufacturing process. Vapor is a gas whose temperature is close to the ambient temperature and which may be condensed under certain conditions. Some additional pollutants are pollen grains from vegetation (causing hay fever), bacterial growth, and combinations of chemical pollutants (smog).

The reduction of air pollution may be accomplished in a number of ways, depending on the foreign matter involved. To remove the larger particles of dust, soot, and smoke, screens and centrifugal force may be used, while adhesives on special screens trap some particles. These special screens may be corrugated metal formed to create a wavy path for the air, causing particles within the air to strike adhesive-coated surfaces. Screens are removable and may be cleaned with solvent. Another type of screen is an electronic device in which particles are electrically charged and adhere to oppositely charged surfaces as they pass through the screen. Liquids can be removed by absorbent blankets or chemicals, settlement chambers, and deflector-drip plates. Molecular-sized particles in gases and vapors can be removed by chemical reaction, dilution, or condensation. Removal of about 90% of pollution is normal, and although removal of more, to 100%, is possible, it is expensive and usually not justified. Most filters are replaceable either by removing and cleaning the initial unit or by using throwaway filters, which are replaced completely. Filters are used in a variety of locations but usually at initial-air intake, return-air intake, at furnace warm-air outlets, and in some duct locations.

*Dampers* (Fig. 27-6) and other duct controls are also necessary to balance a system for efficient operation. Adjustable dampers are installed inside the duct and have an axle extending through the duct side. A lever arm and setscrew allows the damper to be set at any position to provide the most efficient operation and gives a visual indication of that position as well. Fire dampers are installed in ducts to prevent fire spreading through the duct as though it were a flue system. The fire damper is usually hinged at the top and retained in an "open" position with a fusible link. Installation is such that if the fusible link parts as a result of overtemperature, the fire door will swing down and the flow of air will force it into a "closed" position. In a few cases, the fire door may be vertical-sliding, with the fusible link in the duct connected to the door by a pulley-and-wire arrangement. Springs or counterweights may also be used to retain dampers in the "closed" position.

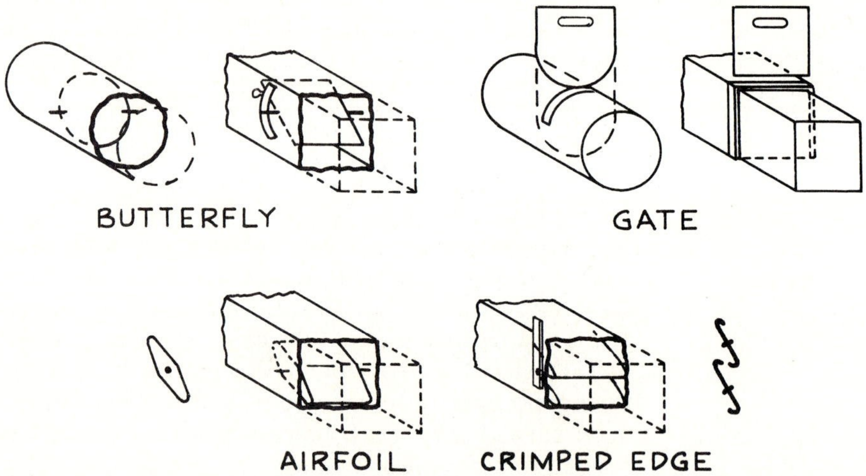


**Figure 27-6** Duct dampers.

Air has weight, and once set in motion it tends to continue in the direction it is moving. Air is also compressible. These two properties create a problem when a stream of air is changed in direction. It also requires energy to make a direction change. When airflow is turned, it tends to crowd to the outside of an elbow, as shown in Fig. 27-7, and loses about 10 times the pressure of an equal length of straight duct. To reduce this loss and also to reduce any turbulence created, *vanes* are often installed in the elbow or connector, as shown. Most vanes are fixed at the radius of the elbow, but some vanes in right-angle connectors may be adjustable with a rod and setscrew similar to that used for dampers. Long sweep elbows reduce this effect, but physical construction may prevent their use.

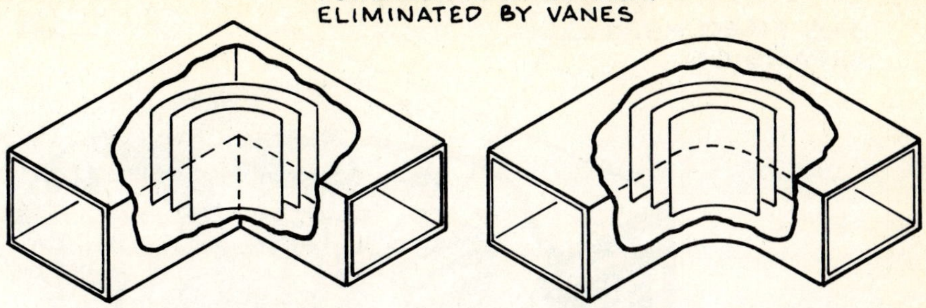


**Figure 27-7** Duct vanes.

## 27.9 AIR-CONDITIONING SYSTEMS

An air-conditioning system is really two systems, one to supply heat and another to supply cooling. Interaction between these two systems results in the "air-conditioning" of a space. The components of either system may be assembled as a unit to provide a separate heating furnace and a cooling package, or may be combined in one piece of equipment to provide both heating and cooling. Selection of the heating equipment depends a great deal on the type of fuel to be used, while the selection of the cooling equipment is determined by temperature and humidity conditions. Fuel for heating systems may be most anything combustible, but natural gas, propane, liquefied petroleum gas, artificial gas, and fuel oil are the most commonly used. Other fuels, such as wood, coal, or heavy oils, are more difficult to handle and have lower fuel ratings and thus are used only when other fuels are not readily available. Electricity is another possible source of heat, but the high cost per kilowatt often eliminates this possibility.

More detailed information regarding furnaces and refrigeration units is contained in other chapters of this book. The major difference between a residential installation and a commercial/industrial installation is usually one of size, influenced by building construction, type of occupancy or work being done, number of people in the space used, and of course the outdoor weather.

Package units for air-conditioning usually provide both heating and cooling facilities by having two separate coil systems within the same unit fixture. Two distribution systems are in common use: a single-zone system, and a multizone system. In the single-zone system (Fig. 27-8), one area and one zone of control is provided. This means that the air supplied to the area is regulated by one thermostat, usually at a central location, and the entire area has the same air conditions. In some cases, where additional heat is required, an electric strip heater may be used in the duct. Additional cooling, however, is more difficult to provide.

Where a number of individual rooms require different amounts of air or different temperatures, a multizone system may be used. This system includes separate heating and cooling coils, each provided with motor-controlled dampers to regulate the various amounts of heat or cooling necessary. The discharge from the central unit then takes on the aspects of single-zone operation, as each separate area is controlled by a thermostat and may be different

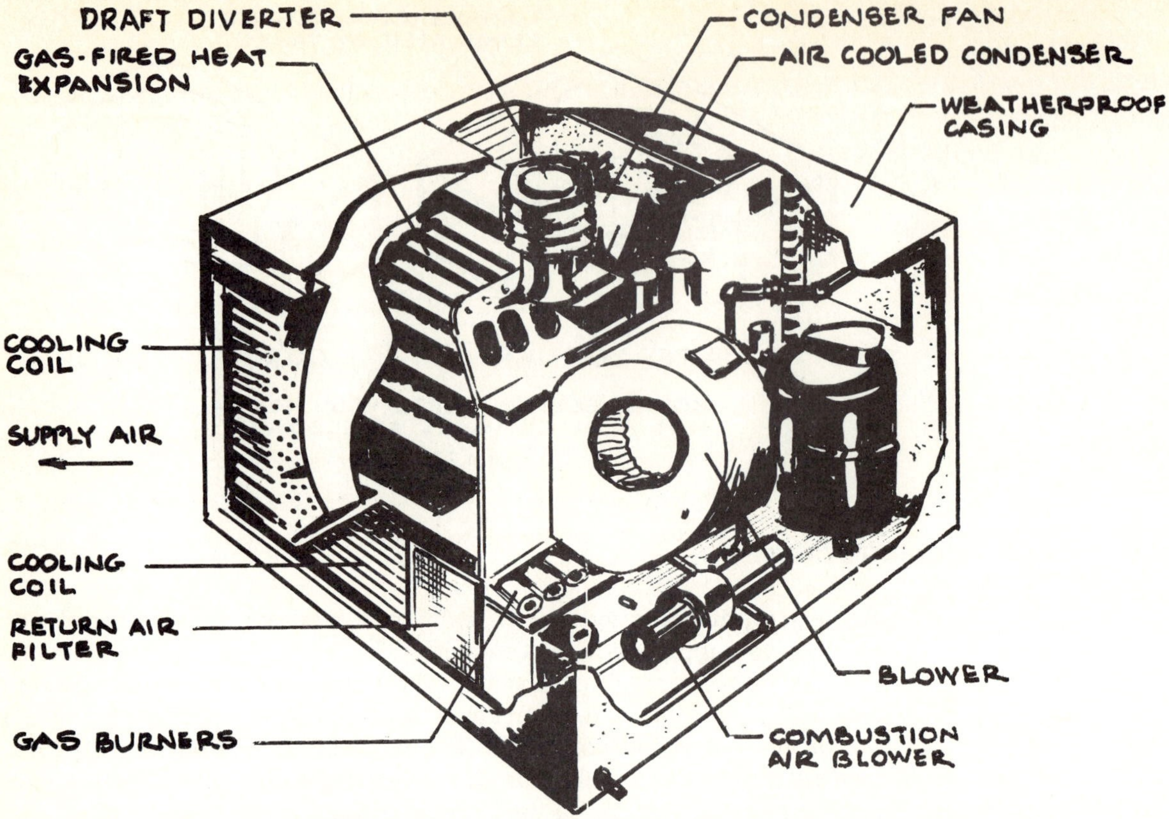


**Figure 27-8** Single-zone packaged unit. (*Courtesy of Lenox Industries Inc.*)

from any other zone. Most multizone systems use direct expansion or chilled water coils for cooling and hot water or steam coils for heating. Equipment is available to provide from about 2000 to 50,000 CFM/min and may be capable of providing air-conditioning to about 30,000 ft$^2$ of space per unit.

The intelligent planning of duct location may provide either additional heating or cooling where needed or eliminate requirements for air-conditioning in some spaces. Halls and corridors are good locations for ductwork, as they are generally centrally located in regard to other spaces and require little conditioning. Some of this conditioning may be provided by seepage or leaks from the ducts installed in the corridor ceiling space. High-velocity dual-duct systems may also help to solve the problem by providing up to 6000 ft/min and thus requiring smaller duct sizes. For small installations such as motel rooms, a fan-coil system may be installed in the space above the bathroom or hall. These units are designed for 100 to 2000 CFM/min and use hot water for heating and chilled water from a central source for cooling. Each unit is a "package" and controlled by a local thermostat.

## 27.10 SUMMARY

The design of proper air-conditioning systems is not work for amateurs but requires a thorough knowledge of construction methods and the physics of air and heat flow, as well as experience in evaluating the client's interpretation of the word "comfort." What may be comfortable to the designer may not be acceptable to the user, or vice versa. Despite such complications, the use of air-conditioning is here to stay and more buildings of every type are designed to include this type of equipment.

The calculation of heat gain or loss is not necessarily a complicated procedure but may become tiresome if accurate results are necessary for ultimate design. The methods outlined in this chapter can be used to determine approximate requirements easily and are probably sufficient for the average building designer, even though they do not include all the factors that might be encountered. With a general knowledge of the methods and reasons for some actions, a building designer should be able to "look over the shoulder" of the mechanical systems engineer and understand what he is doing and why. For those who specialize in residential work, this chapter may provide adequate information for air-conditioning such buildings. Many good books on the design and installation of air-conditioning systems have been written and include all requirements. Manufacturers of furnaces, refrigeration, unit packages, ducts, and other parts are also a good source for information. In general, if cooling requirements can be met, the duct sizes, openings, controls, and other equipment will function adequately for heating.

# 28

# Conveying Systems

## 28.1 INTRODUCTION

Modern building requirements, affected by changes in population, living requirements, and business, have created a situation that requires more rapid and easier vertical and horizontal transportation for both people and freight handling. As land prices continue to rise and buildings are designed in greater height to conserve land, and as office space as well as apartment spaces are stacked higher and higher, vertical transportation in the form of elevator and escalator systems is now a must. This requirement may add as much as 10 to 15% to the cost of building construction but, conversely, may provide better rent returns, lower vacancy rates, more "outside" exposure of the building, and shorter distances within the building. Just as the design of the building is affected by the interior transportation requirements, the design of the transportation system is determined by the design and type of the building.

In addition to vertical transportation in the form of elevators and escalators, other types of conveying systems are necessary in a variety of building forms. Horizontal transportation in the form of moving sidewalks and ramps are common in airports and similar locations. The great number of systems in common use for the handling of freight in manufacturing processes include horizontal tracks with rollers, overhead trolleys, and both open and enclosed chutes. Cranes, from simple equipment designed to handle small parts and boxes to giant overhead units for moving of heavy material, may also be required as a part of some construction. In a few situations where small objects are required to be moved over short distances, a pneumatic tube or hydraulic system may also be needed. Various types of dumbwaiters and individual transportation units, particularly for physically handicapped, may also be required. Conveying systems, then, include a number of different types of equipment, and a general knowledge of these systems is necessary for proper building design.

## 28.2 ELEVATORS

Elevators provide vertical services in a hoistway, at various speeds and by different methods (see Fig. 28-1). The two principal methods used are hydraulic rams, with a vertical height limit of about 150 ft, and cable-hoist elevators, which are theoretically possible to hundreds of feet. *Hydraulic elevators* operate by pumping oil or other liquid into a vertical piston, which in turn raises or lowers a vertical shaft beneath the elevator car. One of the major disadvantages of hydraulic systems is that a hole must be dug or drilled into the ground for at least the same depth as the operating height, unless the shaft is telescoped. *Cable-hoist elevators* are raised and lowered by a system of steel cables coiled around drums and rolled up or allowed to unroll by connected motors. Both systems use a counterweight to partially equalize the car weight.

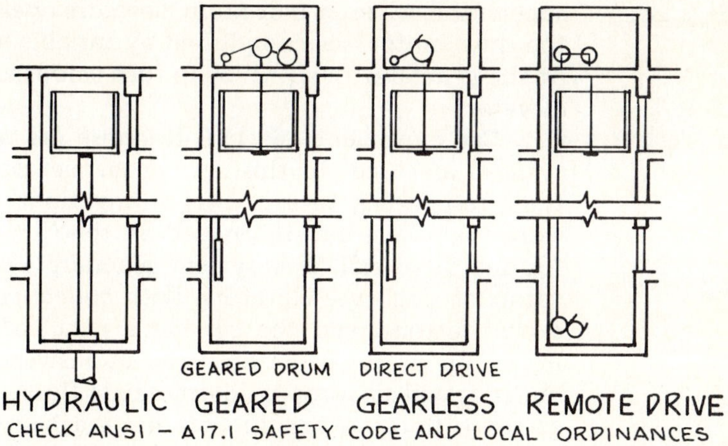


**Figure 28-1** Elevator types.

Elevators are also called as either passenger or freight elevators, which has a considerable influence on the design of the elevators and their capacity and speed. Passenger elevators have a lower weight capacity, smaller car sizes, and operate at higher speeds. Freight elevators may have capacities of many tons, have larger platforms or "cars," operate at relatively slow speeds, and in most cases are not installed in buildings of any great height.

The demand for vertical transportation, or elevators, varies greatly with the type of building use, number of people occupying the building, time of day or night, and location of elevators within a building. During early-morning periods, the traffic in apartment-building elevators is predominantly *down*, while during the same general period traffic in office buildings is primarily up. The reverse is true in the late afternoon, when people leave their work and return to their apartments. Traffic in hospitals usually has an up peak during surgery and doctors' visits in midmorning, and a corresponding up-and-down peak in midafternoon and early evening when patients are visited. Traffic patterns are usually measured in 5-minute increments, as this period allows travel from one position to another in most installations and still enables a possible measure of peak traffic. Desirable intervals between elevators traveling in the same direction are given in Table 28-1.

In the early days of elevator transportation, an "elevator starter," a person in uniform, pushed buttons to control the elevator from the main floor, or an operator sat in the elevator car and controlled the operation of the car in response to indicator lights showing on a panel. Today most systems are automatic and often computerized to take full advantage of the

**TABLE 28-1**
Recommended Elevator Intervals of Service (sec)

| | |
|---|---|
| Apartment buildings | 50-70 |
| College residential | 50-70 |
| Department stores | 30-50 |
| Hospital | 30-50 |
| Hotels, multistory motel | 30-50 |
| Industrial buildings | 25-35 |
| Office buildings | 20-35 |
| Professional buildings | 30-50 |
| Self-parking garages | 40-50 |

entire elevator installation. Generally, this automation is accomplished by three interrelated functions: (1) power control of the drive motor, (2) elevator control of the movement of the car and hoistway doors, and (3) group supervision when two or more elevators operate as a group. Power control of the drive motor is accomplished by variable voltages that change the strength of the generator field, which in turn smooths out the changes in speed of the elevator.

Car movement and call response are relatively simple when *single-call systems* are used. In this system an elevator car serves only one call at a time, responding to a call from any floor and moving to another floor in response to the button pressed on the car panel without stopping en route for any other call. This system generally works satisfactorily on freight elevators or single-use elevators. With *collective operation*, each floor has up or down buttons, and the car has a panel of floor buttons. As the elevator moves up or down the hoistway in answer to buttons pressed by car occupants, it will also stop at intermediate floors when buttons have been pressed at those floors. When the car has reached the highest floor called, it automatically reverses and answers all calls in the other direction. Hoistway and

**TABLE 28-2**
Elevator Requirements for Specific Usage

| *Building type* | *Size* | *Capacity (lb)* | *Speed (ft/min)* |
|---|---|---|---|
| Apartment-hotel | 5′-0″ × 4′-0″ | 1,200 | 200-300 |
| | 7′-0″ × 5′-0″ | 2,500 | 250-300 |
| | 7′-0″ × 6′-0″ | 3,500 | 300-350 |
| | 6′-0″ × 8′-0″ | 4,000 | 300-350 |
| Freight | 5′-6″ × 7′-0″ | 2,500 | 250 |
| | 6′-4″ × 8′-0″ | 3,000 | 200 |
| | 8′-4″ × 12′-0″ | 10,000 | 200 |
| | 10′-4″ × 14′-0″ | 16,000 | 200 |
| Garage-passenger | 6′-6″ × 4′-6″ | 2,000 | 100-200 |
| | 7′-6″ × 6′-0″ | 3,000 | 250-300 |
| | 8′-4″ × 7′-6″ | 3,500 | 350-500 |
| Hospital | 5′-6″ × 18′-10″ | 3,500 | 200-250 |
| | 5′-8″ × 9′-6″ | 4,000 | 250-350 |
| | 5′-8″ × 10′-0″ | 4,500 | 350-500 |
| Office | 7′-0″ × 5′-0″ | 2,500 | 200-250 |
| | 7′-0″ × 5′-6″ | 3,000 | 250 |
| | 7′-2″ × 6′-3″ | 3,500 | 350 |
| | 8′-0″ × 6′-3″ | 4,000 | 350-500 |
| Store | 7′-10″ × 5′-0″ | 2,500 | 200-250 |
| | 7′-0″ × 5′-6″ | 3,000 | 250 |
| | 2′-2″ × 6′-3″ | 3,500 | 350 |
| | 8′-0″ × 6′-3″ | 4,000 | 350-500 |

car doors operate automatically as the car levels at each floor, and doors are usually provided with automatic reversal cushion edges to allow passengers to reverse closing doors.

Elevator size and speed (duty) are subject to the type of building (low-rise or high-rise), population within the building, type of occupancy, and method of ground-level horizontal transportation. In the latter case, the arrival of building occupants via individual autos in relatively small numbers versus the arrival of hundreds from trains or subways will definitely influence the elevator-system design. The cars should be large enough to handle about 75% of anticipated *peak* load and not smaller than about 7′-6″ × 5′-0″ (2500-lb capacity). Residential or small motel or apartment cars and small hydraulic cars may be approximately 5′-0″ × 4′-0″ but must have a limited capacity (1200 lb) and less frequent use (see Table 28-2).

Faster gearless elevators are preferred for longer life and better floor-to-floor performance, and hydraulic elevators are often preferred for lower buildings up to about 12 floors. Combinations of high-speed "express" elevators in high-rise buildings with groups of slower-traveling elevators for intermediate floors and their required "transfer" floors require careful consideration for flexibility and elevator control. All manufacturers of elevator equipment provide full design service and should be contacted before layouts are proposed which may not function.

## 28.3 ESCALATORS

Escalators are another important method of vertical transportation and are used in all types of buildings to about eight or nine floors. After reaching this height, the efficiency of the elevator is generally superior. Installation patterns may be crisscross or parallel from floor to floor (see Fig. 28-2) and

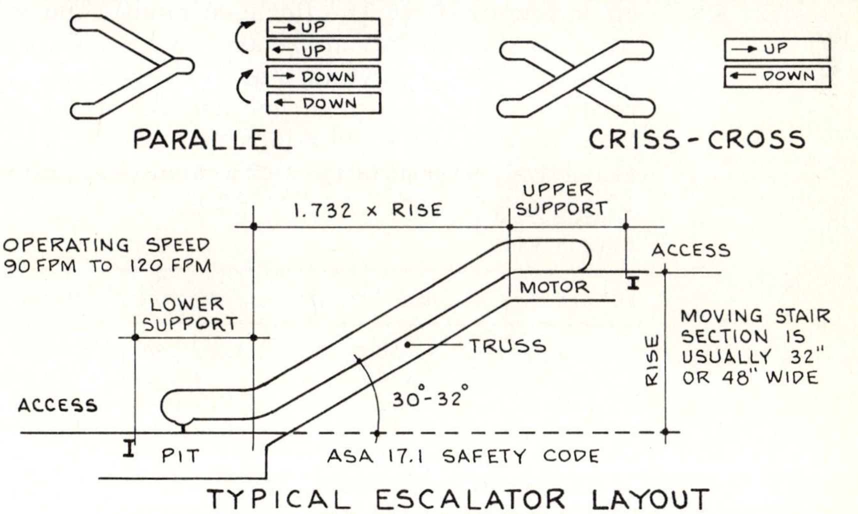


ESCALATORS

| WIDTH | SPEED (fpm) | MAX. CAPACITY | 75% CAPACITY |
|---|---|---|---|
| 32" (1.25 PERSON PER STEP) | 90 | 425/5 MIN. 5,000 HR. | 319/5 MIN. 3750 HR. |
| | 120 | 566/5 MIN. 6,700 HR. | 425/5 MIN. 5025 HR. |
| 48" (2 PERSONS PER STEP) | 90 | 680/5 MIN. 8,000 HR. | 510/5 MIN. 6000 HR. |
| | 120 | 891/5 MIN. 10,000 HR. | 668/5 MIN. 8025 HR. |

**Figure 28-2** Escalators.

may be adjacent to each other or widely separated. Most escalators operate from one floor to another, but in a number of installations, express escalators have been installed to skip several intermediate floors in much the same manner as is used for express elevators. The number of people that may be served by escalators depends on the number of units used, the size or width of the step, the speed of movement, and of course the traffic available. One of the most important values of escalators is that they may channel or direct people into a particular pattern in line with the escalator or adjacent to it.

The speed of an escalator is constant and is related to the amount of incline. Standard speeds are 90 ft/min and 120 ft/min. Higher speeds are possible but not popular, as it becomes difficult for users to get on or off with ease. Step width is also standardized at 32 in. (1.25 persons), which will handle about 5000 persons per hour at 90 ft/min or at 48 in. (2.0 persons), which will handle about 8000 persons per hour at 90 ft/min. A speed of 120 ft/min will increase this capacity to 6700 and 10,000, respectively. These estimates are optimum; in actual use, about 65 to 85% of these figures would be expected.

The space required for an escalator may be separated into three major parts: (1) the upper portion or motor space, (2) the lower portion housing the pit, and (3) the intermediate section including the moving-stair section. Normal incline is at 30°, which makes the horizontal run a product of 1.732 multiplied by the floor-to-floor rise. Power requirements are generally about 44 A starting and 32 A running, 208 V three-phase 60-Hz, but vary slightly with incline, step width, and make of equipment. Heat developed while in operation is approximately 15,000 Btu/hr for the 15-hp motor.

## 28.4 MOVING WALKS

Moving walks are similar to escalators except that they may be installed with an incline of 0° to 15° (inclined ramp). The width of moving walks and ramps is standardized at 32 and 48 in. and operation at 0 to 8° is at 180 ft/min, while the speed at 8 to 15° is generally 140 ft/min. Capacity is also related to the angle of incline and the speed of travel, as shown in Table 28-3. Moving walks and ramps are normally used over longer distances and at relatively level surfaces, such as at airports, subways, and similar locations.

**TABLE 28-3**
Moving Walks

| *Incline (deg)* | *Speed (ft/min)* | *Maximum capacity* | *75% capacity* |
|---|---|---|---|
| 0 | 180 | 1200/5 min (14,400/hr) | 900/5 min (10,000/hr) |
| 5 | 140 | 932/5 min (11,180/hr) | 700/5 min (8,400/hr) |
| 10 | 130 | 867/5 min (10,400/hr) | 650/5 min (7,800/hr) |
| 15 | 125 | 833/5 min (10,000/hr) | 625/5 min (7,500/hr) |

## 28.5 DUMBWAITERS

Dumbwaiters are either of traction or drum type and must have a platform area not greater than 9 ft$^2$ and no more than 4 ft high. Platforms over that size are considered freight elevators and must comply with different safety and installation requirements. Load capacity is from about 50 to 100 lb at speeds of 45 to about 100 ft/min. Dumbwaiters require a hoistway similar to that required for elevators, with traction-type equipment housed overhead and drum machines located in either a pit or overhead. Control may be simple *up* or *down* buttons or more selective floor control when several stations are to be served. Doors may be manually operated or automatically controlled by the position of the car and must be constructed in a fire-resistant manner similar to elevator doors. A water sprinkler is usually required at the head of the hoistway.

## 28.6 INDUSTRIAL CONVEYORS

Industrial conveying systems include a wide variety of equipment installed to provide an easier way of handling products during their manufacture, packaging, and distribution (see Fig. 28-3). For the most part these systems are basically horizontal but may have some small inclines, and may be floor-mounted or overhead-supported. The most common types are (1) gravity package conveyors, (2) powered package conveyors, (3) belt conveyors, (4) magnetic conveyors, (5) towline systems, and (6) special-purpose conveyors. Many combinations are possible to establish a system that will fit almost any requirements, and most systems are engineered and installed by the conveyor manufacturer.

**Figure 28-3** Conveyor types.

*Gravity conveyors* are of two major types: wheel conveyors and roller conveyors. *Wheel conveyors* are almost always the first choice for material that is firm, smooth surfaced, and not too heavy, as capacity is limited to about 125 lb/ft. Available in sections of 5 and 10 ft, and widths of 12 to 24 in., the steel or aluminum ball-bearing wheels are mounted on axles in a number of patterns within a steel frame. Plastic wheels may also be used, but the possible load is reduced to about 40 lb/ft. Gravity *roller conveyors* offer load-carrying capacities to 400 lb/ft and widths to 36 in. Rollers have continuous-width frames and from $1\frac{1}{4}$ to $2\frac{1}{2}$ in. in diameter and are set in frames in a manner similar to wheel conveyors. The pitch of the system is from about 2 to 4% and curves to 90% are possible and common.

*Powered conveyors* are of roller type driven from beneath with an adjustable belt and with a capacity to about 150 lb/ft. Widths are available from 12 to 54 in. and speeds are variable from 5 to 100 ft/min. Changes in direction of 90 to 180° and S bends are possible, together with combinations of turntables, roller transfers from one line to another, and a number of angular diverters and junction sections.

*Belt conveyors* are used for many bulk products, such as gravel, powdered materials, and a wide range of loose or irregular-shaped items. The belt is supported by a series of rollers with a load capacity to approximately 150 lb/ft, belt widths from 8 to 48 in., and speeds from 5 to 100 ft/min. In some types of belt systems the roller supports are replaced with a solid steel base or "table," which facilitates easy movement of packages on and off and allows belt conveyors to be used on inclines to about 25°. Belts are smooth-surfaced or ribbed to handle different materials, inclines to 400 ft are possible, and belts may be reversible for direction of travel. Belt conveyors are also manufactured as mobile units for use in stores and warehouses to assist in loading or unloading trucks and freight cars, and in some uses cantilevered booms may extend into trucks or cars for a distance of 35 ft to facilitate easy handling.

For use in locations where small ferrous metal parts are manufactured or handled, a permanent-*magnet* type of conveyor may be required. This type may be fixed or movable and the incline is usually adjustable from 45 to 75°. Used to transfer machined parts or scrap from processing, the magnetized belt, in widths from 8 to 24 in., may be operated at speeds from 50 to 75 ft/min. Discharge may be to a bin or to another conveyor system. Unfortunately, this conveyor will not handle nonferrous metals, paper, wood, or similar materials.

*Towline systems* also are of two major types, in-floor and overhead trolley, both of which require special carts to connect into the system. In-floor towlines employ a guide track buried flush with the floor and enclosing a moving chain into which a toggle or towpin is inserted from the cart at each floor level. Overhead trolley systems employ a similar but inverted arrangement which requires an extended contact towmast from the cart to the overhead trolley. Each system may be installed complete with turns and switches, low-degree inclines, and automatic controls and have a towing capacity of about 3000 lb at speeds to 100 ft/min.

In addition to the conveying systems outlined, there is also a wide range of equipment for handling multiple heavy parts by overhead chain systems, heavy pallet conveyors, gravity-flow live storage racks, automated vertical-life equipment for use with other conveyor types, and special designs for special uses. Most of this equipment is used in manufacturing but may also profitably be installed and operated where physical handling of any material is necessary and where adequate labor is either not available or too expensive.

## 28.7 CRANES

Cranes are another type of conveying system that is generally limited to manufacturing processes or the handling of heavy or bulky objects (see Fig. 28-4). For the most part cranes in their several recognized forms are used to

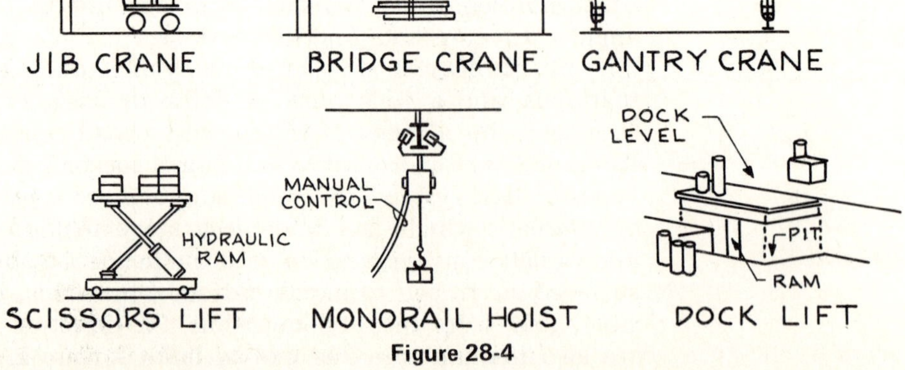


**Figure 28-4**

build buildings, not to complete them, but occasionally a project must include some form of crane, or at least make provision for future installation of a unit, if the building is to function properly. The first questions to be considered are: What floor area of the building is the crane supposed to cover? What material is to be handled? Instead of a crane, perhaps a simple monorail or conveyor would do the work. Other questions relate to the capacity (20 tons or more?), the maximum required span for a bridge crane, how the crane is operated, the height and travel of material lifted, and how often the crane will be used.

Small bridge cranes are of two major types: top-riding cranes and underslung cranes. In a *"bridge"-type crane* the crane is essentially a horizontal beam, supported by and moving on parallel tracks at each end, usually with a movable hoisting mechanism included on the bridge beam moving counter to the end tracks. *Top-riding cranes* ride on top of the side rails with grooved wheels fitting the track, capacities to 10 tons, a span of 15 to 50 ft, with the bridge beam 12 to 30 in. deep. An underslung crane has channel or tee end rails supported from above, on which wheels fabricated with a yoke fitting ride on the bottom flange and in turn support the horizontal bridge team. Crane units more than about 20 ft above the floor line are difficult to operate or move manually, so motorized movement, both horizontally and vertically, is generally included.

## 28.8 HOISTS

Hoists perform some of the same functions that are required of cranes except that hoisting loads seldom exceed 5 or 6 tons and the hoist operates on a suspended monorail from a trolley system (Fig. 28-4). The rail is normally an I section with wheels each side connected by a yoke from which the hoisting mechanism is suspended. Hoist rated capacity may vary from 1000 to 12,000 lb, lift height from about 18 ft for the heaviest load to as much as 150 ft for light loads. Monorail size is nominally 6½ to 24 in. Operation may be manually by chain or motorized, both vertically and horizontally. Many other types of stationary hoists, both chain-operated and motorized, are also available to handle small loads at fixed locations but are most often supplied by the owner or leesee as a part of the "equipment."

## 28.9 LIFTS

Scissors lifts and dock loading lifts (Fig. 28-4) are two major types used to facilitate easy handling of heavy material loads or to provide limited vertical transportation at loading docks. *Scissors lifts* provide capacity from 1500 to 15,000 lb, depending mostly on the maximum vertical travel from 32 to 84 in., with table sizes from 24 in. × 48 in. to 60 in. × 120 in., and at lifting speeds from 7 to 10 ft/min. This type of life is hydraulically operated and a great many special "tables," including rollers, vee troughs, expandable tables, and drop leaves, are available. Installation requires some shallow pits to contain the ram and scissors arms, electrical outlets for 220 to 240 V, and protection of the pit edge with metal angles.

*Dock loading*, or *leveling, lifts* are manufactured to operate in a similar manner to scissors lifts and are usually installed at docks to equalize the difference between ground level and dock height, truck-bed height and dock level, or dock level and conveyor height. Portable equipment capacity varies from 1000 to 10,000 lb with raised height to about 62 in. at a speed of about 7 to 12 ft/min. Lifting capacity of fixed docks is available from 10,000 to 20,000 lb, and at speeds of 30 to 60 seconds for a normal lift of about 60 in.

## 28.10 PNEUMATIC SYSTEMS

Pneumatic conveying is a system of transporting dry bulk material through a pipeline by air pressure. Air pressure may be positive or negative. As the normal pressure of air (14.7 psia) is increased, it becomes heavier and its power to carry more or heavier material is increased. In industrial applications pneumatic systems are successfully used to convey such items as cereals, sawdust, beans, cotton, powder of various types, and some kinds of metal particles. In business installations pneumatic piping moves bank pouches, sales reports, interoffice papers, waste paper, and some forms of trash. Hospitals and other institutions also use this system to move soiled laundry, dry waste, and small packaged items as well as reports and other paperwork.

Several systems of negative pressure (vacuum) or positive pressure are in current use, and vary from 12 psig pressure to 12 in. mercury vacuum. High-vacuum systems are used to move grain, sawdust, cereals, and other dry bulk materials. Pressure systems are used to transport paperwork, laundry, and a number of manufactured products, such as pulverized ore, fertilizer, and granulated sugar. Vacuum systems are capable of velocities from 4000 to 8000 ft/min and pressure systems from about 2700 to 4500 ft/min.

Design of pneumatic systems includes so many different factors that standard design formula have not been fully developed. The essential components include an air compressor or vacuum pump, piping, air dryer, air locks at openings, and other control devices. The principal advantages of pneumatic conveying are easier handling of large quantities of dry bulk material, low loss in handling transfers, higher safety against accidents or fire, and ability to operate in almost any location that the conveyor pipe or tube can be located. Major disadvantages are higher initial installation costs, inability to adjust easily to different materials to be transported, high power costs, and limited distances that systems may operate. Recent improvements have, however, made vacuum systems practical up to about 1500 ft in length and pressure systems to as much as 5200 ft.

## 28.11 SUMMARY

Conveying systems of every kind require expert knowledge of each method in order to design the best installation for any particular use and location. Fortunately, manufacturers of elevators, escalators, and other conveying equipment have qualified engineers and designers who can provide proper advice to ensure that their systems are adequate for the service required and will operate properly under the conditions encountered. The major problem faced by the architect or the building contractor is to provide proper hoistways, pits, supports, and anchors for the different systems.

With taller buildings, more occupants, improvement in manufacturing or distribution, and more paper records of all of this "progress," more rapid and efficient transportation and communication must be provided. Conveying systems in many ways determine when a building will look tall or flat; the interior traffic pattern for people, manufacturing, or sales; and the power requirements for electricity or other power. The actual function of a building may be determined by the type and location of conveying systems, and once installed, most of these systems are difficult or impossible to relocate.

# 29

# Sound Control

## 29.1 INTRODUCTION

Sound control is a comparatively new but important factor in the design and operation of buildings. Together with more requirements for lighting, air-conditioning, and space requirements, occupants of buildings are demanding more sound control. Sound control, especially in industrial/commercial construction, occurs in two major areas: the proper combination of materials to reduce transfer of sound through or over walls, ceilings, and floors, and a reduction of undesirable or internal sound created by people, equipment, or by other personal-comfort systems, such as lighting, air-conditioning, and plumbing.

Exposure to high sound levels has been proven to be detrimental to a person's hearing. Conversely, persons working or living in sound-controlled areas produce more work and are more pleasant to be around. Buildings with good sound control command higher rental fees. In addition, a sound-controlled building often has lower operating costs, as the sound insulation also functions as thermal insulation. Even though sound control may be incorporated in wall, floor, or ceiling construction, special conditions may require additional control for acoustical reasons. Basic principles of both types of sound control are included in this chapter.

## 29.2 NATURE OF SOUND

Sound is created by vibration. A vibrating body creates a high-pressure layer of the conductor (usually air) next to the vibration source. This causes the layer to compress the molecules in the next layer, while the molecules in the first layer tend to return to their original position. The molecules of the second layer in turn squeeze those in the third layer and then return to their original position. During the return a low-pressure area is created in each case. Finally, the sound waves reach the human ear, the eardrum vibrates in turn, and the sensation of hearing is accomplished. The number of times this cycle occurs in 1 second determines the frequency of the sound wave, as measured in hertz.

The *wavelength* of a sound is the distance between the same point in two consecutive waves. Since the speed of sound in any material is constant in that material (air = 1128 ft/sec), the frequency and the speed determine the wavelength. As the frequency increases, the wavelength decreases, as more vibrations are leaving the source. The intensity of a sound is represented by its amplitude, which is created by maximum displacement of the sound layers as the waves travel through the conductor. The measure of intensity is the *decibel* (dB), which gives a ratio of loud sounds to softer ones. A change of approximately 3 dB causes the smallest change in loudness that the human ear can detect. Some comparable decibels are given in Table 29-1.

The human ear detects change of frequency as pitch in the range from 20 to 20,000 Hz. The greater number of frequencies per second, the higher the pitch, with the ear most sensitive to sounds with frequencies between 1000 and 4000 Hz. The quality of a tone is determined by the mixture of various frequencies and their distribution determines the timbre of the sound. The construction of the ear and the automatic adjustment it makes to sound waves determines the loudness of a sound, and this is obviously different with different people.

**TABLE 29-1**
Intensity of Sounds from Common Sources

| *Apparent loudness* | *Decibels* | *Intensity* |
|---|---|---|
| Deafening | | |
| Jet aircraft | 140 | 100,000,000,000,000 |
| Threshold of feeling | 130 | 10,000,000,000,000 |
| Very loud | | |
| Elevated train, thunder | 120 | 1,000,000,000,000 |
| Subway train, nearby riveter | 110 | 100,000,000,000 |
| Noisy industrial plant | 100 | 10,000,000,000 |
| Office with tabulating machines, loud street noise | 90 | 1,000,000,000 |
| Noisy office, vacuum cleaner | 80 | 100,000,000 |
| Loud | | |
| Average street noise | 70 | 10,000,000 |
| Average office, congested department store | 60 | 1,000,000 |
| Moderate | | |
| Moderate restaurant, average two-person talk | 50 | 100,000 |
| School classroom, private office | 40 | 10,000 |
| Faint | | |
| Bedroom | 30 | 1,000 |
| Rustling leaves | 20 | 100 |
| Very faint | | |
| Normal breathing | 10 | 10 |
| Threshold of audibility | 0 | 1 |

U.S. Gypsum Co.

## 29.3 SOUND TRANSMISSION

Sound is transmitted in many ways, but probably the most common is through the atmosphere. When people talk to each other, sound is transmitted from one person to another. Automobiles, airplanes, and other street noises are transmitted through the air and in many cases penetrate a building through open windows. However, almost everyone is also aware that sound is transmitted through walls, ceilings, and floors of a building. This transmission of sound, or noise, has caused the design of building construction to include various methods to eliminate or reduce the possibility of sound transfer from one area to another.

Ideally, the control of unwanted sound should begin with site selection and orientation of the building. Although property near a railroad yard may be desirable because rail shipping is readily available, the site may be undesirable because of the noise created by the trains. This is a rather obvious situation, but a site in the country may also have objectionable noise levels created by farm animals or agricultural machines. The land may be so flat that sounds created miles away are clearly audible, or rolling hills may cut off such sound. Orientation of the building on the site and the actual shape of the building should also be considered. Long, narrow buildings facing each other on opposite sides of the street tend to reflect noises from one to another and from the street. Sound reflects from curved or vee-shaped buildings and may build up to a considerable volume. Rooms that require more quiet should be located in the area away from the sound source. Where an advantageous orientation is not possible, it is often desirable to build artificial hills, berms, or planted areas on the site to block or dissipate the sound partially. This option may also be more economical than the extensive building construction required to control unwanted sound.

Another form of sound is impact sound, created by footsteps, pounding, falling objects, and movement in general. Impact sound travels through the mass of the structure, although some of the sound is airborne. The more rigid and solidly connected structure will transmit more impact sound. Pipes, ductwork, and other continuous systems will distribute impact sound throughout the entire building. Control of such sound may be accomplished by isolation, absorption, or elimination of bypaths. Design to provide "limp" construction—a building not firmly joined or one with lots of flexibility—is another option but is not usually possible. The mass of the various parts is also a help because vibration is more difficult and airborne sound has less effect; however, impact sound may be greater.

**TABLE 29-2**
Sound Levels (Decibels)

| | |
|---|---|
| Aircraft | |
| Near plane | 140 |
| Near residence | 80-90 |
| Elevated train | 120 |
| Industrial plant | 100 |
| Auto horn at 25 ft | 100 |
| Jazz band | 85-100 |
| Loud stereo | 90 |
| Person shouting | 85 |
| Noisy household | 75+ |
| Truck at 25 ft | 75 |
| Busy street at curb | 70 |
| Dense traffic at 200 ft | 65 |
| Office conversation | 60 |
| Average conversation | 50 |
| Quiet household | 40 |
| Outdoor background noise | 20-60 |

U.S. Gypsum Co.

## 29.4 SOUND LEVELS

Reasonably accurate measurements have been made of most of sound-producing sources. These have a direct relationship to the amount of airborne sound that may be present or anticipated. A guide to the principal sound producers is given in Table 29-2, but if combinations or otherwise high sound levels are suspected, the level of such sound should be checked with appropriate measuring equipment both day and night.

Sound *attenuation* levels are necessary as a guide for control of sound affecting various areas. Just as the amount of light (footcandles) varies for different spaces and uses, so does the amount of acceptable sound. The sound level in a manufacturing plant may be such that aircraft noise from overhead adds little to the internal acceptable level. This same aircraft noise or impact noise from adjacent spaces may entirely disrupt the function of an office or cause psychological distress in a hospital or school. Some values for average transmission are given in Table 29-3.

**TABLE 29-3**
Sound Attenuation Performance

| *Building type* | *Room under consideration* | *Adjacent room* | *Average transmission* |
|---|---|---|---|
| Residential, motels, apartments, hospitals | Bedroom | Bedroom | 50 |
| | | Living room | 55 |
| | | Kitchen, bathroom | 55 |
| | | Corridors, lobby | 50 |
| | | Mechanical room | 60 |
| | Living room | Living room | 45 |
| | | Kitchen, bathroom | 50 |
| | | Corridor | 45 |
| | | Lobby | 55 |
| | | Mechanical room | 60 |
| | Kitchen or bathroom | Kitchen, bathroom | 45 |
| | | Corridor | 40 |
| | | Lobby | 50 |
| | | Mechanical room | 65 |
| Business | Office | Office | 50 |
| | | General area | 45 |
| | | Corridor | 45 |
| | | Washroom, kitchen | 50 |
| | | Conference room | 50 |
| | Conference room | General area | 45 |
| | | Corridor | 40 |
| | | Washroom | 45 |
| | | Kitchen | 50 |
| | General area | Corridor | 40 |
| | | Washroom | 45 |
| | | Kitchen | 50 |
| School | Classroom | Classroom | 45 min |
| | | Laboratory | 45 min |
| | | Corridor | 40 min |
| | | Shop | 55 min |
| | | Music room | 60 min |
| | | Washroom | 45 min |
| | Music room | Laboratory | 45 min |
| | | Shop, mechanical room | 50 min |
| | | Music room | 55 min |
| | | Recreational area | 50 min |

U.S. Gypsum Co.

Background noise is present in almost every situation. The amount is related to the tolerance of the occupants and the "work" they are performing. Rather heavy background noise in an automobile factory may be accepted without comment or harm, but the same amount or type of noise may completely disrupt an office. On the other hand, an overcontrolled space may need some added sound to make it a comfortable area in which to work. Some suggested levels for background sound are indicated in Table 29-4. Control of background sound is usually an acoustical problem rather than a structural problem.

**TABLE 29-4**
Optimum Background Noise Levels (Decibels)

| | |
|---|---|
| Broadcast studios | 20-30 |
| Schoolrooms | 40 |
| Conference rooms | 35 |
| Homes, apartments, hotels | |
| Sleeping rooms | 35-45 |
| Living rooms | 40-45 |
| Theaters, movie | 40 |
| Hospitals | 35 |
| Libraries | 35-45 |
| Churches | 30 |
| Private offices | 35-45 |
| Large general offices | 45 |
| Large engineering and drafting rooms | 40-50 |
| Restaurants | 45 |
| Stenographic offices | 50-60 |

U.S. Gypsum Co.

## 29.5 MEASUREMENT

The number of decibels of sound lost through a wall or other surface is called *transmission loss*. To find this figure, the loss attributed to the wall construction must be known or calculated from the various materials used and the decibel value on either side must be known. Such a test may be made in accordance with American Society for Testing and Materials (ASTM) designation E-90.) This type of test is often impractical to perform accurately because the exact frequency may be difficult or impossible to establish.

To simplify sound ratings three major definitions have been established. The most common is the *Sound Transmission Class* (STC), which is a single-number designator for airborne sound transmission evaluating the effectiveness of various combinations of construction materials. The second definition is the *Noise Reduction Coefficient* (NRC), which is a mathematical average of sound absorption coefficients recorded at the frequencies of 250, 500, 1000, and 2000 Hz. The third measurement definition is the *Impact Isolation Class* (IIC), a whole-positive-number rating for evaluating assemblies in isolating impact sound transmission. One additional measurement may be the *Impact Sound Pressure Level* (ISPL), which is derived from the transmission of sound through a floor assembly and measured by a standard "tapping" machine, which produces variable continuous "taps" on a floor sample for measurement of sound.

## 29.6 IMPACT SOUND CONTROL

A number of methods may be utilized to reduce impact sound or the transmission of sound through floors, walls, or ceilings. If it is possible to isolate surface receiving impact sound, such as floors, the sound will not be transmitted to adjoining surfaces. With floors this may not always be easy, but ceilings may be isolated by using resilient clips to hold the drywall or plaster away from the structural framing. Such isolation methods will usually not completely reduce the transmission of sound from overhead floors, but if an adequate material, such as insulation board or carpet padding, is used beneath applied flooring, the sound can be adequately dampened. Another possible option is to mount motors, water closets, kitchen garbage disposal units, and similar equipment on resilient pads or isolation fastenings. Special attention to provide nonconductive joints in pipes and ductwork, although often overlooked, will also reduce sound transmission throughout a building. Pipes and ducts should be fabricated of material heavy enough to prevent the material itself from vibrating, thus creating more sound to be controlled.

Walls may be isolated by providing a caulking bead beneath the floor plate, stopping the wall construction just short (¼ to ½ in.) of contact with a ceiling or floor above, or by a variety of staggered stud systems installed so that opposite wall surfaces are carried on different stud arrangements. Medicine cabinets, electrical switches, and outlet boxes should be installed with an air space between back-to-back units, and all holes for pipe or conduit through plates should be completely caulked.

Although it may reduce construction costs to some extent, the common practice of designing and building toilet rooms and bathrooms back to back may be a serious source of impact sound. Plumbing fixtures and pipe are solid materials that transmit sound efficiently and the sound of running water and its attendant pipe vibration are hard to overcome with such construction.

There are a number of accepted methods to control sound transmission through walls, floors, and ceilings, depending almost entirely on isolation, mass of wall assembly, or combinations of these. The wood-stud wall covered each side with gypsum board is about the poorest barrier possible but is the one most often used. Figure 29-1 illustrates comparative values for various types of wall construction.

Isolation of floors and ceilings is a bit more difficult than for walls, as these surfaces generally either rest on walls or carry walls that rest on them. Complete separation, or isolation, is also difficult because these portions of a building are usually calculated as horizontal diaphragms which resist forces that may rack or twist a building in earthquakes or heavy winds. Here again the simplest and most common construction, joists that carry a floor above and a directly attached ceiling below, is the most economical but provides the least sound control. Many combinations of wood joist, truss joist, concrete slabs, and isolated ceilings are shown in Fig. 29-2.

Isolation of walls, floors, and ceilings will not alone control all airborne or impact sound in a building. Where drop ceilings are designed into a building to act as plenums, the walls are often only taken to the ceiling line. This provides a plenum but also provides an unobstructed space through which sound from one room is easily transmitted to another, and also provides an easy path for fire or noxious gases. For ideal construction, every

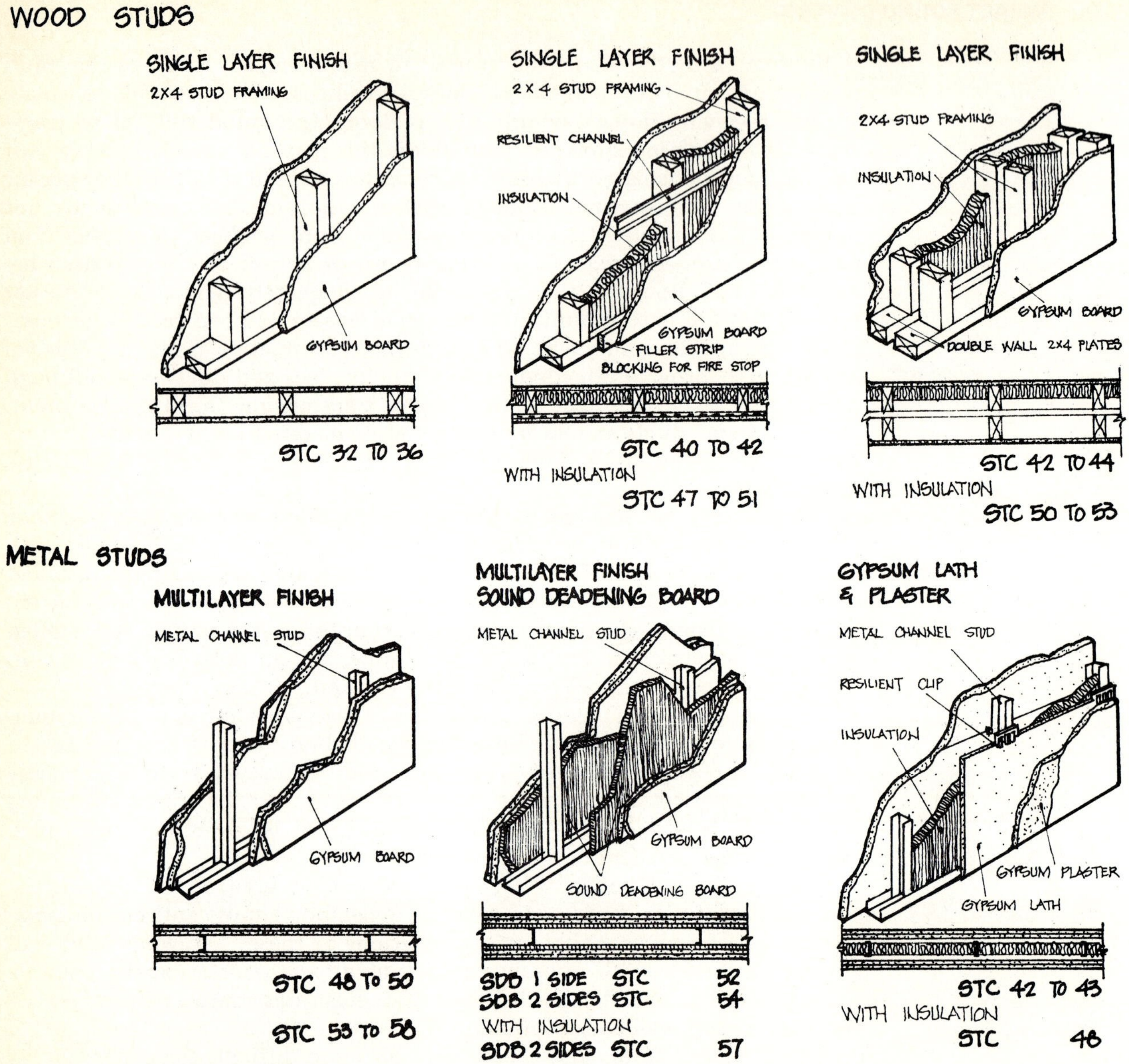


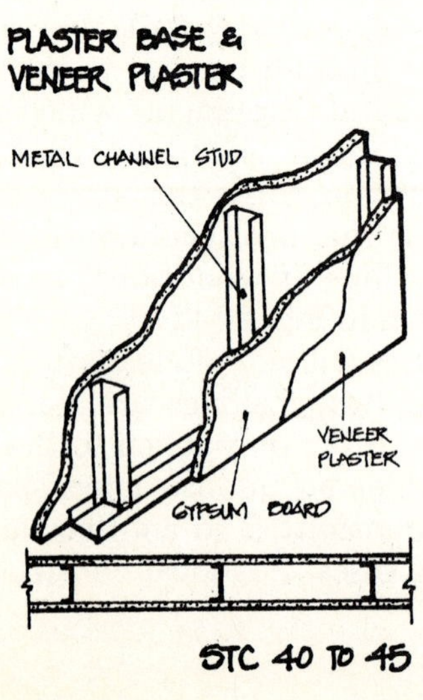


**Figure 29-1** Sound transmission coefficient of walls. (*From the Architectural Draftsman Reference Handbook.*)

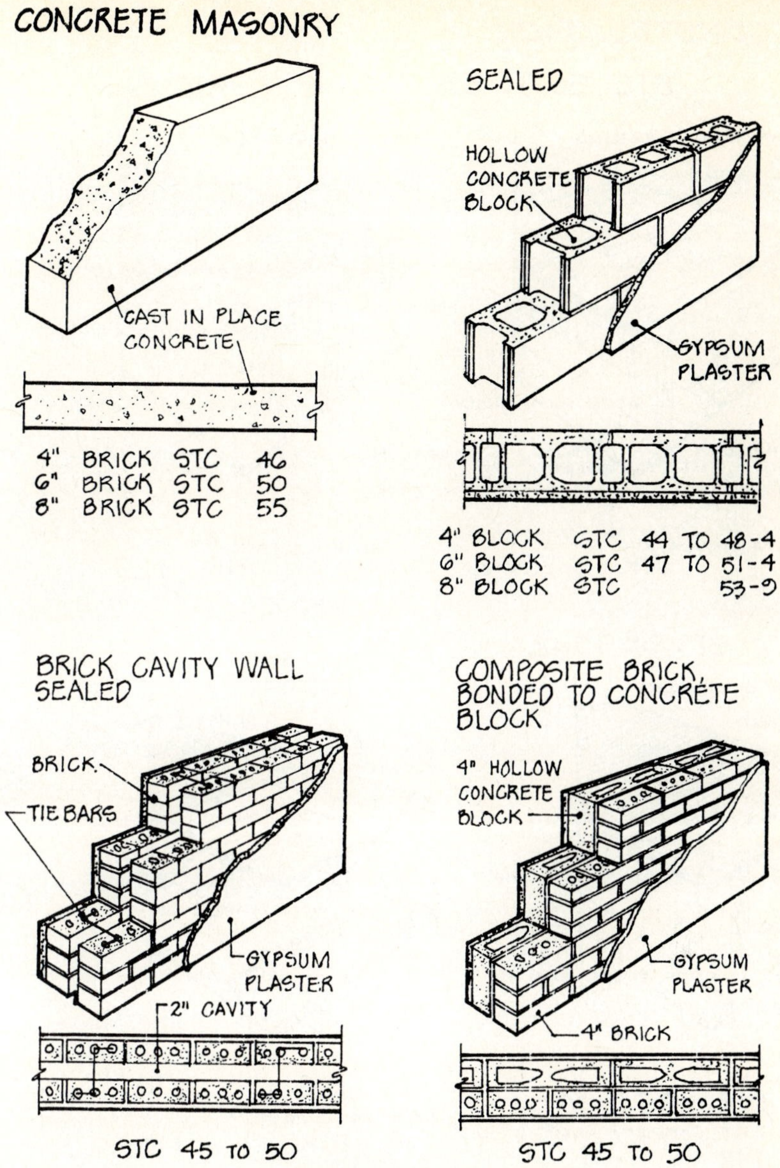


**Figure 29-1** (Continued)

opening, however small, should be sealed; stair stringers should not be fastened to adjacent wall studs; all doors should fit tightly and perhaps even have gaskets at contact with frames; windows should be double-glazed; and mechanical and electrical systems should be isolated as much as possible.

## 29.7 ACOUSTICS

Generally, *acoustics* refers to all sound, but acoustics in buildings or rooms usually means the control of sound within a space to obtain the best hearing possibilities. This means elimination or reduction of unwanted sound or noise and amplification of wanted sounds. Room use is the major criterion for the proper design of acoustic qualities. Some uses dictate that a room be "live" for proper reverberation times, while other rooms be classified as "dead" to absorb as much noise as possible. The shape of the space, the wall

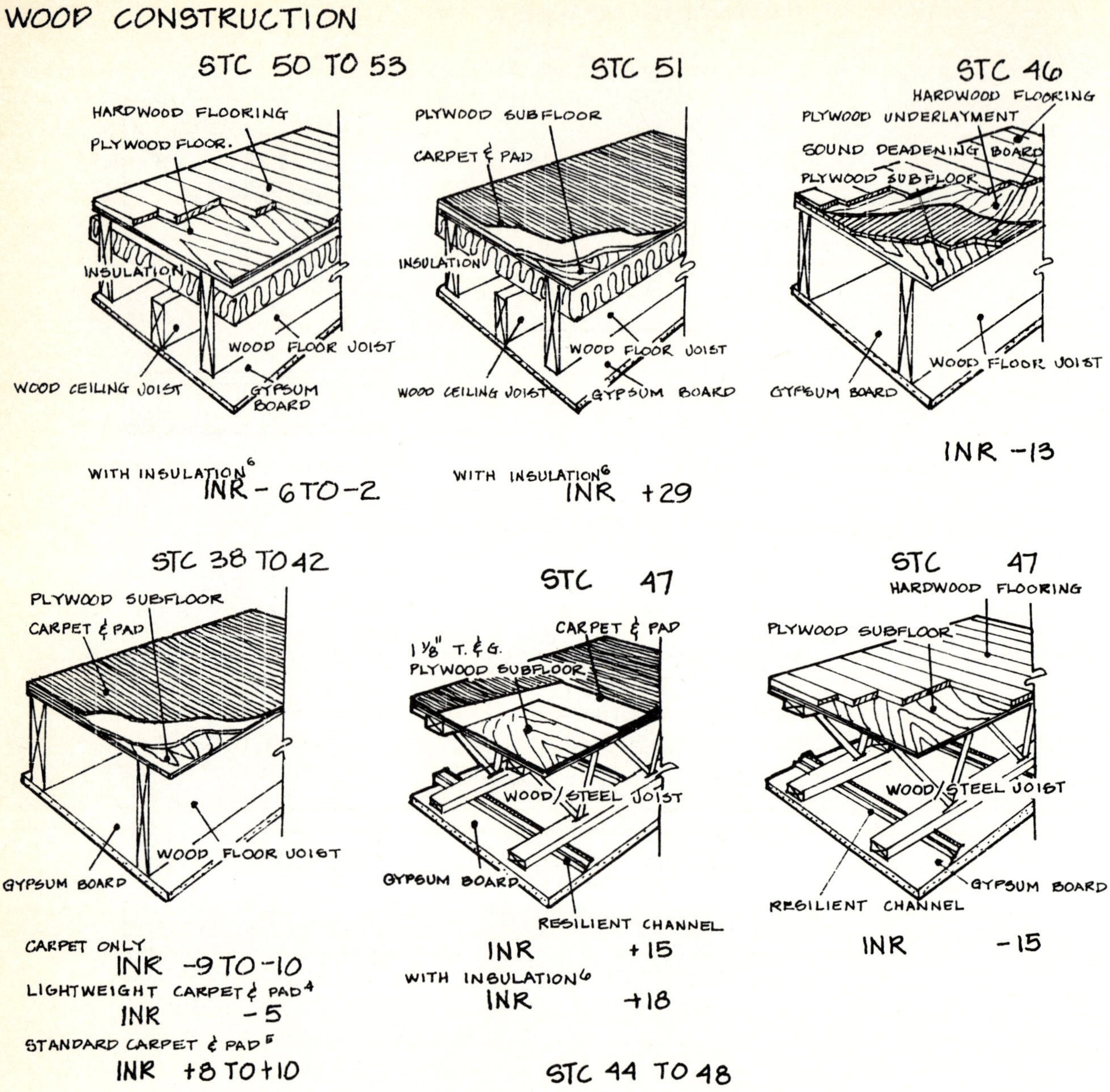


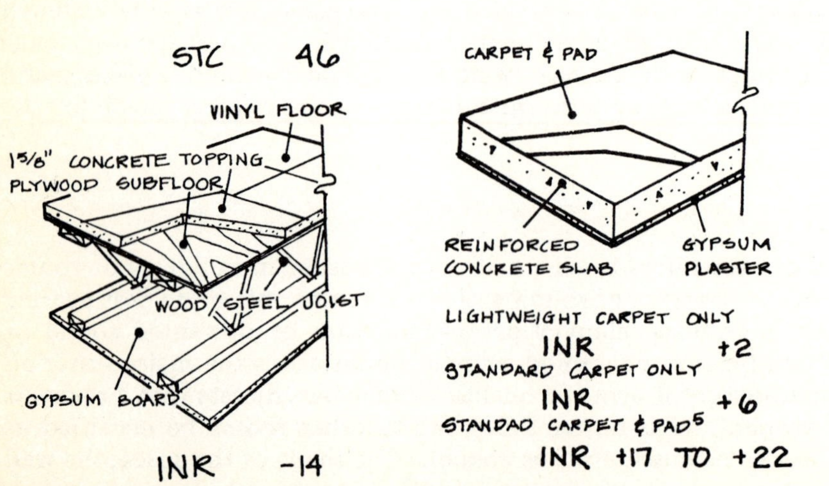


**Figure 29-2** Sound transmission coefficient of floors. (*From the Architectural Draftsman Reference Handbook.*)

and ceiling construction, and the surface material all contribute to acoustical control.

*Reverberation* of a sound is a basic acoustical factor in a room. The time interval between an original sound and its reverberation determines if the sound is heard clearly or slurred or overlapped. In general, the larger the room, the longer the lapse between original sound and reverberation because the sound waves must travel farther. Any original sound reflects off a surface, then bounces to another surface, and to still another, each time losing some energy until it becomes inaudible. The reverberation time of a room is the time in seconds required for a sound to be reduced to an intensity of one-millionth of the intensity at source, or about a drop of 60 dB.

*Absorption* of sound is required so that a direct sound may be heard before its reflected sound. This is accomplished by creating energy loss with surface materials which will soak up reflected sound to as much as 50% before returning it to the room (see Fig. 29-3). The unit of absorption is the *sabin*, which is equal to one square foot of totally absorbent surface, with a perfectly absorbent surface at a value of 1 and a totally reflective surface with a value of zero. Plotting of values of different materials at frequencies of 125, 250, 500, 1000, and 2000 Hz provide the number rating used for Noise Reduction Coefficients (NRC) of those materials.

*Echos* may occur in large rooms where the reflected sound is heard at least $^1/_{17}$ of a second after the original sound. Generally, this means that the reflected sound must travel at least 65 ft farther than the direct sound. Curved or domed ceilings produce louder echos and have a tendency to focus the reflected sound, and rarely may be controlled with absorbent materials. Reflection time in normal residential rooms should be about 0.5 second, but in large rooms where speech or music is to be properly heard may be up to 1.0 second. As a rough guide, sound absorption units in normal rooms should cover from 25 to 30% of the total exposed wall and ceiling area.

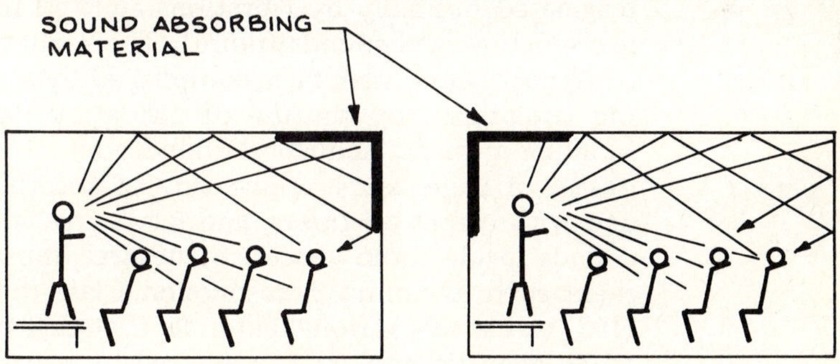


**Figure 29-3** Sound reflection.

## 29.8 ACOUSTIC GUIDELINES

"How much of what?" will confront the occasional designer of sound control, especially in regard to acoustics in buildings that are already constructed. Several rules of thumb may be helpful in average conditions.

The ceiling of a room is usually the largest and most readily available surface for installation of absorbing units. Where ceiling heights are 8 to 15 ft, acoustic material placed on this surface only may bring the NRC down to an acceptable level of 0.60. Small rooms with less ceiling area may require some wall treatment. Rooms with hard-surfaced parallel walls also may require some wall treatment to reduce sound reflection from these walls.

Acoustic control requires a balance of desirable reverberation of sound with amount of absorption installed. Some desirable sound frequencies in the range of normal male speaking (500 Hz) are given in Table 29-5.

The amount of existing sound absorption provided by wall, ceiling, and floor construction of a building may be calculated by formulas to determine the amount of sabins (S) that may exist. By using $S = a \times A$, where $a$ is the coefficient of material and $A$ is the surface area, the total number of sabins is found. Further calculation for desired total absorption is found by multiplying total sabins existing by a constant of 3.0 represent improvement, then subtracting existing sabins from this result. The final step is to divide desired absorption value by surface area to be treated to obtain a fractional value for coefficient of the area treated (see Chapter 2).

**TABLE 29-5**
Reverberation of Sound

| *Location* | *Time (sec)* | *Speech* | *Music* |
|---|---|---|---|
| Small office | Below 0.75 | 0.50-0.75 | Dead |
| Classroom, music rehearsal | 0.75-1.00 | 0.75-1.00 (fair) | 0.80-1.00 (acceptable) |
| Chamber music, general office | 1.00-1.50 | 1.00-1.50 (good) | 1.00-1.50 (fair) |
| Orchestra-choir, workrooms | 1.50-2.00 | 1.50-2.00 (fair) | 1.50-2.00 (good) |
| Large organ, liturgical choir | Above 2.00 | Poor | 2.20-2.25 |

## 29.9 SUMMARY

Sound may be either, or both, airborne or structurally transmitted and is originated normally by vibrations emitted from voices, music, machines, or nonsynchronized sounds (noise). The reduction or elimination of structural or impact sound may be accomplished by separating various parts of a building (isolation), by the use of insulation barriers, or dampening sound by building in mass. Acoustical sound control may be accomplished by proper design of room sizes, application of absorbing material on walls or ceilings and with carpet on floors, and has a great deal to do with proper hearing of sounds inside room spaces. Structural impact sound control uses measurement such as Sound Transmission Class (STC) and Impact Isolation Class (IIC) to classify various materials that may be used in building construction, while acoustic control is measured by wavelength, reflection, and intensity of airborne sound, and uses a measurement termed Noise Reduction Coefficient (NRC) of various materials and combinations.

The amount of acceptable sound, measured in decibels, has a definite relationship with existing or vagrant sound (auto, airplanes), and background sound that may exist from people working or similar sources. The amount of acoustical treatment is relatively easy to calculate if careful attention is given to materials used and their coefficients, and each installation will normally require a different solution.

# 30

# Natural Energy Sources

## 30.1 INTRODUCTION

The title of this chapter may be a misnomer in the absolute sense, as all energy is "natural" if traced back to an original source. However, there are several emerging sources for thermal and electrical power that should be considered before the book is closed. Foremost of these is *solar energy*, which has been used for thousands of years but only in recent times has attained significant importance. Solar energy may produce heat, cooling, or electricity when properly harnessed. Another "natural" source of energy is the taming of *geysers*, which is being done on a commercial basis in California, New Zealand, Italy, and several other places. Still a third source is the *wind*, which through the use of windmills, has been pumping water and grinding grain for centuries and now may produce electricity as well. All these sources are supposedly "free," but the design of equipment and its cost may be a monumental problem. The location of facilities to make proper use of these energy sources is also a problem as adequate "sun-days" for solar energy may not be common in some areas, geysers and other subsurface hot water is relatively limited in geographic location, and wind of sufficient velocity and duration is subject to numerous weather and topographic determinants. At this time, then, large commercial applications to utilize these energy sources are not widespread, but significant experimental and limited-use projects are functioning to warrant discussion and review.

## 30.2 SOLAR ENERGY PRINCIPLES

Solar energy utilization is by two major methods: (1) *passive solar*, which operates without mechanical fans or pumps, and (2) *active solar*, which augments passive solar with pumps or fans. Both systems depend on natural energy "sources" and "sinks" at the building site. An energy *source* is anything in the natural environment that adds heat, and a *sink* is anything that will absorb heat. Passive system heat is transferred by radiation, conduction, and convection. Major on-site sources are (1) solar radiation (direct and diffused), (2) outside air above 75°F, and (3) internal heat generated by people, lights, cooking, and similar activities. The principal sinks are (1) the night sky and upper atmosphere, (2) outside air with temperature below 75°F, and (3) wet surfaces because heat is absorbed when water evaporates.

In a majority of cases direct thermal solar is the most practical method for space heating and cooling. The energy storage and transfer system is usually an element of the structure itself, but may not completely satisfy all the requirements for temperature or quality. Indirect or active solar space-heating systems collect the solar heat outside the building and transfer it inside the building through ducts and pipes using fans or pumps. Solar energy is not a cure-all but needs help from several other items that are necessary for effective results at minimal cost. Five basic factors are (1) thorough insulation, (2) proper orientation of the long axis of the building, (3) location of most windows on the south side for maximum heat gain, (4) provide adequate roof overhang for summer shading, and (5) use of light-colored roof material to reflect sun heat.

The material used for construction, the sun angle, and day length are other important factors in solar heating and cooling. The heating and cooling potential may be increased by increasing the thermal mass. "Solid" building materials, such as adobe, concrete, masonry, rocks, earth, and water, have a high capacity for storing heat. These masses hold heat and radiate it into the building on a delayed schedule. This simple phenomenon is explored further in following sections. Location north or south of the earth's equator will determine the angle of the sun as well as the duration of the sun's availability. When the sun angle is high in summer, the building should have maximum shading while a low angle in winter allows the sun rays to enter the building and provide more heat. These principles have been used for centuries and most southwestern U.S. native Indian tribes used thick

**Figure 30-1** Mesa Verde, Colorado. (*Courtesy of the U.S. National Park Service.*)

massive walls of stone or adobe to construct buildings beneath overhanging rock cliffs which provided the sun-angle shading necessary to keep their buildings warm in winter and cool in summer (see Fig. 30-1).

It has been proven by extensive experiments that all materials used in present buildings have the ability to collect solar energy. The first solar air system was built in Boulder, Colorado, in 1945 on an existing house after about three years of development. Houses in Dover, Massachusetts, Denver and Ft. Collins, Colorado, Ooleillo, France, Atascadero, California, and other locations followed but not until about 1960 with the St. George's School, Wallasey, England, were the real possibilities of direct heating and cooling realized. In 1970 thirty-two "solar buildings" were in use in the United States. In the next eight years this number had increased to more than 500 recorded and an unknown number not recorded.

## 30.3 PASSIVE SOLAR ENERGY

The average residential unit of 1450 to 1500 $ft^2$ requires about 12,500 to 13,000 Btu per day for each degree Fahrenheit of difference between the inside and outside temperatures. A Btu is the amount of heat required to raise one pound of water one degree Fahrenheit. If the outdoor temperature is 32°F and the indoor temperature is to be maintained at 65°F, the following formula will show that the house will require a considerable amount of heat per day:

$$P \times (A - B) = \text{Btu/day}$$

$$13{,}000 \times (65^\circ - 32^\circ) = 429{,}000 \text{ Btu/day}$$

This corresponds to about the useful energy in 5 gal of gasoline, but the heat provided by the sun on a winter day exceeds this amount by approximately threefold, so there is plenty of opportunity to provide adequate heat.

If it were possible to build a building with enough mass and without the presence of humidity, the mass would absorb heat during the day and release heat at night in the same proportion, so that the interior of the building would always remain the same temperature. This is not possible for a number of reasons, so the value of mass and insulation of buildings must be considered. Experiments have shown that two things are necessary for a material to be considered for mass construction: *heat capacity* (the ability to store heat) and *low heat conductivity* (the ability to resist heat flow). Several materials or combinations of materials have approximately the same values when tested for these qualities: (1) 4 in. of water with 1 in. of insulation on exterior surface, (2) 8 in. of concrete with 1 in. of insulation on either inside or outside face, and (3) 3 in. of solid wood with 1 in. of insulation on inside face. Of these the water-insulation combination is more efficient but also harder to install. Passive solar heat considers that there will be acceptable indoor temperature swings inherent in the system and that these swings may well be from 65 to 80°F. Changes of clothing or type of work, then, is often necessary to maintain a reasonable level of comfort. In addition, almost every passive solar installation has required some backup heating in cold weather.

Many "systems" for *passive solar* construction have been used and a number of useful books are available on the subject, so the exact reasons and assumptions for each system will not be investigated thoroughly in this book. The general arrangement and function of a number of the more common systems will be considered, however. *Direct-gain systems* (Fig. 30-2) rely on sun heat entering south-facing windows during the day, storage of heat in building mass (walls, floors, ceilings) during the day, and transfer of stored heat from the mass to the room at night or during cloudy weather. For cooling the direct solar gain is prevented from being absorbed by use of overhangs, shades, or other blocking and the space is ventilated by cool air drawn in at night through adjustable openings. If enough cool air is introduced at night, the mass of the building may absorb part of it and transfer that cooling to the room air during the day. If the mass is properly balanced with ventilation, it is possible, at least theoretically, to maintain an even comfortable interior temperature at all times. Automatic or manually operated shading and venting facilities may make this system expensive or may be inconvenient to use to achieve proper balance.

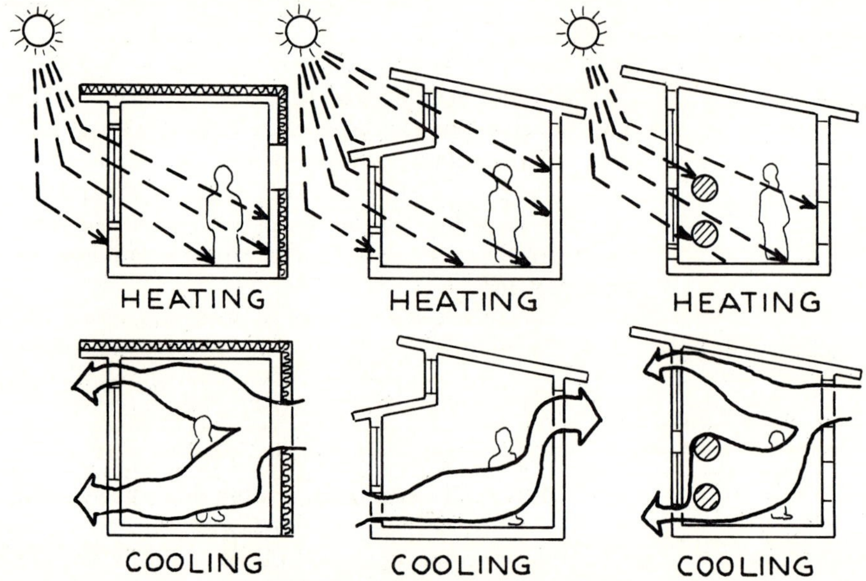


**Figure 30-2** Solar direct-gain systems.

*Indirect solar systems* exclude direct exposure of the room interior to the sun by the construction of solid mass walls on the sunny side. If the exterior face of the wall is of a dark color, the sun's heat will be absorbed by the mass, retain the heat during the day, and radiate the heat into the room at night. Often the wall is vented at top and bottom to create a thermosiphon loop during the day, with the vents closed at night to prevent cold air from flowing into the interior. Cooling is accomplished with overhangs or shades and air may be drawn from another wall into the exterior space outside the mass wall and into the night. One of the major features of this method is the construction of a thermal mass wall, with an exterior glass wall and intermediate air space (Fig. 30-3). Aesthetically, a solid black wall may not be acceptable to an owner or designer and a solid wall may block out a view toward the south. Mullions and other framing of the glass wall will appear more conspicuous, and apertures in the mass wall may provide a high contrast when looking from inside. If water is used for the mass wall, the need for the thermosiphon is diminished but an additional wall of insulation and ventilation is installed on the interior face to provide a siphoning action.

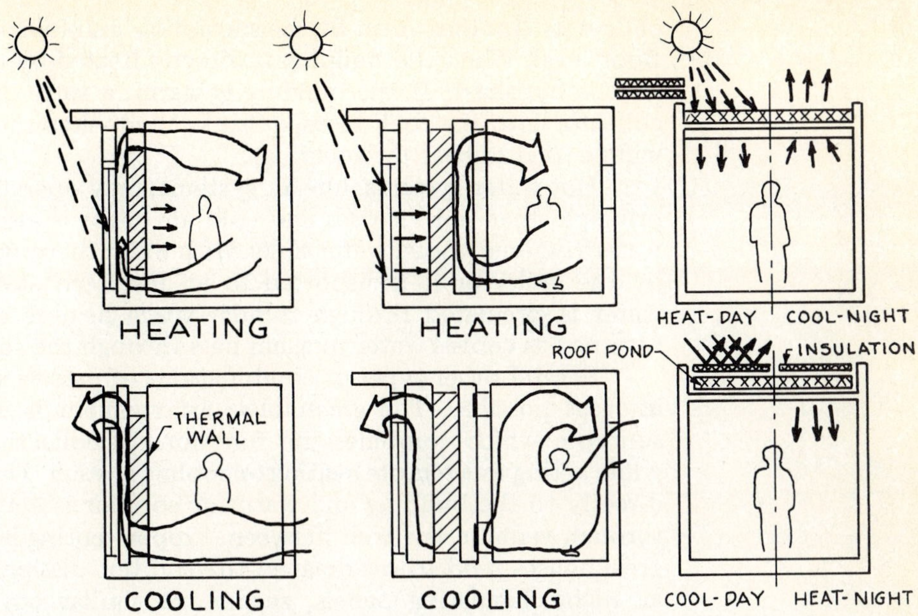


**Figure 30-3** Solar thermal wall systems.

Still another form of passive solar heating/cooling is by *thermosiphon* using the earth and is usually isolated from the actual building. This system takes advantage of the fact that air or liquids can pump themselves as they are heated or cooled. As air is more easily handled than a liquid, most thermosiphon systems use air for heating and cooling. In thermosiphon operation (Fig. 30-4) the air is heated by a glazed, insulated flat-plate collector installed at or below the level of the inlet and outlet vents. As cool air is drawn down behind the collector, warmer air rises across the face of the collector and heats the building. In the most simple arrangement another duct from the north or sheltered side of the building brings in cooler air underground to cool the building when shades or covers are placed over the solar

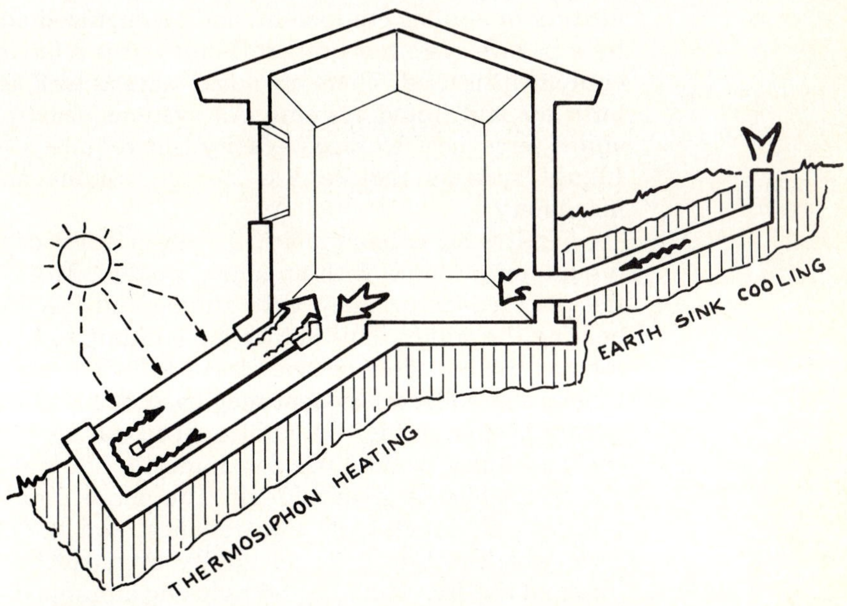


**Figure 30-4** Thermosiphon. (*Courtesy of the California Energy Commission.*)

collector. Another form of thermosiphon employs a rock "sink" below the floor level. When the building is directly heated by the collector, the storage sink is bypassed. If the building is warm, a valve directs the heat from the collector into the rock storage "sink," and when the building cools this heat will be used to heat the building.

Hot-water systems, like air systems, may operate as thermosiphons. The principle that hot air rises and cold air drops is also true of water. Warmer water rises and cooler water sinks. With water thermosiphons the water heated by the collector is transported to an insulated storage tank, then the hot water is circulated through tubular space heaters or a heat exchanger and returned as cooled water to again pass through the solar collector.

Several other types or combinations of passive solar energy are in use or at least possible. The *greenhouse* arrangement is a modified glass-covered sunroom which combines the functions of collection, storage, and transfer while acting as a remote heating or cooling system. The greenhouse is attached directly to the building and warm or cool air is circulated into the building through vents in the wall between. Proper venting is absolutely necessary, as greenhouses tend to overheat in the sun but dissipate that heat very rapidly at night. Insulating panels, shades, or similar covers help to accomplish proper control. Another system employs *elliptical section tanks* filled with water which are rotated to allow the flattened side to have more exposure, and more heat, or turned so that the edge surface allows the sun to enter the building directly. One of the older applications is the use of *water-filled bags* installed on a roof surface and controlled with movable insulated cover panels. Heated by the sun during the day, the bags insulate the rooms below with a layer of water but when covered at night with insulated covers, the heat stored in the bags reverses to heat the building below.

## 30.4 INDIRECT (ACTIVE) SOLAR ENERGY

The essential differences between passive solar energy and indirect (active) solar energy is the use in indirect (active) systems of fans, pumps, auxiliary heating or cooling equipment, and mechanized controls. Circulation of water by a passive system is quite difficult but is relatively simple when mechanical equipment is used. There are advantages as well as disadvantages in the use of both air and liquid systems. Air systems usually require large duct systems and a large heat storage capacity but require a minimum of control valves. Liquid systems require less storage volume and smaller pipes but more machinery.

Unless air is being used, the orientation of the building is not of paramount importance with an active system. Most water systems depend on a flat-plate collector lined with tubing and installed on the building roof for heating the water. This allows the building to be oriented in any direction as long as the collector can be placed to face south. In some commercial installations the collector system may even be installed away from the building to obtain proper facing. An additional advantage in the use of a water system is the possibility of dual use for domestic hot water or pool heating if the capacity for heat is great enough. The major disadvantage is the initial cost of pumping equipment, its maintenance, and the cost of operation, while the greatest advantage is the possibility of retrofit installations in existing buildings and the freedom allowed by using a pumped system.

Several systems are used in *active solar* heating and cooling, but the most common one uses a collector plate, adequate connecting piping, and pumps to force the water through the system (Fig. 30-5). Collector plates are

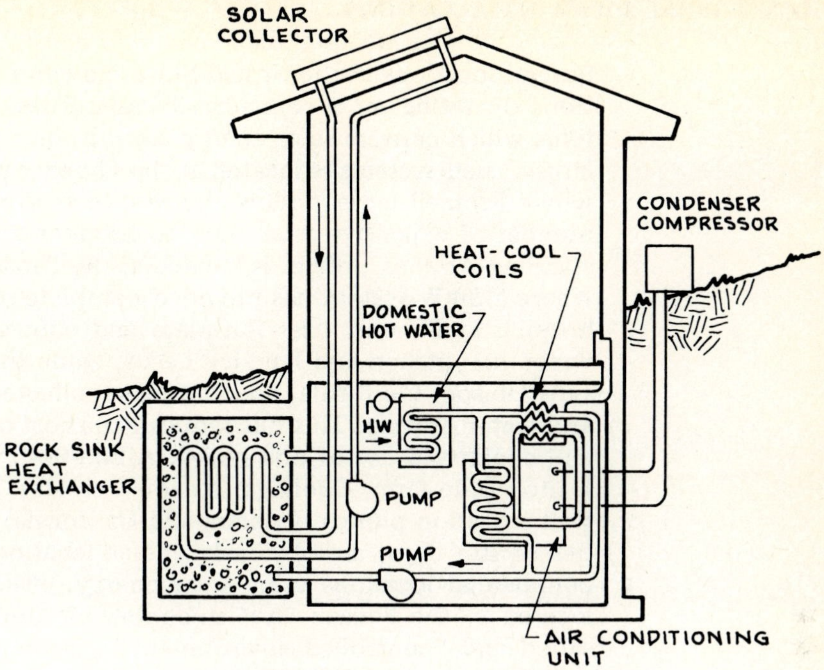


**Figure 30-5** Indirect (active) solar system.

of several types but essentially are tubes of copper or aluminum installed in a case over a thick layer of insulation and covered with water-white glass. Between the glass and the receiver plate is a layer of air. The interior of the box is painted flat black to absorb the sun's heat and raise the temperature of the water in the tubes. A pump is required to cause the water to flow through the collector, as most collectors are roof-mounted. The heated water then returns to a storage tank, heat exchanger, or into heating units. As the heat is extracted, the cooler water is recirculated back through the collector plate. Domestic hot water may be taken from the system as shown in Fig. 30-5 or may constitute the primary use of the system. Swimming pool heating may also use the same system either circulating the pool water or by use of a heat exchanger. Heat exchangers are preferred, as most pool water also contains quantities of chlorine, which is very corrosive to the copper or aluminum tubing used in the collector.

Cooling is accomplished as a liquid refrigerant evaporates, and heat is rejected as the refrigerant liquefies in a condenser. In a solar system the heated water passes through a heat exchanger. The heated solution passes to an absorber-evaporator, where chilled water is further cooled and distributed into a room cooling unit. Another system derives heat from the plate collector, stores heated water, and passes it through a heat exchanger. The resulting heated solution is then condensed and passes through a heat-pump compressor and condenser-evaporator to give a cooling cycle.

In locations where freezing of the water might occur in winter, a heat exchanger is a good solution. The collector system is filled with antifreeze and the heat exchanger is isolated from the hot-water system. The heated antifreeze from the collector passes through an immersed heating coil in the tank and transfers its heat to the water. Depending on the relative location of the tank and the collector, a thermosiphon system may be used or a pump is installed to control the flow of antifreeze. On cloudy days when the collector does not produce enough heat, an auxiliary heating source is applied at the cold-water inlet side of the tank.

## 30.5 COMMERCIAL SOLAR INSTALLATIONS

Several locations in the United States now have very important solar installations operating on a commercial basis. These are mostly photovoltaic systems which convert solar energy directly into usable electricity. One of the largest such systems is located at the Phoenix (Arizona) Sky Harbor Airport, where Fresnell lenses follow the sun, increasing efficiency and reducing the number of expensive solar cells, to produce 225 kW of power. Another outstanding Arizona project is located at the Papago Indian village at Schuchuli, where a 3-kW system has provided complete power for the village. The state hospital at Phoenix uses flat-plate and concentrating collectors to heat all water for kitchen and laundry needs, while the Sierra Vista hospital at San Luis Obispo, California, uses flat-plate collectors to preheat water for boilers and water heating. This has reduced the cost of water heating by about 30% and eliminated use of one boiler. A 23,000-ft$^2$ parabolic trough collector at Dalton Cole farm, Coolidge, Arizona, creates 150 kW to operate three deep-well irrigation pumps. Arizona and California, with their considerable number of sun days, are obviously choice locations for solar energy use. Other commercial locations using solar energy, either for heat or electricity, are located in the Persian Gulf shiekdom of Abu Dhabi, where seawater is desalted and "controlled environment" greenhouses provide vegetables. Solar radiation is used by Universitario de Sonora, Mexico, to improve algae feed for commercial shrimp. The Madeira School, Greenway, Virginia, uses 4500-ft$^2$ sloped roof collector plates and a special oil to avoid freezing to provide hot water and an active forced-air heating system. Located at Walt Disney World, Orlando, Florida, is the Reedy Creek Utilities Co. subsidiary building, which uses 16 horizontal parabolic collectors to provide both hot water and chilled water. Extensive and continuing experimentation and research is being conducted by Arizona State University and Colorado State University as well as many other educational institutions and commercial corporations.

## 30.6 GEYSERS

Geysers occur in some areas of the world and have been tapped successfully to produce steam, hot water, and electricity. The geyser phenomenon is created by underground rivers or lakes coming into contact with natural underground heat with sufficient capacity to heat the water until it rises to the surface through fissures in the earth crust. Hot mineral springs occur in many locations but are not suitable as an energy source because they contain large amounts of soluble minerals and usually are not available in enough quantity to serve as an energy source. In Rotarua, New Zealand, springs provide enough hot water to heat a great number of homes and provide adequate hot water for community laundries and bathing facilities. Steam is not generated, however, so no power is available to generate electricity.

Use of geothermal steam is not a new idea. Reference to the hot water and steam of the Larderello region of central Italy is made as early as the third century A.D. However, industrial activity in this region did not start until 1777, when boric acid was discovered in bubbling pools. In 1827 an ingenious collector was built over a steam vent and the energy captured was used to refine the boric acid. The first "real" power plant was constructed in 1913 with a capacity of 250 kW. Increased activity now enables this field to produce approximately 390,000 kW. In 1949 the New Zealand government approved an investigation program for the Wairakei area in the center of the North Island. Stage I was opened in 1958, producing 69,000 kW, followed in 1959 by Stage II, which increased total capacity to 192,600 kW. The Wairaki generation plant (Fig. 30-6) supplies approximately 18% of the North Island electrical requirements. The average depth of wells is about 3200 ft, with steam produced at about 100 psi. The nearby Kawerau field provides 200,000 lb/hr to Tasman Pulp and Paper Co. Ltd., about one-fourth of their power needs.

**Figure 30-6** Wairakei geothermal area. (*Courtesy of the New Zealand Tourist Office.*)

The Geysers power plants (Fig. 30-7) in Sonoma County, California, are the only geothermal power plants in operation in the United States. In the early 1920s some shallow wells were drilled to provide steam to heat a nearby resort, but attempts to expand the system were unsuccessful until in 1958 the Pacific Gas and Electric Co. joined with Magma-Thermal to drill steam wells and build a power plant. Unit 1 went into service in 1960 and produced 11,000 kW. Subsequent wells and generators have now increased the total output to 910,000 kW from 15 power plants, now first in total geothermal production in the world. Geothermal areas also exist in southern and eastern California, northwestern Nevada, southern Oregon, western Idaho, and northern New Mexico, but have not been commercially developed. Oregon Institute of Technology, Klamath, Oregon, heats approximately 450,000 $ft^2$ of floor space from hot-water wells at temperatures of 190°F at an average of 2.8 million Btu per hour. Japan, Mexico, the Soviet Union and Iceland have also begun to use the energy of geothermal sources. The first Japanese experiments were in 1940s but produced only 25 to 30 kW of power. In 1966 a plant was built at Matsukawa with an initial output of 22,000 kW, which has now been expanded to 72,000 kW. Plants located at Onikobe

Figure 30-7 "The Geysers" – Plant #6. (*Courtesy of Pacific Gas and Electric Company.*)

(25,000 kW), Hatchobaru (50,000 kW), and Otake (11,000 kW) are also in operation. Mexico has a geothermal operation at Cerro Prieto, Baja California, which went into operation in 1973. In early 1963 the Soviet Union erected a plant at Kamchatka Peninsula (5000 kW). It is estimated that 50 to 60% of the Soviet Union is underlaid with thermal water with temperatures suitable for heating and low-pressure electrical generation. Extensive experiments have also been made by injecting cold water into hot bedrock and extracting hot water under pressure. The city of Reykjavik, Iceland, uses natural heat to serve 11,000 homes with 90,000 inhabitants. Initial drilling was done in 1928, and at present 385 million watts of power is distributed over most of the Reykjavik area.

## 30.7 WIND POWER

Wind power is available in mechanical form, so is more easily converted to useful energy than is true in some of the other systems we have discussed. The familiar agricultural windmill in farmers' pastures has for many years been a source of energy for pumping water for livestock. Today many windmills are still used for direct mechanical operations, but there is increasing interest in electrical generation, of both direct current and alternating current, and some installations actually generate enough electricity to sell some back to the utility company. The siting of windmills and the types of blades used determine the generating potential. In general, areas where the wind measured at approximately 30 ft above the earth surface is below 10 miles per hour estimated mean annual wind speed are not adequate for windmill operation. Good locations require a mean wind speed of 10 to 15 mph, and excellent sites require 15 mph or greater to generate 300 W per square meter of windmill frontal area annually. Anemometers (wind-speed sensors) are used to gauge the wind speed, and continuous recording of the wind speed will give the annual mean power. The wind power available is directly related to the cube of the wind speed; that is, if the wind speed is doubled, eight times the power is available. The shape of the blades, and their number and location in relation to wind direction, also determine the speed of rotation and the efficiency of the windmill. Some of the major types of windmill blades are shown in Fig. 30-8, and the relationship among wind speed, wind power, and rotor diameter in Fig. 30-9.

**Figure 30-8** Types of wind machine blades.

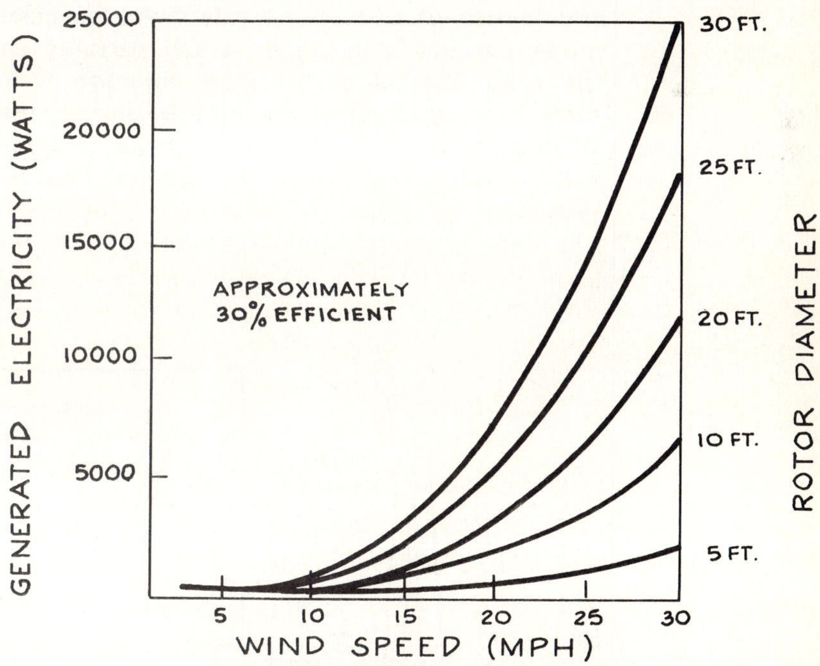


**Figure 30-9** Wind speed and power.

Windmills slow the wind speed down to about one-third of the original speed. Wind speed is difficult to convert directly, but it can be converted by using the formula

$$P = \tfrac{1}{2} D A v^3$$

where $P$ is the available power in watts, $D$ the air density in kg/m$^3$, $A$ the cross-sectional area facing wind flow in m$^2$, and $v$ the instantaneous wind speed in m/sec. It may be more convenient to consider wind power in terms of power per unit area facing the wind, written as follows:

$$\frac{P}{A} = \tfrac{1}{2}Dv^3$$

Farm windmills used to pump water have many flat vanes which rotate rather slowly and are designed to lift water directly from deep wells. Blades for windcharges are more propeller-shaped because they do not have to start against heavy static loads as do farm water windmills. If water is to be pumped, it is now often pumped by electric motors connected to wind generators. Generators are constructed in a manner similar to that used for any other generator (see Chapter 7) and may provide direct current or alternating current via *inverters*. Each horsepower of pump capacity requires about 1 kW of electrical power. Other forms of energy from windmills are heat generated by splashing water, which in turn develops heat by friction, and by direct drive of heat pump compressors. Unfortunately, most of these must also be dual-connected to utility company power to continue operation when wind power fails.

In the past wind systems have been used primarily to charge batteries. Now that utility companies have extended their facilities to more remote areas, battery charging is used mostly for microwave repeaters, fire service lookouts, and similar service. With the introduction of inverters for direct-current generators, some power companies are now cooperating with individual generating systems to accept any surplus power generated. Such *grid-connected systems* provide some financial return on the high cost of modern windmill installations and partially eliminate complete reliance on the wind. Several of the more common systems illustrated in Fig. 30-10 show how windmill power may be interconnected with the utility source. Obviously, safety is one of the major factors, as surges or discrepancies in voltage, frequency, or phase must not be allowed to creep into the utility system.

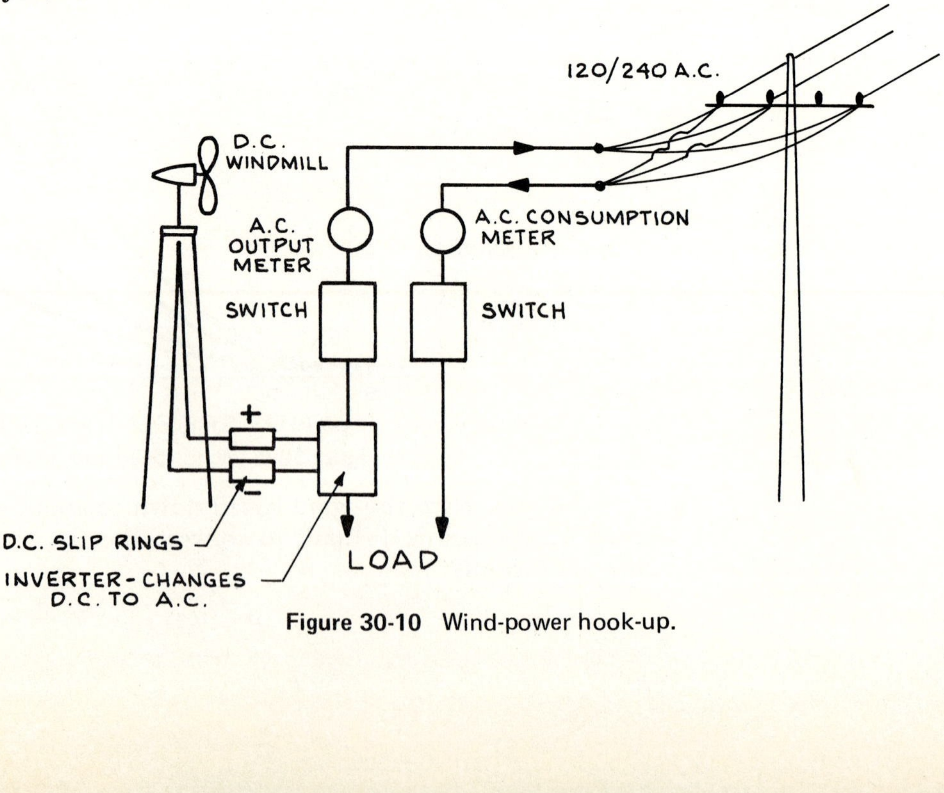


**Figure 30-10** Wind-power hook-up.

## 30.8 SUMMARY

In addition to the natural resources discussed here, many others are being considered: sea-tide forces, heat from incineration, fuel from garbage, and the by-products of useful gases, salvaged metals, and other "reusables." Solar, geothermal, and wind power use are only in their infancy, but in the next decade all of these will have to be developed more fully if we are to continue to use energy extravagantly. Perhaps by that time the ability to profitably harness lightning, air thermals, and volcanoes will also be developed. It's not impossible!

# 31

# Solid-Waste Disposal

## 31.1 INTRODUCTION

The collection and disposal of waste materials created by human habitation have long been a major expense and a troublesome problem for municipalities. Waste falls into two principal categories: (1) *liquid waste*, such as sewage, storm drainage, industrial waste, oil and liquefied chemicals, and animal waste, and (2) *solid waste*, including household trash, cans, bottles, street sweepings, dead animals, and similar material which contains little or no moisture. The handling and treatment of liquid sewage has been covered in Chapter 17, and the collection and disposal of solid waste is the subject of this chapter.

Solid-waste disposal is not limited to actual buildings, although most of such waste material is generated by people in those buildings. The scope of solid waste is so extensive that it staggers the imagination, and only a cursory review can be included in this chapter. Only solid waste is generally considered as *refuse*, but dewatered sewage sludge may become a solid, fly ash from industrial smokestacks becomes a solid, and dust, smoke, and fumes from gases may even become solids under certain conditions. Refuse comprises all the solid waste of the community, as well as some semisolid slurries, pastes, and market garbage. Disposal of this refuse is by sanitary landfill, incineration, and recycling and in some cases requires sorting or other pretreatment at the original source.

## 31.2 SOLID-WASTE STATISTICS

Conservative estimates placed the annual cost of solid-waste collection, treatment, and final disposal in the United States in 1974 at about $5 billion. Of this amount about 80% was for collection and delivery to disposal sites, and about 60% of this figure ($2.4 billion) was spent for residential collection. National surveys conducted by the U.S. Public Health Service indicate that about 5.3 lb of solid waste per person per day was generated in 1968, for a total figure of about 190 million tons per year. At an annual increase of 4 to 5% the average will be about 8 to 10 lb per capita per day in 1985 or more than 350 million tons per year.

A recent survey made for Los Angeles County (California) waste management indicated that the present load is about 25,000 tons per day with a serviced population of 9.2 million people but this is expected to jump to 34,000 tons per day by 1990 when the population is anticipated to be approximately 12.6 million persons, an increase of 4 percent in less than 10 years.

**TABLE 31-1**
Solid Waste Increase in U.S.

| | *Solid waste* | | *U.S. population* | |
|---|---|---|---|---|
| | *Lb/person/day* | *Percent increase* | *Thousands* | *Percent increase* |
| 1920 | 2.75 | — | 106,421.6 | — |
| 1970 | 5.30 | 96.3 | 203,184.7 | 91.5 |
| 1980 | 8.00 | 50.9 | 243,291.0 | 19.7 |

U.S. Bureau of Census.

These figures represent only the amount handled by public and private collectors and do not consider the 10 to 15% that is self-collected in households and industry and handled separately from municipal systems. If this additional amount of waste is added to that handled by organized collectors, the total for 1981 approaches 400 million tons. Further investigation reveals that about 60% of solid waste originates in households, 30% in industrial or commercial, 3.7% comes from construction activities, and about 6.3% is miscellaneous. With an increase of about 30% in the U.S. population since 1950, the waste disposal load has increased about 90% in the past 30 years.

An increasing problem for waste disposal is the proliferation of various nondegradable plastics used for containers, insulation, equipment casings, and similar products. Estimates indicated that in 1980 the share of solid waste provided by plastics was about 2.8% over 1970, or over 6.5 million tons. A roadside litter survey by the National Research Council indicated that about 1 cubic yard of litter (paper cups, cigarette packs, beer cans, bottles, and general trash) was accumulated per month for each mile of major U.S. highways. The annual total of solid waste and litter in 1971 included 40 million tons of paper, 3.8 million tons of plastics, 71 billion metal

**TABLE 31-2**
Types of Solid Waste

| | *1960* | *1965* | *1970* | *1975* | *1980* |
|---|---|---|---|---|---|
| Gross waste lb/person/day | 86.9 | 104.8 | 131.1 | 135.6 | 155.6 |
| Paper (percent) | 18.8 | 15.8 | 16.0 | 15.5 | 20.6 |
| Glass | 1.4 | 1.2 | 1.3 | 2.7 | 4.9 |
| Metal | 0.5 | 1.0 | 1.2 | 4.4 | 4.1 |
| Aluminum | — | — | 1.3 | 8.7 | 14.8 |
| Rubber | 5.7 | 12.5 | 8.2 | 6.9 | 3.9 |

U.S. Environmental Protection Agency.

**TABLE 31-3**
Industrial Sources of Waste (Thousands short tons)

| | *1975* | *1980* |
|---|---|---|
| Total waste | 139,051 | 149,873 |
| Food process | 11,501 | 14,410 |
| Lumber/wood | 8,143 | 5,893 |
| Paper | 9,123 | 12,251 |
| Chemicals | 38,695 | 43,431 |
| Clay/glass | 11,345 | 13,336 |
| Primary metals | 42,708 | 37,523 |
| Electrical equipment | 1,500 | 2,148 |
| Transportation equipment | 3,785 | 4,168 |
| Petroleum | 2,038 | 4,873 |

U.S. Bureau of Census.

cans, 38 billion glass bottles and jars, 7.6 million television sets, and more than 7 million cars and trucks. The average cost of waste management in 1972 was about $6 billion ($27 per ton) and is expected to increase to more than $15 billion by the year 2000.

## 31.3 TYPES OF SOLID WASTE

Solid waste is present in many forms. The terms used to describe this type of waste differ with different organizations, but knowledge of the origin of the waste is not of much value in any case. The properties of waste have more to do with the collection, processing, and disposal than with the type of material. Any combination of waste containing animal or vegetable matter is usually termed *refuse* or *garbage*. Waste that does not contain garbage is usually termed *rubbish* or *trash*. Street refuse consisting of paper, leaves, occasional cans or bottles, dirt, and similar solids is most often collected by street maintenance crews, and leaves, tree trimmings, and park floral debris is often composted for later use as a soil additive for city parks and landscaping work. Abandoned vehicles and dead animals require special attention and equipment for removal and disposition by police or health departments and usually require no special consideration in the design or equipment for buildings dealing with refuse or trash disposal.

*Garbage* is animal and vegetable waste resulting from preparing, cooking, and serving food. Garbage is produced in home kitchens, restaurants, stores, hotel dining room/kitchens, and similar locations. Although most garbage is moist it does not contain enough moisture to require special handling except to prevent dripping on floors and other dry areas. Garbage decomposes rapidly and may produce a disagreeable odor and attract vermin and rats. Garbage does not include waste from meat-packing plants or canneries. Much household garbage today bypasses garbage-collection services by being ground up in disposal units and flushed down the sanitary sewer system, creating an overloading problem in many of the older sewage-treatment plants. *Market refuse* is composed of trimmings from vegetables, meat scrap, spoiled fruit or vegetables, cartons, boxes, and crates, and does not usually include cooked food. This refuse, like garbage, has a high factor for putrescibility, so must be treated in a manner similar to garbage treatment. Wooden crates and cardboard cartons are often separated and reused or recycled.

*Rubbish* is composed of burnable material such as paper, rags, wood, rubber, dry grass and leaves, plastics, and other combustible items. Rubbish may also consist of noncombustible material such as cans, glass bottles, ceramics, discarded metal, ashes, and dirt or rocks. Combustible rubbish is usually reduced by incineration, but noncombustible rubbish may be used for sanitary landfill. In many communities all types of rubbish are used in landfill, as separation at the original source is difficult and inconvenient, and later separation would require extensive equipment and considerable expense.

Other types of solid waste are not very common in direct connection with design, construction, or operation of buildings but include street refuse, dead animals, abandoned vehicles, trees, construction and demolition refuse, and special types of explosive, toxic, caustic, or radioactive waste. Most of these are handled with special equipment and have special disposal requirements. Residue and sludge from water or sewage treatment plants, pumpings from septic tanks, and agricultural waste and manure also require additional treatment and disposal and may be incinerated, chemically treated, or in a few cases buried in landfill sites after minimal chemical exposure.

## 31.4 BUILDING REQUIREMENTS FOR SOLID WASTE

Liquid waste in the form of storm drainage is disposed of by proper size and type of roof drains, scuppers, gutters, or downspouts in buildings and by proper grading, underground culverts and drains, or ditches for surface drainage. Collection and disposal of "gray" or "black" water and sewage is accomplished by a system of pipes in the building and connection to municipal sewers or private septic-tank systems. Unfortunately, most buildings do not have adequate facilities for temporary storage of solid waste, sorting by economical methods, or on-site disposal abilities. This simply means that most disposal systems must be specialized and either publically or privately owned and operated for the sole purpose of waste disposal.

Some thought should certainly be given to economical and convenient methods for solid waste collection in buildings. In residential planning the use of a *garbage disposal unit* in connection with kitchen plumbing will usually take care of meat scraps, vegetable trimmings, and other cooking and serving waste. Even though most of this material might be termed "solid waste," it is included as part of liquid sewage when properly ground. Many modern homes are now being equipped with *trash compactors* (Fig. 31-1) installed in the kitchen counter space and capable of compacting refuse into a neat watertight package about 8 in. × 15 in. × 15 in. The weight of the package depends on the material compacted. Most compactors are manufactured freestanding or as built-in counter units and require about 500 W of electricity per cycle. General capacity is often advertised as "equal to three 20-gal garbage cans," but this of course varies with the contents.

Apartment houses and condominiums may have the same type disposal units and compactors as those used in a single-family residence. If this is true and all occupants use them properly, there is no problem with kitchen garbage, and space for temporary storage of compacted-trash packages is the major consideration required. Commercial trash bins approximately 48 in. × 54 in. × 72 in. with hinged lids and large wheels are provided by most private collection agencies and some municipal services and will generally hold average solid rubbish from about six families for each week. Adequate space in a basement or other location must be provided for storage, access, and handling of each unit on a ratio of at least twice the size of each con-

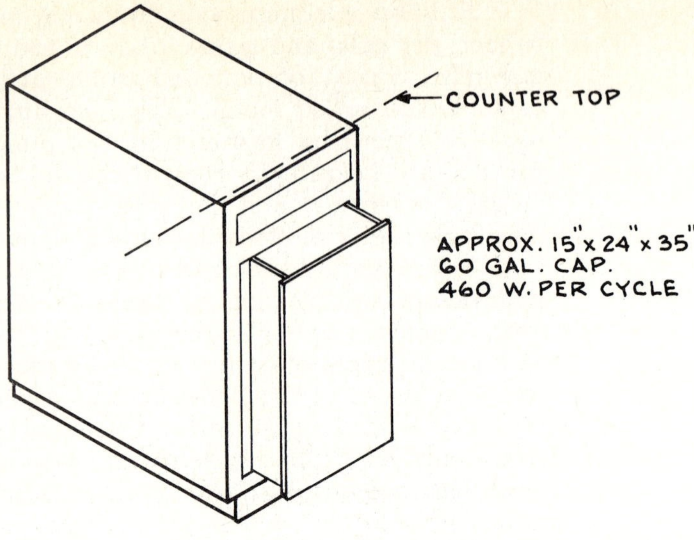


**Figure 31-1** Trash compactor.

tainer. In multistory apartment buildings, especially older ones without trash compactors, a metal *trash chute*, with access doors in the corridor at each floor, is often installed. Trash chutes are usually made of 14- or 16-gauge steel, 14 to 48 in. square or round, with single or interacting doors and are vented to free air at the top and equipped with automatic fire-sprinkler heads. The bottom must be adequately supported to withstand the shock of a trash package dropped from several floors above and should be at the proper height to dump its contents automatically into the refuse container. Obviously, such a chute must be properly enclosed with fire-resistant walls and provided with fire dampers when required by state or local fire codes.

Multistory buildings have an entirely different problem regarding solid-waste disposal. In most cases the refuse from offices is primarily paper. This is also true of some types of retail stores, where cardboard cartons are the principal waste product. The matter of collection is simpler because employees or custodians usually collect and empty waste baskets or similar containers each night. The problem is what to do with all this waste after it has been collected. In many cases this material, mostly paper, is baled and sold, which, at least partly, pays for the process. Small balers may be hand-operated or machine-operated, require about 36 in. × 48 in. of area for the baler and at least 10 times that area for unbaled waste, access to equipment, and minimum baled storage. Bales are customarily about 22 in. × 32 in. × 70 in. and weigh about 450 lb, so dollies or forklift equipment may be required for handling. Where large quantities of paper are regularly the major part of solid waste, large hydraulically operated balers may be used to handle up to 10 tons per hour.

Solid waste from groceries, meat markets, restaurants, and similar operations is a different type of operation. In a great many cases meat bones, fat, and other scrap is collected by independent processors and rendered to retrieve the tallow and fat remaining. This material, in turn, is used in the production of cosmetics, plastics, and many other products. Bones are ground up and used in commercial fertilizers. Vegetable waste was formerly gathered by farmers and used as feed for hogs, but most states now require all vegetable garbage to be cooked before it is used as animal feed, so this practice is limited.

## 31.5 SANITARY LANDFILLS

A sanitary landfill is an engineered burial of solid waste and is the most common method of solid waste disposal in the United States and many other countries. In the United States approximately 80% of cities with populations larger than 25,000 utilize landfills. Only about 9% of solid waste is incinerated; the rest is disposed of by other processes or in open dumps. Most sanitary landfills are made on land not generally considered as very valuable in their original state, such as swamps, marshlands, canyons, or man-made depressions such as quarries and sand pits. The final resulting land obtained by landfill operations is usually used for recreational purposes, parks, and golf courses but not for construction, as it must be obvious that good building foundation systems cannot be established on questionably compacted fill.

The landfill process is relatively simple and certainly the least expensive of any solid-waste disposal method. Essentially, the process is one of successive and repeated layers of waste each covered with earth until the final fill elevation is reached. The density of the compacted waste, "boxing" of the layers with earth dikes on all sides to prevent the spread of possible fire, relief of generated methane gas, and control of vermin, rodents, and flies are of primary consideration. Several methods to provide all these requirements are in common practice (see Fig. 31-2). All methods must start with a depression in the earth, either a natural condition or one made by early grading. Where the normal land is reasonably flat, the *area method* may be used. In this method the compacted waste from collection trucks is dumped in wide, flat rows, spread and further compacted with bulldozers, and covered with earth on sides and top at the end of each day or more often. In the *trench method* a wide trench is dug, the compacted waste is deposited in the trench, and the waste is covered with earth excavated from an adjoining trench as that trench is dug. A third variation is the *ramp method*, in which the waste is deposited against a slope, compacted, and then covered with earth from the bottom of the slope. Compaction of the solid waste is of utmost importance and may occur at the building generating the waste, in collection trucks equipped with compactor units, or at the disposal site using heavy equipment. Several municipalities, primarily in England, have converted incinerator plants to collector plants where loose waste is compacted into high-density bales before it is trucked to a disposal site.

Like most every other process, landfill has some definite advantages as well as some disadvantages. Some of the major advantages are:

1. Low initial capital investment compared to other methods.
2. Acceptability of all types of solid waste.
3. Minimum equipment and personnel required.
4. Equipment and personnel can be used flexibly from one site or operation to another.
5. Complete and final disposal and use of marginal land.

**Figure 31-2** Sanitary landfill.

The disadvantages are related primarily to community and weather:

1. Sites are usually outside city limits and may create governmental control problems.
2. Hauling distances may cause high costs per ton.
3. Freezing and/or snow conditions in winter may create problems.
4. Communities may object to the added traffic of collection trucks, blowing dust and loose paper, and the assumed devaluation of property near a dump.

Both the American Public Works Association and the U.S. Public Health Service have worked out an acceptable formula to determine the landfill area required. Most fill is compacted to between 40 and 50% of original volume, and approximately 1 acre of land per year is required for each 10,000 people. Earth cover is about 15 to 20% for deep fills and the average density of refuse is about 300 to 350 lb per cubic yard. Excluding the original costs of land, landfill disposal costs vary from about \$0.50 per ton in the cities to about \$1.25 per ton for disposal in outlying areas. In 1968–1969 the city of Los Angeles processed about 4000 tons per day with an operating force of 18 at an average cost of \$0.67 per ton. The area of Los Angeles County, which has a tributary population of more than 7 million, collectively disposes of more than 10 million tons of solid waste per year, nearly 8 lb per person per day.

## 31.6 INCINERATION

The incineration of solid waste is a combustion process by which the hydrocarbons in the waste combine with oxygen in the air to form carbon dioxide and water. The metals and other noncombustible waste form a residue or ash, and the carbon dioxide and water pass into the atmosphere as a vapor. Under proper control conditions this process creates a high heating value that may be utilized to generate steam or electricity. Proper sorting of solid waste is a prime requirement for incineration, as some materials will not burn or may cause equipment malfunction if introduced into the furnace. Continuous feeding is also desirable, as is preheating to drive off surface moisture that may be present in the solid waste.

The proper design and construction of municipal incineration plants (see Fig. 31-3) is a complicated and specialized field, but a brief description

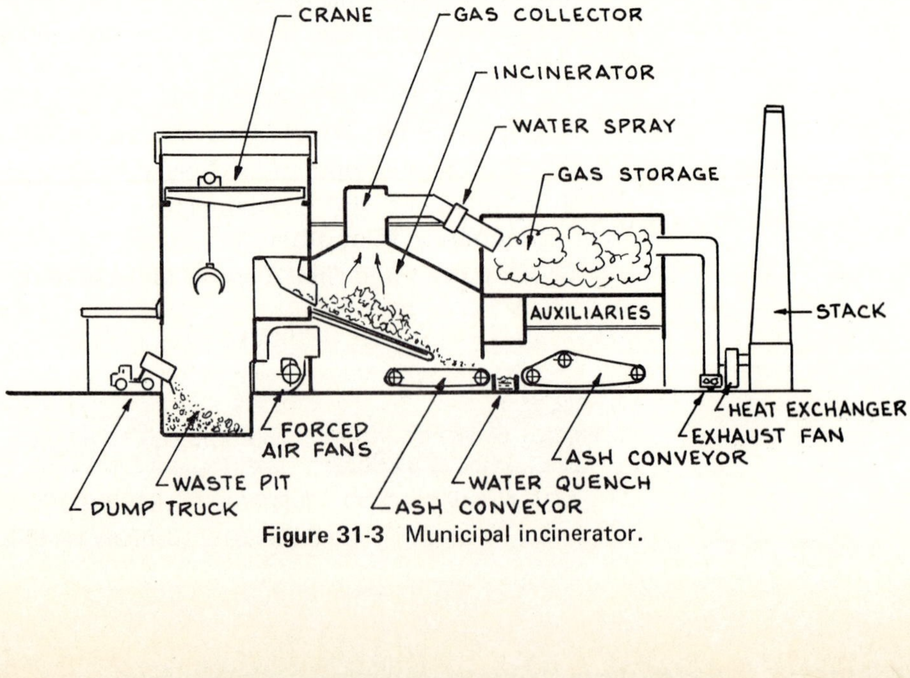


**Figure 31-3** Municipal incinerator.

of the basic principles should be helpful. Most materials that truly burn must be converted to a gas by heat. Solid waste contains moisture which must be removed before the waste will ignite and burn. Most incinerators expose the raw waste to hot gases or other heat at the bottom of the feed chute before the waste drops into the burning chamber and begins to vaporize in contact with oxygen contained in a forced-air draft. The intense heat generated by this vaporization is passed back over the raw waste, thereby drying the waste in a continuous process. The burning waste is carried along by stoker, moving grates, or other means, and finally the noncombustible parts drop into a cooling system and are disposed of, while the hot gases are usually used to heat boilers. The steam thus generated in the boilers may be used for heating, driving electric generating turbines, or other uses. The various arrangements for presorting, drying, combustion, and final disposal of residue are reasonably standard.

The American Public Works Association has classified six types of solid waste and the Incinerator Institute of America has used that classification to establish parallel types of incinerators. Type 0 waste ("a mixture of highly combustible waste such as paper, wood boxes, . . . 10% plastic bags, oily rags or rubber scraps") has a heating value of 8500 Btu/lb. Other types provide heating values from 6500 Btu/lb for waste containing 25% moisture (type 1) to garbage containing 70% moisture (type 3), which has a heating value of only 2500 Btu/lb. The national annual average composition of municipal refuse indicates that about 50% of total waste is paper, 19% is food waste, 10% is metal, 10% is glass, and the remainder is composed of plastics, leather, textile waste, and wood.

Incineration has advantages and disadvantages that affect selection, location, and operation. Some of the advantages are:

1. Less land is required than for landfill.
2. An incinerator can often be located near the source of waste, so less hauling is required.
3. Residue is minimal and contains no organic matter if properly processed.
4. Flexibility allows for handling various quantities of waste.

Some of the disadvantages are:

1. An incinerator plant is relatively expensive, costing from \$5 to \$9 per ton of material handled.
2. Skilled workers are required.
3. Maintenance and repair costs are high.
4. An incinerator is often objectionable to the community due to smoke and possible devaluation of nearby property.

Heat from incineration may be utilized by heat transfer through water-tube boilers, water-tube wall combustion chambers, or combinations of these. Steam may be produced at a ratio of 1 to 3.5 lb of steam per pound of incinerated waste. Despite this high return on "wasted" heat, most municipal incinerators use such heat only for the immediate building heat, hot water, or waste drying; only about 5% generates electricity. One plant operating at about 200 tons of waste per day produces about 75,000 lb of steam per hour at 400 psig. Metal recovery from incinerator residue includes aluminum, ferrous metal, zinc, copper, and miscellaneous other metals. These metals are best separated before incineration but may be processed from residue. Separation may or may not be economical, depending on the market value of the salvaged metal.

## 31.7 WASTE RECYCLING

Today the entire world is facing not only an acute shortage of energy but also a steadily increasing amount of waste. In the United States the waste per person per year has jumped about 50% since 1968, while in the United Kingdom it has increased by 1% per year. About 2% of total U.S. energy demand could be saved by recycling ferrous metals, paper, and aluminum. However, waste recycling presents some problems regarding collection, sorting, transportation, and reuse. At the present time the recycling of automobiles and other large metal objects is carried on extensively but has little to do with the operation of buildings. The salvage of paper, some metals, glass, and other solid waste may become a factor in the construction and operation of buildings as more and more possible shortages are indicated. At the present time most recycling of metal, glass, or paper occurs at plants producing or handling large quantities of those materials. Wood and textile waste is almost entirely salvaged at the mills and reprocessed into other products, but plastics (18 billion pounds produced in 1970) remain the major nonsalvageable material in use today.

At first glance, recycling seems to be entirely feasible and should eventually reduce the costs of energy in remanufacturing, but several currently proven factors dispute this premise. Further research is needed to provide a method for easy separation of various classes of solid waste so that materials that could be recycled will not be mixed with other refuse. In addition, fluctuating markets for the salvaged material make the cost of recycling a doubtful investment. Freight rates for transporting recycled material and high labor costs for handling as well as present restrictive laws and codes also tend to discourage the process.

## 31.8 SUMMARY

Much of the solid waste produced by modern society consists of replaceable or renewable energy-rich materials. Many schemes and methods have been proposed and tried in practice, but the processes of separation, transportation, and final disposal have often been halted or slowed by government, labor, and the very communities that would profit by having such facilities. Landfill is still the most common method of disposal because it uses existing collection and transportation facilities and is the cheapest method available. Proponents of landfill also contend that the resulting land area may be more useful than a marsh, canyon, or abandoned quarry and certainly is more attractive. The incineration of waste is expensive, both in initial cost and in operation of the plant. Sorting of waste is minimal if a little effort is made at salvage before treatment, but the final disposal of ash and residue may present a problem. At present, recycling is carried on primarily by industries producing materials which have a by-product or waste material that may be reused. Other disposal methods, such as composting and dumping at sea, are costly, inefficient, polluting, and fortunately not widespread. In-house solid waste disposal for living units, offices, retail stores, and similar nonmanufacturing operations is at present limited to some slight separation at the option of the inhabitant and generally means that waste that cannot be handled by the sewer system will be collected and disposed of by public or private collectors.

# Glossary

A

**ABS:** Acrylonitrile-butadiene-styrene plastic.

**Absolute humidity:** Amount of moisture in the air, measured in grains per cubic foot.

**Absolute pressure:** Gauge pressure plus atmospheric pressure of 14.7 psi.

**Absolute zero temperature:** Minus 460°F or minus 275°C.

**Absorption:** A process of assimilation. For example, an absorption refrigerator creates low temperatures by absorption of refrigeration by chemicals.

**Accumulator:** A storage tank for liquid refrigerant from an evaporator that prevents the refrigerant from entering the suction line.

**Acoustics:** The branch of physics that relates to the laws of sound. The study of materials and methods for the control of sound and its transmission.

**ACP:** Asbestos-cement pipe.

**Activated carbon:** Carbon that has been specially treated. Usually used to clean air in filters.

**Aerator:** A device that adds air to water. Fills flowing water with bubbles.

**AGA:** American Gas Association.

**Air:** A mixture of dry air and saturated water with a density of 0.075 lb/ft$^3$ and an absolute viscosity of $0.379 \times 10^{-5}$ lb per foot of mass. Dry air at 70°F has a barometric pressure of 29.92 in.

**Air break:** A piping arrangement that provides an air space above the flood level of the receptor.

**Air change:** A measure of the amount of air moved into, or extracted from, a room or other space. Usually measured in cubic feet per minute.

**Air cleaner:** A device that removes airborne contaminants from exhaust air.

*Terms in italic type are defined elsewhere in the Glossary.

**Air-conditioning:** A system that controls temperature, humidity, motion, odors, distribution, gases, and related factors that affect human health or comfort.

**Air cooler:** Equipment designed to lower the temperature of air that passes through it.

**Air lock:** Air trapped within a pipe which restricts or blocks the flow of liquid.

**Alternating current:** Electric current that regularly alternates its direction of flow at a fixed frequency. Most electric power is alternating current and may be increased or decreased in voltage by means of a transformer. Abbreviation: ac.

**Alternator:** A machine that converts mechanical energy into alternating current.

**Ambient temperature:** The temperature of outside air or water surrounding an object (building, person) on all sides.

**Ammeter:** A measuring instrument designed to indicate the flow of electric amperes.

**Ammonia:** A refrigerant composed of nitrogen and hydrogen ($NH_2$) and identified as R-117.

**Ampere:** A unit indicating the amount of electricity passing a given point per second. Abbreviation: A or amp.

**Amplifier:** An electrical device that increases the flow of electrons in a circuit. An electronic amplifier controls the volume of sound in a communication system.

**Anaerobic:** A microorganism that flourishes without oxygen. The active bacteria in a septic tank.

**Anemometer:** An instrument that measures the rate of flow of wind or air.

**Anode:** The positive (+) connection post of a battery.

**Antihammer device:** An air chamber designed to absorb the shock of a rapidly closed valve.

**Armature:** The center rotating portion of an electric generator or motor.

**ASHRAE:** American Society of Heating, Refrigeration and Air-Conditioning Engineers.

**Aspiration:** Movement of gas or liquid produced by suction.

**ASTM:** American Society for Testing and Materials.

**Atmospheric pressure:** The pressure that air exerts on the earth's surface measured in pounds per square inch (psi).

**Atom:** The smallest particle of matter. A minuscule amount.

**AWWA:** American Water Works Association.

## B

**Backflow:** Flow of liquids into the pure water supply caused by pressure drop in the water-supply piping.

**Backpressure:** Compressed trapped air that resists the flow of water in its normal course.

**Back-siphon:** Backflow situation caused by siphonage.

**Balance point:** The point at which the output of a heat pump is equal to the heat requirements of the structure as compared to the outside temperature.

**Barometer:** An instrument for measuring the atmospheric pressure. An aneroid barometer operates by pressure on a sealed metal container. A mercurial barometer indicates pressure by a column of mercury contained in a glass tube.

**Battery:** A self-contained unit (wet or dry) in which electrical direct current is generated by chemical action on chemical or metal plates or containers.

**Bell and spigot:** Type of pipe connection in which a straight end (spigot) fits into an enlarged hub (bell) of the adjoining pipe. This is one type of joint for cast-iron pipe and is caulked watertight with packing (oakum) and molten lead.

**Bench mark:** A fixed location of known elevation from which other measurements may be taken.

**Blast gate:** A sliding damper.

**Blower:** A fan used to increase air pressure.

**BOCA:** Building Officials Conference of America.

**Boyle's law:** Law of physics stating that the volume of a gas varies as pressure varies, as long as the temperature does not change.

**Brake horsepower:** One horsepower equals 33,000 ft-lb/min.

**Branch:** An addition or change of direction of a pipe or duct by intersection.

**Brazing:** The joining of metal with an alloy having a melting point higher than solder but less than the melting point of the metal being joined.

**Break:** Opening or discontinuation of an electrical circuit by accident or by a switch.

**Breaker panel:** An electrical distribution panel using ampere-rated automatic disconnects to protect the various circuits.

**Brine:** Water saturated with a salt. Used in commercial refrigeration systems as a heat-transfer material.

**Brush:** A graphite or carbon bar that makes continuous contact with the rotating commutator of a generator or motor and conducts the current into the rest of the circuit.

**Btu:** British thermal unit. The quantity of heat required to raise the temperature of one pound of water one degree Fahrenheit.

**Bus bar:** A heavy bar used to conduct primary electricity where a wire would be too light.

C

**Cable:** A stranded single conductor of large size, bare or insulated.

**Calorie:** The amount of heat required to raise the temperature of one gram of water one degree Celsius.

**Calorimeter:** A device used to measure amount of heat or to determine specific heat.

**Capacitor:** Two conductors separated by an insulator. A capacitor blocks the flow of direct current and partially impedes the flow of alternating current. Impedence decreases as current frequency increases.

**Capillarity:** The action by which a liquid from one source is moved or absorbed, usually vertically, into another source.

**Carbon dioxide:** A chemical compound of oxygen and carbon ($CO_2$), identified as refrigerant R-744. A dangerous gas when concentrated but comprising about 0.04% of the atmosphere.

**Carcinogen:** A substance that causes cancer.

**Cascade system:** A refrigeration system having two or more circuits where the evaporator of one cools the condenser of the other.

**Cathode:** The negative (–) terminal of an electrical device. Opposite terminal of anode (+).

**Caulking:** Method used to make a joint or opening tight. Bell-and-spigot pipe connections are caulked with oakum and molten lead. Cracks in buildings are caulked with various pastelike products.

**Celsius:** The temperature recording system using the metric system. The freezing point of water is taken as 0°, the boiling point as 100°. Abbreviation: °C. The older term for this system is "centigrade."

**Cesspool:** A deep pit into which liquid waste is deposited to be absorbed by the earth.

**Charge:** The quantity of electricity held by a capacitor. The amount of liquid refrigerant used to fill a system.

**Charles's law:** The volume of a given mass of gas at a constant pressure varies according to its temperature.

**Chase:** A space or shaft, usually vertical, in which pipes or ducts are placed or carried from one floor to another.

**Check valve:** A device that permits flow in one direction but automatically closes upon pressure from the other direction.

**Circuit:** A system of pipe, ducts, or electrical wiring which permits flow from an energy source, through various pieces of equipment, and back to the original source.

**Circuit vent:** A branch plumbing vent that serves two or more traps, extending from in front of the last fixture connection to the vent stack. Also called a loop vent.

**Cleanout:** A removable drainage plug which allows access to the inside of a pipe for cleaning purposes. Usually installed at the ends of a house sewer and at each major change in sewer direction.

**Coil:** A heating or cooling unit made of pipe or tubing. An electrical conductor wound with wire which creates a magnetic field with electrical passage.

**Common vent:** A plumbing vent connecting at the junction of two fixtures and serving both fixtures.

**Commutator:** A series of separately insulated segments that connect the windings of an armature coil to the brushes of a motor or electric generator.

**Compound:** A combination of chemical elements or a combination of ingredients before being processed into a final product.

**Compressor:** A type of pump that handles gases. Used in refrigeration to compress low pressure from the cooling side into high pressure for the condensing phase.

**Condensate:** Water retrieved from steam as it loses heat. Fluid that forms on a refrigeration evaporator. Any liquid obtained by condensing a heated gas.

**Conductance:** Opposite of resistance. The ability of a material to transfer heat, electricity, or another type of energy from one point to another without physical motion or change of shape.

**Conduit:** A pipe or duct containing a fluid or containing electric wiring. Conduit-wire relationships with regard to conduit size and allowed number and type of wires is indicated in the *National Electrical Code®*.

**Contaminant:** Any material foreign to the base material. Dirt, oil, and moisture are contaminants of refrigerants. Dust, metallic particles, and gases are contaminants in ventilation.

**Continuous vent:** A vertical plumbing vent that is a continuation of the drain to which it is connected.

**Convection:** The transfer of heat by means of movement of a liquid or gas.

**Corporation stop:** A plumbing valve installed in the building water-service line at the water meter or main.

**Coulomb:** An electrical unit. One coulomb per second equals one ampere or $6.25 \times 10^{18}$ electrons per second past a given point.

**CPVC:** Chlorinated polyvinyl chloride. A plastic compound used to manufacture pipe capable of carrying hot water.

**Creep:** Dimensional change of a plastic pipe under load over a long period of time (cold flow).

**Cryogenic:** A substance that exists as a liquid or gas at 250°F or lower.

**Curb box:** (Buffalo box): A fitting placed over a corporation stop to allow it to be operated from ground level.

**Cycle:** A series of events repeated in the same order. A repeated series of mechanical operations.

## D

**Dalton's law:** The total vapor pressure of a mixture of gases is equal to the sum of the individual vapor pressures of the gases contained in the mixture.

**Damper:** A valve for controlling airflow or controlling distribution.

**Defrosting:** A process of removing accumulated frost from refrigeration evaporators.

**Degree-day:** A unit representing one degree of difference from a given point in average outdoor temperature of one day. One method for figuring fuel requirements. Degree-days are calculated from averages. If the average temperature is 65°F and the average degree-day 75°F, the factor is 75 − 65 or 10.

**Dehumidifier:** Any device used to remove moisture from air in an enclosed space.

**Dehydration:** The removal of water vapor from air by absorbing materials. Removal from animals of moisture caused by excessive heat. A dehydrator is a drier used to dry "wet" or saturated air.

**Delta connection:** Connection at the generating source of a three-phase electrical system forming a triangle (the shape of the Greek capital letter delta). When in the proper position, one phase will have full current and the other two will have half current.

**Density:** The mass of a unit of a substance. Density is the weight per unit volume of a material; the ratio of mass of a substance to the volume of the substance.

**Desiccant:** Any material, liquid or solid, which will absorb water or water vapor from another material.

**Dew point:** Temperature at which vapor at 100% humidity begins to condense into a liquid.

**Dielectric:** The matter between the plates of a capacitor and used to store electrostatic energy; an insulator or nonconductor.

**Direct current:** Electric current that flows steadily in one direction and is usually provided by batteries. Where direct current is rectified from alternating current generators it may have pulsating characteristics. Abbreviation: dc.

**Disposal field:** A system of buried piping in connection with a septic tank, to provide absorption of liquid waste into the earth. Also called a leach bed.

**Double-pole:** A switch or other device that connects electric service into one group of circuits while disconnecting service to an opposite group of circuits.

**Downspout:** A vertical pipe that carries storm water from a roof gutter to the ground or into a storm drain system.

**Drainage system:** The complete set of pipes and fittings that carry sewage from fixtures to the building main drain.

**Drum trap:** Plumbing trap designed as a vertical cylinder with the inlet near the bottom and the outlet at the top. The cylinder holds water between the outlets, thus providing a water seal to prevent sewer gas from entering the building.

**Dry ice:** An ice substitute made of solid carbon dioxide. The carbon dioxide gas changes directly from a gas to a solid at a temperature of $-109°F$.

**Dry vent:** Any plumbing vent that does not contain water.

**Dry well:** A well filled with rock or gravel and designed to permit water to be absorbed into the earth. A dry well is not a cesspool but generally receives storm water or purified septic-tank effluent.

**Duct:** An air conductor, usually of metal or fiber and either rectangular or round in cross section, used for conducting low-pressure air or gases.

**Ductility:** The property of a metal that allows it to be formed into thin sheets or other shapes without breaking.

**DWV:** A plumbing abbreviation meaning drain–waste–vent.

**Dynamometer:** A mechanical device for measuring power input or output. Usually consists of a unit arranged to measure rotation of a shaft or flywheel which records electronically.

## E

**E:** Mathematical symbol for voltage.

**Easement:** Right of a government or utility company to use another's land for access to utility service systems.

**Ebullator:** A device inserted in flooded evaporator tubes to prevent the water from becoming oil-bound or refrigerant from becoming quiescent at a pressure lower than the boiling point.

**Effluent:** The outflow of a liquid, usually in connection with sewage.

**Eighth bend:** A plumbing pipe fitting causing a 45° change of direction.

**Ejector:** A plumbing device used to pump sewage. An air device used to force contaminated air into a hood or receiver system of ventilation.

**Elastic limit:** The greatest stress a piece of material can withstand and still return to original shape when the stress is removed.

**Electrode:** The plates of a storage battery. In electroplating, current enters the bath through the anode electrode (+) and leaves through the cathode electrode (–), causing plating metal to be deposited on the base metal.

**Electrolyte:** A liquid or paste compound that is capable of conducting electricity.

**Electromagnet:** A magnet created by a coil winding on a metal core. The magnet operates only during a current flow.

**Enthalpy:** The total amount of heat in one pound of a substance calculated from an accepted temperature base.

**Entropy:** The energy in a system. A mathematical factor used in engineering.

**Enzyme:** An organic substance that speeds up the spoiling of food by changes in the food cells and is slowed by refrigeration.

**Equivalent direct radiation:** A unit of heat delivery of 240 Btu/hr. Abbreviation: EDR.

**Erosion:** Gradual wearing away of a material by continuous abrasive action.

**Escutcheon:** A plate or ring used to conceal open spaces around a pipe or electrical device installed in a wall or similar surface.

**Eutectic:** A mixture that melts or freezes at constant temperature with constant composition and whose melting point is the lowest possible for a given solution.

**Evaporation:** A change from a liquid to a gas, usually caused by heating.

**Evasé:** A section of duct in an exhaust stack that gradually increases in diameter.

**Exfiltration:** The opposite of *infiltration*. The condition when inside air passes through cracks or other openings into the outside air.

**Expansion valve:** A valve used in refrigeration systems which maintains pressure differences between high side and low side.

**Extrusion:** A method of forcing a semiplastic substance through a die shaped like the cross section required. Possible with plastic or aluminum but not with steel.

**Factor of safety:** The ratio of ultimate stress to working stress. A calculated overload included in designs to protect against failures of a product.

**Fahrenheit:** A measurement of temperature in which 32° indicates freezing and 212° the boiling point of water at a sea-level pressure of 14.696 psi. Abbreviation: °F.

**Fail-safe control:** A device that opens an electrical circuit or operates valves or other equipment to stop action if normal controls fail to function.

**Farad:** A unit of capacitance storage of one coulomb when one volt is applied.

**Filter:** Equipment that removes foreign matter from a liquid or gas.

**Fittings:** The couplings, elbows, and other units (but not valves) that join pipe together.

**Fixture:** Plumbing fixtures are sinks, tubs, water closets, and similar units. Electrical fixtures are various lighting devices designed for lamps or fluorescent tubes.

**Flapper valve:** A type of valve that allows gas or liquid to flow in only one direction. Valves are made of thin metal or other flexible material and are opened or closed by pressure. Also known as a reed valve.

**Flare fitting:** Pipe connections that are not threaded, soldered, or welded. The end of the pipe, usually copper, is expanded at about 45° and firmly gripped by a recessed ring of the connecting fitting to form a leak-tight joint.

**Flash point:** Temperature at which a fuel oil gives off enough vapor to support a flash flame but will not support continuous combustion.

**Flood level:** The point in a plumbing fixture where water overflows the rim if the drain is plugged.

**Fluorescent lamp:** A glass tube coated inside with phosphors and filled with mercury vapor. An electric charge causes the vapor to actuate a glow in the phosphor coating and emit light.

**Fluorocarbon:** A material made by the polymerization of monomers of carbon and fluorine.

**Flushometer:** A plumbing valve connected directly to the water supply and designed to provide a measured amount of water to flush a water closet or urinal. Also called a flush valve.

**Flush valve:** A *flushometer.*

**Foaming:** Formation of froth or suds caused by turbulence of sewage or other liquids in a tank. Foam created by rapid boiling-out of refrigerant of an oil–refrigerant mixture when the pressure is suddenly reduced.

**Footcandle:** A unit of measure for illumination expressed in lumens. The number of footcandles for a given surface is found by dividing the number of lumens on the surface by the area in square feet. Abbreviation: fc.

**Foot-lambert:** A unit of brightness expressed in lumens per square foot. One foot-lambert equals one lumen per square foot.

**Foot-pounds:** The work done in moving a substance one foot against a force or resistance of one pound.

**Freeze:** To convert a liquid to a solid by extracting heat, usually by refrigeration. To preserve meat or vegetables by causing the contained moisture to become a solid, thereby preventing decay in the living cells.

**Freezer burn:** Drying or discoloration of products, principally meat, caused by poor wrapping, long exposure, or inadequate control of refrigeration.

**Freon:** Trade name for a series of refrigerants produced by du Pont de Nemours, Inc.

**Frequency:** The number of complete cycles or vibrations in a unit of time. Alternating electric current generally has a frequency of 60 hertz (cycles per second) or a change of flow direction once every $1/60$ of a second.

**Friction loss:** Pressure loss due to friction in ducts or pipes, usually expressed in percent of total efficiency.

**Fumes:** Smoke, odors, vapor, fog, and similar airborne material over a very wide range that may be seen, smelled, or be invisible and that may range from harmless to toxic.

**Furnace:** Any unit burning fuel (coal, oil, gas) or using electricity to produce heat in a warm-air or air-conditioning system.

**Fuse:** A device in an electric circuit designed to melt and disconnect the flow of electricity in cases of overload.

**Fusible link:** A piece of metal with a known melting point used to connect control devices for doors, dampers, and similar equipment. When the temperature surrounding the link exceeds the melting point of the link, the link parts and springs, or gravity, close the damper or other device against a fire.

## G

**Galvanic:** Electric current generated chemically when two dissimilar metals are immersed in a proper electrolyte.

**Galvanometer:** An instrument for measuring small amounts of direct current.

**Gate valve:** A type of valve in which a disk or machined wedge is moved vertically at right angles to the flow to open or close the flow. The valve may be opened to full bore of the pipe.

**Gauge:** Also gage. Any instrument used to measure pressure or liquid level. Sheet metal, wire, and a number of other materials have their properties of thickness, diameters, and other qualities established by various gauges.

**Generator:** A machine that converts mechanical energy into electrical energy, or alternating current. Also called an alternator.

**Globe valve:** A valve in which the liquid flow is controlled by a disk operating vertically, but parallel, to the liquid flow. The flow is controlled by the pressure of the disk against the seat and normal to the axis of the pipe to which the valve is connected.

**Gooseneck:** A type of faucet with a return-bend at the top often used in pantries or on chemical sinks. A short length of flexible pipe used to connect water-service pipe to the water main.

**GPM:** Abbreviation for gallons per minute.

**Grab sample:** A sample of material collected over a very short period of time. A sample "grabbed" rather than carefully measured.

**Grease trap:** A type of plumbing device installed in the drain line of a fixture likely to be loaded with grease from cooking or garbage. The grease trap floats out the grease by a series of baffles, the water is drained off, and the floating grease is periodically removed.

**Grille:** A louvered or perforated cover unit for a heating or air-conditioning outlet in a wall, ceiling, or floor. Grilles may be of a single thickness or multiple thickness, stationary or adjustable. Large louvered grilles are also used for air inlets to many forms of commercial chillers, condensers, and water-cooling towers.

**Gross output:** A boiler rating at which the boiler will deliver total quantity of heat while meeting all testing and rating requirements.

**Ground:** A deliberate or accidental connection between an electric circuit and the earth. A point of zero electrical potential. The neutral or "ground" wire in a three-wire 220/240-V supply.

**Groundwater:** Water beneath the earth's surface held at some level by a layer of clay, rock, or other impervious material.

## H

**Head:** The sum of static and velocity pressures at a point of measurement. Static head is the pressure of fluid in terms of height of a column of water or mercury.

**Heat:** The form of energy transferred due to temperature change. Latent heat is a change of enthalpy during a change of state and is absorbed or rejected at constant pressure. Sensible heat changes only the temperature of the substance involved. Specific heat is the heat absorbed or released by a mass when temperature is increased or decreased.

**Heat exchanger:** Equipment designed to transfer heat from a warm surface to a cooler surface. Refrigeration evaporators and condensers are heat exchangers. High-temperature water surrounding a piping grid transfers heat to cooler water inside the pipes. Hot water passing through pipes in a warm-air heating system warms the surrounding air.

**Heating value:** The amount of heat, expressed in Btu, contained in unit quantities of various fuels.

**Heat loss:** The amount of heat, expressed in Btu, which is lost or removed from a space in 24 hours.

**Heat pump:** A pump that operates on a compression cycle and can supply either heat or cooling to a temperature-controlled space.

**Heat transmission coefficient:** The values given to building materials used in the calculation of heat transmission by conductance, convection, or radiation. Abbreviation: HTC.

**Henry:** A basic unit of inductance. One henry results from the current variation of one ampere per second and produces one volt. Abbreviation: H.

**High side:** The parts of a refrigeration system which are under high pressure. The condensing side of a system.

**Hood:** A general ventilation term for any booth, canopy, or open receiver designed to exhaust contaminated air.

**Horsepower:** A unit of power indicating work done at a rate of 550 ft-lb/sec or 33,000 ft-lb/min. One horsepower equals 747 W.

**Hose bibb:** A faucet outside a building to which a hose may be attached. Also called a sill cock.

**House drain:** The lowest horizontal piping of a sewage system designed to collect waste from all fixtures and conduct it outside the building; the building drain.

**Hubless joint:** A type of connection for piping which uses a sleeve covering the adjoining ends of the pipe and is held in place with bands or clamps.

**Humidistat:** An electrical air-conditioning control device actuated by differences in air humidity.

**Humidity:** The percent of moisture in the air in relation to totally dry air. Absolute humidity is the weight of water per unit volume.

**Hydrocarbon:** A material composed of carbon and hydrogen only.

**Hydronics:** The science of heating or cooling liquids.

**Hygrometer:** An instrument responsive to the relative humidity of the atmosphere.

**Hygroscopic:** The ability of a substance to absorb and retain moisture and to change physical shape as the moisture content changes.

## I

**IAPMO:** International Association of Plumbing and Mechanical Officials.

**ICBOA:** International Conference of Building Officials Association.

**ICC:** Interstate Commerce Commission.

**Ice melting equivalent:** The amount of heat absorbed by melting ice at 32°F is 144 Btu/hr of ice or 288,000 Btu/ton. Abbreviation: IME.

**Illumination:** The volume of light thrown upon an object by a luminous body; specifically, the density of light on a surface per unit of area.

**Impedence:** Total opposition to alternating current flow. Impedence is indicated as $Z$, and the unit is *ohms*.

**Impeller:** The rotating portion of a centrifugal pump.

**Inch of mercury:** A unit used in measuring pressure. One inch of mercury column equals 0.491 psi or 13.6 in. of water column.

**Inch of water:** Also used as a measure of pressure. One inch of water column equals 0.578 oz/in.$^2$.

**Indirect vent:** A plumbing vent that connects above the fixture served or terminates in the open air.

**Indirect waste:** A plumbing waste pipe that does not connect directly into the main waste system but empties into another fixture which is directly connected.

**Induced draft:** An artificial draft in a firebox or other location created by a fan or arrangement of dampers.

**Inductance:** The change in a magnetic field created by a change in current flow of alternating current. The change in magnetic field induces a counter-voltage which tends to hold current so that it may be used to retard current or to augment it.

**Infiltration:** Air leakage into a building caused by wind or temperature differences. Leakage is primarily through cracks and other small openings rather than through large open areas.

**Input rate:** The amount of heat, or cold, supplied to an appliance and usually expressed as Btu per hour or cubic feet per hour.

**Insulation:** A material used to decrease or retard the flow of heat through a wall or partition or to retain the heat inside a furnace, duct, pipe, or building.

**Interlock:** A safety device that requires a series of functions in proper order so that an electric circuit or a machine may function.

**Invert:** The elevation of the inside of a horizontal pipe at the bottom.

***IR* drop:** Voltage drop resulting from current flow through a resistor.

**Irradiation:** The quantity of radiant energy on a surface per unit of time per unit of area. A method of exposing food to certain wavelengths to kill certain bacteria.

**J**

**J box:** An electrical junction box where conduit, wire, or cable are interconnected. J boxes are not usually used as fixture boxes. A junction box.

**Joint runner:** An incombustible rope used to retain the molten lead in a bell-and-spigot plumbing joint. After the lead has partly cooled, the runner is removed and the lead is packed by using a yarning iron.

**Joule:** The unit of work resulting from one ampere of current flowing through one ohm of resistance in one second.

**Jumper:** A conductor or other device used to make temporary connections between two electrical circuits or appliances.

**K**

**Kelvin:** The absolute scale of temperature based on the average kinetic energy per molecule of a perfect gas. Zero is equal to −459°F.

**Kilometer:** A metric unit of measure equaling 1000 m; about ⅝ mile or 3280.83 ft. Abbreviation: km.

**Kilovolt-ampere:** Unit of apparent power as different from kilowatts or true power. Abbreviation: kVA.

**Kilowatt:** Unit of electric power equal to 1000 W, or 3413 Btu/hr. Abbreviation: kW.

**Kilowatt-hour:** One kilowatt of power used for one hour. Abbreviation: kWh.

**Kirchhoff's law:** Kirchhoff's current law states that the sum of all currents flowing from a given point must be equal to the sum of all currents flowing into a point. Kirchhoff's voltage law states that the sum of all the voltage rises in a complete circuit must equal the sum of all the voltage drops in the circuit.

**L**

**Lag:** The time between waves in alternating electric current. The time in a radiant heating system that heat is continued after the heat source is turned off.

**Latent heat:** Heat energy absorbed in the process of changing the form of a substance without change in temperature or pressure.

**Lateral:** A secondary plumbing pipe. A sewer or water line from the street main to the building.

**Lavatory:** A bathroom plumbing fixture designed for washing hands and face.

**Leach bed:** A septic-tank disposal field. The leach bed is designed to allow liquid from the septic tank to be absorbed into the surrounding soil.

**Leader:** A pipe, usually of sheet metal, which conducts rainwater from the gutter to the ground or storm sewer.

**Line voltage:** Voltage at normal outlets, usually either 110/120 V or 220/240 V. Not low voltage below 30 V.

**Liquefaction:** The process of changing a substance, ordinarily gaseous, to a liquid by means other than condensation.

**Liquefied natural gas:** Natural gas that has been cooled until liquefied. Abbreviation: LNG.

**Liquefied petroleum gas:** Any gas fuel primarily composed of hydrocarbons, such as propane, propylene, butane, isobutane, or butylene. Abbreviation: LPG.

**Liter:** A metric unit equal to 0.26 U.S. gallon.

**Litharge:** A lead powder mixed with glycerine and used to seal threaded pipe joints.

**Load:** The amount of heat per unit of time, kilowatts of electricity, amount of conditioned air, or other factors imposed upon a system.

**Load factor:** The ratio of mean actual load to a maximum load or maximum production capacity in a given period.

$$\text{load factor} = \frac{\text{kWh supplied in period}}{\text{Peak kW} \times \text{hours in period}}$$

**Loop vent:** A circuit vent. A branch plumbing vent serving two or more traps, extending from the last fixture to the main vent but not connected at intervening traps.

**Louver:** A ventilation device constructed with overlapping slats, fixed or adjustable, designed to control ventilation and shed rainwater.

**Low side:** Portion of a refrigeration system at the evaporator or low-pressure side. Opposite of *high side*.

**Low voltage:** Voltage lower than 30 V. Usually direct current supplied by batteries or a transformer and used for control or communication systems.

**Lumen:** The unit for measurement of the time rate of the flow of light from a standard candle.

## M

**Magnetic circuit:** The closed path of magnetic lines of force.

**Main:** The primary carrier of electricity, steam, hot or cold water or sewage. The wire, pipe, or duct that carries energy to branches. The principal conduit or conductor.

**Main vent:** The principal vent to which plumbing branch or fixtures vents are connected.

**Makeup:** Material added to a system for filling or adding to the operational level. Air may be supplied to a building to compensate for air lost or water may be added to a boiler for loss by escape or condensation.

**Malleable:** Capable of being shaped by hammering, rolling, or pressure.

**Manhole:** An opening through which entrance may be gained to a sewer, boiler, tank, or other enclosed area for inspection or repair.

**Manifold:** A main pipe or duct with many branches connected to it. The feeder line for multiple gas or oil burners.

**Manometer:** A U-shaped tube device used to balance pressure of gases against a column of mercury.

**Mass:** An assembly of matter that collectively makes one quantity. The quantity of matter in a body that is resistant to a force.

**Mechanical equivalent:** An energy ratio conversion in which 1 Btu equals 778.2 ft-lb.

**Mega:** Prefix for one million. One megavolt equals one million volts.

**Mercury switch:** A type of silent electrical switch in which a minute tube of mercury is used to complete a circuit or break it.

**Metabolism:** The physical and chemical processes by which a living organism converts other materials into living tissue, energy, and waste.

**Meter:** Metric unit of measure that equals 39.37 in. A device for measuring amounts of liquids, gas, or electricity. Various specialized instruments such as barometer, ammeter, voltmeter, manometer, used to measure energy, pressure, or other force.

**Methane:** The hydrocarbon gas $CH_4$, which is the principal component of natural gas.

**Micrometer (Micron):** The thousandth part of one millimeter in the metric system. Abbreviation: $\mu$m.

**Milli:** An expression of one one-thousandth part, such as millivolt, which is one one-thousandth of a volt.

**Modulation:** A control by increments rather than simply on or off. The control of sound regarding volume or clarity.

**Moisture content:** The amount of moisture contained in a material as compared to total dry condition, expressed as a percentage.

**Molecule:** The smallest part of a compound or element having all of the properties of the compound or element.

**Monel:** A low-corrosive metal compounded from a copper–nickel alloy. Used extensively for sinks, commercial vats, tanks and some pipe.

**Monochlorodifluoromethane:** A refrigerant better known as Freon 12 or R-22 and having the chemical formula $CHClF_2$.

**Multiple system:** A system in which several units are connected together. A refrigeration system may have multiple evaporators. A heating system may have multiple main pipes or duct runs.

## N

*National Electrical Code®:* Code of electrical rules and requirements for interior wiring, primarily concerned with fire resistance, overloading, and electrical safety.

**National Plumbing Code:** Code of plumbing rules and requirements primarily concerned with waste, venting, and water installations.

**Natural draft:** A circulation of air created by natural, nonmechanical means such as adjustment of doors, windows, dampers, chimneys, and similar objects.

**Natural gas:** Nonmanufactured fuel gas generally obtained in conjunction with the production of petroleum.

**NBFU:** National Bureau of Fire Underwriters. A national organization that establishes fire insurance rules and requirements for construction to meet such rules.

**Needle valve:** A valve with a very small orifice using a sharpened "needle-like" part to control flow. Much used in oil burners and similar equipment.

**Negative:** Opposite terminal in a battery from *positive*. Electrode with excessive electrons.

**Neoprene:** A synthetic rubber which is resistant to hydrocarbons, oil and gas.

**Net rating:** The quantity of heat available in Btu/hr for the connected load in a boiler.

**Neutron:** An electrically neutral part of an atom, having the weight of a proton.

**NFPA:** National Fire Protection Association.

**Nipple:** A length of pipe, usually less than 12 in., with male threads on both ends, and used to join pipe or fittings together.

**Nitrates:** Native sodium salts. To treat with nitric acid and change into a nitrate. Fertilizer consisting of sodium and potassium nitrate. A possible sewage by-product.

**Nonferrous:** A metal other than iron or steel. Nonferrous metals used in support systems are aluminum, copper, lead, and zinc.

**Nozzle:** A fitting at the end of a pipe or hose which causes the stream to become a variable spray as the nozzle control is operated. The end of an air jet used to clear dust or induce suction in a ventilating system.

## O

**Oakum:** Stranded or combed sisal fiber that is further treated with oil or other water-resistant material. Oakum is used for caulking a bell-and-spigot cast-iron pipe joint.

**Odorant:** A substance added to an otherwise odorless or tasteless gas to give warning of a leak of that gas.

**Ohm:** A unit of electrical resistance. The resistance that allows one ampere to flow with the potential of one volt. Abbreviation: $\Omega$.

**Ohm's law:** Current (amperes) is directly proportional to voltage and inversely proportional to resistance (ohms).

**Organic compound:** A chemical compound produced by a life process. Any compound containing carbon.

**Orifice:** An accurate-sized opening for controlling fluid flow.

**Outlet:** A connection in an electrical circuit wherever a switch, light fixture, convenience outlet, or similar device is included in the circuit. The end of a warm-air duct, usually covered by a grille or register.

**Output:** Capacity, performance rate, or net refrigeration.

**Overload:** Any load greater than the rated or designed capacity. Overload protection is provided by electric circuit breakers or fuses and by pressure-relief valves in liquid or gas systems.

**Oxygen:** A gas comprising approximately 21% of the atmosphere by volume, and 50% of all materials above the earth's crust. Its chemical symbol is O.

**Ozone:** An acrid, unstable gas created by electric discharges in the air or by high-energy solar radiation in the upper atmosphere.

## P

**Panel board:** A single unit or a group of units with fuses, switches, or circuit breakers designed to protect electric circuits against overloads. Usually installed in a cabinet on small projects or freestanding in industrial installations.

**Panel heat:** A type of heating system in which heat is transferred by radiation and convection from panel units in wall or ceiling. Hot-water and electric radiant heating are examples of panel heating.

**Parallel:** An electrical circuit designed with two or more paths to follow. A system where flow is divided between two or more channels equally. See *series*.

**Particulate:** Airborne material that has a relatively fixed shape and volume such as dust, mist, or smoke.

**Pascal's law:** The pressure imposed on a fluid is transmitted equally in all directions.

**Pasteurization:** A process of controlled heating of beer and milk at about 130 to 150°F which destroys bacteria and arrests spoiling but does not change the flavor. Discovered by Louis Pasteur (1822–1895), a French scientist.

**Pathogens:** Disease-producing bacteria generated when proper conditions exist to expose raw sewage to free air.

**Peak load:** The maximum load carried by equipment or a system over an extended period of time.

**Performance factor:** The ratio of input energy to work or other output of a system or equipment.

**Perm:** The unit for measuring permanence. One perm equals one grain per square foot (per hour or per inch of mercury vapor difference).

**Petcock:** A small valve often used as a drain cock or air control exhaust valve. Usually used with copper tubing.

**Phase:** A distinguishable part of a cycle, as one of the states of a liquid, solid, gas, or electric current. Phases usually have a time frequency, heat requirement, or other catalyst. Angular relationship between waves and expressed in degrees.

**Phosphates:** Salts of phosphoric acid.

**pH value:** A measure of the hydrogen-ion concentration of a substance to determine its acidity or alkalinity. A pH value of 6 or less indicates acidity, 7 indicates neutrality, and 8 indicates alkalinity.

**Pilot light:** A small, continuously burning flame which acts to ignite other, larger burners.

**Plenum:** A chamber or space used to conduct low-pressure air. Enclosed spaces above the ceiling are often used as warm-air plenum chambers.

**Plug:** A stopper. In a plumbing system, a plug is a metal object several inches long with male threads on the outside, which is screwed into a pipe to close it. A *cleanout* is a plumbing plug installed at the ends of the sewer lines to facility cleaning of the pipe.

**Polarity:** The electrical condition indicating positive or negative, magnetic north pole and south pole, or direction of current flow.

**Polyethylene:** A plastic used for manufacturing pipe for gas installations. A sheet plastic used for damp-proofing.

**Polymer:** A product resulting from a chemical change involving the successive addition of a large amount of relatively small molecules.

**Polyphase:** Multiphase. Two or more alternating electric currents that differ in phase.

**Polystyrene:** A plastic often used as an insulating material.

**Positive:** Having a relatively high potential. Having a deficiency of electrons. Opposite of negative.

**Potable water:** Water fit for human consumption. Water that is not contaminated.

**Power:** The rate of doing work. The time rate at which energy is transferred.

**Pressure:** A force distributed over a surface, expressed as units of force per unit area. The normal force exerted by a homogeneous liquid or gas on the walls of a container per unit of area.

**Primary air:** Combustion air mixed with the gas before it reaches the burner and expressed as a percent of combustion air.

**Primary voltage:** The voltage on the supply side of a transformer, almost always the high-voltage side.

**Privy:** An early method of sewage disposal consisting of a hole in the earth covered by a small building that contains seating facilities.

**Propane:** Hydrocarbon gas fuel obtained from crude petroleum and natural gas.

**Psi:** Pounds per square inch.

**Psia:** Pounds per square inch absolute, which equals gauge pressure plus atmospheric pressure.

**Psig:** Pounds per square inch gauge, which means the pressure actually shown on the pressure gauge.

**Psychrometer:** An instrument including a wet-bulb thermometer and a dry-bulb thermometer and used to determine the relationship between the two.

**Purging:** Cleaning of pipes, especially refrigerant lines, by compressed air to remove contaminants.

**PVC:** Polyvinyl chloride, a plastic used in manufacture of pipe and many other thermosetting products.

**Pyrometer:** An instrument used to measure high temperatures.

**Q**

**Quarter bend:** A plumbing pipe, usually cast iron sewer pipe, which makes a 90° turn. Termed an elbow in water piping.

**Quench:** To cool a piece of hot metal quickly by immersion in cold water.

**R**

**Radiant heat:** A type of heat obtained by direct radiation from panels in the wall, ceiling, or floor. Hot water in small pipes may be used in all locations, and electric resistance wiring is used in wall or ceiling installations.

**Radiation:** The process of sending out rays from a central source. Radiation in connection with heating means the stream of heat given off by a radiator into a clear space around it.

**Radiator:** The room heating element connected to the heat source. Radiators for steam or hot-water systems are often freestanding cast-iron units. Radiators for electric heat are usually resistance coils.

**Rankin scale:** Name given to the absolute temperature scale, on which zero equals −460°F.

**Reactance:** Opposition to alternating electric current by either inductance or capacitance, or both. Indicated by the symbol $X$ and measured in *ohms.*

**Recovery rate:** Time required for a water heater to heat cold water to the desired, or calculated, temperature.

**Rectifier:** An electrical device that changes alternating current into direct current. A refrigeration externally cooled heat exchanger on the high side for condensing absorbent before passing refrigerant to the condenser.

**Reed valve:** A thin metal valve operated solely by pressure. Same as flapper valve.

**Refrigerant:** A substance that produces a refrigeration effect by absorption of heat during expansion or vaporization as it changes from a liquid to a gas, and releases its heat by condensing from a gas back to a liquid. Most refrigerants have long chemical names which have been simplified by using R-number designations, as follows:

| | |
|---|---|
| R-11 | Trichloromonofluoromethane: a low-pressure chemical |
| R-12 | Dichlorodifluoromethane: Freon 12 |
| R-22 | Monochlorodifluoromethane: synthetic chemical |
| R-40 | Methyl chloride: early-1930s refrigerant |
| R-113 | Trichlorotrifluoromethane: synthetic chemical |
| R-160 | Ethyl chloride: obsolete chemical refrigerant |
| R-170 | Ethane: low-temperature refrigerant |
| R-290 | Propane: low-temperature refrigerant; gas fuel |
| R-600 | Butane: low-temperature refrigerant; gas fuel |
| R-611 | Methyl format: low-pressure refrigerant |
| R-717 | Ammonia: acrid industrial refrigerant |
| R-764 | Sulfur dioxide: obsolete low-pressure refrigerant |

**Relay:** Electrical device that uses small current in the control circuit to operate a valve or switch in the operating current.

**Relief valve:** A safety valve. A valve designed to open before a dangerous condition of overpressure or overtemperature develops in a piece of equipment.

**Resin:** An organic compound, usually synthetic, which is used as a base material or binder in the manufacture of another product.

**Resistance:** Opposition to current flow by a physical conductor when electric current is applied. Any force tending to hinder motion.

**Return:** A pipe, wire, duct, or other device used to carry a substance from a terminal position back to its original source.

**Rheostat:** An adjustable electrical resistor.

**Ringelmann scale:** A device for measuring smoke density.

**Riser:** The vertical portion of a system. Risers usually extend from a horizontal main (water, sewer, or vent) upward to other floors or the roof. Electric risers usually extend from a main panel to secondary panels.

**Rough-in:** The earliest stages of construction of a system. This stage usually includes installation of work to be concealed later, such as water and sewer pipe, ducts, electric conduit and boxes, and some equipment, and is often the major portion of the entire installation.

## S

**Saddle fitting:** A fitting that "saddles" an existing pipe when a new branch is to be added and no existing plug or fitting is available. A hole is made in the existing pipe and the saddle fitting installed over it with clamps.

**Safety valve:** A relief valve to prevent over pressure or temperature.

**Sanitary sewer:** A common name for the plumbing system designed to carry away sewage. A *storm drain* is not a sanitary sewer.

**Saponify:** To create a soapy compound by accidentally combining oil with alkali or alkaline refrigerant.

**Saturation:** A condition where a substance contains the maximum amount of another substance at a given temperature and pressure.

**Seal:** Any material used to firmly close a pipe, crack, or passage. A plumbing seal provided by the water in a trap. A material used to close a building crack. A device or material used to prevent leakage.

**Sensible heat:** Heat that causes a change in temperature of a substance.

**Septic tank:** A watertight tank in a private waste disposal system which receives sewage and reduces the sewage to sludge and water by the *anaerobic* action of bacteria.

**Series:** An arrangement where related things are placed in succession. Electrical devices in which the current flows through each in order. A heating system where hot water or steam passes through each radiator one after the other.

**Service drop:** The portion of an electrical system from the last pole to the first connection at the building.

**Sewage:** Any liquid waste that contains animal or vegetable matter in suspension. Water and waterborne waste in a waste-disposal system.

**Shunt:** An electrical device connected in parallel with another part.

**Sill cock:** A faucet outside a building to which a hose may be attached. Also called a hose bibb.

**Sink:** A kitchen plumbing fixture used in the preparation of food and cleaning. A depository for heat generated by a solar collector.

**Sludge:** The semisolid waste developed in a septic tank by removal of water. Pasty refuse.

**Soil pipe:** Any pipe that carries sewage. Cast-iron bell-and-spigot pipe.

**Soil stack:** The vertical sewage pipe of a drain-waste-vent plumbing system.

**Solenoid:** An electrical device consisting of a plunger actuated by a magnetic coil.

**Specific gravity:** Weight of a substance compared to water with an assigned weight of 1.0.

**Specific heat:** Ratio of the quantity of heat required to raise the temperature of a substance one degree to that required to raise an equal mass of water one degree.

**Squirrel cage:** A type of fan where the blades are arranged in a circle and perpendicular to the air flow.

**Standard air density:** The density of air at *standard conditions.* Density equals 0.075 lb/ft.

**Standard atmosphere:** The condition where air equals 29.92 in. of mercury at 68°F.

**Standard conditions:** Used in air-conditioning and ventilation where the temperature of the air is assumed to be at 68°F, the relative humidity at 50%, and the pressure at 29.92 in. of mercury.

**Stator:** The stationary part of a motor.

**Steam:** Water in a vapor state, above 212°F.

**Storm drain:** A piping system designed to collect and carry off rainwater and similar surface or subsurface water. Not a sewer.

**Sublimation:** The condition where a substance changes from a solid to a gas without becoming a liquid.

**Sump:** A pit constructed below the low point of a system that handles liquids. The sump is not connected to a drain but is occasionally pumped dry.

**Surge tank:** A tank that accommodates increases in liquid by compressing the enclosed air and later releasing it as the volume of liquid decreases.

**Switch:** A mechanical device designed to make or break an electrical circuit. Any device that can divert a substance from one position or use to another.

## T

**Tee:** A pipe or duct fitting shaped like a letter "T" that is usually used to form a branch from a main.

**Teflon:** A synthetic rubberlike material used for packing.

**Temperature:** Measurement of heat or cold in degrees; measurement of speed of motion of molecules. One degree Fahrenheit equals 5/9 degree Celsius. See *Celsius* and *Fahrenheit.*

**Terminal:** The point in an electrical circuit at which it is started or stopped. The connections on a battery.

**Therm:** A heat unit with a value of 100,000 Btu.

**Thermal conductivity:** The ability of a material to transmit heat. The reciprocal of *thermal resistance.*

**Thermal radiation:** Transmission of heat from a warm surface to a cooler surface by invisible electromagnetic waves.

**Thermal resistance:** Resistance to the transmission of heat. Insulating materials have high thermal resistance while metals have low thermal resistance.

**Thermistor:** A material that has qualities between an insulator and a conductor (semiconductor) and has electrical resistance that varies with temperature.

**Thermocouple:** A device consisting of two different metals welded together, then heated to produce an electric current at open ends.

**Thermodynamics:** The science that deals with relations of heat and work.

**Thermometer:** A device that indicates *ambient temperature.*

**Thermoplastic:** A plastic capable of being repeatedly softened and hardened by temperature.

**Thermoset:** A plastic which, when properly cured, cannot be changed by heat or chemical means.

**Thermostat:** A temperature-sensing device that automatically controls an energy device.

**Ton-day of refrigeration:** Heat removed by 1 ton of refrigeration per day. Approximately equal to *latent heat* in melting 1 ton of ice at 32°F, or 288,000 Btu.

**Tons of refrigeration:** Useful refrigeration equal to 12,000 Btu/hr or 200 Btu/min.

**Transducer:** A device actuated by power from one system and supplies power in another form to a second system.

**Transformer:** An electrical device that may increase or decrease voltage on one side over line voltage on the other side. Consists of an iron core with windings on each side. Transformers are usually used for doorbells, chimes, and low-voltage controls with voltage requirements from 6 to 30 V. When voltage decreases, the current increases.

**Transmission:** The system of conductors used to carry electricity from the generating plant to the distribution or substation.

**Trap:** A plumbing fitting connected in the drain line of a fixture (water closets have a built-in trap) to prevent sewer gas from entering the building from the sewer. The usual trap (P trap) provides a water barrier in the loop, continuously maintained by each drain.

**Tubing:** Thin-wall pipe, usually copper, that is connected with flared or soldered fittings.

U

**Ultraviolet:** Invisible radiation rays having wavelengths shorter than those of visible violet but longer than those of x-rays.

**Underwriters' Laboratories, Inc.:** Laboratories maintained for the testing of electrical equipment, safety devices, and equipment. Abbreviation: UL.

**Union:** A plumbing fitting used to join two pipes that may also be disconnected without cutting. The unit consists of two parts connected to the pipe ends and connected together by a threaded ring at the meeting of the two end pieces.

**Updraft:** Low pressure at upper terminal of a smokestack which tends to increase velocity and volume of the smoke.

**Upfeed system:** A water or heating system in which the main supply line is below the distribution units and the feed is upward.

**Urethane:** A synthetic chemical compound that may be foamed as an insulation. A synthetic substitute for varnish.

**Urinal:** A plumbing fixture designed for the reception of urine. Usually installed in buildings other than residential and may be wall-hung or floor-supported.

**Utility gas:** Liquefied petroleum gas, manufactured gas, or natural gas or their mixtures.

V

**Vacuum:** A space devoid, or nearly devoid, of matter. A reduction of the pressure in a space below atmospheric pressure.

**Valve:** A device for controlling the flow of fluids. *Gate valves*, *globe valves*, and *check valves* are the most common types among the thousands available.

**Vapor:** The gaseous form of any substance that, under other conditions of pressure, temperature, or both, is a solid or a liquid.

**Vapor lock:** Blockage of a pipeline when liquid is trapped in the line and vapor cannot flow.

**Vector:** A line that represents the direction and magnitude of a quantity.

**Velocity:** A vector representing the rate of change of position of a body. The speed at which a material passes a given point measured as feet per second or feet per minute.

**Vent:** A plumbing pipe system designed to provide an avenue of escape for sewer gas to the outside air. A pipe or chimney from a unit heater to carry off any escaping gas fumes.

**Viscosity:** The property of a fluid to resist flow, or to change the arrangement of its molecules. A measurement of the ability of a fuel oil to flow.

**Vitrified clay pipe:** Pipe formed of clay, then fired for permanence. Used in septic-tank disposal fields, foundation drains, and similar locations.

**Volatile liquid:** A liquid that evaporates at low temperature and pressure.

**Volt:** The electromotive force (EMF) that will send one ampere through one ohm of resistance; the unit of electrical potential or pressure. Abbreviation: V. Indicated as *E* or *V* in calculations.

**Volt-ampere:** Volts multiplied by amperes. Abbreviation: VA.

**Volume:** A measure of quantity in terms of space occupied and designated by cubic inches, cubic feet, cubic centimeters, liters, quarts, pints, and similar terms.

**VTR:** Drawing indication for "vent through roof."

W

**Waste:** Liquid that does not contain fecal matter and is discharged from a plumbing fixture, condensing system, or similar equipment.

**Water closet:** A plumbing fixture designed to receive human excrement and discharge it into the sewer system with a charge of water. Erroneously called a toilet.

**Water column:** A unit expressing pressure. One inch of water column equals a pressure of 0.578 oz/in.$^2$.

**Water conditioner:** A plumbing device used to remove dissolved minerals from water. A water "softener" that reduces the amount of calcium and other minerals in water.

**Water hammer:** The banging sound in water piping when valves are operated with vigor. See *antihammer device*.

**Water tower:** A structure designed to cool refrigerant by allowing cooler water to flow over the piping arranged in a "tower."

**Watt:** The electrical unit of power or the rate of energy transfer equivalent to one ampere flowing under the pressure of one volt. Abbreviation: W. One *horsepower* equals 746 W. Watts are indicated as *W* in general electrical formulas and as *P* in power calculations.

**Wet bulb:** Part of a *psychrometer* and used together with a dry-bulb thermometer to determine the relative humidity in the air.

**Wet cell:** A type of battery in which combinations of chemical plates are immersed in *electrolyte* liquid to generate direct electrical current.

**Wet vent:** A plumbing vent pipe that serves as both a drain and vent.

## Y

**Yoke:** A collar by which a lead trap is secured. A strap that is bolted around a vertical pipe to hold it from falling.

## Z

**Zone:** An area of a building in which a portion of the heating system operates.

**Zone system:** A heating system divided into several zones, each controlled by a separate thermostat or control valve, to permit independent control of room or area.

# Bibliography

## GENERAL


Andrews, F. I., *Architects Guide to Mechanical Systems.* New York, N.Y.: Reinhold Publishing Corporation, 1966.

Brady, Constance, *Right Where You Live.* Millbrae, Calif.: Celestial Arts Press, 1979.

Glorioso, Robert M., and Hill, Francis, *Introduction to Engineering.* Englewood Cliffs, N.J.: Prentice-Hall, Inc., 1975.

Goodban, William T., and Hayslett, Jack J., *Architectural Drawing and Planning.* New York, N.Y.: McGraw-Hill Book Company, 1972.

Griffin, C. W., *Energy Conservation in Buildings.* Washington, D.C.: Construction Specifications Institute, 1974.

Kenward, Michael, *Potential Energy.* Cambridge, England: Cambridge University Press, 1976.

Landsberg, Jack R., and Stewart, Ronald, *Improving Energy Efficiency in Buildings.* Albany, N.Y.: State University of New York Press, 1980.

Lewis, Jack R., *Architectural Draftsman's Reference Handbook.* Englewood Cliffs, N.J.: Prentice-Hall, Inc., 1982.

Merkel, James A., *Basic Engineering Principles.* Westport, Conn.: AVI Publishing Co. Inc., 1974.

Olin, Harold, Schmidt, John I., and Lewis, Walter, *Construction Principles, Materials and Methods.* Chicago, Ill.: Interstate Press, 1980.

Schwenk, Harry C., Shannon, Robert H., and Skrotzki, B. G. A., *Nuclear Power Engineering.* New York, N.Y.: McGraw-Hill Book Company, 1957.

Seale, R. L., and Sierka, R. A., *Energy Needs and the Environment.* Tucson, Ariz.: The University of Arizona Press, 1973.

Spence, William P., *Architecture, Design Engineering Drawing.* Bloomington, Ill.: McKnight Publishing Co., 1979.

Watson, Donald, *Energy Conservation through Building Design.* New York, N.Y.: McGraw-Hill Book Company, 1979.

## ELECTRICAL


Andersen, Edwin P., *Wiring Diagrams for Lighting and Power*. Indianapolis, Ind.: Theodore Audel & Co., 1975.

Crow, Leonard, *Learning Electrical Essentials.* Indianapolis, Ind.: Howard W. Sams & Co., Inc., Publishers, 1975.

Dagostino, Frank, *Mechanical and Electrical Systems in Construction and Architecture.* Reston, Va.: Reston Publishing Co., Inc., 1978.

Key, Eugene, *Principles of Electricity*. New York, N.Y.: Barnes & Noble Books, 1975.

McGuiness, W. J., and Stein, Benjamin, *Mechanical and Electrical Equipment for Buildings.* New York, N.Y.: John Wiley & Sons, Inc., 1980.

Mix, Floyd M., *House Wiring Simplified*. South Holland, Ill.: Goodheart-Wilcox Co., Inc., 1981.

Mullin, Ray C., *Electric Wiring—Residential*. Albany, N.Y.: Delmar Publishers, Inc., 1981.

*National Electric Code®*. Quincy, Mass.: National Fire Protection Association, 1981.

Shrader, Robert L., *Electrical Transmission and Distribution.* New York, N.Y.: McGraw-Hill Book Company, 1969.

Skrotzki, Bernhardt, *Electrical Transmission and Distribution.* New York: N.Y.: McGraw-Hill Book Company, 1954.

Timbie, William H., and Pike, Arthur L., *Essentials of Electricity*, 3rd ed. New York, N.Y.: John Wiley & Sons, Inc., 1963.

Waddicor, H., *Principles of Electrical Power Transmission.* London, England: Chapman & Hall Ltd., 1964.


## PLUMBING


Blankenbaker, E. K., *Modern Plumbing*. South Holland, Ill.: Goodheart-Willcox Company, 1978.

Fair, Gordon, Geyer, John, and Okun, Daniel A., *Water and Wastewater Engineering.* New York, N.Y.: John Wiley & Sons, Inc., 1966.

Nielsen, Louis S., *Standard Plumbing Engineering Design*. New York, N.Y.: McGraw-Hill Book Company, 1963.

Powell, Sheppard, *Water Conditioning for Industry*. New York, N.Y.: McGraw-Hill Book Company, 1954.

Sanks, Robert L., *Water Treatment Plant Design*. Ann Arbor, Mich.: Ann Arbor Science Publishers, 1978.

Slater, Harry, and Smith, Lee, *Basic Plumbing*. Albany, N.Y.: Delmar Publishers, Inc., 1979.

Sunset Magazine, *Basic Plumbing Illustrated*. Menlo Park, Calif.: Lane Publishing Co., 1975.

Tebbutt, T. H. Y., *Water Science and Technology.* London, England; John Murray Ltd., 1973.

Uniform Plumbing Code. Los Angeles, Calif.: International Association of Plumbing and Mechanical Officials, 1979.


## HVAC


Air Conditioning and Refrigeration Institute, *Refrigeration and Air Conditioning*. Englewood Cliffs, N.J.: Prentice-Hall, Inc., 1979.

Althouse, Andrew D., Turnquist, Carl, and Bracciano, Alfred, *Modern Refrigeration and Air Conditioning*. South Holland, Ill.: Goodheart-Willcox Company, 1979.

Brumbaugh, J. E., *Heating, Ventilating and Air-Conditioning Library*. Indianapolis, Ind.: Howard W. Sams & Co., Inc., Publishers, 1976.

Dixon, A. E., and Leslie, J. D., *Solar Energy Conversion*. Toronto, Canada: Pergamon Press, 1979.

Gunther, Raymond C., *Refrigeration, Air Conditioning and Cold Storage*. Philadelphia, Pa.: Chilton Book Company, 1969.

Hemeon, W. C. L., *Plant and Process Ventilation*. New York, N.Y.: Industrial Press, Inc., 1955.

Lang, V. Paul, *Principles of Air Conditioning*. Albany, N.Y.: Delmar Publishers, Inc., 1979.

McDermott, Henry J., *Handbook for Ventilation for Contaminant Control*. Ann Arbor, Mich.: Ann Arbor Science Publishers, 1976.

Weaver, Michael, and Kirkpatrick, James M., *Environmental Control*. New York, N.Y.: Harper & Row, Publishers, Inc., 1974.

## MISCELLANEOUS

Adler, Rodney R., *Vertical Transportation for Buildings*. New York, N.Y.: Elsevier Publishing Co., Inc., 1970.

American Public Works Association, *Solid Waste Collection Practice*. Chicago, Ill.: APWA, 1975.

Baum, Richard, and Parker, Charles, *Solid Waste Disposal*. Ann Arbor, Mich.: Ann Arbor Science Publishers, 1973.

*California Solar Information Packet*. Sacramento, Calif.: California Energy Commission, 1978.

*Common Sense Wind Energy*. Sacramento, Calif.: California Energy Commission, 1981.

Freese, Stanley, *Windmills and Millwrights*. Cambridge, England: Davis & Charles, 1974.

Niles, Philip, and Haggard, Kenneth, *Passive Solar Handbook*. Sacramento, Calif.: California Energy Commission, 1980.

Smith, Paul Slee, *Recycling Waste*. Bruzley, England: Scientific Publishing Ltd., 1976.

Strakosch, George, *Elevators and Escalators*. New York, N.Y.: John Wiley & Sons, Inc., 1967.

# Appendix

## Engineering Societies and Associations

Acoustical Society of America
335 East 45th Street, New York, NY 10017

Air-Conditioning and Refrigeration Institute
1815 North Fort Meyer Drive, Arlington, VA 22208

Air Conditioning Contractors of America
1228 17th Street NW, Washington, DC 20036

American Boiler Manufacturers Association
1500 Wilson Boulevard, Suite 700, Arlington, VA 22209

American Consulting Engineers Council
1155 15th Street NW, Washington, DC 20005

American Gas Association
1515 Wilson Boulevard, Arlington, VA 22209

American Institute of Plant Engineers
3975 Erie Avenue, Cincinnati, OH 45208

American National Standards Institute
1430 Broadway, New York, NY 10018

American Society for Testing and Materials
1916 Race Street, Philadelphia, PA 19103

American Society of Gas Engineers
290 Temple Avenue, Long Beach CA 90803

American Society of Heating, Refrigerating, and Air-Conditioning Engineers, Inc.
1791 Tullie Circle NE, Atlanta, GA 30329; United Engineering Center, 345 47th Street, New York, NY 10017

American Society of Plumbing Engineers
15233 Ventura Boulevard, Sherman Oaks, CA 91403

Architectural Acoustics Society
464 Armour Circle, Atlanta, GA 30324

Associated Builders and Contractors, Inc.
444 North Capitol Street NW, Suite 409, Washington, DC 20001

Associated General Contractors of America
1857 E Street, NW, Washington, DC 20006

Canadian Institute of Plumbing and Heating
784 Plymouth Avenue, Suite 305, Montreal, Quebec, Canada

Commercial Refrigerator Manufacturers Association
1101 Connecticut Avenue, Suite 700, Washington, DC 20038

Construction Specifications Institute of Canada
1027 Vonge Street, Suite 105, Toronto, Ontario, Canada

Construction Specifications Institute
601 Madison Street, Alexandria, VA 22314

Edison Electric Institute
1111 18th Street NW, Washington, DC 20036

Heating, Refrigeration and Air-Conditioning Institute of Canada
5468 Dundas Street, Suite 226, West Islington, Ontario, Canada

Illuminating Engineering Society of North America
345 East 47th Street, New York, NY 10017

International Association of Plumbing and Mechanical Officials
5032 Alhambra Avenue, Los Angeles, CA 90032

International Solar Energy Society
P.O. Box 26, Highett, Victoria, Australia

Mechanical Contractors Association of America
5530 Wisconsin Avenue NW, Suite 750, Washington, DC 20015

Mechanical Constractors Association of Canada
116 Albert Street, Suite 408, Ottawa, Ontario, Canada

National Electrical Contractors Association
7315 Wisconsin Avenue, Washington, DC 20014

National Electrical Manufacturers Association
2101 L Street NW, Washington, DC 20037

National Fire Protection Association
Batterymarch Park, Quincy, MA 02269

National Society of Professional Engineers
2029 K Street NW, Washington, DC 20006

Plumbing, Heating, Cooling Information Bureau
111 East Wacker Drive, Suite 1407, Chicago, IL 60601

Producers Council
1717 Massachusetts Avenue NW, Washington, DC 20036

Refrigeration Research Foundation
7315 Wisconsin Avenue, Bethesda, MD 20014

Sheet Metal and Air-Conditioning Contractors National Association
8224 Old Courthouse Road, Tysons Corner, Vienna, VA 22180

Solar Energy Industries Association, Inc.
1001 Connecticut Avenue NW, Washington, DC 20036

# ELECTRICAL CONDUIT AND WIRING

| CONDUIT INFORMATION | | | | | | | | | |
|---|---|---|---|---|---|---|---|---|---|
| | NORMAL EXTERIOR DIAMETER | | | NORMAL INTERIOR DIAMETER | | | WEIGHT LBS. PER 100 LIN. FT. | | |
| TRADE SIZE INCHES | STEEL CONDUIT | ALUM. CONDUIT | ELECT. MET. TUBE | STEEL CONDUIT | ALUM. CONDUIT | ELECT. MET. TUBE | STEEL CONDUIT | ALUM. CONDUIT | ELECT. MET. TUBE |
| ½ | 0.840 | 0.840 | 0.706 | 0.622 | 0.622 | 0.622 | 79.0 | 27.4 | 28.5 |
| ¾ | 1.050 | 1.050 | 0.922 | 0.824 | 0.824 | 0.824 | 105.0 | 36.4 | 43.5 |
| 1 | 1.315 | 1.315 | 1.163 | 1.049 | 1.049 | 1.049 | 153.0 | 53.0 | 64.0 |
| 1¼ | 1.660 | 1.660 | 1.510 | 1.380 | 1.380 | 1.380 | 201.0 | 69.6 | 95.0 |
| 1½ | 1.900 | 1.900 | 1.740 | 1.610 | 1.610 | 1.610 | 249.0 | 86.2 | 110.0 |
| 2 | 2.375 | 2.375 | 2.197 | 2.067 | 2.067 | 2.067 | 332.0 | 115.7 | 140.0 |
| 2½ | 2.875 | 2.875 | 2.875 | 2.469 | 2.469 | 2.731 | 527.0 | 182.5 | 220.0 |
| 3 | 3.500 | 3.500 | 3.500 | 3.068 | 3.068 | 3.356 | 682.6 | 238.9 | 260.0 |
| 3½ | 4.000 | 4.000 | 4.000 | 3.548 | 3.548 | 3.834 | 831.0 | 287.7 | 365.0 |
| 4 | 4.500 | 4.500 | 4.500 | 4.026 | 4.026 | 4.334 | 972.3 | 340.0 | 400.0 |
| 5 | 5.563 | 5.563 | | 5.047 | 5.047 | | 1313.6 | 465.4 | |
| 6 | 6.625 | 6.625 | | 6.065 | 6.065 | | 1745.3 | 612.5 | |

| WIRE INFORMATION | | | | | | | |
|---|---|---|---|---|---|---|---|
| TYPE LETTER | TRADE NAME | MAX. OP. TEMP. | USE | TYPE LETTER | TRADE NAME | MAX. OP. TEMP. | USE |
| AF | ASBESTOS HEAT RESIST. | 150°C 302°F | FIXTURE | SA | SILICONE ASBESTOS | 90°C 194°F | DRY |
| AVA | ASBESTOS VAR. CAMBRIC | 110°C 230°F | DRY | SF-1 | SILICONE RUBBER | 200°C 392°F | FIXTURE 300V. MAX. |
| AVB | ASBESTOS VAR. CAMBRIC | 90°C 194°F | DRY | SF-2 | SILICONE RUBBER | 200°C 392°F | FIXTURE 7 STRAND |
| AVL | ASBESTOS VAR. CAMBRIC | 110°C 230°F | WET & DRY | T | THERMOPLASTIC | 60°C 140°F | DRY |
| MI | MINERAL INSULATED | 85°C 185°F | WET & DRY | TA | THERMOPLASTIC ASBESTOS | 90°C 194°F | SW'BD. |
| RF-1 | RUBBER COVER | 60°C 140°F | FIXTURE 300V MAX. | TF | THERMOPLASTIC SOLID OR STRAND | 60°C 140°C | FIXTURE |
| RF-2 | RUBBER COVER | 60°C 140°F | FIXTURE 7 STRAND | TFN | THERMOPLASTIC HEAT RESIST. | 90°C | FIXTURE |
| RFH-1 | RUBBER COVER HEAT RESIST. | 75°C 167°F | FIXTURE 300V. MAX. | THHN | THERMOPLASTIC HEAT RESIST. | 90°C 194°F | DRY |
| RFH-2 | RUBBER COVER HEAT RESIST. | 75°C 167°F | FIXTURE 7 STRAND | THW | THERMO, MOIST. & HEAT RESIST. | 75°C 167°F | WET & DRY |
| RH | RUBBER COVER HEAT RESIST. | 75°C 167°F | DRY | THWN | THERMO, MOIST. & HEAT RESIST. | 75°C 167°F | WET & DRY |
| RHH | RUBBER COVER HEAT RESIST. | 90°C 194°F | DRY | TW | THERMOPLASTIC MOIST. RESIST. | 60°C 140°F | WET & DRY |
| RHW | RUBBER, MOIST, & HEAT RESIST. | 75°C 167°F | WET & DRY | V | VARNISH CAMBRIC | 85°C 185°F | DRY |
| RUH | LATEX HEAT RESIST. | 75°C | DRY | XRHW | THERMO.-POLY X-LINK-RESIST. | 75°C 167°F | WET & DRY |
| RUW | LATEX MOIST. RESIST. | 60°C 146°F | WET & DRY | | PAPER COVER | 85°C 185°F | U'GRND. |

# ELECTRICAL CONDUIT AND WIRING

NEW WORK OR REWIRING: TYPES RF-2, RFH-2, R, RH, RW, RHN, RHW, RH-RW
NEW WORK: FEP, FEBP, RUN, RUW, T, TF, THHN, THW, THWN, TW

| WIRE SIZE | MAXIMUM NUMBER OF CONDUCTORS IN CONDUIT AND TUBING | | | | | | | | | | | |
|---|---|---|---|---|---|---|---|---|---|---|---|---|
| | 1/2 | 3/4 | 1 | 1 1/4 | 1 1/2 | 2 | 2 1/2 | 3 | 3 1/2 | 4 | 5 | 6 |
| 18 | 7 | 12 | 20 | 35 | 49 | 80 | 115 | 176 | | | | |
| 16 | 6 | 10 | 17 | 30 | 41 | 68 | 98 | 150 | | | | |
| 14 | 4 | 6 | 10 | 18 | 25 | 41 | 58 | 90 | 121 | 155 | | |
| 12 | 3 | 5 | 8 | 15 | 21 | 34 | 50 | 76 | 103 | 132 | 208 | |
| 10 | 1 | 4 | 7 | 13 | 17 | 29 | 41 | 64 | 86 | 110 | 173 | |
| 8 | 1 | 3 | 4 | 7 | 10 | 17 | 25 | 38 | 52 | 67 | 105 | 152 |
| 6 | 1 | 1 | 3 | 4 | 6 | 10 | 15 | 23 | 32 | 41 | 64 | 93 |
| 4 | 1 | 1 | 1 | 3 | 5 | 8 | 12 | 17 | 24 | 31 | 49 | 72 |
| 3 | | 1 | 1 | 3 | 4 | 7 | 10 | 16 | 21 | 28 | 44 | 63 |
| 2 | | 1 | 1 | 3 | 3 | 6 | 9 | 14 | 19 | 24 | 38 | 55 |
| 1 | | 1 | 1 | 1 | 3 | 4 | 7 | 10 | 14 | 18 | 29 | 42 |
| 0 | | | 1 | 1 | 2 | 4 | 6 | 9 | 12 | 16 | 25 | 37 |
| 00 | | | 1 | 1 | 1 | 3 | 5 | 8 | 11 | 14 | 22 | 32 |
| 000 | | | 1 | 1 | 1 | 3 | 4 | 7 | 9 | 12 | 19 | 27 |
| 0000 | | | | 1 | 1 | 2 | 3 | 6 | 8 | 10 | 16 | 23 |
| 250 | | | | 1 | 1 | 1 | 3 | 5 | 6 | 8 | 13 | 19 |
| 300 | | | | 1 | 1 | 1 | 3 | 4 | 5 | 7 | 11 | 16 |
| 350 | | | | 1 | 1 | 1 | 1 | 3 | 5 | 6 | 10 | 15 |
| 400 | | | | | 1 | 1 | 1 | 3 | 4 | 6 | 9 | 13 |
| 500 | | | | | 1 | 1 | 1 | 3 | 4 | 5 | 8 | 11 |

REWIRING: TYPES TF, T, THW, TW, RUH, RUW

| WIRE SIZE | MAXIMUM NUMBER OF CONDUCTORS IN CONDUIT OR TUBING | | | | | | | | | | | |
|---|---|---|---|---|---|---|---|---|---|---|---|---|
| | 1/2 | 3/4 | 1 | 1 1/4 | 1 1/2 | 2 | 2 1/2 | 3 | 3 1/2 | 4 | 5 | 6 |
| 18 | 13 | 24 | 38 | 68 | 93 | 152 | | | | | | |
| 16 | 11 | 19 | 31 | 55 | 75 | 123 | 176 | 270 | | | | |
| 14 | 5 | 10 | 16 | 29 | 40 | 65 | 93 | 143 | 192 | | | |
| 12 | 4 | 8 | 13 | 24 | 32 | 53 | 76 | 117 | 157 | 202 | | |
| 10 | 4 | 6 | 11 | 19 | 26 | 43 | 61 | 95 | 127 | 163 | 257 | |
| 8 | 1 | 4 | 6 | 11 | 15 | 25 | 36 | 56 | 75 | 96 | 152 | 219 |
| 6 | 1 | 2 | 4 | 7 | 10 | 16 | 23 | 36 | 48 | 62 | 97 | 141 |
| 4 | 1 | 1 | 3 | 5 | 7 | 12 | 17 | 27 | 36 | 46 | 73 | 106 |
| 3 | 1 | 1 | 2 | 4 | 6 | 10 | 15 | 23 | 31 | 40 | 63 | 91 |
| 2 | 1 | 1 | 1 | 4 | 5 | 9 | 15 | 20 | 27 | 34 | 54 | 78 |
| 1 | | 1 | 1 | 2 | 4 | 6 | 9 | 14 | 19 | 25 | 39 | 57 |
| 0 | | 1 | 1 | 2 | 3 | 5 | 8 | 12 | 16 | 21 | 33 | 48 |
| 00 | | 1 | 1 | 1 | 3 | 4 | 7 | 10 | 14 | 18 | 28 | 41 |
| 000 | | | 1 | 1 | 2 | 4 | 5 | 9 | 12 | 15 | 24 | 35 |
| 0000 | | | 1 | 1 | 1 | 3 | 5 | 7 | 10 | 13 | 20 | 29 |
| 250 | | | | 1 | 1 | 2 | 4 | 6 | 8 | 10 | 16 | 23 |
| 300 | | | | 1 | 1 | 2 | 3 | 5 | 7 | 9 | 14 | 20 |
| 350 | | | | 1 | 1 | 1 | 3 | 4 | 6 | 8 | 12 | 18 |
| 400 | | | | | 1 | 1 | 2 | 4 | 5 | 7 | 11 | 16 |
| 500 | | | | | 1 | 1 | 1 | 3 | 4 | 6 | 9 | 14 |

## SOLID ANNEALED COPPER WIRE SIZES •

| GAGE NO. | DIAMETER INCHES | DIAMETER MILS | SECTION AREA INCH$^2$ | SECTION AREA CIR. MILS | $\Omega$ PER 1000 FT. AT 77°F. |
|---|---|---|---|---|---|
| 4/0 | 0.528 | 460.0 | 0.166 | 212000 | 0.0500 |
| 3/0 | 0.470 | 410.0 | 0.132 | 167800 | 0.0630 |
| 2/0 | 0.418 | 364.8 | 0.105 | 133100 | 0.0795 |
| 1/0 | 0.372 | 325.0 | 0.083 | 106000 | 0.1000 |
| 1 | 0.332 | 289.3 | 0.0657 | 83700 | 0.1260 |
| 2 | 0.292 | 258.0 | 0.0521 | 66370 | 0.1590 |
| 3 | 0.260 | 229.4 | 0.0413 | 52600 | 0.2010 |
| 4 | 0.232 | 204.3 | 0.0328 | 41700 | 0.2530 |
| 5 | 0.208 | 181.9 | 0.0260 | 33100 | 0.3190 |
| 6 | 0.184 | 162.0 | 0.0206 | 26300 | 0.4030 |
| 7 | 0.1562 | 144.0 | 0.0164 | 20800 | 0.5080 |
| 8 | 0.1285 | 128.0 | 0.0130 | 16500 | 0.6410 |
| 9 | 0.1152 | 114.0 | 0.0103 | 13100 | 0.8080 |
| 10 | 0.1019 | 102.0 | 0.0082 | 10400 | 1.020 |
| 11 | 0.0913 | 91.0 | 0.00647 | 8230 | 1.280 |
| 12 | 0.0808 | 80.8 | 0.00513 | 6530 | 1.620 |
| 13 | 0.0720 | 72.0 | 0.00407 | 5180 | 2.040 |
| 14 | 0.0641 | 64.1 | 0.00323 | 4110 | 2.580 |
| 15 | 0.0571 | 57.0 | 0.00256 | 3260 | 3.250 |
| 16 | 0.0508 | 51.0 | 0.00203 | 2580 | 4.090 |
| 18 | 0.0403 | 40.3 | 0.00128 | 1620 | 6.510 |
| 20 | 0.0321 | 32.0 | 0.00081 | 1020 | 10.400 |

# ELECTRICAL LIGHTING

| EFFECTIVE CEILING CAVITY REFLECTANCE (CCR) = 70%<br>EFFECTIVE WALL CAVITY REFLECTANCE (WCR) = 50, 30, 10%<br>EFFECTIVE FLOOR CAVITY REFLECTANCE (FCR) = 20% | TYPICAL DISTRIBUTION AND PERCENT LAMP LUMENS | | COEFFICIENT OF UTILIZATION WCR | | |
|---|---|---|---|---|---|
| TYPICAL LUMINAIRE | MAXIMUM S/MH GUIDE | RCR ↓ | 50 | 30 | 10 |
| 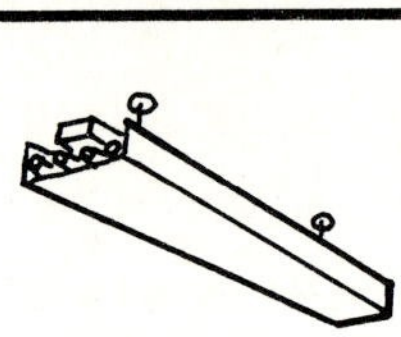 | 1.4/1.2 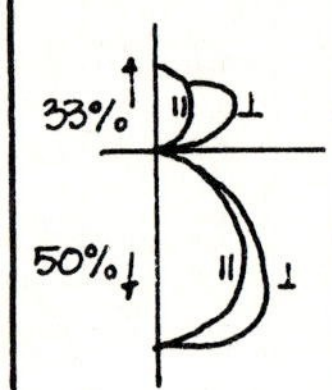 | 0 | .84 | .84 | .84 |
| | | 1 | .75 | .73 | .73 |
| | | 2 | .67 | .63 | .63 |
| | | 3 | .60 | .55 | .50 |
| | | 4 | .53 | .48 | .43 |
| | | 5 | .48 | .42 | .37 |
| | | 6 | .43 | .37 | .33 |
| PRISMATIC BOTTOM AND SIDES, OPEN TOP, 4 LAMP SUSPENDED UNIT—MULTIPLY BY 1.05 FOR 2 LAMPS | | 7 | .39 | .33 | .28 |
| | | 8 | .35 | .29 | .25 |
| | | 9 | .31 | .26 | .23 |
| | | 10 | .28 | .23 | .19 |
| 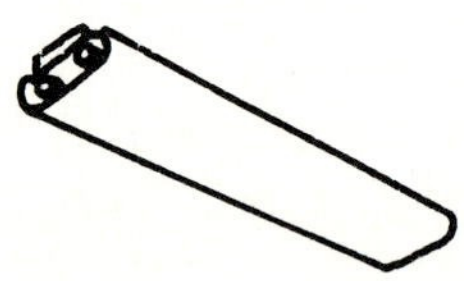 | 1.2 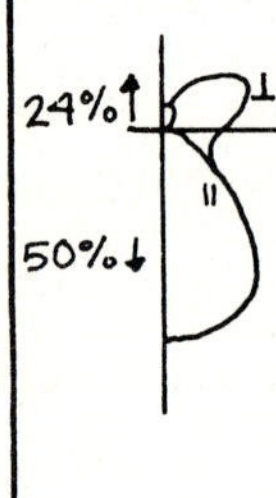 | 0 | .77 | .77 | .77 |
| | | 1 | .67 | .65 | .62 |
| | | 2 | .59 | .55 | .52 |
| | | 3 | .53 | .45 | .44 |
| | | 4 | .45 | .42 | .38 |
| | | 5 | .43 | .37 | .33 |
| | | 6 | .39 | .32 | .29 |
| 2 LAMP PRISMATIC WRAPAROUND—MULTIPLY BY 0.95 FOR 4 LAMPS | | 7 | .35 | .39 | .26 |
| | | 8 | .32 | .36 | .23 |
| | | 9 | .29 | .23 | .20 |
| | | 10 | .26 | .21 | .18 |
| 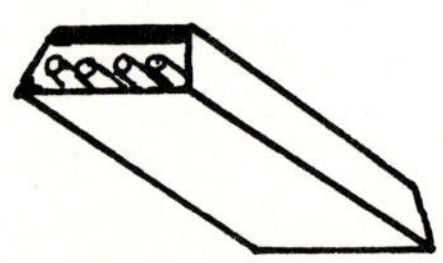 | 1.4/1.2 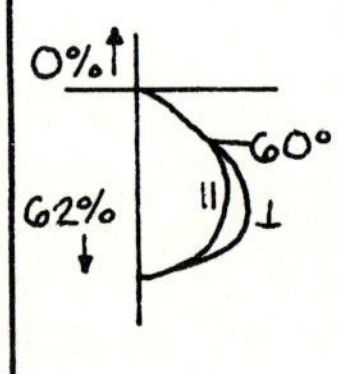 | 0 | .72 | .72 | .72 |
| | | 1 | .65 | .63 | .61 |
| | | 2 | .58 | .54 | .52 |
| | | 3 | .52 | .48 | .44 |
| | | 4 | .48 | .42 | .38 |
| | | 5 | .41 | .37 | .33 |
| | | 6 | .37 | .32 | .29 |
| | | 7 | .33 | .29 | .26 |
| FLUORESCENT UNIT WITH FLAT PRISMATIC LENS, 4 LAMP 2' WIDE—MULTIPLY BY 1.10 FOR 2 LAMP | | 8 | .30 | .25 | .22 |
| | | 9 | .27 | .22 | .19 |
| | | 10 | .24 | .20 | .16 |
| CEILING CAVITY REFLECTANCE = 50% 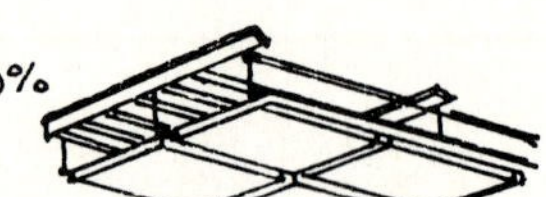 | | 1 | .58 | .50 | .54 |
| | | 2 | .51 | .47 | .43 |
| | | 3 | .45 | .41 | .36 |
| $\rho$CC FROM BELOW ~65% | | 4 | .39 | .35 | .31 |
| | | 5 | .35 | .30 | .26 |
| DIFFUSING PLASTIC OR GLASS | | 6 | .31 | .26 | .23 |
| 1) CEILING EFFICIENCY ~60%, DIFFUSER TRANSMITTANCE ~50%, DIFFUSER REFLECTANCE ~40%. CAVITY WITH MIN. OBSTRUCTIONS AND PAINTED WITH 80% REFLECTANCE PAINT—USE $\rho_c$ = 70. | | 7 | .28 | .23 | .20 |
| | | 8 | .25 | .20 | .17 |
| | | 9 | .23 | .18 | .15 |
| | | 10 | .21 | .16 | .13 |
| 2) FOR LOWER REFLECTANCE PAINT OR OBSTRUCTIONS—USE $\rho_c$ = 50. | | | | | |

# ELECTRICAL LIGHTING

| EFFECTIVE CEILING CAVITY REFLECTANCE (CCR) = 70%<br>EFFECTIVE WALL CAVITY REFLECTANCE (WCR) = 30,50,10%<br>EFFECTIVE FLOOR CAVITY REFLECTANCE (FCR) = 20% | TYPICAL DISTRIBUTION AND PER CENT LAMP LUMENS | | COEFFICIENT OF UTILIZATION WCR | | |
|---|---|---|---|---|---|
| | | | 50 | 30 | 10 |
| TYPICAL LUMINAIRE | MAXIMUM S/MH GUIDE | RCR[c] ↓ | | | |
| 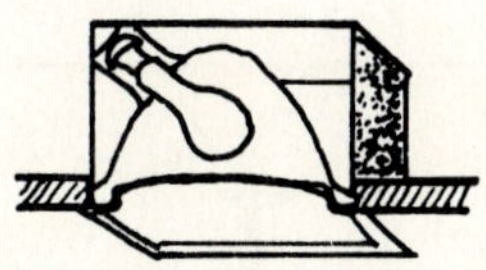<br>MEDIUM DISTRIBUTION UNIT WITH LENS PLATE AND INSIDE FROST LAMP. | 1.0<br>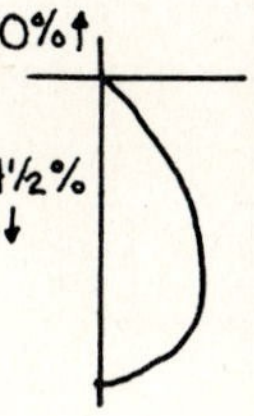<br>0%↑<br>54½%↓ | 0 | .63 | .63 | .63 |
| | | 1 | .58 | .57 | .56 |
| | | 2 | .54 | .52 | .50 |
| | | 3 | .50 | .47 | .45 |
| | | 4 | .47 | .44 | .41 |
| | | 5 | .43 | .40 | .38 |
| | | 6 | .40 | .37 | .35 |
| | | 7 | .37 | .34 | .32 |
| | | 8 | .35 | .31 | .30 |
| | | 9 | .32 | .30 | .27 |
| | | 10 | .30 | .27 | .24 |
| 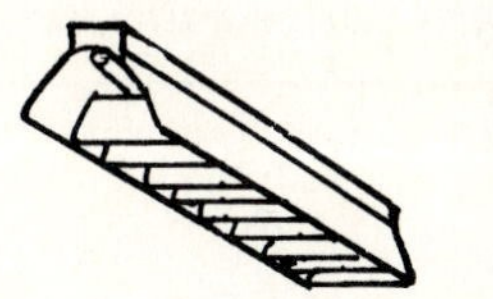<br>1" WIDE ALUMINUM TROFFER WITH 40° CW x 45° LW SHIELDING AND SINGLE EXTRA-HIGH-OUTPUT LAMP | 1.1/0.8<br>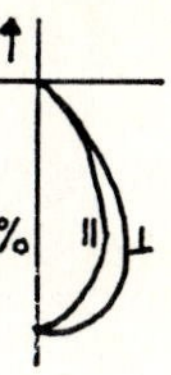<br>0%↑<br>42½%↓ | 0 | .49 | .49 | .49 |
| | | 1 | .45 | .44 | .43 |
| | | 2 | .42 | .40 | .38 |
| | | 3 | .39 | .36 | .34 |
| | | 4 | .35 | .33 | .31 |
| | | 5 | .33 | .30 | .28 |
| | | 6 | .30 | .28 | .26 |
| | | 7 | .28 | .25 | .23 |
| | | 8 | .26 | .23 | .21 |
| | | 9 | .24 | .21 | .19 |
| | | 10 | .22 | .19 | .17 |
| <br>2-LAMP, SURFACE MOUNTED, BARE LAMP UNIT—PHOTOMETRY WITH 18" WIDE PANEL ABOVE LUMINAIRE (LAMPS ON 6" CENTERS) | 1.3<br>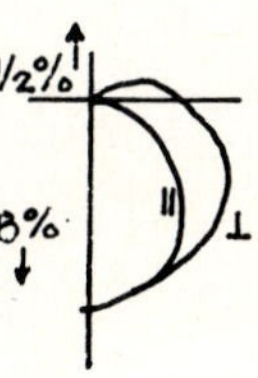<br>9½%↑<br>78%↓ | 0 | .98 | .98 | .98 |
| | | 1 | .85 | .79 | .75 |
| | | 2 | .71 | .65 | .60 |
| | | 3 | .62 | .55 | .49 |
| | | 4 | .55 | .47 | .41 |
| | | 5 | .48 | .40 | .34 |
| | | 6 | .43 | .35 | .29 |
| | | 7 | .38 | .30 | .25 |
| | | 8 | .34 | .27 | .22 |
| | | 9 | .31 | .23 | .18 |
| | | 10 | .28 | .21 | .16 |
| 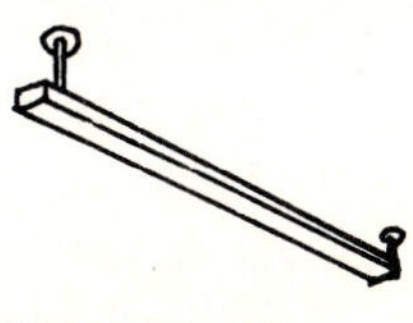<br>LUMINOUS BOTTOM SUSPENDED UNIT WITH EXTRA-HIGH-OUTPUT LAMP | 1.5<br>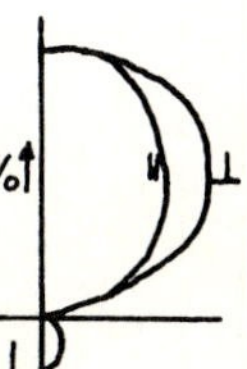<br>66%↑<br>12%↓ | 0 | .67 | .67 | .67 |
| | | 1 | .59 | .57 | .54 |
| | | 2 | .51 | .48 | .45 |
| | | 3 | .45 | .41 | .37 |
| | | 4 | .40 | .35 | .31 |
| | | 5 | .35 | .30 | .27 |
| | | 6 | .32 | .27 | .23 |
| | | 7 | .28 | .23 | .30 |
| | | 8 | .25 | .21 | .17 |
| | | 9 | .23 | .18 | .15 |
| | | 10 | .21 | .16 | .13 |

CONSTRUCTION SPECIFICATIONS INSTITUTE

# ELECTRICAL LIGHTING

## COEFFICIENTS OF UTILIZATION

| EFFECTIVE CEILING CAVITY REFLECTANCE (CCR) = 70% EFFECTIVE WALL CAVITY REFLECTANCE (WCR) = 50, 30, 10% EFFECTIVE FLOOR CAVITY REFLECTANCE (FCR) = 20% | TYPICAL DISTRIBUTION AND PER CENT LAMP LUMENS | | COEFFICIENT OF UTILIZATION WCR | | |
|---|---|---|---|---|---|
| TYPICAL LUMINAIRE | MAXIMUM S/MH GUIDE[d] | RCR[c] ↓ | 50 | 30 | 10 |
| 1  PENDANT DIFFUSING SPHERE WITH INCANDESCENT LAMP | 1.5 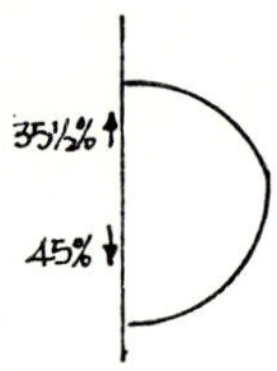  | 0 | .81 | .81 | .81 |
| | | 1 | .66 | .62 | .59 |
| | | 2 | .56 | .50 | .46 |
| | | 3 | .48 | .42 | .37 |
| | | 4 | .42 | .36 | .30 |
| | | 5 | .37 | .30 | .25 |
| | | 6 | .33 | .26 | .21 |
| | | 7 | .29 | .23 | .18 |
| | | 8 | .27 | .20 | .16 |
| | | 9 | .24 | .18 | .14 |
| | | 10 | .22 | .16 | .12 |
| 2 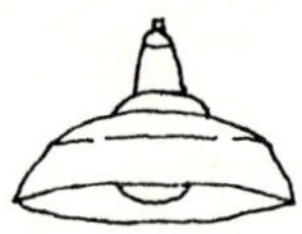 PORCELAIN-ENAMELED VENTILATED STANDARD DOME WITH INCANDESCENT LAMP | 1.3 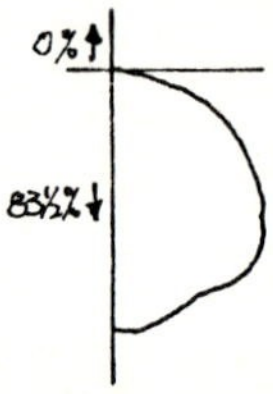  | 0 | .97 | .97 | .97 |
| | | 1 | .86 | .83 | .81 |
| | | 2 | .76 | .72 | .67 |
| | | 3 | .67 | .61 | .57 |
| | | 4 | .60 | .53 | .48 |
| | | 5 | .53 | .46 | .41 |
| | | 6 | .47 | .40 | .35 |
| | | 7 | .42 | .35 | .30 |
| | | 8 | .38 | .31 | .26 |
| | | 9 | .34 | .27 | .23 |
| | | 10 | .31 | .24 | .20 |
| 3 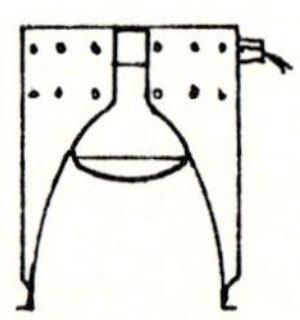 R40 FOOD WITH SPECULAR ANODIZED REFLECTOR SKIRT; 45° CUTOFF | 0.7 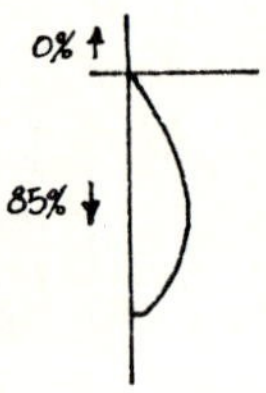  | 0 | .98 | .98 | .98 |
| | | 1 | .94 | .92 | .91 |
| | | 2 | .90 | .87 | .85 |
| | | 3 | .86 | .83 | .81 |
| | | 4 | .82 | .79 | .77 |
| | | 5 | .79 | .75 | .73 |
| | | 6 | .76 | .72 | .70 |
| | | 7 | .73 | .69 | .66 |
| | | 8 | .70 | .66 | .63 |
| | | 9 | .67 | .63 | .60 |
| | | 10 | .64 | .60 | .58 |
| 4 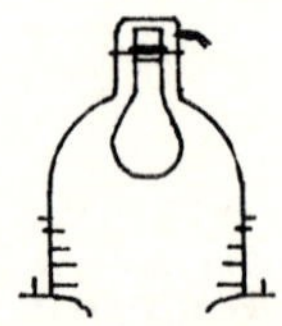 REFLECTOR DOWNLIGHT WITH BAFFLES AND INSIDE FROSTED LAMP | 0.7 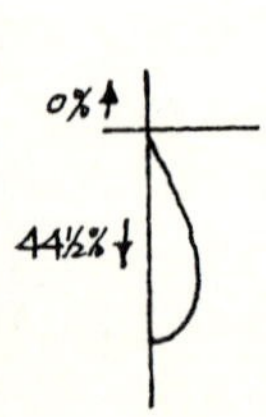  | 0 | .52 | .52 | .52 |
| | | 1 | .50 | .49 | .48 |
| | | 2 | .48 | .48 | .45 |
| | | 3 | .46 | .45 | .43 |
| | | 4 | .44 | .43 | .42 |
| | | 5 | .43 | .41 | .40 |
| | | 6 | .41 | .40 | .38 |
| | | 7 | .40 | .38 | .37 |
| | | 8 | .38 | .37 | .36 |
| | | 9 | .37 | .35 | .34 |
| | | 10 | .36 | .34 | .38 |

# PLUMBING PIPE SIZES

## COMPARATIVE PIPE SIZES

| NOM. PIPE SIZE | GALVANIZED STEEL | | | COPPER PIPE | | | COPPER TUBING TYPE L | | | CAST IRON PIPE STANDARD WT. BELL AND SPIGOT | | |
|---|---|---|---|---|---|---|---|---|---|---|---|---|
| | OUT DI. | IN DI. | WT/FT. | OUT DI. | IN DI. | WT/FT | OUT DI. | IN DI. | WT./FT. | OUT DI. | IN DI. | WT./FT. |
| 1/4 | .540 | .364 | .42 | .540 | .375 | .46 | .375 | .315 | .126 | | | |
| 1/2 | .640 | .622 | .85 | .840 | .625 | .957 | .625 | .545 | .285 | | | |
| 3/4 | 1.050 | .824 | 1.13 | 1.050 | .822 | 1.30 | .875 | .785 | .455 | | | |
| 1 | 1.315 | 1.049 | 1.68 | 1.315 | 1.062 | 1.83 | 1.125 | 1.025 | .655 | | | |
| 1 1/4 | 1.660 | 1.380 | 2.27 | 1.660 | 1.365 | 2.69 | 1.375 | 1.265 | .884 | | | |
| 1 1/2 | 1.900 | 1.610 | 2.72 | 1.900 | 1.600 | 3.20 | 1.625 | 1.505 | 1.14 | | | |
| 2 | 2.375 | 2.067 | 3.65 | 2.375 | 2.062 | 4.23 | 2.125 | 1.985 | 1.75 | 2.38 | 2.00 | 3.5 |
| 2 1/2 | 2.875 | 2.469 | 5.79 | 2.875 | 2.500 | 6.14 | 2.265 | 2.105 | 2.48 | | | |
| 3 | 3.500 | 3.178 | 7.58 | 3.500 | 3.062 | 8.75 | 3.125 | 2.945 | 3.33 | 3.50 | 3.00 | 4.5 |
| 3 1/2 | 4.000 | 3.645 | 9.11 | 4.000 | 3.500 | 11.41 | 3.625 | 3.425 | 4.29 | | | |
| 4 | 4.500 | 4.126 | 10.79 | 4.500 | 4.000 | 12.94 | 4.125 | 3.905 | 5.38 | 4.50 | 4.00 | 6.5 |
| 5 | 5.563 | 5.147 | 14.62 | 5.563 | 5.062 | 15.21 | 5.125 | 4.875 | 7.61 | 5.50 | 5.00 | 8.4 |
| 6 | 6.625 | 6.065 | 18.97 | 6.625 | 6.125 | 19.41 | 6.125 | 5.845 | 10.2 | 6.50 | 6.00 | 10.4 |
| 8 | 8.625 | 7.981 | 28.55 | 8.625 | 8.000 | 31.63 | 8.125 | 7.725 | 19.3 | 8.63 | 8.00 | 17.0 |
| 10 | 10.75 | 10.02 | 40.48 | 10.75 | 10.02 | 46.22 | 10.13 | 9.625 | 30.1 | 10.75 | 10.00 | 23.0 |
| 12 | 12.75 | 12.00 | 49.56 | | | | 12.13 | 11.56 | 40.4 | 12.75 | 12.00 | 35.0 |

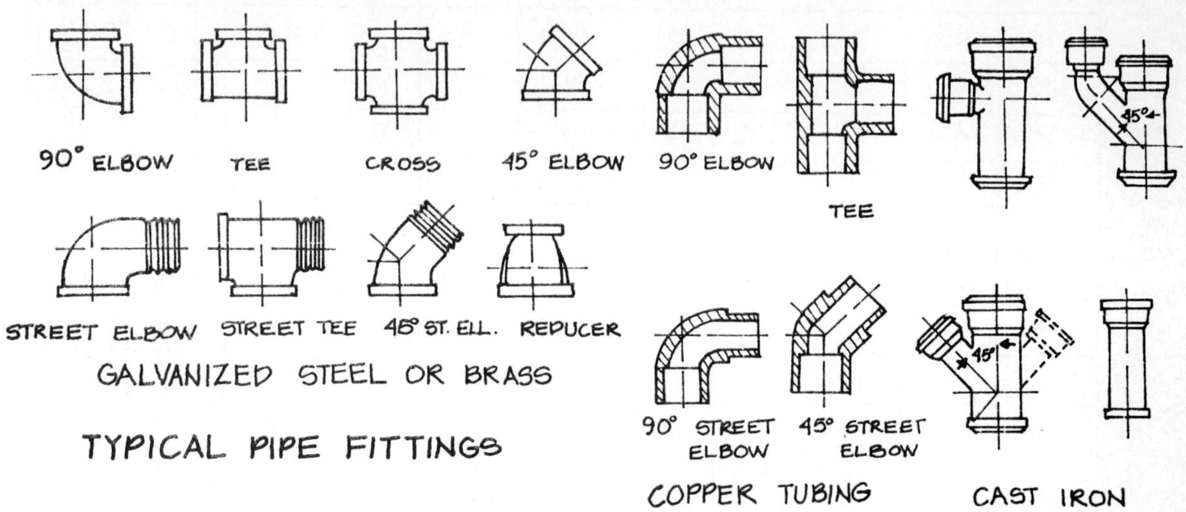


TYPICAL PIPE FITTINGS

# CAST IRON PIPE FITTINGS

125 LB. PER SQ. IN. BOLTED FLANGE

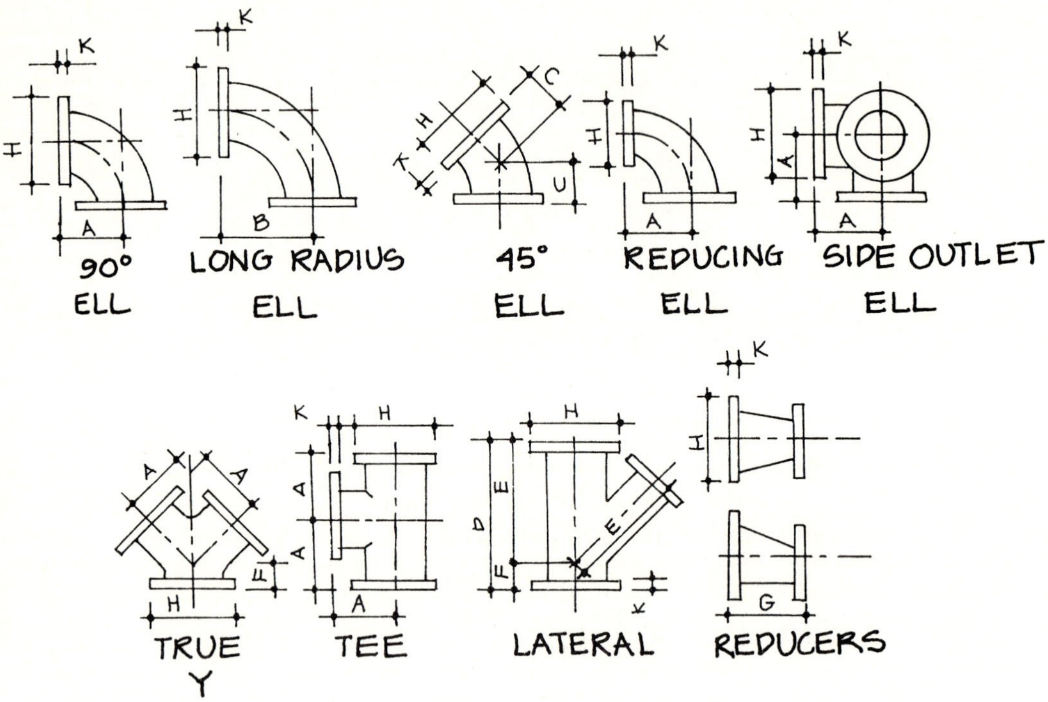


| NOM. SIZE | A | B | C | D | E | F | G | H | K | L | M |
|---|---|---|---|---|---|---|---|---|---|---|---|
| 1 | 3½ | 5 | 1¾ | 7½ | 5¾ | 1¾ | — | 4¼ | 7/16 | 3⅛ | 5/8 |
| 1¼ | 3¾ | 5½ | 2 | 8 | 6¼ | 1¾ | — | 4 5/8 | ½ | 3½ | 5/8 |
| 1½ | 4 | 6 | 2¼ | 9 | 7 | 2 | — | 5 | 9/16 | 3⅞ | 5/8 |
| 2 | 4½ | 6½ | 2½ | 10½ | 8 | 2½ | 5 | 6 | 5/8 | 4¾ | 3/4 |
| 2½ | 5 | 7 | 3 | 12 | 9½ | 2½ | 5½ | 7 | 11/16 | 5½ | 3/4 |
| 3 | 5½ | 7¾ | 3 | 13 | 10 | 3 | 6 | 7½ | 3/4 | 6 | 3/4 |
| 3½ | 6 | 8½ | 3½ | 14½ | 11½ | 3 | 6½ | 8½ | 13/16 | 7 | 3/4 |
| 4 | 6½ | 9 | 4 | 15 | 12 | 3 | 7 | 9 | 15/16 | 7½ | 3/4 |
| 5 | 7½ | 10¼ | 4½ | 17 | 13½ | 3½ | 8 | 10 | 15/16 | 8½ | 7/8 |
| 6 | 8 | 11½ | 5 | 18 | 14½ | 3½ | 9 | 11 | 1 | 9½ | 7/8 |
| 8 | 9 | 14 | 5½ | 22 | 17½ | 4½ | 11 | 13½ | 1⅛ | 11¾ | 7/8 |
| 10 | 11 | 16½ | 6½ | 25½ | 20½ | 5 | 12 | 16 | 1 3/16 | 14¼ | 1 |
| 12 | 12 | 19 | 7½ | 30 | 24½ | 5½ | 14 | 19 | 1¼ | 17 | 1 |

# CAPACITY OF PIPE

## CONTENTS OF PIPE

CAPACITIES IN CUBIC FEET AND IN UNITED STATES GALLONS (231 CUBIC INCHES) PER FOOT OF LENGTH

| DIAMETER, INCHES | DIAMETER, FEET | FOR 1 FOOT LENGTH | | DIAMETER, INCHES | DIAMETER, FEET | FOR 1 FOOT LENGTH | | DIAMETER, INCHES | DIAMETER, FEET | FOR 1 FOOT LENGTH | |
|---|---|---|---|---|---|---|---|---|---|---|---|
| | | CUBIC FT., ALSO AREA IN SQ. FEET | U.S. GAL. (231 CU. IN.) | | | CUBIC FT., ALSO AREA IN SQ. FEET | U.S. GAL. (231 CU. IN.) | | | CUBIC FT., ALSO AREA IN SQ. FEET | U.S. GAL. (231 CU. IN.) |
| 1/4 | .0208 | .0003 | .0026 | 6.75 | .5625 | .2485 | 1.859 | 19.0 | 1.583 | 1.969 | 14.73 |
| 5/16 | .0260 | .0005 | .0040 | 7.00 | .5833 | .2673 | 1.999 | 19.5 | 1.625 | 2.074 | 15.52 |
| 3/8 | .0313 | .0008 | .0057 | 7.25 | .6042 | .2868 | 2.144 | 20.0 | 1.666 | 2.182 | 16.32 |
| 7/16 | .0365 | .0010 | .0078 | 7.50 | .6250 | .3068 | 2.295 | 20.5 | 1.708 | 2.292 | 17.15 |
| 1/2 | .0417 | .0014 | .0102 | 7.75 | .6458 | .3275 | 2.450 | 21.0 | 1.750 | 2.405 | 17.99 |
| 9/16 | .0469 | .0017 | .0129 | 8.00 | .6667 | .3490 | 2.611 | 21.5 | 1.792 | 2.521 | 18.86 |
| 5/8 | .0521 | .0021 | .0159 | 8.25 | .6875 | .3713 | 2.777 | 22.0 | 1.833 | 2.640 | 19.65 |
| 11/16 | .0573 | .0026 | .0193 | 8.50 | .7083 | .3940 | 2.948 | 22.5 | 1.875 | 2.761 | 20.75 |
| 3/4 | .0625 | .0031 | .0230 | 8.75 | .7292 | .4175 | 3.125 | 23.0 | 1.917 | 2.885 | 21.58 |
| 13/16 | .0677 | .0036 | .0270 | 9.00 | .7500 | .4418 | 3.305 | 23.5 | 1.958 | 3.012 | 22.53 |
| 7/8 | .0729 | .0042 | .0312 | 9.25 | .7708 | .4668 | 3.492 | 24.0 | 2.000 | 3.142 | 23.50 |
| 15/16 | .0781 | .0048 | .0359 | 9.50 | .7917 | .4923 | 3.682 | 25.0 | 2.083 | 3.409 | 25.50 |
| 1.00 | .0833 | .0055 | .0408 | 9.75 | .8125 | .5185 | 3.879 | 26.0 | 2.166 | 3.687 | 27.58 |
| 1.25 | .1042 | .0085 | .0638 | 10.00 | .8333 | .5455 | 4.081 | 27.0 | 2.250 | 3.976 | 29.74 |
| 1.50 | .1250 | .0123 | .0918 | 10.25 | .8542 | .5730 | 4.286 | 28.0 | 2.333 | 4.276 | 31.99 |
| 1.75 | .1458 | .0168 | .1250 | 10.50 | .8750 | .6013 | 4.498 | 29.0 | 2.416 | 4.587 | 34.31 |
| 2.00 | .1667 | .0218 | .1632 | 10.75 | .8958 | .6303 | 4.714 | 30.0 | 2.500 | 4.909 | 36.72 |
| 2.25 | .1875 | .0276 | .2066 | 11.00 | .9167 | .6600 | 4.937 | 31.0 | 2.583 | 5.241 | 39.21 |
| 2.50 | .2083 | .0341 | .2550 | 11.25 | .9375 | .6903 | 5.163 | 32.0 | 2.666 | 5.585 | 41.78 |
| 2.75 | .2292 | .0413 | .3085 | 11.50 | .9583 | .7213 | 5.397 | 33.0 | 2.750 | 5.940 | 44.43 |
| 3.00 | .2500 | .0491 | .3673 | 11.75 | .9792 | .7530 | 5.633 | 34.0 | 2.833 | 6.305 | 47.17 |
| 3.25 | .2708 | .0576 | .4310 | 12.00 | 1.000 | .7854 | 5.876 | 35.0 | 2.916 | 6.681 | 49.98 |
| 3.50 | .2917 | .0668 | .4998 | 12.50 | 1.042 | .8523 | 6.375 | 36.0 | 3.000 | 7.069 | 52.88 |
| 3.75 | .3125 | .0767 | .5738 | 13.00 | 1.083 | .9218 | 6.895 | 37.0 | 3.083 | 7.486 | 55.86 |
| 4.00 | .3333 | .0873 | .6528 | 13.50 | 1.125 | .9940 | 7.435 | 38.0 | 3.166 | 7.876 | 58.92 |
| 4.25 | .3542 | .0985 | .7370 | 14.00 | 1.167 | 1.069 | 7.997 | 39.0 | 3.250 | 8.296 | 62.06 |
| 4.50 | .3750 | .1105 | .8263 | 14.50 | 1.208 | 1.147 | 8.578 | 40.0 | 3.333 | 8.728 | 65.29 |
| 4.75 | .3958 | .1231 | .9205 | 15.00 | 1.250 | 1.227 | 9.180 | 41.0 | 3.416 | 9.168 | 68.58 |
| 5.00 | .4167 | .1364 | 1.020 | 15.50 | 1.292 | 1.310 | 9.801 | 42.0 | 3.500 | 9.620 | 71.96 |
| 5.25 | .4375 | .1503 | 1.124 | 16.00 | 1.333 | 1.396 | 10.44 | 43.0 | 3.583 | 10.084 | 75.43 |
| 5.50 | .4583 | .1650 | 1.234 | 16.50 | 1.375 | 1.485 | 11.11 | 44.0 | 3.666 | 10.560 | 79.00 |
| 5.75 | .4792 | .1803 | 1.349 | 17.00 | 1.417 | 1.576 | 11.79 | 45.0 | 3.750 | 11.044 | 82.62 |
| 6.00 | .5000 | .1963 | 1.469 | 17.50 | 1.458 | 1.670 | 12.50 | 46.0 | 3.833 | 11.540 | 86.32 |
| 6.25 | .5208 | .2130 | 1.594 | 18.00 | 1.500 | 1.767 | 13.22 | 47.0 | 3.916 | 12.048 | 90.12 |
| 6.50 | .5417 | .2305 | 1.724 | 18.50 | 1.542 | 1.867 | 13.97 | 48.0 | 4.000 | 12.566 | 94.02 |

1 CUBIC FOOT OF WATER WEIGHS 62.35 POUNDS; 1 GALLON (U.S.) WEIGHS 8.335 POUNDS.

CAST IRON PIPE RESEARCH ASSOCIATION

# PRESSURE OF WATER

PRESSURES IN POUNDS PER SQUARE INCH,
CORRESPONDING TO HEADS OF WATER IN FEET

| HEAD FT. | 0 | 1 | 2 | 3 | 4 | 5 | 6 | 7 | 8 | 9 |
|---|---|---|---|---|---|---|---|---|---|---|
| 0 | | 0.433 | 0.866 | 1.299 | 1.732 | 2.165 | 2.598 | 3.031 | 3.464 | 3.987 |
| 10 | 4.330 | 4.763 | 5.196 | 5.629 | 6.062 | 6.495 | 6.928 | 7.361 | 7.794 | 8.277 |
| 20 | 8.660 | 9.093 | 9.526 | 9.959 | 10.392 | 10.825 | 11.258 | 11.691 | 12.124 | 12.557 |
| 30 | 12.990 | 13.423 | 13.856 | 14.289 | 14.722 | 14.722 | 15.588 | 16.021 | 16.454 | 16.877 |
| 40 | 17.320 | 17.753 | 18.186 | 18.619 | 19.052 | 19.052 | 19.918 | 20.351 | 20.784 | 21.217 |
| 50 | 21.650 | 22.083 | 22.516 | 22.949 | 23.382 | 23.815 | 24.248 | 24.681 | 25.114 | 25.547 |
| 60 | 25.980 | 26.413 | 26.846 | 27.279 | 27.712 | 28.145 | 28.578 | 29.001 | 29.444 | 29.877 |
| 70 | 30.310 | 30.743 | 31.176 | 31.609 | 32.042 | 32.475 | 32.908 | 33.341 | 33.774 | 34.207 |
| 80 | 34.640 | 35.073 | 35.506 | 35.939 | 36.372 | 36.805 | 37.238 | 37.671 | 38.104 | 38.537 |
| 90 | 38.970 | 39.403 | 39.836 | 40.269 | 40.702 | 41.135 | 41.568 | 42.001 | 42.436 | 42.867 |

HEADS OF WATER IN FEET CORRESPONDING TO PRESSURES
IN POUNDS PER SQUARE INCH

| PRESSURE LBS. PER SQ. IN. | 0 | 1 | 2 | 3 | 4 |
|---|---|---|---|---|---|
| 0 | | 2.309 | 4.619 | 6.928 | 9.238 |
| 10 | 23.095 | 25.404 | 27.714 | 30.023 | 32.333 |
| 20 | 46.189 | 48.499 | 50.808 | 53.118 | 55.427 |
| 30 | 69.284 | 71.594 | 73.903 | 76.213 | 78.522 |
| 40 | 92.379 | 94.688 | 96.998 | 99.307 | 101.62 |
| 50 | 115.47 | 117.78 | 120.69 | 122.40 | 124.71 |
| 60 | 138.57 | 140.88 | 143.19 | 145.50 | 147.81 |
| 70 | 161.66 | 163.97 | 166.28 | 168.59 | 170.90 |
| 80 | 184.76 | 187.07 | 189.38 | 191.69 | 194.00 |
| 90 | 207.85 | 210.16 | 212.47 | 214.78 | 217.09 |
| | **5** | **6** | **7** | **8** | **9** |
| 0 | 11.547 | 13.857 | 16.166 | 18.476 | 20.785 |
| 10 | 34.642 | 36.952 | 39.261 | 41.570 | 43.880 |
| 20 | 57.737 | 60.046 | 62.356 | 64.665 | 66.975 |
| 30 | 80.831 | 83.141 | 85.450 | 87.760 | 90.069 |
| 40 | 103.93 | 106.24 | 108.55 | 110.85 | 113.16 |
| 50 | 127.02 | 129.33 | 131.64 | 133.95 | 136.26 |
| 60 | 150.12 | 152.42 | 154.73 | 157.04 | 159.35 |
| 70 | 173.21 | 175.52 | 177.83 | 180.14 | 182.45 |
| 80 | 197.31 | 198.61 | 200.92 | 203.23 | 205.54 |
| 90 | 219.40 | 221.71 | 224.02 | 226.33 | 228.64 |

AT 62° F., 1 FOOT HEAD = 0.433 LB. PER SQUARE INCH; 0.433 x 144 = 0.433 x 144 = 62.355 LB. PER CUBIC FOOT. 1 LB. PER SQUARE INCH = 2.30947 FEET HEAD. 1 ATMOSPHERE = 14.7 LB. PER SQUARE INCH = 33.94 FEET HEAD.

CAST IRON PIPE ASSOCIATION

# COLOR REFLECTIVE VALUES

| COLOR | APPROX. PERCENT OF REFLECTION |
|---|---|
| WHITE, DULL OR FLAT | 75 – 85 |
| WHITE, GLOSS | 85 – 90 |
| **LIGHT TINTS** | |
| CREAM OR EGGSHELL | 79 |
| IVORY | 75 |
| PALE PINK AND PALE YELLOW | 75 – 80 |
| LIGHT GREEN, LIGHT BLUE, LIGHT ORCHID | 70 – 75 |
| SOFT PINK, LIGHT PEACH | 69 |
| LIGHT BEIGE, PALE GREY | 70 |
| **MEDIUM TONES** | |
| APRICOT | 56 – 62 |
| PINK | 64 |
| TAN, YELLOW GOLD | 55 |
| LIGHT GREYS | 35 – 50 |
| MEDIUM TOURQUOISE | 44 |
| MEDIUM LIGHT BLUE | 42 |
| YELLOW GREEN | 45 |
| OLD GOLD, PUMPKIN | 34 |
| ROSE | 29 |
| **DEEP TONES** | |
| COCOA BROWN, MAUVE | 24 |
| MEDIUM GREEN, MEDIUM BLUE | 21 |
| | 20 |
| **UNSUITABLE DARK COLORS** | |
| DARK BROWN, DARK GREY | 10 – 15 |
| OLIVE GREEN | 12 |
| DARK BLUE, BLUE GREEN | 5 – 10 |
| FOREST GREEN | 7 |
| **NATURAL WOOD TONES** | |
| BIRCH AND BEECH | 35 – 50 |
| LIGHT MAPLE | 25 – 35 |
| LIGHT OAK | 25 – 35 |
| DARK OAK, CHERRY | 10 – 15 |
| REDWOOD | 10 – 15 |
| BLACK WALNUT, MAHOGANY | 5 – 15 |

RECOMMENDED CEILING VALUES SHOULD BE IN THE RANGE OF 60-90%. FLOOR REFLECTION VALUES SHOULD BE IN THE RANGE OF 15-35%. OVERALL REFLECTION VALUES OF A ROOM SHOULD BE IN THE 35-60% RANGE.

**TABLE A9**

Friction Loss in Pipe

(a) Allowance for friction loss in copper valves and fittings expressed as equivalent length of tube (ft)

| Diameter of fitting (in.) | Standard Ls | | 90° T | | | | |
|---|---|---|---|---|---|---|---|
| | 90° | 45° | Side branch | Straight run | Coupling | Gate valve | Globe valve |
| 3/8 | 0.5 | 0.3 | 0.75 | 0.15 | 0.15 | 0.1 | 4 |
| 1/2 | 1 | 0.6 | 1.5 | 0.3 | 0.3 | 0.2 | 7.5 |
| 3/4 | 1.25 | 0.75 | 2 | 0.4 | 0.4 | 0.25 | 10 |
| 1 | 1.5 | 1.0 | 2.5 | 0.45 | 0.45 | 0.3 | 12.5 |
| 1 1/4 | 2 | 1.2 | 3 | 0.6 | 0.6 | 0.4 | 18 |
| 1 1/2 | 2.5 | 1.5 | 3.5 | 0.8 | 0.8 | 0.5 | 23 |
| 2 | 3.5 | 2 | 5 | 1 | 1 | 0.7 | 28 |
| 2 1/2 | 4 | 2.5 | 6 | 1.3 | 1.3 | 0.8 | 33 |
| 3 | 5 | 3 | 7.5 | 1.5 | 1.5 | 1 | 40 |
| 3 1/2 | 6 | 3.5 | 9 | 1.8 | 1.8 | 1.2 | 50 |
| 4 | 7 | 4 | 10.5 | 2 | 2 | 1.4 | 63 |
| 5 | 9 | 5 | 13 | 2.5 | 2.5 | 1.7 | 70 |
| 6 | 10 | 6 | 15 | 3 | 3 | 2 | 84 |

Copper Development Association Inc.

[a]Allowances are for streamlined soldered fittings and recessed threaded fittings. For threaded fittings, double the allowances show in the table.

(b) Allowance for friction loss in valves and threaded fittings expressed as equivalent length of galvanized pipe (ft)

| Diameter of fitting (in.) | Standard L | | | | | | |
|---|---|---|---|---|---|---|---|
| | 90° | 45° | 90° side T | Coupling or straight run | Gate valve | Globe valve | Angle valve |
| 3/8 | 1 | 0.6 | 1.5 | 0.3 | 0.2 | 8 | 4 |
| 1/2 | 2 | 1.2 | 3 | 0.6 | 0.4 | 15 | 8 |
| 3/4 | 2.5 | 1.5 | 4 | 0.8 | 0.5 | 20 | 12 |
| 1 | 3 | 1.8 | 5 | 0.9 | 0.6 | 25 | 15 |
| 1 1/4 | 4 | 2.4 | 6 | 1.2 | 0.8 | 35 | 18 |
| 1 1/2 | 5 | 3 | 7 | 1.5 | 1.0 | 45 | 22 |
| 2 | 7 | 4 | 10 | 2 | 1.3 | 55 | 28 |
| 2 1/2 | 8 | 5 | 12 | 2.5 | 1.6 | 65 | 34 |
| 3 | 10 | 6 | 15 | 3 | 2 | 80 | 40 |
| 3 1/2 | 12 | 7 | 18 | 3.6 | 2.4 | 100 | 50 |
| 4 | 14 | 8 | 21 | 4 | 2.7 | 125 | 55 |
| 5 | 17 | 10 | 25 | 5 | 3.3 | 140 | 70 |
| 6 | 20 | 12 | 30 | 6 | 4 | 165 | 80 |

Environmental Protection Agency.

**TABLE A10**

Climatic Conditions for the United States[a]

| Col. 1 | Col. 2 | Col. 3 | Col. 4 | Winter[b] Col. 5 | | Summer[c] Col. 6 | | | Col. 7 | Col. 8 | | |
|---|---|---|---|---|---|---|---|---|---|---|---|---|
| | | | | Design dry-bulb | | Design dry-bulb and mean coincident wet-bulb | | | Mean daily range | Design wet-bulb | | |
| State and station | Latitude[d] | Longitude[d] | Elevation[e] Ft | 99% | 97.5% | 1% | 2.5% | 5% | | 1% | 2.5% | 5% |
| Alabama | | | | | | | | | | | | |
| Alexander City | 33°0′ | 86°0′ | 660 | 18 | 22 | 96/77 | 93/76 | 91/76 | 21 | 79 | 78 | 78 |
| Anniston AP | 33°4′ | 85°5′ | 599 | 18 | 22 | 97/77 | 94/76 | 92/76 | 21 | 79 | 78 | 78 |
| Auburn | 32°4′ | 85°3′ | 730 | 18 | 22 | 96/77 | 93/76 | 91/76 | 21 | 79 | 78 | 78 |
| Birmingham AP | 33°3′ | 86°5′ | 610 | 17 | 21 | 96/74 | 94/75 | 92/74 | 21 | 78 | 77 | 76 |
| Decatur | 34°4′ | 87°0′ | 580 | 11 | 16 | 95/75 | 93/74 | 91/74 | 22 | 78 | 77 | 76 |
| Dothan AP | 31°2′ | 85°2′ | 321 | 23 | 27 | 94/76 | 92/76 | 91/76 | 20 | 80 | 79 | 78 |
| Florence AP | 34°5′ | 87°4′ | 528 | 17 | 21 | 97/74 | 94/74 | 92/74 | 22 | 78 | 77 | 76 |
| Gadsden | 34°0′ | 86°0′ | 570 | 16 | 20 | 96/75 | 94/75 | 92/74 | 22 | 78 | 77 | 76 |
| Huntsville AP | 34°4′ | 86°4′ | 619 | 11 | 16 | 95/75 | 93/74 | 91/74 | 23 | 78 | 77 | 76 |
| Mobile AP | 30°4′ | 88°2′ | 211 | 25 | 29 | 95/77 | 93/77 | 91/76 | 18 | 80 | 79 | 78 |
| Mobile CO | 30°4′ | 88°1′ | 119 | 25 | 29 | 95/77 | 93/77 | 91/76 | 16 | 80 | 79 | 78 |
| Montgomery AP | 32°2′ | 86°2′ | 195 | 22 | 25 | 96/76 | 95/76 | 93/76 | 21 | 79 | 79 | 78 |
| Selma-Craig AFB | 32°2′ | 87°0′ | 207 | 22 | 26 | 97/78 | 95/77 | 93/77 | 21 | 81 | 80 | 79 |
| Talladega | 33°3′ | 86°1′ | 565 | 18 | 22 | 97/77 | 94/76 | 92/76 | 21 | 79 | 78 | 78 |
| Tuscaloosa AP | 33°1′ | 87°4′ | 170r | 20 | 23 | 98/75 | 96/76 | 94/76 | 22 | 79 | 78 | 77 |
| Alaska | | | | | | | | | | | | |
| Anchorage AP | 61°1′ | 150°0′ | 90 | −23 | −18 | 71/59 | 68/58 | 66/56 | 15 | 60 | 59 | 57 |
| Barrow (S) | 71°2′ | 156°5′ | 22 | −45 | −41 | 57/53 | 53/50 | 49/47 | 12 | 54 | 50 | 47 |
| Fairbanks AP (S) | 64°5′ | 147°5′ | 436 | −51 | −47 | 82/62 | 78/60 | 75/59 | 24 | 64 | 62 | 60 |
| Juneau AP | 58°2′ | 134°4′ | 17 | −4 | 1 | 74/60 | 70/58 | 67/57 | 15 | 61 | 59 | 58 |
| Kodiak | 57°3′ | 152°3′ | 21 | 10 | 13 | 69/58 | 65/56 | 62/55 | 10 | 60 | 58 | 56 |
| Nome AP | 64°3′ | 165°3′ | 13 | −31 | −27 | 66/57 | 62/55 | 59/54 | 10 | 58 | 56 | 55 |
| Arizona | | | | | | | | | | | | |
| Douglas AP | 31°3′ | 109°3′ | 4,098 | 27 | 31 | 98/63 | 95/63 | 93/63 | 31 | 70 | 69 | 68 |
| Flagstaff AP | 35°1′ | 111°4′ | 6,973 | −2 | 4 | 84/55 | 82/55 | 80/54 | 31 | 61 | 60 | 59 |
| Fort Huachuca AP (S) | 31°3′ | 110°2′ | 4,664 | 24 | 28 | 95/62 | 92/62 | 90/62 | 27 | 69 | 68 | 67 |
| Kingman AP | 35°2′ | 114°0′ | 3,446 | 18 | 25 | 103/65 | 100/64 | 97/64 | 30 | 70 | 69 | 69 |
| Nogales | 31°2′ | 111°0′ | 3,800 | 28 | 32 | 99/64 | 96/64 | 94/64 | 31 | 71 | 70 | 69 |
| Phoenix AP (S) | 33°3′ | 112°0′ | 1,117 | 31 | 34 | 109/71 | 107/71 | 105/71 | 27 | 76 | 75 | 75 |
| Prescott AP | 34°4′ | 112°3′ | 5,014 | 4 | 9 | 96/61 | 94/60 | 92/60 | 30 | 66 | 65 | 64 |
| Tucson AP (S) | 32°1′ | 111°0′ | 2,584 | 28 | 32 | 104/66 | 102/66 | 100/66 | 26 | 72 | 71 | 71 |
| Winslow AP | 35°0′ | 110°4′ | 4,880 | 5 | 10 | 97/61 | 95/60 | 93/60 | 32 | 66 | 65 | 64 |
| Yuma AP | 32°4′ | 114°4′ | 199 | 36 | 39 | 111/72 | 109/72 | 107/71 | 27 | 79 | 78 | 77 |
| Arkansas | | | | | | | | | | | | |
| Blytheville AFB | 36°0′ | 90°0′ | 264 | 10 | 15 | 96/78 | 94/77 | 91/76 | 21 | 81 | 80 | 78 |

**TABLE A10 Continued**

| Col. 1 | Col. 2 | Col. 3 | Col. 4 | Col. 5 (Winter[b]) | | Col. 6 (Summer[c]) | | | Col. 7 | Col. 8 | | |
|---|---|---|---|---|---|---|---|---|---|---|---|---|
| | | | | Design dry-bulb | | Design dry-bulb and mean coincident wet-bulb | | | Mean daily | Design wet-bulb | | |
| State and station | Latitude[d] | Longitude[d] | Elevation[e] Ft | 99% | 97.5% | 1% | 2.5% | 5% | range | 1% | 2.5% | 5% |
| Camden | 33°4′ | 92°5′ | 116 | 18 | 23 | 98/76 | 96/76 | 94/76 | 21 | 80 | 79 | 78 |
| El Dorado AP | 33°1′ | 92°5′ | 252 | 18 | 23 | 98/76 | 96/76 | 94/76 | 21 | 80 | 79 | 78 |
| Fayetteville AP | 36°0′ | 94°1′ | 1,253 | 7 | 12 | 97/72 | 94/73 | 92/73 | 23 | 77 | 76 | 75 |
| Fort Smith AP | 35°2′ | 94°2′ | 449 | 12 | 17 | 101/75 | 98/76 | 95/76 | 24 | 80 | 79 | 78 |
| Hot Springs | 34°3′ | 93°1′ | 535 | 17 | 23 | 101/77 | 97/77 | 94/77 | 22 | 80 | 79 | 78 |
| Jonesboro | 35°5′ | 90°4′ | 345 | 10 | 15 | 96/78 | 94/77 | 91/76 | 21 | 81 | 80 | 78 |
| Little Rock AP (S) | 34°4′ | 92°1′ | 257 | 15 | 20 | 99/76 | 96/77 | 94/77 | 22 | 80 | 79 | 78 |
| Pine Bluff AP | 34°1′ | 92°0′ | 204 | 16 | 22 | 100/78 | 97/77 | 95/78 | 22 | 81 | 80 | 80 |
| Texarkana AP | 33°3′ | 94°0′ | 361 | 18 | 23 | 98/76 | 96/77 | 93/76 | 21 | 80 | 79 | 78 |
| California | | | | | | | | | | | | |
| Bakersfield AP | 35°2′ | 119°0′ | 495 | 30 | 32 | 104/70 | 101/69 | 98/68 | 32 | 73 | 71 | 70 |
| Barstow AP | 34°5′ | 116°5′ | 2,142 | 26 | 29 | 106/68 | 104/68 | 102/67 | 37 | 73 | 71 | 70 |
| Blythe AP | 33°4′ | 114°3′ | 390 | 30 | 33 | 112/71 | 110/71 | 108/70 | 28 | 75 | 75 | 74 |
| Burbank AP | 34°1′ | 118°2′ | 699 | 37 | 39 | 95/68 | 91/68 | 88/67 | 25 | 71 | 70 | 69 |
| Chico | 39°5′ | 121°5′ | 205 | 28 | 30 | 103/69 | 101/68 | 98/67 | 36 | 71 | 70 | 68 |
| Concord | 38°0′ | 122°0′ | 195 | 24 | 27 | 100/69 | 97/68 | 94/67 | 32 | 71 | 70 | 68 |
| Covina | 34°0′ | 117°5′ | 575 | 32 | 35 | 98/69 | 95/68 | 92/67 | 31 | 73 | 71 | 70 |
| Crescent City AP | 41°5′ | 124°0′ | 50 | 31 | 33 | 68/60 | 65/59 | 63/58 | 18 | 62 | 60 | 59 |
| Downey | 34°0′ | 118°1′ | 116 | 37 | 40 | 93/70 | 89/70 | 86/69 | 22 | 72 | 71 | 70 |
| El Cajon | 32°4′ | 117°0′ | 525 | 42 | 44 | 83/69 | 80/69 | 78/68 | 30 | 71 | 70 | 68 |
| El Centro AP (S) | 32°5′ | 115°4′ | −30 | 35 | 38 | 112/74 | 110/74 | 108/74 | 34 | 81 | 80 | 78 |
| Escondido | 33°0′ | 117°1′ | 660 | 39 | 41 | 89/68 | 85/68 | 82/68 | 30 | 71 | 70 | 69 |
| Eureaka/Arcata AP | 41°0′ | 124°1′ | 217 | 31 | 33 | 68/60 | 65/59 | 63/58 | 11 | 62 | 60 | 59 |
| Fairfield-Travis AFB | 38°2′ | 122°0′ | 72 | 29 | 32 | 99/68 | 95/67 | 91/66 | 34 | 70 | 68 | 67 |
| Fresno AP (S) | 36°5′ | 119°4′ | 326 | 28 | 30 | 102/70 | 100/69 | 97/68 | 34 | 72 | 71 | 70 |
| Hamilton AFB | 38°0′ | 122°3′ | 3 | 30 | 32 | 89/68 | 84/66 | 80/65 | 28 | 72 | 69 | 67 |
| Laguna Beach | 33°3′ | 117°5′ | 35 | 41 | 43 | 83/68 | 80/68 | 77/67 | 18 | 70 | 69 | 68 |
| Livermore | 37°4′ | 122°0′ | 545 | 24 | 27 | 100/69 | 97/68 | 93/67 | 24 | 71 | 70 | 68 |
| Lompoc, Vandenburg AFB | 34°4′ | 120°3′ | 552 | 35 | 38 | 75/61 | 70/61 | 67/60 | 20 | 63 | 61 | 60 |
| Long Beach AP | 33°5′ | 118°1′ | 34 | 41 | 43 | 83/68 | 80/68 | 77/67 | 22 | 70 | 69 | 68 |
| Los Angeles AP (S) | 34°0′ | 118°2′ | 99 | 41 | 43 | 83/68 | 80/68 | 77/67 | 15 | 70 | 69 | 68 |
| Los Angeles CO (S) | 34°0′ | 118°1′ | 312 | 37 | 40 | 93/70 | 89/70 | 86/69 | 20 | 72 | 71 | 70 |
| Merced-Castle AFB | 37°2′ | 120°3′ | 178 | 29 | 31 | 102/70 | 99/69 | 96/68 | 36 | 72 | 71 | 70 |
| Modesto | 37°4′ | 121°0′ | 91 | 28 | 30 | 101/69 | 98/68 | 95/67 | 36 | 71 | 70 | 69 |
| Monterey | 36°4′ | 121°5′ | 38 | 35 | 38 | 75/63 | 71/61 | 68/61 | 20 | 64 | 62 | 61 |
| Napa | 38°2′ | 122°2′ | 16 | 30 | 32 | 100/69 | 96/68 | 92/67 | 30 | 71 | 69 | 68 |
| Needles AP | 34°5′ | 114°4′ | 913 | 30 | 33 | 112/71 | 110/71 | 108/70 | 27 | 75 | 75 | 74 |

| | | | | | | | | | | | | |
|---|---|---|---|---|---|---|---|---|---|---|---|---|
| Oakland AP | 37°4′ | 122°1′ | 3 | 34 | 36 | 85/64 | 80/63 | 75/62 | 19 | 66 | 64 | 63 |
| Oceanside | 33°1′ | 117°2′ | 30 | 41 | 43 | 83/68 | 80/68 | 77/67 | 13 | 70 | 69 | 68 |
| Ontario | 34°0′ | 117°36′ | 995 | 31 | 33 | 102/70 | 99/69 | 96/67 | 36 | 74 | 72 | 71 |
| Oxnard | 34°1′ | 119°1′ | 43 | 34 | 36 | 83/66 | 80/64 | 77/63 | 19 | 70 | 68 | 67 |
| Palmdale AP | 34°4′ | 118°1′ | 2,517 | 18 | 22 | 103/65 | 101/65 | 98/64 | 35 | 69 | 67 | 66 |
| Palm Springs | 33°5′ | 116°4′ | 411 | 33 | 35 | 112/71 | 110/70 | 108/70 | 35 | 76 | 74 | 73 |
| Pasadena | 34°1′ | 118°1′ | 864 | 32 | 35 | 98/69 | 95/68 | 92/67 | 29 | 73 | 71 | 70 |
| Petaluma | 38°1′ | 122°4′ | 27 | 26 | 29 | 94/68 | 90/66 | 87/65 | 31 | 72 | 70 | 68 |
| Pomona CO | 34°0′ | 117°5′ | 871 | 28 | 30 | 102/70 | 99/69 | 95/68 | 36 | 74 | 72 | 71 |
| Redding AP | 40°3′ | 122°1′ | 495 | 29 | 31 | 105/68 | 102/67 | 100/66 | 32 | 71 | 69 | 68 |
| Redlands | 34°0′ | 117°1′ | 1,318 | 31 | 33 | 102/70 | 99/69 | 96/68 | 33 | 74 | 72 | 71 |
| Richmond | 38°0′ | 122°2′ | 55 | 34 | 36 | 85/64 | 80/63 | 75/62 | 17 | 66 | 64 | 63 |
| Riverside-March AFB (S) | 33°5′ | 117°2′ | 1,511 | 29 | 32 | 100/68 | 98/68 | 95/67 | 37 | 72 | 71 | 70 |
| Sacramento AP | 38°3′ | 121°3′ | 17 | 30 | 32 | 101/70 | 98/70 | 94/69 | 36 | 72 | 71 | 70 |
| Salinas AP | 36°4′ | 121°4′ | 74 | 30 | 32 | 74/61 | 70/60 | 67/59 | 24 | 62 | 61 | 59 |
| San Bernardino, Norton AFB | 34°1′ | 117°1′ | 1,125 | 31 | 33 | 102/70 | 99/69 | 96/68 | 38 | 74 | 72 | 71 |
| San Diego AP | 32°4′ | 117°1′ | 19 | 42 | 44 | 83/69 | 80/69 | 78/68 | 12 | 71 | 70 | 68 |
| San Fernando | 34°1′ | 118°3′ | 977 | 37 | 39 | 95/68 | 91/68 | 88/67 | 38 | 71 | 70 | 69 |
| San Francisco AP | 37°4′ | 122°2′ | 8 | 35 | 38 | 82/64 | 77/63 | 73/62 | 20 | 65 | 64 | 62 |
| San Francisco CO | 37°5′ | 122°3′ | 52 | 38 | 40 | 74/63 | 71/62 | 69/61 | 14 | 64 | 62 | 61 |
| San Jose AP | 37°2′ | 122°0′ | 70r | 34 | 36 | 85/66 | 81/65 | 77/64 | 26 | 68 | 67 | 65 |
| San Luis Obispo | 35°2′ | 120°4′ | 315 | 33 | 35 | 92/69 | 88/70 | 84/69 | 26 | 73 | 71 | 70 |
| Santa Ana AP | 33°4′ | 117°5′ | 115r | 37 | 39 | 89/69 | 85/68 | 82/68 | 28 | 71 | 70 | 69 |
| Santa Barbara MAP | 34°3′ | 119°5′ | 10 | 34 | 36 | 81/67 | 77/66 | 75/65 | 24 | 68 | 67 | 66 |
| Santa Cruz | 37°0′ | 122°0′ | 125 | 35 | 38 | 75/63 | 71/61 | 68/61 | 28 | 64 | 62 | 61 |
| Santa Maria AP (S) | 34°5′ | 120°3′ | 238 | 31 | 33 | 81/64 | 76/63 | 73/62 | 23 | 65 | 64 | 63 |
| Santa Monica CO | 34°0′ | 118°3′ | 57 | 41 | 43 | 83/68 | 80/68 | 77/67 | 16 | 70 | 69 | 68 |
| Santa Paula | 34°2′ | 119°0′ | 263 | 33 | 35 | 90/68 | 86/67 | 84/66 | 36 | 71 | 69 | 68 |
| Santa Rosa | 38°3′ | 122°5′ | 167 | 27 | 29 | 99/68 | 95/67 | 91/66 | 34 | 70 | 68 | 67 |
| Stockton AP | 37°5′ | 121°2′ | 28 | 28 | 30 | 100/69 | 97/68 | 94/67 | 37 | 71 | 70 | 68 |
| Ukiah | 39°1′ | 122°4′ | 620 | 27 | 29 | 99/69 | 95/68 | 91/67 | 40 | 70 | 68 | 67 |
| Visalia | 36°2′ | 119°1′ | 354 | 28 | 30 | 102/70 | 100/69 | 97/68 | 38 | 72 | 71 | 70 |
| Yreka | 41°4′ | 122°4′ | 2,625 | 13 | 17 | 95/65 | 92/64 | 89/63 | 38 | 67 | 65 | 64 |
| Yuba City | 39°1′ | 121°4′ | 70 | 29 | 31 | 104/68 | 101/67 | 99/66 | 36 | 71 | 69 | 68 |
| Colorado | | | | | | | | | | | | |
| Alamosa AP | 37°3′ | 105°5′ | 7,536 | −21 | −16 | 84/57 | 82/57 | 80/57 | 35 | 62 | 61 | 60 |
| Boulder | 40°0′ | 105°2′ | 5,385 | −2 | 8 | 93/59 | 91/59 | 89/59 | 27 | 64 | 63 | 62 |
| Colorado Springs AP | 38°5′ | 104°4′ | 6,173 | −3 | 2 | 91/58 | 88/57 | 86/57 | 30 | 63 | 62 | 61 |
| Denver AP | 39°5′ | 104°5′ | 5,283 | −5 | 1 | 93/59 | 91/59 | 89/59 | 28 | 64 | 63 | 62 |
| Durango | 37°1′ | 107°5′ | 6,550 | −1 | 4 | 89/59 | 87/59 | 85/59 | 30 | 64 | 63 | 62 |
| Fort Collins | 40°4′ | 105°0′ | 5,001 | −10 | −4 | 93/59 | 91/59 | 89/59 | 28 | 64 | 63 | 62 |
| Grand Junction AP (S) | 39°1′ | 108°3′ | 4,849 | 2 | 7 | 96/59 | 94/59 | 92/59 | 29 | 64 | 63 | 62 |
| Greeley | 40°3′ | 104°4′ | 4,648 | −11 | −5 | 96/60 | 94/60 | 92/60 | 29 | 65 | 64 | 63 |
| La Junta AP | 38°0′ | 103°3′ | 4,188 | −3 | 3 | 100/68 | 98/68 | 95/67 | 31 | 72 | 70 | 69 |
| Leadville | 39°2′ | 106°2′ | 10,177 | −8 | −4 | 84/52 | 91/51 | 78/50 | 30 | 56 | 55 | 54 |
| Pueblo AP | 38°2′ | 104°2′ | 4,639 | −7 | 0 | 97/61 | 95/61 | 92/61 | 31 | 67 | 66 | 65 |
| Sterling | 40°4′ | 103°1′ | 3,939 | −7 | −2 | 95/62 | 93/62 | 90/62 | 30 | 67 | 66 | 65 |

**TABLE A10 Continued**

| Col. 1 | Col. 2 | Col. 3 | Col. 4 | Col. 5 (Winter[b]) | | Col. 6 (Summer[c]) | | | Col. 7 | Col. 8 | | |
|---|---|---|---|---|---|---|---|---|---|---|---|---|
| | | | | Design dry-bulb | | Design dry-bulb and mean coincident wet-bulb | | | Mean daily | Design wet-bulb | | |
| State and station | Latitude[d] | Longitude[d] | Elevation[e] Ft | 99% | 97.5% | 1% | 2.5% | 5% | range | 1% | 2.5% | 5% |
| Trinidad AP | 37°2′ | 104°2′ | 5,746 | −2 | 3 | 93/61 | 91/61 | 89/61 | 32 | 66 | 65 | 64 |
| Connecticut | | | | | | | | | | | | |
| Bridgeport AP | 41°1′ | 73°1′ | 7 | 6 | 9 | 86/73 | 84/71 | 81/70 | 18 | 75 | 74 | 73 |
| Hartford, Brainard Field | 41°5′ | 72°4′ | 15 | 3 | 7 | 91/74 | 88/73 | 85/72 | 22 | 77 | 75 | 74 |
| New Haven AP | 41°2′ | 73°0′ | 6 | 3 | 7 | 88/75 | 84/73 | 82/72 | 17 | 76 | 75 | 74 |
| New London | 41°2′ | 72°1′ | 60 | 5 | 9 | 88/73 | 85/72 | 83/71 | 16 | 76 | 75 | 74 |
| Norwalk | 41°1′ | 73°3′ | 37 | 6 | 9 | 86/73 | 84/71 | 81/70 | 19 | 75 | 74 | 73 |
| Norwich | 41°3′ | 72°0′ | 20 | 3 | 7 | 89/75 | 86/73 | 83/72 | 18 | 76 | 75 | 74 |
| Waterbury | 41°3′ | 73°0′ | 605 | −4 | 2 | 88/73 | 85/71 | 82/70 | 21 | 75 | 74 | 72 |
| Windsor Locks, Bradley Field (S) | 42°0′ | 72°4′ | 169 | 0 | 4 | 91/74 | 88/72 | 85/71 | 22 | 76 | 75 | 73 |
| Delaware | | | | | | | | | | | | |
| Dover AFB | 39°0′ | 75°3′ | 38 | 11 | 15 | 92/75 | 90/75 | 87/74 | 18 | 79 | 77 | 76 |
| Wilmington AP | 39°4′ | 75°3′ | 78 | 10 | 14 | 92/74 | 89/74 | 87/73 | 20 | 77 | 76 | 75 |
| District of Columbia | | | | | | | | | | | | |
| Andrews AFB | 38°5′ | 76°5′ | 279 | 10 | 14 | 92/75 | 90/74 | 87/73 | 18 | 78 | 76 | 75 |
| Washington National AP | 38°5′ | 77°0′ | 14 | 14 | 17 | 93/75 | 91/74 | 89/74 | 18 | 78 | 77 | 76 |
| Florida | | | | | | | | | | | | |
| Belle Glade | 26°4′ | 80°4′ | 16 | 41 | 44 | 92/76 | 91/76 | 89/76 | 16 | 79 | 78 | 78 |
| Cape Kennedy AP | 28°3′ | 80°3′ | 16 | 35 | 38 | 90/78 | 88/78 | 87/78 | 15 | 80 | 79 | 79 |
| Daytona Beach AP | 29°1′ | 81°0′ | 31 | 32 | 35 | 92/78 | 90/77 | 88/77 | 15 | 80 | 79 | 78 |
| Fort Lauderdale | 26°0′ | 80°1′ | 13 | 42 | 46 | 92/78 | 91/78 | 90/78 | 15 | 80 | 79 | 79 |
| Fort Myers AP | 26°4′ | 81°5′ | 13 | 41 | 44 | 93/78 | 92/78 | 91/77 | 18 | 80 | 79 | 79 |
| Fort Pierce | 27°3′ | 80°2′ | 10 | 38 | 42 | 91/78 | 90/78 | 89/78 | 15 | 80 | 79 | 79 |
| Gainesville AP (S) | 29°4′ | 82°2′ | 155 | 28 | 31 | 95/77 | 93/77 | 92/77 | 18 | 80 | 79 | 78 |
| Jacksonville AP | 30°3′ | 81°4′ | 24 | 29 | 32 | 96/77 | 94/77 | 92/76 | 19 | 79 | 79 | 78 |
| Key West AP | 24°3′ | 81°5′ | 6 | 55 | 57 | 90/78 | 90/78 | 89/78 | 9 | 80 | 79 | 79 |
| Lakeland CO (S) | 28°0′ | 82°0′ | 214 | 39 | 41 | 93/76 | 91/76 | 89/76 | 17 | 79 | 78 | 78 |
| Miami AP (S) | 25°5′ | 81°5′ | 7 | 44 | 47 | 91/77 | 90/77 | 89/77 | 15 | 79 | 79 | 78 |
| Miami Beach CO | 25°5′ | 80°1′ | 9 | 45 | 48 | 90/77 | 89/77 | 88/77 | 10 | 79 | 79 | 78 |
| Ocala | 29°1′ | 82°1′ | 86 | 31 | 34 | 95/77 | 93/77 | 92/76 | 18 | 80 | 79 | 78 |
| Orlando AP | 28°3′ | 81°2′ | 106r | 35 | 38 | 94/76 | 93/76 | 91/76 | 17 | 79 | 78 | 78 |
| Panama City, Tyndall AFB | 30°0′ | 85°4′ | 22 | 29 | 33 | 92/78 | 90/77 | 89/77 | 14 | 81 | 80 | 79 |
| Pensacola CO | 30°3′ | 87°1′ | 13 | 25 | 29 | 94/77 | 93/77 | 91/77 | 14 | 80 | 79 | 79 |
| St. Augustine | 29°5′ | 81°2′ | 15 | 31 | 35 | 92/78 | 89/78 | 87/78 | 16 | 80 | 79 | 79 |
| St. Petersburg | 28°0′ | 82°4′ | 35 | 36 | 40 | 92/77 | 91/77 | 90/76 | 16 | 79 | 79 | 78 |
| Sanford | 28°5′ | 81°2′ | 14 | 35 | 38 | 94/76 | 93/76 | 91/76 | 17 | 79 | 78 | 78 |

| | | | | | | | | | | | | |
|---|---|---|---|---|---|---|---|---|---|---|---|---|
| Sarasota | 27°2′ | 82°3′ | 30 | 39 | 42 | 93/77 | 92/77 | 90/76 | 17 | 79 | 79 | 78 |
| Tallahassee AP (S) | 30°2′ | 84°2′ | 58 | 27 | 30 | 94/77 | 92/76 | 90/76 | 19 | 79 | 78 | 78 |
| Tampa AP (S) | 28°0′ | 82°3′ | 19 | 36 | 40 | 92/77 | 91/77 | 90/76 | 17 | 79 | 79 | 78 |
| West Palm Beach AP | 26°4′ | 80°1′ | 15 | 41 | 45 | 92/78 | 91/78 | 90/78 | 16 | 80 | 79 | 79 |
| Georgia | | | | | | | | | | | | |
| Albany, Turner AFB | 31°3′ | 84°1′ | 224 | 25 | 29 | 97/77 | 95/76 | 93/76 | 20 | 80 | 79 | 78 |
| Americus | 32°0′ | 84°2′ | 476 | 21 | 25 | 97/77 | 94/76 | 92/75 | 20 | 79 | 78 | 77 |
| Athens | 34°0′ | 83°2′ | 700 | 18 | 22 | 94/74 | 92/74 | 90/74 | 21 | 78 | 77 | 76 |
| Atlanta AP (S) | 33°4′ | 84°3′ | 1,005 | 17 | 22 | 94/74 | 92/74 | 90/73 | 19 | 77 | 76 | 75 |
| Augusta AP | 33°2′ | 82°0′ | 143 | 20 | 23 | 97/77 | 95/76 | 93/76 | 19 | 80 | 79 | 78 |
| Brunswick | 31°1′ | 81°3′ | 14 | 29 | 32 | 92/78 | 89/78 | 87/78 | 18 | 80 | 79 | 79 |
| Columbus, Lawson AFB | 32°3′ | 85°0′ | 242 | 21 | 24 | 95/76 | 93/76 | 91/75 | 21 | 79 | 78 | 77 |
| Dalton | 34°5′ | 85°0′ | 720 | 17 | 22 | 94/76 | 93/76 | 91/76 | 22 | 79 | 78 | 77 |
| Dublin | 32°3′ | 83°0′ | 215 | 21 | 25 | 96/77 | 93/76 | 91/75 | 20 | 79 | 78 | 77 |
| Gainesville | 34°2′ | 83°5′ | 1,254 | 16 | 21 | 93/74 | 91/74 | 89/73 | 21 | 77 | 76 | 75 |
| Griffin (S) | 33°1′ | 84°2′ | 980 | 18 | 22 | 93/76 | 90/75 | 88/74 | 21 | 78 | 77 | 76 |
| La Grange | 33°0′ | 85°0′ | 715 | 19 | 23 | 94/76 | 91/75 | 89/74 | 21 | 78 | 77 | 76 |
| Macon AP | 32°4′ | 83°4′ | 356 | 21 | 25 | 96/77 | 93/76 | 91/75 | 22 | 79 | 78 | 77 |
| Marietta, Dobbins AFB | 34°0′ | 84°3′ | 1,016 | 17 | 21 | 94/74 | 92/74 | 90/74 | 21 | 78 | 77 | 76 |
| Moultrie | 31°1′ | 83°4′ | 340 | 27 | 30 | 97/77 | 95/77 | 92/76 | 20 | 80 | 79 | 78 |
| Rome AP | 34°2′ | 85°1′ | 637 | 17 | 22 | 94/76 | 93/76 | 91/76 | 23 | 79 | 78 | 77 |
| Savannah-Travis AP | 32°1′ | 81°1′ | 52 | 24 | 27 | 96/77 | 93/77 | 91/77 | 20 | 80 | 79 | 78 |
| Valdosta-Moody AFB | 31°0′ | 83°1′ | 239 | 28 | 31 | 96/77 | 94/77 | 92/76 | 20 | 80 | 79 | 78 |
| Waycross | 31°2′ | 82°2′ | 140 | 26 | 29 | 96/77 | 94/77 | 91/76 | 20 | 80 | 79 | 78 |
| Hawaii | | | | | | | | | | | | |
| Hilo AP (S) | 19°4′ | 155°1′ | 31 | 61 | 62 | 84/73 | 83/72 | 82/72 | 15 | 75 | 74 | 74 |
| Honolulu AP | 21°2′ | 158°0′ | 7 | 62 | 63 | 87/73 | 86/73 | 85/72 | 12 | 76 | 75 | 74 |
| Kaneohe Bay MCAS | 21°2′ | 157°5′ | 18 | 65 | 66 | 85/75 | 84/74 | 83/74 | 12 | 76 | 76 | 75 |
| Wahiawa | 21°3′ | 158°0′ | 900 | 58 | 59 | 96/73 | 85/72 | 84/72 | 14 | 75 | 74 | 73 |
| Idaho | | | | | | | | | | | | |
| Boise AP (S) | 43°3′ | 116°1′ | 2,842 | 3 | 10 | 96/65 | 94/64 | 91/64 | 31 | 68 | 66 | 65 |
| Burley | 42°3′ | 113°5′ | 4,180 | −3 | 2 | 99/62 | 95/61 | 92/66 | 35 | 64 | 63 | 61 |
| Coeur d'Alene AP | 47°5′ | 116°5′ | 2,973 | −8 | −1 | 89/62 | 86/61 | 83/60 | 31 | 64 | 63 | 61 |
| Idaho Falls AP | 43°3′ | 112°0′ | 4,730r | −11 | −6 | 89/61 | 87/61 | 84/59 | 38 | 65 | 63 | 61 |
| Lewiston AP | 46°2′ | 117°0′ | 1,413 | −1 | 6 | 96/65 | 93/64 | 90/63 | 32 | 67 | 66 | 64 |
| Moscow | 46°4′ | 117°0′ | 2,660 | −7 | 0 | 90/63 | 87/62 | 84/61 | 32 | 65 | 64 | 62 |
| Mountain Home AFB | 43°0′ | 115°5′ | 2,992 | 6 | 12 | 99/64 | 97/63 | 94/62 | 36 | 66 | 65 | 63 |
| Pocatello AP | 43°0′ | 112°4′ | 4,444 | −8 | −1 | 94/61 | 91/60 | 89/59 | 35 | 64 | 63 | 61 |
| Twin Falls AP (S) | 42°3′ | 114°3′ | 4,148 | −3 | 2 | 99/62 | 95/61 | 92/60 | 34 | 64 | 63 | 61 |
| Illinois | | | | | | | | | | | | |
| Aurora | 41°5′ | 88°2′ | 744 | −6 | −1 | 93/76 | 91/76 | 88/75 | 20 | 79 | 78 | 76 |
| Belleville, Scott AFB | 38°3′ | 89°5′ | 447 | 1 | 6 | 94/76 | 92/76 | 89/75 | 21 | 79 | 78 | 76 |
| Bloomington | 40°3′ | 89°0′ | 775 | −6 | −2 | 92/75 | 90/74 | 88/73 | 21 | 78 | 76 | 75 |
| Carbondale | 37°5′ | 89°1′ | 380 | 2 | 7 | 95/77 | 93/77 | 90/76 | 21 | 80 | 79 | 77 |
| Champaign/Urbana | 40°0′ | 88°2′ | 743 | −3 | 2 | 95/75 | 92/74 | 90/73 | 21 | 78 | 77 | 75 |
| Chicago, Midway AP | 41°5′ | 87°5′ | 610 | −5 | 0 | 94/74 | 91/73 | 88/72 | 20 | 77 | 75 | 74 |
| Chicago, O'Hare AP | 42°0′ | 87°5′ | 658 | −8 | −4 | 91/74 | 89/74 | 86/72 | 20 | 77 | 76 | 74 |

**TABLE A10 Continued**

| Col. 1 | Col. 2 | Col. 3 | Col. 4 | Col. 5 (Winter[b]) | | Col. 6 (Summer[c]) | | | Col. 7 | Col. 8 | | |
|---|---|---|---|---|---|---|---|---|---|---|---|---|
| | | | | Design dry-bulb | | Design dry-bulb and mean coincident wet-bulb | | | Mean daily | Design wet-bulb | | |
| State and station | Latitude[d] | Longitude[d] | Elevation[e] Ft | 99% | 97.5% | 1% | 2.5% | 5% | range | 1% | 2.5% | 5% |
| Chicago CO | 41°5′ | 87°4′ | 594 | −3 | 2 | 94/75 | 91/74 | 88/73 | 15 | 79 | 77 | 75 |
| Danville | 40°1′ | 87°4′ | 558 | −4 | 1 | 93/75 | 90/74 | 88/73 | 21 | 78 | 77 | 75 |
| Decatur | 39°5′ | 88°5′ | 670 | −3 | 2 | 94/75 | 91/74 | 88/73 | 21 | 78 | 77 | 75 |
| Dixon | 41°5′ | 89°3′ | 696 | −7 | −2 | 93/75 | 90/74 | 88/73 | 23 | 78 | 77 | 75 |
| Elgin | 42°0′ | 88°2′ | 820 | −7 | −2 | 91/75 | 88/74 | 86/73 | 21 | 78 | 77 | 75 |
| Freeport | 42°2′ | 89°4′ | 780 | −9 | −4 | 91/74 | 89/73 | 87/72 | 24 | 77 | 76 | 74 |
| Galesburg | 41°0′ | 90°3′ | 771 | −7 | −2 | 93/75 | 91/75 | 88/74 | 22 | 78 | 77 | 75 |
| Greenville | 39°0′ | 89°2′ | 563 | −1 | 4 | 94/76 | 92/75 | 89/74 | 21 | 79 | 78 | 76 |
| Joliet | 41°3′ | 88°1′ | 588 | −5 | 0 | 93/75 | 90/74 | 88/73 | 20 | 78 | 77 | 75 |
| Kankakee | 41°1′ | 87°5′ | 625 | −4 | 1 | 93/75 | 90/74 | 88/73 | 21 | 78 | 77 | 75 |
| La Salle/Peru | 41°2′ | 89°1′ | 520 | −7 | −2 | 93/75 | 91/75 | 88/74 | 22 | 78 | 77 | 75 |
| Macomb | 40°3′ | 90°4′ | 702 | −5 | 0 | 95/76 | 92/76 | 89/75 | 22 | 79 | 78 | 76 |
| Moline AP | 41°3′ | 90°3′ | 582 | −9 | −4 | 93/75 | 91/75 | 88/74 | 23 | 78 | 77 | 75 |
| Mt. Vernon | 38°2′ | 88°5′ | 500 | 0 | 5 | 95/76 | 92/75 | 89/74 | 21 | 79 | 78 | 76 |
| Peoria AP | 40°4′ | 89°4′ | 652 | −8 | −4 | 91/75 | 89/74 | 87/73 | 22 | 78 | 76 | 75 |
| Quincy AP | 40°0′ | 91°1′ | 762 | −2 | 3 | 96/76 | 93/76 | 90/76 | 22 | 80 | 78 | 77 |
| Rantoul, Chanute AFB | 40°2′ | 88°1′ | 740 | −4 | 1 | 94/75 | 91/74 | 89/73 | 21 | 78 | 77 | 75 |
| Rockford | 42°1′ | 89°0′ | 724 | −9 | −4 | 91/74 | 89/73 | 87/72 | 24 | 77 | 76 | 74 |
| Springfield AP | 39°5′ | 89°4′ | 587 | −3 | 2 | 94/75 | 92/74 | 89/74 | 21 | 79 | 77 | 76 |
| Waukegan | 42°2′ | 87°5′ | 680 | −6 | −3 | 92/76 | 89/74 | 87/73 | 21 | 78 | 76 | 75 |
| Indiana | | | | | | | | | | | | |
| Anderson | 40°0′ | 85°4′ | 847 | 0 | 6 | 95/76 | 92/75 | 89/74 | 22 | 79 | 78 | 76 |
| Bedford | 38°5′ | 86°3′ | 670 | 0 | 5 | 95/76 | 92/75 | 89/74 | 22 | 79 | 78 | 76 |
| Bloomington | 39°1′ | 86°4′ | 820 | 0 | 5 | 95/76 | 92/75 | 89/74 | 22 | 79 | 78 | 76 |
| Columbus, Bakalar AFB | 39°2′ | 85°5′ | 661 | 3 | 7 | 95/76 | 92/75 | 90/74 | 22 | 79 | 78 | 76 |
| Crawfordsville | 40°0′ | 86°5′ | 752 | −2 | 3 | 94/75 | 91/74 | 88/73 | 22 | 79 | 77 | 76 |
| Evansville AP | 38°0′ | 87°3′ | 381 | 4 | 9 | 95/76 | 93/75 | 91/75 | 22 | 79 | 78 | 77 |
| Fort Wayne AP | 41°0′ | 85°1′ | 791 | −4 | 1 | 92/73 | 89/72 | 87/72 | 24 | 77 | 75 | 74 |
| Goshen AP | 41°3′ | 85°5′ | 823 | −3 | 1 | 91/73 | 89/73 | 86/72 | 23 | 77 | 75 | 74 |
| Hobart | 41°3′ | 87°2′ | 600 | −4 | 2 | 91/73 | 88/73 | 85/72 | 21 | 77 | 75 | 74 |
| Huntington | 40°4′ | 85°3′ | 802 | −4 | 1 | 92/73 | 89/72 | 87/72 | 23 | 77 | 75 | 74 |
| Indianapolis AP (S) | 39°4′ | 86°2′ | 793 | −2 | 2 | 92/74 | 90/74 | 87/73 | 22 | 78 | 76 | 75 |
| Jeffersonville | 38°2′ | 85°5′ | 455 | 5 | 10 | 95/74 | 93/74 | 90/74 | 23 | 79 | 77 | 76 |
| Kokomo | 40°3′ | 86°1′ | 790 | −4 | 0 | 91/74 | 90/73 | 88/73 | 22 | 77 | 75 | 74 |
| Lafayette | 40°2′ | 86°5′ | 600 | −3 | 3 | 94/74 | 91/73 | 88/73 | 22 | 78 | 76 | 75 |
| La Porte | 41°3′ | 86°4′ | 810 | −3 | 3 | 93/74 | 90/74 | 87/73 | 22 | 78 | 76 | 75 |
| Marion | 40°3′ | 85°4′ | 791 | −4 | 0 | 91/74 | 90/73 | 88/73 | 23 | 77 | 75 | 74 |

| | | | | | | | | | | | | |
|---|---|---|---|---|---|---|---|---|---|---|---|---|
| Muncie | 40°1′ | 85°2′ | 955 | −3 | 2 | 92/74 | 90/73 | 87/73 | 22 | 76 | 76 | 75 |
| Peru, Bunker Hill AFB | 40°4′ | 86°1′ | 804 | −6 | −1 | 90/74 | 88/73 | 86/73 | 22 | 77 | 75 | 74 |
| Richmond AP | 39°5′ | 84°5′ | 1,138 | −2 | 2 | 92/74 | 90/74 | 87/73 | 22 | 78 | 76 | 75 |
| Shelbyville | 39°3′ | 85°5′ | 765 | −1 | 3 | 93/74 | 91/74 | 88/73 | 22 | 78 | 76 | 75 |
| South Bend AP | 41°4′ | 86°2′ | 773 | −3 | 1 | 91/73 | 89/73 | 86/72 | 22 | 77 | 75 | 74 |
| Terre Haute AP | 39°3′ | 87°2′ | 601 | −2 | 4 | 95/75 | 92/74 | 89/73 | 22 | 79 | 77 | 76 |
| Valparaise | 41°2′ | 87°0′ | 801 | −3 | 3 | 93/74 | 90/74 | 87/73 | 22 | 78 | 76 | 75 |
| Vincennes | 38°4′ | 87°3′ | 420 | 1 | 6 | 95/75 | 92/74 | 90/73 | 22 | 79 | 77 | 76 |
| Iowa | | | | | | | | | | | | |
| Ames (S) | 42°0′ | 93°4′ | 1,004 | −11 | −6 | 93/75 | 90/74 | 87/73 | 23 | 78 | 76 | 75 |
| Burlington AP | 40°5′ | 91°1′ | 694 | −7 | −3 | 94/74 | 91/75 | 88/73 | 22 | 78 | 77 | 75 |
| Cedar Rapids AP | 41°5′ | 91°4′ | 863 | −10 | −5 | 91/76 | 88/75 | 86/74 | 23 | 78 | 77 | 75 |
| Clinton | 41°5′ | 90°1′ | 595 | −8 | −3 | 92/75 | 90/75 | 87/74 | 23 | 78 | 77 | 75 |
| Council Bluffts | 41°2′ | 95°5′ | 1,210 | −8 | −3 | 94/76 | 91/75 | 88/74 | 22 | 78 | 77 | 75 |
| Des Moines AP | 41°3′ | 93°4′ | 948r | −10 | −5 | 94/75 | 91/74 | 88/73 | 23 | 78 | 77 | 75 |
| Dubuque | 42°2′ | 90°4′ | 1,065 | −12 | −7 | 90/74 | 88/73 | 86/72 | 22 | 77 | 75 | 74 |
| Fort Dodge | 42°3′ | 94°1′ | 1,111 | −12 | −7 | 91/74 | 88/74 | 86/72 | 23 | 77 | 75 | 74 |
| Iowa City | 41°4′ | 91°3′ | 645 | −11 | −6 | 92/76 | 89/76 | 87/74 | 22 | 80 | 78 | 76 |
| Keokuk | 40°2′ | 91°2′ | 526 | −5 | 0 | 95/75 | 92/75 | 89/74 | 22 | 79 | 77 | 76 |
| Marshalltown | 42°0′ | 92°5′ | 898 | −12 | −7 | 92/76 | 90/75 | 88/74 | 23 | 78 | 77 | 75 |
| Mason City AP | 43°1′ | 93°2′ | 1,194 | −15 | −11 | 90/74 | 88/74 | 85/72 | 24 | 77 | 75 | 74 |
| Newton | 41°4′ | 93°0′ | 946 | −10 | −5 | 94/75 | 91/74 | 88/73 | 23 | 78 | 77 | 75 |
| Ottumwa AP | 41°1′ | 92°2′ | 842 | −8 | −4 | 94/75 | 91/74 | 88/73 | 22 | 78 | 77 | 75 |
| Sioux City AP | 42°2′ | 96°2′ | 1,095 | −11 | −7 | 95/74 | 92/74 | 89/73 | 24 | 78 | 77 | 75 |
| Waterloo | 42°3′ | 92°2′ | 868 | −15 | −10 | 91/76 | 89/75 | 86/74 | 23 | 78 | 77 | 75 |
| Kansas | | | | | | | | | | | | |
| Atchison | 39°3′ | 95°1′ | 945 | −2 | 2 | 96/77 | 93/76 | 91/76 | 23 | 81 | 79 | 77 |
| Chanute AP | 34°4′ | 95°3′ | 977 | 3 | 7 | 100/74 | 97/74 | 94/74 | 23 | 78 | 77 | 76 |
| Dodge City AP (S) | 37°5′ | 100°0′ | 2,594 | 0 | 5 | 100/69 | 97/69 | 95/69 | 25 | 74 | 73 | 71 |
| El Dorado | 37°5′ | 96°5′ | 1,282 | 3 | 7 | 101/72 | 98/73 | 96/73 | 24 | 77 | 76 | 75 |
| Emporia | 38°2′ | 96°1′ | 1,209 | 1 | 5 | 100/74 | 97/74 | 94/73 | 25 | 78 | 77 | 76 |
| Garden City AP | 38°0′ | 101°0′ | 2,882 | −1 | 4 | 99/69 | 96/69 | 94/69 | 28 | 74 | 73 | 71 |
| Goodland AP | 39°2′ | 101°4′ | 3,645 | −5 | 0 | 99/66 | 96/65 | 93/66 | 31 | 71 | 70 | 68 |
| Great Bend | 38°2′ | 98°5′ | 1,940 | 0 | 4 | 101/73 | 98/73 | 95/73 | 28 | 78 | 76 | 75 |
| Hutchinson AP | 38°0′ | 97°5′ | 1,524 | 4 | 8 | 102/72 | 99/72 | 97/72 | 28 | 77 | 75 | 74 |
| Liberal | 37°0′ | 101°0′ | 2,838 | 2 | 7 | 99/68 | 96/68 | 94/68 | 28 | 73 | 72 | 71 |
| Manhattan, Fort Riley (S) | 39°0′ | 96°5′ | 1,076 | −1 | 3 | 99/75 | 95/75 | 92/74 | 24 | 78 | 77 | 76 |
| Parsons | 37°2′ | 95°3′ | 908 | 5 | 9 | 100/74 | 97/74 | 94/74 | 23 | 79 | 77 | 76 |
| Russell AP | 38°5′ | 98°5′ | 1,864 | 0 | 4 | 101/73 | 98/73 | 95/73 | 29 | 78 | 76 | 75 |
| Salina | 38°5′ | 97°4′ | 1,271 | 0 | 5 | 103/74 | 100/74 | 97/73 | 26 | 78 | 77 | 75 |
| Topeka AP | 39°0′ | 95°4′ | 877 | 0 | 4 | 99/75 | 96/75 | 93/74 | 24 | 79 | 78 | 76 |
| Wichita AP | 37°4′ | 97°3′ | 1,321 | 3 | 7 | 101/72 | 98/73 | 96/73 | 23 | 77 | 76 | 75 |
| Kentucky | | | | | | | | | | | | |
| Ashland | 38°3′ | 82°4′ | 551 | 5 | 10 | 94/76 | 91/74 | 89/73 | 22 | 78 | 77 | 75 |
| Bowling Green AP | 37°0′ | 86°3′ | 535 | 4 | 10 | 94/77 | 92/75 | 89/74 | 21 | 79 | 77 | 76 |
| Corbin AP | 37°0′ | 84°1′ | 1,175 | 4 | 9 | 94/73 | 92/73 | 89/72 | 23 | 77 | 76 | 75 |
| Covington AP | 39°0′ | 84°4′ | 869 | 1 | 6 | 92/73 | 90/72 | 88/72 | 22 | 77 | 75 | 74 |

**TABLE A10 Continued**

| Col. 1 | Col. 2 | Col. 3 | Col. 4 | Col. 5 (Winter[b]) | | Col. 6 (Summer[c]) | | | Col. 7 | Col. 8 | | |
|---|---|---|---|---|---|---|---|---|---|---|---|---|
| | | | | Design dry-bulb | | Design dry-bulb and mean coincident wet-bulb | | | Mean daily | Design wet-bulb | | |
| State and station | Latitude[d] | Longitude[d] | Elevation[e] Ft | 99% | 97.5% | 1% | 2.5% | 5% | range | 1% | 2.5% | 5% |
| Hopkinsville, Campbell AFB | 36°4′ | 87°3′ | 540 | 4 | 10 | 94/77 | 92/75 | 89/74 | 21 | 79 | 77 | 76 |
| Lexington AP (S) | 38°0′ | 84°4′ | 979 | 3 | 8 | 93/73 | 91/73 | 88/72 | 22 | 77 | 76 | 75 |
| Louisville AP (S) | 38°1′ | 85°4′ | 474 | 5 | 10 | 95/74 | 93/74 | 90/74 | 23 | 79 | 77 | 76 |
| Madisonville | 37°2′ | 87°3′ | 439 | 5 | 10 | 96/76 | 93/75 | 90/75 | 22 | 79 | 78 | 77 |
| Owensboro | 37°5′ | 87°1′ | 420 | 5 | 10 | 97/76 | 94/75 | 91/75 | 23 | 79 | 78 | 77 |
| Paducah AP | 37°0′ | 88°4′ | 398 | 7 | 12 | 98/76 | 95/75 | 92/75 | 20 | 79 | 78 | 77 |
| Louisiana | | | | | | | | | | | | |
| Alexandria AP | 31°2′ | 92°2′ | 92 | 23 | 27 | 95/77 | 94/77 | 92/77 | 20 | 80 | 79 | 78 |
| Baton Rouge AP | 30°3′ | 91°1′ | 64 | 25 | 29 | 95/77 | 93/77 | 92/77 | 19 | 80 | 80 | 79 |
| Bogalusa | 30°5′ | 89°5′ | 103 | 24 | 28 | 95/77 | 93/77 | 92/77 | 19 | 80 | 80 | 79 |
| Houma | 29°3′ | 90°4′ | 13 | 31 | 35 | 95/78 | 93/78 | 92/77 | 15 | 81 | 80 | 79 |
| Lafayette AP | 30°1′ | 92°0′ | 38 | 26 | 30 | 95/78 | 94/78 | 92/78 | 18 | 81 | 80 | 79 |
| Lake Charles AP (S) | 30°1′ | 93°1′ | 14 | 27 | 31 | 95/77 | 93/77 | 92/77 | 17 | 80 | 79 | 79 |
| Minden | 32°4′ | 93°2′ | 250 | 20 | 25 | 99/77 | 96/76 | 94/76 | 20 | 79 | 79 | 78 |
| Monroe AP | 32°3′ | 92°0′ | 78 | 20 | 25 | 99/77 | 96/76 | 94/76 | 20 | 79 | 79 | 78 |
| Natchitoches | 31°5′ | 93°0′ | 120 | 22 | 26 | 97/77 | 95/77 | 93/77 | 20 | 80 | 79 | 78 |
| New Orleans AP | 30°0′ | 90°2′ | 3 | 29 | 33 | 93/78 | 92/78 | 90/77 | 16 | 81 | 80 | 79 |
| Shreveport AP (S) | 32°3′ | 93°5′ | 252 | 20 | 25 | 99/77 | 96/76 | 94/76 | 20 | 79 | 79 | 78 |
| Maine | | | | | | | | | | | | |
| Augusta AP | 44°2′ | 69°5′ | 350 | −7 | −3 | 88/73 | 85/70 | 82/68 | 22 | 74 | 72 | 70 |
| Bangor, Dow AFB | 44°5′ | 68°5′ | 162 | −11 | −6 | 86/70 | 83/68 | 80/67 | 22 | 73 | 71 | 69 |
| Caribou AP (S) | 46°5′ | 68°0′ | 624 | −18 | −13 | 84/69 | 81/67 | 78/66 | 21 | 71 | 69 | 67 |
| Lewiston | 44°0′ | 70°1′ | 182 | −7 | −2 | 88/73 | 85/70 | 82/68 | 22 | 74 | 72 | 70 |
| Millinocket AP | 45°4′ | 68°4′ | 405 | −13 | −9 | 87/69 | 83/68 | 80/66 | 22 | 72 | 70 | 68 |
| Portland (S) | 43°4′ | 70°2′ | 61 | −6 | −1 | 87/72 | 84/71 | 81/69 | 22 | 74 | 72 | 70 |
| Waterville | 44°3′ | 69°4′ | 89 | −8 | −4 | 87/72 | 84/69 | 81/68 | 22 | 74 | 72 | 70 |
| Maryland | | | | | | | | | | | | |
| Baltimore AP | 39°1′ | 76°4′ | 146 | 10 | 13 | 94/75 | 91/75 | 89/74 | 21 | 78 | 77 | 76 |
| Baltimore CO | 39°2′ | 76°3′ | 14 | 14 | 17 | 92/77 | 89/76 | 87/75 | 17 | 80 | 78 | 76 |
| Cumberland | 39°4′ | 78°5′ | 945 | 6 | 10 | 92/75 | 89/74 | 87/74 | 22 | 77 | 76 | 75 |
| Frederick AP | 39°3′ | 77°3′ | 294 | 8 | 12 | 94/76 | 91/75 | 88/74 | 22 | 78 | 77 | 76 |
| Hagerstown | 39°4′ | 77°4′ | 660 | 8 | 12 | 94/75 | 91/74 | 89/74 | 22 | 77 | 76 | 75 |
| Salisbury (S) | 38°2′ | 75°3′ | 52 | 12 | 16 | 93/75 | 91/75 | 88/74 | 18 | 79 | 77 | 76 |
| Massachusetts | | | | | | | | | | | | |
| Boston AP (S) | 42°2′ | 71°0′ | 15 | 6 | 9 | 91/73 | 88/71 | 85/70 | 16 | 75 | 74 | 72 |
| Clinton | 42°2′ | 71°4′ | 398 | −2 | 2 | 90/72 | 87/71 | 84/69 | 17 | 75 | 73 | 72 |
| Fall River | 41°4′ | 71°1′ | 190 | 5 | 9 | 87/72 | 84/71 | 81/69 | 18 | 74 | 73 | 72 |

| | | | | | | | | | | | | |
|---|---|---|---|---|---|---|---|---|---|---|---|---|
| Framingham | 42°2′ | 71°3′ | 170 | 3 | 6 | 89/72 | 86/71 | 83/69 | 17 | 74 | 73 | 71 |
| Gloucester | 42°3′ | 70°4′ | 10 | 2 | 5 | 89/73 | 86/71 | 83/70 | 15 | 75 | 74 | 72 |
| Greenfield | 42°3′ | 72°4′ | 205 | −7 | −2 | 88/72 | 85/71 | 82/69 | 23 | 74 | 73 | 71 |
| Lawrence | 42°4′ | 71°1′ | 57 | −6 | 0 | 90/73 | 87/72 | 84/70 | 22 | 76 | 74 | 73 |
| Lowell | 42°3′ | 71°2′ | 90 | −4 | 1 | 91/73 | 88/72 | 85/70 | 21 | 76 | 74 | 73 |
| New Bedford | 41°4′ | 71°0′ | 70 | 5 | 9 | 85/72 | 82/71 | 80/69 | 19 | 74 | 73 | 72 |
| Pittsfield AP | 42°3′ | 73°2′ | 1,170 | −8 | −3 | 87/71 | 84/70 | 81/68 | 23 | 73 | 72 | 70 |
| Springfield, Westover AFB | 42°1′ | 72°3′ | 247 | −5 | 0 | 90/72 | 87/71 | 84/69 | 19 | 75 | 73 | 72 |
| Taunton | 41°5′ | 71°1′ | 20 | 5 | 9 | 89/73 | 86/72 | 83/70 | 18 | 75 | 74 | 73 |
| Worcester AP | 42°2′ | 71°5′ | 986 | 0 | 4 | 87/71 | 84/70 | 81/68 | 18 | 73 | 72 | 70 |
| Michigan | | | | | | | | | | | | |
| Adrian | 41°5′ | 84°0′ | 754 | −1 | 3 | 91/73 | 88/72 | 85/71 | 23 | 76 | 75 | 73 |
| Alpena AP | 45°0′ | 83°3′ | 689 | −11 | −6 | 89/70 | 85/70 | 83/69 | 27 | 73 | 72 | 70 |
| Battle Creek AP | 42°2′ | 85°2′ | 939 | 1 | 5 | 92/74 | 88/72 | 85/70 | 23 | 76 | 74 | 73 |
| Benton Harbor AP | 42°1′ | 86°3′ | 649 | 1 | 5 | 91/72 | 88/72 | 85/70 | 20 | 75 | 74 | 72 |
| Detroit | 42°3′ | 83°0′ | 633 | 3 | 6 | 91/73 | 88/72 | 86/71 | 20 | 76 | 74 | 73 |
| Escanaba | 45°4′ | 87°0′ | 594 | −11 | −7 | 87/70 | 83/69 | 80/68 | 17 | 73 | 71 | 69 |
| Flint AP | 42°0′ | 83°4′ | 766 | −4 | 1 | 90/73 | 87/72 | 85/71 | 25 | 76 | 74 | 72 |
| Grand Rapids AP | 42°5′ | 85°3′ | 681 | 1 | 5 | 91/72 | 88/72 | 85/70 | 24 | 75 | 74 | 72 |
| Holland | 42°5′ | 86°1′ | 612 | 2 | 6 | 88/72 | 86/71 | 83/70 | 22 | 75 | 73 | 72 |
| Jackson AP | 42°2′ | 84°2′ | 1,003 | 1 | 5 | 92/74 | 88/72 | 85/70 | 23 | 76 | 74 | 73 |
| Kalamazoo | 42°1′ | 85°3′ | 930 | 1 | 5 | 92/74 | 88/72 | 85/70 | 23 | 76 | 74 | 73 |
| Lansing AP | 42°5′ | 84°4′ | 852 | −3 | 1 | 90/73 | 87/72 | 84/70 | 24 | 75 | 74 | 72 |
| Marquette CO | 46°3′ | 87°3′ | 677 | −12 | −8 | 84/70 | 81/69 | 77/66 | 18 | 72 | 70 | 68 |
| Mt. Pleasant | 43°4′ | 84°5′ | 796 | 0 | 4 | 91/73 | 87/72 | 84/71 | 24 | 76 | 74 | 72 |
| Muskegon AP | 43°1′ | 86°1′ | 627 | 2 | 6 | 86/72 | 84/70 | 82/70 | 21 | 75 | 73 | 72 |
| Pontiac | 42°4′ | 83°2′ | 974 | 0 | 4 | 90/73 | 87/72 | 85/71 | 21 | 76 | 74 | 73 |
| Port Huron | 43°0′ | 82°3′ | 586 | 0 | 4 | 90/73 | 87/72 | 83/71 | 21 | 76 | 74 | 73 |
| Saginaw AP | 43°3′ | 84°1′ | 662 | 0 | 4 | 91/73 | 87/72 | 84/71 | 23 | 76 | 74 | 72 |
| Sault Ste. Marie AP (S) | 46°3′ | 84°2′ | 721 | −12 | −8 | 84/70 | 81/69 | 77/66 | 23 | 72 | 70 | 68 |
| Traverse City AP | 44°4′ | 85°4′ | 618 | −3 | 1 | 89/72 | 86/71 | 83/69 | 22 | 75 | 73 | 71 |
| Ypsilanti | 42°1′ | 83°3′ | 777 | 1 | 5 | 92/72 | 89/71 | 86/70 | 22 | 75 | 74 | 72 |
| Minnesota | | | | | | | | | | | | |
| Albert Lea | 43°4′ | 93°2′ | 1,235 | −17 | −12 | 90/74 | 87/72 | 84/71 | 24 | 77 | 75 | 73 |
| Alexandria AP | 45°5′ | 95°2′ | 1,421 | −22 | −16 | 91/72 | 88/72 | 85/70 | 24 | 76 | 74 | 72 |
| Bemidji AP | 47°3′ | 95°0′ | 1,392 | −31 | −26 | 88/69 | 85/69 | 81/67 | 24 | 73 | 71 | 69 |
| Brainerd | 46°2′ | 94°2′ | 1,214 | −20 | −16 | 90/73 | 87/71 | 84/69 | 24 | 75 | 73 | 71 |
| Duluth AP | 46°5′ | 92°1′ | 1,426 | −21 | −16 | 85/70 | 82/68 | 79/66 | 22 | 72 | 70 | 68 |
| Fairbault | 44°2′ | 93°2′ | 1,190 | −17 | −12 | 91/74 | 88/72 | 85/71 | 24 | 77 | 75 | 73 |
| Fergus Falls | 46°1′ | 96°0′ | 1,210 | −21 | −17 | 91/72 | 88/72 | 85/70 | 24 | 76 | 74 | 72 |
| International Falls AP | 48°3′ | 93°2′ | 1,179 | −29 | −25 | 85/68 | 83/68 | 80/66 | 26 | 71 | 70 | 68 |
| Mankato | 44°1′ | 94°0′ | 785 | −17 | −12 | 91/72 | 88/72 | 85/70 | 24 | 77 | 75 | 73 |
| Minneapolis/St. Paul AP | 44°5′ | 93°1′ | 822 | −16 | −12 | 92/75 | 89/73 | 86/71 | 22 | 77 | 75 | 73 |
| Rochester AP | 44°0′ | 92°3′ | 1,297 | −17 | −12 | 90/74 | 87/72 | 84/71 | 24 | 77 | 75 | 73 |
| St. Cloud AP (S) | 45°4′ | 94°1′ | 1,034 | −15 | −11 | 91/74 | 88/72 | 85/70 | 24 | 76 | 74 | 72 |
| Virginia | 47°3′ | 92°3′ | 1,435 | −25 | −21 | 85/69 | 83/68 | 80/66 | 23 | 71 | 70 | 68 |
| Willmar | 45°1′ | 95°0′ | 1,133 | −15 | −11 | 91/74 | 88/72 | 85/71 | 24 | 76 | 74 | 72 |

**TABLE A10 Continued**

| Col. 1 | Col. 2 | Col. 3 | Col. 4 | Col. 5 (Winter[b]) | | Col. 6 (Summer[c]) | | | Col. 7 | Col. 8 | | |
|---|---|---|---|---|---|---|---|---|---|---|---|---|
| | | | | Design dry-bulb | | Design dry-bulb and mean coincident wet-bulb | | | Mean daily | Design wet-bulb | | |
| State and station | Latitude[d] | Longitude[d] | Elevation[e] Ft | 99% | 97.5% | 1% | 2.5% | 5% | range | 1% | 2.5% | 5% |
| Winona | 44°1′ | 91°4′ | 652 | −14 | −10 | 91/75 | 88/73 | 85/72 | 24 | 77 | 75 | 74 |
| Mississippi | | | | | | | | | | | | |
| Biloxi, Keesler AFB | 30°2′ | 89°0′ | 25 | 28 | 31 | 94/79 | 92/79 | 90/78 | 16 | 82 | 81 | 80 |
| Clarksdale | 34°1′ | 90°3′ | 178 | 14 | 19 | 96/77 | 94/77 | 92/76 | 21 | 80 | 79 | 78 |
| Columbus AFB | 33°4′ | 88°3′ | 224 | 15 | 20 | 95/77 | 93/77 | 91/76 | 22 | 80 | 79 | 78 |
| Greenville AFB | 33°3′ | 91°1′ | 139 | 15 | 20 | 95/77 | 93/77 | 91/76 | 21 | 80 | 79 | 78 |
| Greenwood | 33°3′ | 90°1′ | 128 | 15 | 20 | 95/77 | 93/77 | 91/76 | 21 | 80 | 79 | 78 |
| Hattiesburg | 31°2′ | 89°2′ | 200 | 24 | 27 | 96/78 | 94/77 | 92/77 | 21 | 81 | 80 | 79 |
| Jackson AP | 32°2′ | 90°1′ | 330 | 21 | 25 | 97/76 | 95/76 | 93/76 | 21 | 79 | 78 | 78 |
| Laurel | 31°4′ | 89°1′ | 264 | 24 | 27 | 96/78 | 94/77 | 92/77 | 21 | 81 | 80 | 79 |
| McComb AP | 31°2′ | 90°3′ | 458 | 21 | 26 | 96/77 | 94/76 | 92/76 | 18 | 80 | 79 | 78 |
| Meridian AP | 32°2′ | 88°5′ | 294 | 19 | 23 | 97/77 | 95/76 | 93/76 | 22 | 80 | 79 | 78 |
| Natchez | 31°4′ | 91°3′ | 168 | 23 | 27 | 96/78 | 94/78 | 92/77 | 21 | 81 | 80 | 79 |
| Tupelo | 34°2′ | 88°4′ | 289 | 14 | 19 | 96/77 | 94/77 | 92/76 | 22 | 80 | 79 | 78 |
| Vicksburg CO | 32°2′ | 91°0′ | 234 | 22 | 26 | 97/78 | 95/78 | 93/77 | 21 | 81 | 80 | 79 |
| Missouri | | | | | | | | | | | | |
| Cape Girardeau | 37°1′ | 89°3′ | 330 | 8 | 13 | 98/76 | 95/75 | 92/75 | 21 | 79 | 78 | 77 |
| Columbia AP (S) | 39°0′ | 92°2′ | 778 | −1 | 4 | 97/74 | 94/74 | 91/73 | 22 | 78 | 77 | 76 |
| Farmington AP | 37°5′ | 90°3′ | 928 | 3 | 8 | 96/76 | 93/75 | 90/74 | 22 | 78 | 77 | 75 |
| Hannibal | 39°4′ | 91°2′ | 489 | −2 | 3 | 96/76 | 93/76 | 90/76 | 22 | 80 | 78 | 77 |
| Jefferson City | 38°4′ | 92°1′ | 640 | 2 | 7 | 98/75 | 95/74 | 92/74 | 23 | 78 | 77 | 76 |
| Joplin AP | 37°1′ | 94°3′ | 982 | 6 | 10 | 100/73 | 97/73 | 94/73 | 24 | 78 | 77 | 76 |
| Kansas City AP | 39°1′ | 94°4′ | 742 | 2 | 6 | 99/75 | 96/74 | 93/74 | 20 | 78 | 77 | 76 |
| Kirksville AP | 40°1′ | 92°4′ | 966 | −5 | 0 | 96/74 | 93/74 | 90/73 | 24 | 78 | 77 | 76 |
| Mexico | 39°1′ | 92°0′ | 775 | −1 | 4 | 97/74 | 94/74 | 91/73 | 22 | 78 | 77 | 76 |
| Moberly | 39°3′ | 92°3′ | 850 | −2 | 3 | 97/74 | 94/74 | 91/73 | 23 | 78 | 77 | 76 |
| Poplar Bluff | 36°5′ | 90°3′ | 322 | 11 | 16 | 98/78 | 95/76 | 92/76 | 22 | 81 | 79 | 78 |
| Rolla | 38°0′ | 91°5′ | 1,202 | 3 | 9 | 94/77 | 91/75 | 89/74 | 22 | 78 | 77 | 76 |
| St. Joseph AP | 39°5′ | 95°0′ | 809 | −3 | 2 | 96/77 | 93/76 | 91/76 | 23 | 81 | 79 | 77 |
| St. Louis AP | 38°5′ | 90°2′ | 535 | 2 | 6 | 97/75 | 94/75 | 91/74 | 21 | 78 | 77 | 76 |
| St. Louis CO | 38°4′ | 90°2′ | 465 | 3 | 8 | 98/75 | 94/75 | 91/74 | 18 | 78 | 77 | 76 |
| Sedalia, Whiteman AFB | 38°4′ | 93°3′ | 838 | −1 | 4 | 95/76 | 92/76 | 90/75 | 22 | 79 | 78 | 76 |
| Sikeston | 36°5′ | 89°3′ | 318 | 9 | 15 | 98/77 | 95/76 | 92/75 | 21 | 80 | 78 | 77 |
| Springfield AP | 37°1′ | 93°2′ | 1,265 | 3 | 9 | 96/73 | 93/74 | 91/74 | 23 | 78 | 77 | 75 |
| Montana | | | | | | | | | | | | |
| Billings AP | 45°5′ | 108°3′ | 3,567 | −15 | −10 | 94/64 | 91/64 | 88/63 | 31 | 67 | 66 | 64 |
| Bozeman | 45°5′ | 111°0′ | 4,856 | −20 | −14 | 90/61 | 87/60 | 84/59 | 32 | 63 | 62 | 60 |

| | | | | | | | | | | | | |
|---|---|---|---|---|---|---|---|---|---|---|---|---|
| Butte AP | 46°0′ | 112°3′ | 5,526r | –24 | –17 | 86/58 | 83/56 | 80/56 | 35 | 60 | 58 | 57 |
| Cut Bank AP | 48°4′ | 112°2′ | 3,838r | –25 | –20 | 88/61 | 85/61 | 82/60 | 35 | 64 | 62 | 61 |
| Glasgow AP (S) | 48°1′ | 106°4′ | 2,277 | –22 | –18 | 92/64 | 89/63 | 85/62 | 29 | 68 | 66 | 64 |
| Glendive | 47°1′ | 104°4′ | 2,076 | –18 | –13 | 95/66 | 92/64 | 89/62 | 29 | 69 | 67 | 65 |
| Great Falls AP (S) | 47°3′ | 111°2′ | 3,664r | –21 | –15 | 91/60 | 88/60 | 85/59 | 28 | 64 | 62 | 60 |
| Havre | 48°3′ | 109°4′ | 2,488 | –18 | –11 | 94/65 | 90/64 | 87/63 | 33 | 68 | 66 | 65 |
| Helena AP | 46°4′ | 112°0′ | 3,893 | –21 | –16 | 91/60 | 88/60 | 85/59 | 32 | 64 | 62 | 61 |
| Kalispell AP | 48°2′ | 114°2′ | 2,965 | –14 | –7 | 91/62 | 87/61 | 84/60 | 34 | 65 | 63 | 62 |
| Lewiston AP | 47°0′ | 109°3′ | 4,132 | –22 | –16 | 90/62 | 87/61 | 83/60 | 30 | 65 | 63 | 62 |
| Livingston AP | 45°4′ | 110°3′ | 4,653 | –20 | –14 | 90/61 | 87/60 | 84/59 | 32 | 63 | 62 | 60 |
| Miles City AP | 46°3′ | 105°5′ | 2,629 | –20 | –15 | 98/66 | 95/66 | 92/65 | 30 | 70 | 68 | 67 |
| Missoula AP | 46°5′ | 114°1′ | 3,200 | –13 | –6 | 92/62 | 88/61 | 85/60 | 36 | 65 | 63 | 62 |
| Nebraska | | | | | | | | | | | | |
| Beatrice | 40°2′ | 96°5′ | 1,235 | –5 | –2 | 99/75 | 95/74 | 92/74 | 24 | 78 | 77 | 76 |
| Chadron AP | 42°5′ | 103°0′ | 3,300 | –8 | –3 | 97/66 | 94/65 | 91/65 | 30 | 71 | 69 | 68 |
| Columbus | 41°3′ | 97°2′ | 1,442 | –6 | –2 | 98/74 | 95/73 | 92/73 | 25 | 77 | 76 | 75 |
| Fremont | 41°3′ | 96°3′ | 1,203 | –6 | –2 | 98/75 | 95/74 | 92/74 | 22 | 78 | 77 | 76 |
| Grand Island AP | 41°0′ | 98°2′ | 1,841 | –8 | –3 | 97/72 | 94/71 | 91/71 | 28 | 75 | 74 | 73 |
| Hastings | 40°4′ | 98°3′ | 1,932 | –7 | –3 | 97/72 | 94/71 | 91/71 | 27 | 75 | 74 | 73 |
| Kearney | 40°4′ | 99°1′ | 2,146 | –9 | –4 | 96/71 | 93/70 | 90/70 | 28 | 74 | 73 | 72 |
| Lincoln CO (S) | 40°5′ | 96°5′ | 1,150 | –5 | –2 | 99/75 | 95/74 | 92/74 | 24 | 78 | 77 | 76 |
| McCook | 40°1′ | 100°4′ | 2,565 | –6 | –2 | 98/69 | 95/69 | 91/69 | 28 | 74 | 72 | 71 |
| Norfolk | 42°0′ | 97°3′ | 1,532 | –8 | –4 | 97/74 | 93/74 | 90/73 | 30 | 78 | 77 | 75 |
| North Platte AP (S) | 41°1′ | 100°4′ | 2,779 | –8 | –4 | 97/69 | 94/69 | 90/69 | 28 | 74 | 72 | 71 |
| Omaha AP | 41°2′ | 95°5′ | 978 | –8 | –3 | 94/76 | 91/75 | 88/74 | 22 | 78 | 77 | 75 |
| Scottsbluff AP | 41°5′ | 103°4′ | 3,950 | –8 | –3 | 95/65 | 92/65 | 90/64 | 31 | 70 | 68 | 67 |
| Sidney AP | 41°1′ | 103°0′ | 4,292 | –8 | –3 | 95/65 | 92/65 | 90/64 | 31 | 70 | 68 | 67 |
| Nevada | | | | | | | | | | | | |
| Carson City | 39°1′ | 119°5′ | 4,675 | 4 | 9 | 94/60 | 91/59 | 89/58 | 42 | 63 | 61 | 60 |
| Elko AP | 40°5′ | 115°5′ | 5,075 | –8 | –2 | 94/59 | 92/59 | 90/58 | 42 | 63 | 62 | 60 |
| Ely AP (S) | 39°1′ | 114°5′ | 6,257 | –10 | –4 | 89/57 | 87/56 | 85/55 | 39 | 60 | 59 | 58 |
| Las Vegas AP (S) | 36°1′ | 115°1′ | 2,162 | 25 | 28 | 108/66 | 106/65 | 104/65 | 30 | 71 | 70 | 69 |
| Lovelock AP | 40°0′ | 118°3′ | 3,900 | 8 | 12 | 98/63 | 96/63 | 93/62 | 42 | 66 | 65 | 64 |
| Reno AP (S) | 39°3′ | 119°5′ | 4,404 | 5 | 10 | 95/61 | 92/60 | 90/59 | 45 | 64 | 62 | 61 |
| Reno CO | 39°3′ | 119°5′ | 4,490 | 6 | 11 | 96/61 | 93/60 | 91/59 | 45 | 64 | 62 | 61 |
| Tonopah AP | 38°0′ | 117°1′ | 5,426 | 5 | 10 | 94/60 | 92/59 | 90/58 | 40 | 64 | 62 | 61 |
| Winnemucca AP | 40°5′ | 117°5′ | 4,299 | –1 | 3 | 96/60 | 94/60 | 92/60 | 42 | 64 | 62 | 61 |
| New Hampshire | | | | | | | | | | | | |
| Berlin | 44°3′ | 71°1′ | 1,110 | –14 | –9 | 87/71 | 84/69 | 81/68 | 22 | 73 | 71 | 70 |
| Claremont | 43°2′ | 72°2′ | 420 | –9 | –4 | 89/72 | 86/70 | 83/69 | 24 | 74 | 73 | 71 |
| Concord AP | 43°1′ | 71°3′ | 339 | –8 | –3 | 90/72 | 87/70 | 84/69 | 26 | 74 | 73 | 71 |
| Keene | 43°0′ | 72°2′ | 490 | –12 | –7 | 90/72 | 87/70 | 83/69 | 24 | 74 | 73 | 71 |
| Laconia | 43°3′ | 71°3′ | 505 | –10 | –5 | 89/72 | 86/70 | 83/69 | 25 | 74 | 73 | 71 |
| Manchester, Grenier AFB | 43°0′ | 71°3′ | 253 | –8 | –3 | 91/72 | 88/71 | 85/70 | 24 | 75 | 74 | 72 |
| Portsmouth, Pease AFB | 43°1′ | 70°5′ | 127 | –2 | 2 | 89/73 | 85/71 | 83/70 | 22 | 75 | 74 | 72 |
| New Jersey | | | | | | | | | | | | |
| Atlantic City CO | 39°3′ | 74°3′ | 11 | 10 | 13 | 92/74 | 89/74 | 86/72 | 18 | 78 | 77 | 75 |

**TABLE A10 Continued**

| Col. 1 | Col. 2 | Col. 3 | Col. 4 | Winter[b] Col. 5 | | Summer[c] Col. 6 | | | Col. 7 | Col. 8 | | |
|---|---|---|---|---|---|---|---|---|---|---|---|---|
| | | | | Design dry-bulb | | Design dry-bulb and mean coincident wet-bulb | | | Mean daily | Design wet-bulb | | |
| State and station | Latitude[d] | Longitude[d] | Elevation[e] Ft | 99% | 97.5% | 1% | 2.5% | 5% | range | 1% | 2.5% | 5% |
| Long Branch | 40°2′ | 74°0′ | 20 | 10 | 13 | 93/74 | 90/73 | 87/72 | 18 | 78 | 77 | 75 |
| Newark AP | 40°4′ | 74°1′ | 11 | 10 | 14 | 94/74 | 91/73 | 88/72 | 20 | 77 | 76 | 75 |
| New Brunswick | 40°3′ | 74°3′ | 86 | 6 | 10 | 92/74 | 89/73 | 86/72 | 19 | 77 | 76 | 75 |
| Paterson | 40°5′ | 74°1′ | 100 | 6 | 10 | 94/74 | 91/73 | 88/72 | 21 | 77 | 76 | 75 |
| Phillipsburg | 40°4′ | 75°1′ | 180 | 1 | 6 | 92/73 | 89/72 | 86/71 | 21 | 76 | 75 | 74 |
| Trenton CO | 40°1′ | 74°5′ | 144 | 11 | 14 | 91/75 | 88/74 | 85/73 | 19 | 78 | 76 | 75 |
| Vineland | 39°3′ | 75°0′ | 95 | 8 | 11 | 91/75 | 89/74 | 86/73 | 19 | 78 | 76 | 75 |
| New Mexico | | | | | | | | | | | | |
| Alamagordo, Holloman AFB | 32°5′ | 106°1′ | 4,070 | 14 | 19 | 98/64 | 96/64 | 94/64 | 30 | 69 | 68 | 67 |
| Albuquerque AP (S) | 35°0′ | 106°4′ | 5,310 | 12 | 16 | 96/61 | 94/61 | 92/61 | 27 | 66 | 65 | 64 |
| Artesia | 32°5′ | 104°2′ | 3,375 | 13 | 19 | 103/67 | 100/67 | 97/67 | 30 | 72 | 71 | 70 |
| Carlsbad AP | 32°2′ | 104°2′ | 3,234 | 13 | 19 | 103/67 | 100/67 | 97/67 | 28 | 72 | 71 | 70 |
| Clovis AP | 34°3′ | 103°1′ | 4,279 | 8 | 13 | 95/65 | 93/65 | 91/61 | 28 | 69 | 68 | 67 |
| Farmington AP | 36°5′ | 108°1′ | 5,495 | 1 | 6 | 95/63 | 93/62 | 91/61 | 30 | 67 | 65 | 64 |
| Gallup | 35°3′ | 108°5′ | 6,465 | 0 | 5 | 90/59 | 89/58 | 86/58 | 32 | 64 | 62 | 61 |
| Grants | 35°1′ | 107°5′ | 6,520 | −1 | 4 | 89/59 | 88/58 | 85/57 | 32 | 64 | 62 | 61 |
| Hobbs AP | 32°4′ | 103°1′ | 3,664 | 13 | 18 | 101/66 | 99/66 | 97/66 | 29 | 71 | 70 | 69 |
| Las Cruces | 32°2′ | 107°0′ | 3,900 | 15 | 20 | 99/64 | 96/64 | 94/64 | 30 | 69 | 68 | 67 |
| Los Alamos | 35°5′ | 106°2′ | 7,410 | 5 | 9 | 89/60 | 87/60 | 85/60 | 32 | 62 | 61 | 60 |
| Raton AP | 36°5′ | 104°3′ | 6,379 | −4 | 1 | 91/60 | 89/60 | 87/60 | 34 | 65 | 64 | 63 |
| Roswell, Walker AFB | 33°2′ | 104°3′ | 3,643 | 13 | 18 | 100/66 | 98/66 | 96/66 | 33 | 71 | 70 | 69 |
| Santa Fe CO | 35°4′ | 106°0′ | 7,045 | 6 | 10 | 90/61 | 88/61 | 86/61 | 28 | 63 | 62 | 61 |
| Silver City AP | 32°4′ | 108°2′ | 5,373 | 5 | 10 | 95/61 | 94/60 | 91/60 | 30 | 66 | 64 | 63 |
| Socorro AP | 34°0′ | 106°5′ | 4,617 | 13 | 17 | 97/62 | 95/62 | 93/62 | 30 | 67 | 66 | 65 |
| Tucumcari AP | 35°1′ | 103°4′ | 4,053 | 8 | 13 | 99/66 | 97/66 | 95/65 | 28 | 70 | 69 | 68 |
| New York | | | | | | | | | | | | |
| Albany AP (S) | 42°5′ | 73°5′ | 277 | −6 | −1 | 91/73 | 88/72 | 85/70 | 23 | 75 | 74 | 72 |
| Albany CO | 42°5′ | 73°5′ | 19 | −4 | 1 | 91/73 | 88/72 | 85/70 | 20 | 75 | 74 | 72 |
| Auburn | 43°0′ | 76°3′ | 715 | −3 | 2 | 90/73 | 87/71 | 84/70 | 22 | 75 | 73 | 72 |
| Batavia | 43°0′ | 78°1′ | 900 | 1 | 5 | 90/72 | 87/71 | 84/70 | 22 | 75 | 73 | 72 |
| Binghamton AP | 42°1′ | 76°0′ | 1,590 | −2 | 1 | 86/71 | 83/69 | 81/68 | 20 | 73 | 72 | 70 |
| Buffalo AP | 43°0′ | 78°4′ | 705r | 2 | 6 | 88/71 | 85/70 | 83/69 | 21 | 74 | 73 | 72 |
| Cortland | 42°4′ | 76°1′ | 1,129 | −5 | 0 | 88/71 | 85/71 | 82/70 | 23 | 74 | 73 | 71 |
| Dunkirk | 42°3′ | 79°2′ | 590 | 4 | 9 | 88/73 | 85/72 | 83/71 | 18 | 75 | 74 | 72 |
| Elmira AP | 42°1′ | 76°5′ | 860 | −4 | 1 | 89/71 | 86/71 | 83/70 | 24 | 74 | 73 | 71 |
| Geneva (S) | 42°5′ | 77°0′ | 590 | −3 | 2 | 90/73 | 87/71 | 84/70 | 22 | 75 | 73 | 72 |
| Glen Falls | 42°2′ | 73°4′ | 321 | −11 | −5 | 88/72 | 85/71 | 82/69 | 23 | 74 | 73 | 71 |

| | | | | | | | | | | | | |
|---|---|---|---|---|---|---|---|---|---|---|---|---|
| Gloversville | 43°1′ | 74°2′ | 790 | −8 | −2 | 89/72 | 86/71 | 83/69 | 23 | 75 | 74 | 72 |
| Hornell | 42°2′ | 77°4′ | 1,325 | −4 | 0 | 88/71 | 85/70 | 82/69 | 24 | 74 | 73 | 72 |
| Ithaca (S) | 42°3′ | 76°3′ | 950 | −5 | 0 | 88/71 | 85/71 | 82/70 | 24 | 74 | 73 | 71 |
| Jamestown | 42°1′ | 79°2′ | 1,390 | −1 | 3 | 88/70 | 86/70 | 83/69 | 20 | 74 | 72 | 71 |
| Kingston | 42°0′ | 74°0′ | 279 | −3 | 2 | 91/73 | 88/72 | 85/70 | 22 | 76 | 74 | 73 |
| Lockport | 43°1′ | 78°4′ | 520 | 4 | 7 | 89/74 | 86/72 | 84/71 | 21 | 76 | 74 | 73 |
| Massena AP | 45°0′ | 75°0′ | 202r | −13 | −8 | 86/70 | 83/69 | 80/68 | 20 | 73 | 72 | 70 |
| Newburg-Stewart AFB | 41°3′ | 74°1′ | 460 | −1 | 4 | 90/73 | 88/72 | 85/70 | 21 | 76 | 74 | 73 |
| NYC-Central Park (S) | 40°5′ | 74°0′ | 132 | 11 | 15 | 92/74 | 89/73 | 87/72 | 17 | 76 | 75 | 74 |
| NYC-Kennedy AP | 40°4′ | 73°5′ | 16 | 12 | 15 | 90/73 | 87/72 | 84/71 | 16 | 76 | 75 | 74 |
| NYC-La Guardia AP | 40°5′ | 73°5′ | 19 | 11 | 15 | 92/74 | 89/73 | 87/72 | 16 | 76 | 75 | 74 |
| Niagra Falls AP | 43°1′ | 79°0′ | 596 | 4 | 7 | 89/74 | 86/72 | 84/71 | 20 | 76 | 74 | 73 |
| Olean | 42°1′ | 78°3′ | 1,420 | −2 | 2 | 87/71 | 84/71 | 81/70 | 23 | 74 | 73 | 71 |
| Oneonta | 42°3′ | 75°0′ | 1,150 | −7 | −4 | 86/71 | 83/69 | 80/68 | 24 | 73 | 72 | 70 |
| Oswego CO | 43°3′ | 76°3′ | 300 | 1 | 7 | 86/73 | 83/71 | 80/70 | 20 | 75 | 73 | 72 |
| Plattsburg AFB | 44°4′ | 73°3′ | 165 | −13 | −8 | 86/70 | 83/69 | 80/68 | 22 | 73 | 72 | 70 |
| Poughkeepsie | 41°4′ | 73°5′ | 103 | 0 | 6 | 92/74 | 89/74 | 86/72 | 21 | 77 | 75 | 74 |
| Rochester AP | 43°1′ | 77°4′ | 543 | 1 | 5 | 91/73 | 88/71 | 85/70 | 22 | 75 | 73 | 72 |
| Rome-Griffiss AFB | 43°1′ | 75°3′ | 515 | −11 | −5 | 88/71 | 85/70 | 83/69 | 22 | 75 | 73 | 71 |
| Schenectady (S) | 42°5′ | 74°0′ | 217 | −4 | 1 | 90/73 | 87/72 | 84/70 | 22 | 75 | 74 | 72 |
| Suffolk County AFB | 40°5′ | | 57 | 7 | 10 | 86/72 | 83/71 | 80/70 | 16 | 76 | 74 | 73 |
| Syracuse AP | 43°1′ | 76°1′ | 424 | −3 | 2 | 90/73 | 87/71 | 84/70 | 20 | 75 | 73 | 72 |
| Utica | 43°1′ | 75°2′ | 714 | −12 | −6 | 88/73 | 85/71 | 82/70 | 22 | 75 | 73 | 71 |
| Watertown | 44°0′ | 76°0′ | 497 | −11 | −6 | 86/73 | 83/71 | 81/70 | 20 | 75 | 73 | 72 |
| North Carolina | | | | | | | | | | | | |
| Ashville AP | 35°3′ | 82°3′ | 217r | 10 | 14 | 89/73 | 87/72 | 85/71 | 21 | 75 | 74 | 72 |
| Charlotte AP | 35°0′ | 81°0′ | 735 | 18 | 22 | 95/74 | 93/74 | 91/74 | 20 | 77 | 76 | 76 |
| Durham | 36°0′ | 78°5′ | 406 | 16 | 20 | 94/75 | 92/75 | 90/75 | 20 | 78 | 77 | 76 |
| Elizabeth City AP | 36°2′ | 76°1′ | 10 | 12 | 19 | 93/78 | 91/77 | 89/76 | 18 | 80 | 78 | 78 |
| Fayetteville, Pope AFB | 35°1′ | 79°0′ | 95 | 17 | 20 | 95/76 | 92/76 | 90/75 | 20 | 79 | 78 | 77 |
| Goldsboro, Seymour-Johnson AFB | 35°2′ | 78°0′ | 88 | 18 | 21 | 94/77 | 91/76 | 89/75 | 18 | 79 | 78 | 77 |
| Greensboro | 36°1′ | 80°0′ | 887 | 14 | 18 | 93/74 | 91/73 | 89/73 | 21 | 77 | 76 | 75 |
| Greenville | 35°4′ | 77°2′ | 25 | 18 | 21 | 93/77 | 91/76 | 89/75 | 19 | 79 | 78 | 77 |
| Henderson | 36°2′ | 78°2′ | 510 | 12 | 15 | 95/77 | 92/76 | 90/76 | 20 | 79 | 78 | 77 |
| Hickory | 35°4′ | 81°2′ | 1,165 | 14 | 18 | 92/73 | 90/72 | 88/72 | 21 | 75 | 74 | 73 |
| Jacksonville | 34°5′ | 77°3′ | 24 | 20 | 24 | 92/78 | 90/78 | 88/77 | 18 | 80 | 79 | 78 |
| Lumberton | 34°4′ | 79°0′ | 132 | 18 | 21 | 95/76 | 92/76 | 90/75 | 20 | 79 | 78 | 77 |
| New Bern AP | 35°1′ | 77°0′ | 17 | 20 | 24 | 92/78 | 90/78 | 88/77 | 18 | 80 | 79 | 78 |
| Raleigh/Durham AP (S) | 35°5′ | 78°5′ | 433 | 16 | 20 | 94/75 | 92/75 | 90/75 | 20 | 78 | 77 | 76 |
| Rocky Mount | 36°0′ | 77°5′ | 81 | 18 | 21 | 94/77 | 91/76 | 89/75 | 19 | 79 | 78 | 77 |
| Wilmington AP | 34°2′ | 78°0′ | 30 | 23 | 26 | 93/79 | 91/78 | 89/77 | 18 | 81 | 80 | 79 |
| Winston-Salem AP | 36°1′ | 80°1′ | 967 | 16 | 20 | 94/74 | 91/73 | 89/73 | 20 | 76 | 75 | 74 |
| North Dakota | | | | | | | | | | | | |
| Bismark AP (S) | 46°5′ | 100°5′ | 1,647 | −23 | −19 | 95/68 | 91/68 | 88/67 | 27 | 73 | 71 | 70 |
| Devil's Lake | 48°1′ | 98°5′ | 1,471 | −25 | −21 | 91/69 | 88/68 | 85/66 | 25 | 73 | 71 | 69 |
| Dickinson AP | 46°5′ | 102°5′ | 2,595 | −21 | −17 | 94/68 | 90/66 | 87/65 | 25 | 71 | 69 | 68 |

**TABLE A10 Continued**

| Col. 1 | Col. 2 | Col. 3 | Col. 4 | Col. 5 (Winter[b]) | | Col. 6 (Summer[c]) | | | Col. 7 | Col. 8 | | |
|---|---|---|---|---|---|---|---|---|---|---|---|---|
| | | | | Design dry-bulb | | Design dry-bulb and mean coincident wet-bulb | | | Mean daily | Design wet-bulb | | |
| State and station | Latitude[d] | Longitude[d] | Elevation[e] Ft | 99% | 97.5% | 1% | 2.5% | 5% | range | 1% | 2.5% | 5% |
| Fargo AP | 46°5′ | 96°5′ | 900 | −22 | −18 | 92/73 | 89/71 | 85/69 | 25 | 76 | 74 | 72 |
| Grands Forks AP | 48°0′ | 97°2′ | 832 | −26 | −22 | 91/70 | 87/70 | 84/68 | 25 | 74 | 72 | 70 |
| Jamestown AP | 47°0′ | 98°4′ | 1,492 | −22 | −18 | 94/70 | 90/69 | 87/68 | 26 | 74 | 74 | 71 |
| Minot AP | 48°2′ | 101°2′ | 1,713 | −24 | −20 | 92/68 | 89/67 | 86/65 | 25 | 72 | 70 | 68 |
| Williston | 48°1′ | 103°4′ | 1,877 | −25 | −21 | 91/68 | 88/67 | 85/65 | 25 | 72 | 70 | 68 |
| Ohio | | | | | | | | | | | | |
| Akron-Canton AP | 41°0′ | 81°3′ | 1,210 | 1 | 6 | 89/72 | 86/71 | 84/70 | 21 | 75 | 73 | 72 |
| Ashtabula | 42°0′ | 80°5′ | 690 | 4 | 9 | 88/73 | 85/72 | 83/71 | 18 | 75 | 74 | 72 |
| Athens | 39°2′ | 82°1′ | 700 | 0 | 6 | 95/75 | 92/74 | 90/73 | 22 | 78 | 76 | 74 |
| Bowling Green | 41°3′ | 83°4′ | 675 | −2 | 2 | 92/73 | 89/73 | 86/71 | 23 | 76 | 75 | 73 |
| Cambridge | 40°0′ | 81°4′ | 800 | 1 | 7 | 93/75 | 90/74 | 87/73 | 23 | 78 | 76 | 75 |
| Chillicothe | 39°2′ | 83°0′ | 638 | 0 | 6 | 95/75 | 92/74 | 90/73 | 22 | 78 | 76 | 74 |
| Cincinnati CO | 39°1′ | 84°4′ | 761 | 1 | 6 | 92/73 | 90/72 | 88/72 | 21 | 77 | 75 | 74 |
| Cleveland AP (S) | 41°2′ | 81°5′ | 777r | 1 | 5 | 91/73 | 88/72 | 86/71 | 22 | 76 | 74 | 73 |
| Columbus AP (S) | 40°0′ | 82°5′ | 812 | 0 | 5 | 92/73 | 90/73 | 87/72 | 24 | 77 | 75 | 74 |
| Dayton AP | 39°5′ | 84°1′ | 997 | −1 | 4 | 91/73 | 89/72 | 86/71 | 20 | 76 | 75 | 73 |
| Defiance | 41°2′ | 84°2′ | 700 | −1 | 4 | 94/74 | 91/73 | 88/72 | 24 | 77 | 76 | 74 |
| Findlay AP | 41°0′ | 83°4′ | 797 | 2 | 3 | 92/74 | 90/73 | 87/72 | 24 | 77 | 76 | 74 |
| Fremont | 41°2′ | 83°1′ | 600 | −3 | 1 | 90/73 | 88/73 | 85/71 | 24 | 76 | 75 | 73 |
| Hamilton | 39°2′ | 84°3′ | 650 | 0 | 5 | 92/73 | 90/72 | 87/71 | 22 | 76 | 75 | 73 |
| Lancaster | 39°4′ | 82°4′ | 920 | 0 | 5 | 93/74 | 91/73 | 88/72 | 23 | 77 | 75 | 74 |
| Lima | 40°4′ | 84°0′ | 860 | −1 | 4 | 94/74 | 91/73 | 88/72 | 24 | 77 | 76 | 74 |
| Mansfield AP | 40°5′ | 82°3′ | 1,297 | 0 | 5 | 90/73 | 87/72 | 85/72 | 22 | 76 | 74 | 73 |
| Marion | 40°4′ | 83°1′ | 920 | 0 | 5 | 93/74 | 91/73 | 88/72 | 23 | 77 | 76 | 74 |
| Middletown | 39°3′ | 84°3′ | 635 | 0 | 5 | 92/73 | 90/72 | 87/71 | 22 | 76 | 75 | 73 |
| Newark | 40°1′ | 82°3′ | 825 | −1 | 5 | 94/73 | 92/73 | 89/72 | 23 | 77 | 75 | 74 |
| Norwalk | 41°1′ | 82°4′ | 720 | −3 | 1 | 90/73 | 88/73 | 85/71 | 22 | 76 | 75 | 73 |
| Portsmouth | 38°5′ | 83°0′ | 530 | 5 | 10 | 95/76 | 92/74 | 89/73 | 22 | 78 | 77 | 75 |
| Sandusky CO | 41°3′ | 82°4′ | 606 | 1 | 6 | 93/73 | 91/72 | 88/71 | 21 | 76 | 74 | 73 |
| Springfield | 40°0′ | 83°5′ | 1,020 | −1 | 3 | 91/74 | 89/73 | 87/72 | 21 | 77 | 76 | 74 |
| Steubenville | 40°2′ | 80°4′ | 992 | 1 | 5 | 89/72 | 86/71 | 84/70 | 22 | 74 | 73 | 72 |
| Toledo AP | 41°4′ | 83°5′ | 676r | −3 | 1 | 90/73 | 88/73 | 85/71 | 25 | 76 | 75 | 73 |
| Warren | 41°2′ | 80°5′ | 900 | 0 | 5 | 89/71 | 87/71 | 85/70 | 23 | 74 | 73 | 71 |
| Wooster | 40°5′ | 82°0′ | 1,030 | 1 | 6 | 89/72 | 86/71 | 84/70 | 22 | 75 | 73 | 72 |
| Youngstown AP | 41°2′ | 80°4′ | 1,178 | −1 | 4 | 88/71 | 86/71 | 84/70 | 23 | 74 | 73 | 71 |
| Zanesville AP | 40°0′ | 81°5′ | 881 | 1 | 7 | 93/75 | 90/74 | 87/73 | 23 | 78 | 76 | 75 |
| Oklahoma | | | | | | | | | | | | |
| Ada | 34°5′ | 96°4′ | 1,015 | 10 | 14 | 100/74 | 97/74 | 95/74 | 23 | 77 | 76 | 75 |
| Altus AFB | 34°4′ | 99°2′ | 1,390 | 11 | 16 | 102/73 | 100/73 | 98/73 | 25 | 77 | 76 | 75 |
| Ardmore | 34°2′ | 97°1′ | 880 | 13 | 17 | 100/74 | 98/74 | 95/74 | 23 | 77 | 77 | 76 |
| Bartlesville | 36°5′ | 96°0′ | 715 | 6 | 10 | 101/73 | 98/74 | 95/74 | 23 | 77 | 77 | 76 |

| | | | | | | | | | | | | |
|---|---|---|---|---|---|---|---|---|---|---|---|---|
| McAlester | 34°5′ | 95°5′ | 760 | 14 | 19 | 99/74 | 96/74 | 93/74 | 23 | 77 | 76 | 75 |
| Muskogee AP | 35°4′ | 95°2′ | 610 | 10 | 15 | 101/74 | 98/75 | 95/75 | 23 | 79 | 78 | 77 |
| Norman | 35°1′ | 97°3′ | 1,109 | 9 | 13 | 99/74 | 96/74 | 94/74 | 24 | 77 | 76 | 75 |
| Oklahoma City AP (S) | 35°2′ | 97°4′ | 1,280 | 9 | 13 | 100/74 | 97/74 | 95/73 | 23 | 78 | 77 | 76 |
| Ponca City | 36°4′ | 97°0′ | 996 | 5 | 9 | 100/74 | 97/74 | 94/74 | 24 | 77 | 76 | 76 |
| Seminole | 35°2′ | 96°4′ | 865 | 11 | 15 | 99/74 | 96/74 | 94/73 | 23 | 77 | 76 | 75 |
| Stillwater (S) | 36°1′ | 97°1′ | 884 | 8 | 13 | 100/74 | 96/74 | 93/74 | 24 | 77 | 76 | 75 |
| Tulsa AP | 36°1′ | 95°5′ | 650 | 8 | 13 | 101/74 | 98/75 | 95/75 | 22 | 79 | 78 | 77 |
| Woodward | 36°3′ | 99°3′ | 1,900 | 6 | 10 | 100/73 | 97/73 | 94/73 | 26 | 78 | 76 | 75 |
| Oregon | | | | | | | | | | | | |
| Albany | 44°4′ | 123°1′ | 224 | 18 | 22 | 92/67 | 89/66 | 86/65 | 31 | 69 | 67 | 66 |
| Astoria AP (S) | 46°1′ | 123°5′ | 8 | 25 | 29 | 75/65 | 71/62 | 68/61 | 16 | 65 | 63 | 62 |
| Baker AP | 44°5′ | 117°5′ | 3,368 | −1 | 6 | 92/63 | 89/61 | 86/60 | 30 | 65 | 63 | 61 |
| Bend | 44°0′ | 121°2′ | 3,599 | −3 | 4 | 90/62 | 87/60 | 84/59 | 33 | 64 | 62 | 60 |
| Corvallis (S) | 44°3′ | 123°2′ | 221 | 18 | 22 | 92/67 | 89/66 | 86/65 | 31 | 69 | 67 | 66 |
| Eugene AP | 44°1′ | 123°1′ | 364 | 17 | 22 | 92/67 | 89/66 | 86/65 | 31 | 69 | 67 | 66 |
| Grants Pass | 42°3′ | 123°2′ | 925 | 20 | 24 | 99/69 | 96/68 | 93/67 | 33 | 71 | 69 | 68 |
| Klamath Falls AP | 42°1′ | 121°4′ | 4,091 | 4 | 9 | 90/61 | 87/60 | 84/59 | 36 | 63 | 61 | 60 |
| Medford AP (S) | 42°2′ | 122°5′ | 1,298 | 19 | 23 | 98/68 | 94/67 | 91/66 | 35 | 70 | 68 | 67 |
| Pendleton AP | 45°4′ | 118°5′ | 1,492 | −2 | 5 | 97/65 | 93/64 | 90/62 | 29 | 66 | 65 | 63 |
| Portland AP | 45°4′ | 122°4′ | 21 | 17 | 23 | 89/68 | 85/67 | 81/65 | 23 | 69 | 67 | 66 |
| Portland CO | 45°3′ | 122°4′ | 57 | 18 | 24 | 90/68 | 86/67 | 82/65 | 21 | 69 | 67 | 66 |
| Roseburg AP | 43°1′ | 123°2′ | 505 | 18 | 23 | 93/67 | 90/66 | 87/65 | 30 | 69 | 67 | 66 |
| Salem AP | 45°0′ | 123°0′ | 195 | 18 | 23 | 92/68 | 88/66 | 84/65 | 31 | 69 | 68 | 66 |
| The Dalles | 45°4′ | 121°1′ | 102 | 13 | 19 | 93/69 | 89/68 | 85/66 | 28 | 70 | 68 | 67 |
| Pennsylvania | | | | | | | | | | | | |
| Allentown AP | 40°4′ | 75°3′ | 376 | 4 | 9 | 92/73 | 88/72 | 86/72 | 22 | 76 | 75 | 73 |
| Altoona CO | 40°2′ | 78°2′ | 1,468 | 0 | 5 | 90/72 | 87/71 | 84/70 | 23 | 74 | 73 | 72 |
| Butler | 40°4′ | 80°0′ | 1,100 | 1 | 6 | 90/73 | 87/72 | 85/71 | 22 | 75 | 74 | 73 |
| Chambersburg | 40°0′ | 77°4′ | 640 | 4 | 8 | 93/75 | 90/74 | 87/73 | 23 | 77 | 76 | 75 |
| Erie AP | 42°1′ | 80°1′ | 732 | 4 | 9 | 88/73 | 85/72 | 83/71 | 18 | 75 | 74 | 72 |
| Harrisburg AP | 40°1′ | 76°5′ | 335 | 7 | 11 | 94/75 | 91/74 | 88/73 | 21 | 77 | 76 | 75 |
| Johnstown | 40°2′ | 78°5′ | 1,214 | −3 | 2 | 86/70 | 83/70 | 80/68 | 23 | 72 | 71 | 70 |
| Lancaster | 40°1′ | 76°2′ | 255 | 4 | 8 | 93/75 | 90/74 | 87/73 | 22 | 77 | 76 | 75 |
| Meadville | 41°4′ | 80°1′ | 1,065 | 0 | 4 | 88/71 | 85/70 | 83/69 | 21 | 73 | 72 | 71 |
| New Castle | 41°0′ | 80°2′ | 825 | 2 | 7 | 91/73 | 88/72 | 86/71 | 23 | 75 | 74 | 73 |
| Philadelphia AP | 39°5′ | 75°2′ | 7 | 10 | 14 | 93/75 | 90/74 | 87/72 | 21 | 77 | 76 | 75 |
| Pittsburgh AP | 40°3′ | 80°1′ | 1,137 | 1 | 5 | 89/72 | 86/71 | 84/70 | 22 | 74 | 73 | 72 |
| Pittsburgh CO | 40°3′ | 80°0′ | 749r | 3 | 7 | 91/72 | 88/71 | 86/70 | 19 | 74 | 73 | 72 |
| Reading CO | 40°2′ | 76°0′ | 226 | 9 | 13 | 92/73 | 89/72 | 86/72 | 19 | 76 | 75 | 73 |
| Scranton/Wilkes-Barre | 41°2′ | 75°4′ | 940 | 1 | 5 | 90/72 | 87/71 | 84/70 | 19 | 74 | 73 | 72 |
| State College (S) | 40°5′ | 77°5′ | 1,175 | 3 | 7 | 90/72 | 87/71 | 84/70 | 23 | 74 | 73 | 72 |
| Sunbury | 40°5′ | 76°5′ | 480 | 2 | 7 | 92/73 | 89/72 | 86/70 | 22 | 75 | 74 | 73 |
| Uniontown | 39°5′ | 79°4′ | 1,040 | 5 | 9 | 91/74 | 88/73 | 85/72 | 22 | 76 | 75 | 74 |
| Warren | 41°5′ | 79°1′ | 1,280 | −2 | 4 | 89/71 | 86/71 | 83/70 | 24 | 74 | 73 | 72 |
| West Chester | 40°0′ | 75°4′ | 440 | 9 | 13 | 92/75 | 89/74 | 86/72 | 20 | 77 | 76 | 75 |
| Williamsport AP | 41°1′ | 77°0′ | 527 | 2 | 7 | 92/73 | 89/72 | 86/70 | 23 | 75 | 74 | 73 |

**TABLE A10 Continued**

| Col. 1 | Col. 2 | Col. 3 | Col. 4 | Col. 5 (Winter[b]) | | Col. 6 (Summer[c]) | | | Col. 7 | Col. 8 | | |
|---|---|---|---|---|---|---|---|---|---|---|---|---|
| | | | | Design dry-bulb | | Design dry-bulb and mean coincident wet-bulb | | | Mean daily | Design wet-bulb | | |
| State and station | Latitude[d] | Longitude[d] | Elevation[e] Ft | 99% | 97.5% | 1% | 2.5% | 5% | range | 1% | 2.5% | 5% |
| York | 40°0′ | 76°4′ | 390 | 8 | 12 | 94/75 | 91/74 | 88/73 | 22 | 77 | 76 | 75 |
| Rhode Island | | | | | | | | | | | | |
| Newport (S) | 41°3′ | 71°2′ | 20 | 5 | 9 | 88/73 | 85/72 | 82/70 | 16 | 76 | 75 | 73 |
| Providence AP | 41°4′ | 71°3′ | 55 | 5 | 9 | 89/73 | 86/72 | 83/70 | 19 | 75 | 74 | 73 |
| South Carolina | | | | | | | | | | | | |
| Anderson | 34°3′ | 82°4′ | 764 | 19 | 23 | 94/74 | 92/74 | 90/74 | 21 | 77 | 76 | 75 |
| Charleston AFB (S) | 32°5′ | 80°0′ | 41 | 24 | 27 | 93/78 | 91/78 | 89/77 | 18 | 81 | 80 | 79 |
| Charleston CO | 32°5′ | 80°0′ | 9 | 25 | 28 | 94/78 | 92/78 | 90/77 | 13 | 81 | 80 | 79 |
| Columbia AP | 34°0′ | 81°1′ | 217 | 20 | 24 | 97/76 | 95/75 | 93/75 | 22 | 79 | 78 | 77 |
| Florence AP | 34°1′ | 79°4′ | 146 | 22 | 25 | 94/77 | 92/77 | 90/76 | 21 | 80 | 79 | 78 |
| Georgetown | 33°2′ | 79°2′ | 14 | 23 | 26 | 92/79 | 90/78 | 88/77 | 18 | 81 | 80 | 79 |
| Greenville AP | 34°5′ | 82°1′ | 957 | 18 | 22 | 93/74 | 91/74 | 89/74 | 21 | 77 | 76 | 75 |
| Greenwood | 34°1′ | 82°1′ | 671 | 18 | 22 | 95/75 | 93/74 | 91/74 | 21 | 78 | 77 | 76 |
| Orangeburg | 33°3′ | 80°5′ | 244 | 20 | 24 | 97/76 | 95/75 | 93/75 | 20 | 79 | 78 | 77 |
| Rock Hill | 35°0′ | 81°0′ | 470 | 19 | 23 | 96/75 | 94/74 | 92/74 | 20 | 78 | 77 | 76 |
| Spartanburg AP | 35°0′ | 82°0′ | 816 | 18 | 22 | 93/74 | 91/74 | 89/74 | 20 | 77 | 76 | 75 |
| Sumter-Shaw AFB | 34°0′ | 80°3′ | 291 | 22 | 25 | 95/77 | 92/76 | 90/75 | 21 | 79 | 78 | 77 |
| South Dakota | | | | | | | | | | | | |
| Aberdeen AP | 45°3′ | 98°3′ | 1,296 | −19 | −15 | 94/73 | 91/72 | 88/70 | 27 | 77 | 75 | 73 |
| Brookings | 44°2′ | 96°5′ | 1,642 | −17 | −13 | 95/73 | 92/72 | 89/71 | 25 | 77 | 75 | 73 |
| Huron AP | 44°3′ | 98°1′ | 1,282 | −18 | −14 | 96/73 | 93/72 | 90/71 | 28 | 77 | 75 | 73 |
| Mitchel | 43°5′ | 98°0′ | 1,346 | −15 | −10 | 96/72 | 93/71 | 90/70 | 28 | 76 | 75 | 73 |
| Pierre AP | 44°2′ | 100°2′ | 1,718 | −15 | −10 | 99/71 | 95/71 | 92/69 | 29 | 75 | 74 | 72 |
| Rapid City AP (S) | 44°0′ | 103°0′ | 3,165 | −11 | −7 | 95/66 | 92/65 | 89/65 | 28 | 71 | 69 | 67 |
| Sioux Falls AP | 43°4′ | 96°4′ | 1,420 | −15 | −11 | 94/73 | 91/72 | 88/71 | 24 | 76 | 75 | 73 |
| Watertown AP | 45°0′ | 97°0′ | 1,746 | −19 | −15 | 94/73 | 91/72 | 88/71 | 26 | 76 | 75 | 73 |
| Yankton | 43°0′ | 97°2′ | 1,280 | −13 | −7 | 94/73 | 91/72 | 88/71 | 25 | 77 | 76 | 74 |
| Tennessee | | | | | | | | | | | | |
| Athens | 33°3′ | 84°4′ | 940 | 13 | 18 | 95/74 | 92/73 | 90/73 | 22 | 77 | 76 | 75 |
| Bristol-Tri City AP | 36°3′ | 82°2′ | 1,519 | 9 | 14 | 91/72 | 89/72 | 87/71 | 22 | 75 | 75 | 73 |
| Chattanooga AP | 35°0′ | 85°1′ | 670 | 13 | 18 | 96/75 | 93/74 | 91/74 | 22 | 78 | 77 | 76 |
| Clarksville | 36°4′ | 87°2′ | 470 | 6 | 12 | 95/76 | 93/74 | 90/74 | 21 | 78 | 77 | 76 |
| Columbia | 35°4′ | 87°0′ | 690 | 10 | 15 | 97/75 | 94/74 | 91/74 | 21 | 78 | 77 | 76 |
| Dyersburg | 36°0′ | 89°3′ | 334 | 10 | 15 | 96/78 | 94/77 | 91/76 | 21 | 81 | 80 | 78 |
| Greenville | 35°5′ | 82°5′ | 1,320 | 11 | 16 | 92/73 | 90/72 | 88/72 | 22 | 76 | 75 | 74 |
| Jackson AP | 35°4′ | 88°5′ | 413 | 11 | 16 | 98/76 | 95/75 | 92/75 | 21 | 79 | 78 | 77 |
| Knoxville AP | 35°5′ | 84°0′ | 980 | 13 | 19 | 94/74 | 92/73 | 90/73 | 21 | 77 | 76 | 75 |

| | | | | | | | | | | | | |
|---|---|---|---|---|---|---|---|---|---|---|---|---|
| Memphis AP | 35°0′ | 90°0′ | 263 | 13 | 18 | 98/77 | 95/76 | 93/76 | 21 | 80 | 79 | 78 |
| Murfreesboro | 35°5′ | 86°2′ | 608 | 9 | 14 | 97/75 | 94/74 | 91/74 | 22 | 78 | 77 | 76 |
| Nashville AP (S) | 36°1′ | 86°4′ | 577 | 9 | 14 | 97/75 | 94/74 | 91/74 | 21 | 78 | 77 | 76 |
| Tullahoma | 35°2′ | 86°1′ | 1,075 | 8 | 13 | 96/74 | 93/73 | 91/73 | 22 | 77 | 76 | 75 |
| Texas | | | | | | | | | | | | |
| Abilene AP | 32°3′ | 99°4′ | 1,759 | 15 | 20 | 101/71 | 99/71 | 97/71 | 22 | 75 | 74 | 74 |
| Alice AP | 27°4′ | 98°0′ | 180 | 31 | 34 | 100/78 | 98/77 | 95/77 | 20 | 82 | 81 | 79 |
| Amarillo AP | 35°1′ | 101°4′ | 3,607 | 6 | 11 | 98/67 | 95/67 | 93/67 | 26 | 71 | 70 | 70 |
| Austin AP | 30°2′ | 97°4′ | 597 | 24 | 28 | 100/74 | 98/74 | 97/74 | 22 | 78 | 77 | 77 |
| Bay City | 29°0′ | 96°0′ | 52 | 29 | 33 | 96/77 | 94/77 | 92/77 | 16 | 80 | 79 | 79 |
| Beaumont | 30°0′ | 94°0′ | 18 | 27 | 31 | 95/79 | 93/78 | 91/78 | 19 | 81 | 80 | 80 |
| Beeville | 28°2′ | 97°4′ | 225 | 30 | 33 | 99/78 | 97/77 | 95/77 | 18 | 82 | 81 | 79 |
| Big Springs AP (S) | 32°2′ | 101°3′ | 2,537 | 16 | 20 | 100/69 | 97/69 | 95/69 | 26 | 74 | 73 | 72 |
| Brownsville AP (S) | 25°5′ | 97°3′ | 16 | 35 | 39 | 94/77 | 93/77 | 92/77 | 18 | 80 | 79 | 79 |
| Brownwood | 31°5′ | 99°0′ | 1,435 | 18 | 22 | 101/73 | 99/73 | 96/73 | 22 | 77 | 76 | 75 |
| Bryan AP | 30°4′ | 96°2′ | 275 | 24 | 29 | 98/76 | 96/76 | 94/76 | 20 | 79 | 78 | 78 |
| Corpus Christi AP | 27°5′ | 97°3′ | 43 | 31 | 35 | 95/78 | 94/78 | 92/78 | 19 | 80 | 80 | 79 |
| Corsicana | 32°0′ | 96°3′ | 425 | 20 | 25 | 100/75 | 98/75 | 96/75 | 21 | 79 | 78 | 77 |
| Dallas AP | 32°5′ | 96°5′ | 481 | 18 | 22 | 102/75 | 100/75 | 97/75 | 20 | 78 | 78 | 77 |
| Del Rio, Laughlin AFB | 29°2′ | 101°0′ | 1,072 | 26 | 31 | 100/73 | 98/73 | 97/73 | 24 | 79 | 77 | 76 |
| Denton | 33°1′ | 97°1′ | 655 | 17 | 22 | 101/74 | 99/74 | 97/74 | 22 | 78 | 77 | 76 |
| Eagle Pass | 28°5′ | 100°3′ | 743 | 27 | 32 | 101/73 | 99/73 | 98/73 | 24 | 78 | 78 | 77 |
| El Paso AP (S) | 31°5′ | 106°2′ | 3,918 | 20 | 24 | 100/64 | 98/64 | 96/64 | 27 | 69 | 68 | 68 |
| Fort Worth AP (S) | 32°5′ | 97°0′ | 544r | 17 | 22 | 101/74 | 99/74 | 97/74 | 22 | 78 | 77 | 76 |
| Galveston AP | 29°2′ | 94°5′ | 5 | 31 | 36 | 90/79 | 89/79 | 88/78 | 10 | 81 | 80 | 80 |
| Greenville | 33°0′ | 96°1′ | 575 | 17 | 22 | 101/74 | 99/74 | 97/74 | 21 | 78 | 77 | 76 |
| Harlingen | 26°1′ | 97°4′ | 37 | 35 | 39 | 96/77 | 94/77 | 93/77 | 19 | 80 | 79 | 79 |
| Houston AP | 29°4′ | 95°2′ | 50 | 27 | 32 | 96/77 | 94/77 | 92/77 | 18 | 80 | 79 | 79 |
| Houston CO | 29°5′ | 95°2′ | 158r | 28 | 33 | 97/77 | 95/77 | 93/77 | 18 | 80 | 79 | 79 |
| Huntsville | 30°4′ | 95°3′ | 494 | 22 | 27 | 100/75 | 98/75 | 96/75 | 20 | 78 | 78 | 77 |
| Killeen-Gray AFB | 31°0′ | 97°4′ | 1,021 | 20 | 25 | 99/73 | 97/73 | 95/73 | 22 | 77 | 76 | 75 |
| Lamesa | 32°5′ | 102°0′ | 2,965 | 13 | 17 | 99/69 | 96/69 | 94/69 | 26 | 73 | 72 | 71 |
| Laredo AFB | 27°3′ | 99°3′ | 503 | 32 | 36 | 102/73 | 101/73 | 99/74 | 23 | 78 | 78 | 77 |
| Longview | 32°2′ | 94°4′ | 345 | 19 | 24 | 99/76 | 97/76 | 95/76 | 20 | 80 | 79 | 78 |
| Lubbock AP | 33°4′ | 101°5′ | 3,243 | 10 | 15 | 98/69 | 96/69 | 94/69 | 26 | 73 | 72 | 71 |
| Lufkin AP | 31°1′ | 94°5′ | 286 | 25 | 29 | 99/76 | 97/76 | 94/76 | 20 | 80 | 79 | 78 |
| McAllen | 26°1′ | 98°1′ | 122 | 35 | 39 | 97/77 | 95/77 | 94/77 | 21 | 80 | 79 | 79 |
| Midland AP (S) | 32°0′ | 102°1′ | 2,815r | 16 | 21 | 100/69 | 98/69 | 96/69 | 26 | 73 | 72 | 71 |
| Mineral Wells AP | 32°5′ | 98°0′ | 934 | 17 | 22 | 101/74 | 99/74 | 97/74 | 22 | 78 | 77 | 76 |
| Palestine CO | 31°5′ | 95°4′ | 580 | 23 | 27 | 100/76 | 98/76 | 96/76 | 20 | 79 | 79 | 78 |
| Pampa | 35°3′ | 101°0′ | 3,230 | 7 | 12 | 99/67 | 96/67 | 94/67 | 26 | 71 | 70 | 70 |
| Pecos | 31°2′ | 103°3′ | 2,580 | 16 | 21 | 100/69 | 98/69 | 96/69 | 27 | 73 | 72 | 71 |
| Plainview | 34°1′ | 101°4′ | 3,400 | 8 | 13 | 98/68 | 96/68 | 94/68 | 26 | 72 | 71 | 70 |
| Port Arthur AP | 30°0′ | 94°0′ | 16 | 27 | 31 | 95/79 | 93/78 | 91/78 | 19 | 81 | 80 | 80 |
| San Angelo, Goodfellow AFB | 31°2′ | 100°2′ | 1,878 | 18 | 22 | 101/71 | 99/71 | 97/70 | 24 | 75 | 74 | 73 |
| San Antonio AP (S) | 29°3′ | 98°3′ | 792 | 25 | 30 | 99/72 | 97/73 | 96/73 | 19 | 77 | 76 | 76 |
| Sherman Perrin AFB | 33°4′ | 96°4′ | 763 | 15 | 20 | 100/75 | 98/75 | 95/74 | 22 | 78 | 77 | 76 |

**TABLE A10 Continued**

| Col. 1 | Col. 2 | Col. 3 | Col. 4 | Col. 5 (Winter[b]) | | Col. 6 (Summer[c]) | | | Col. 7 | Col. 8 | | |
|---|---|---|---|---|---|---|---|---|---|---|---|---|
| | | | | Design dry-bulb | | Design dry-bulb and mean coincident wet-bulb | | | Mean daily | Design wet-bulb | | |
| State and station | Latitude[d] | Longitude[d] | Elevation[e] Ft | 99% | 97.5% | 1% | 2.5% | 5% | range | 1% | 2.5% | 5% |
| Snyder | 32°4′ | 101°0′ | 2,325 | 13 | 18 | 100/70 | 98/70 | 96/70 | 26 | 74 | 73 | 72 |
| Temple | 31°1′ | 97°2′ | 675 | 22 | 27 | 100/74 | 99/74 | 97/74 | 22 | 78 | 77 | 77 |
| Tyler AP | 32°2′ | 95°2′ | 527 | 19 | 24 | 99/76 | 97/76 | 95/76 | 21 | 80 | 79 | 78 |
| Vernon | 34°1′ | 99°2′ | 1,225 | 13 | 17 | 102/73 | 100/73 | 97/73 | 24 | 77 | 76 | 75 |
| Victoria AP | 28°5′ | 97°0′ | 104 | 29 | 32 | 98/78 | 96/77 | 94/77 | 18 | 82 | 81 | 79 |
| Waco AP | 31°4′ | 97°0′ | 500 | 21 | 26 | 101/75 | 99/75 | 97/75 | 22 | 78 | 78 | 77 |
| Wichita Falls AP | 34°0′ | 98°3′ | 994 | 14 | 18 | 103/73 | 101/73 | 98/73 | 24 | 77 | 76 | 75 |
| Utah | | | | | | | | | | | | |
| Cedar City AP | 37°4′ | 113°1′ | 5,613 | −2 | 5 | 93/60 | 91/60 | 89/59 | 32 | 65 | 63 | 62 |
| Logan | 41°4′ | 111°5′ | 4,775 | −3 | 2 | 93/62 | 91/61 | 88/60 | 33 | 65 | 64 | 63 |
| Moab | 38°5′ | 109°3′ | 3,965 | 6 | 11 | 100/60 | 98/60 | 96/60 | 30 | 65 | 64 | 63 |
| Ogden AP | 41°1′ | 112°3′ | 4,455 | 1 | 5 | 93/63 | 91/61 | 88/61 | 33 | 66 | 65 | 64 |
| Price | 39°4′ | 110°5′ | 5,580 | −2 | 5 | 93/60 | 91/60 | 89/59 | 33 | 65 | 63 | 62 |
| Provo | 40°1′ | 111°4′ | 4,470 | 1 | 6 | 98/62 | 96/62 | 94/61 | 32 | 66 | 65 | 64 |
| Richfield | 38°5′ | 112°0′ | 5,300 | −2 | 5 | 93/60 | 91/60 | 89/59 | 34 | 65 | 63 | 62 |
| St. George CO | 37°1′ | 113°4′ | 2,899 | 14 | 21 | 103/65 | 101/65 | 99/64 | 33 | 70 | 68 | 67 |
| Salt Lake City AP (S) | 40°5′ | 112°0′ | 4,220 | 3 | 8 | 97/62 | 95/62 | 92/61 | 32 | 66 | 65 | 64 |
| Vernal AP | 40°3′ | 109°3′ | 5,280 | −5 | 0 | 91/61 | 89/60 | 86/59 | 32 | 64 | 63 | 62 |
| Vermont | | | | | | | | | | | | |
| Barre | 44°1′ | 72°3′ | 1,120 | −16 | −11 | 84/71 | 81/69 | 78/68 | 23 | 73 | 71 | 70 |
| Burlington AP (S) | 44°3′ | 73°1′ | 331 | −12 | −7 | 88/72 | 85/70 | 82/69 | 23 | 74 | 72 | 71 |
| Rutland | 43°3′ | 73°0′ | 620 | −13 | −8 | 87/72 | 84/70 | 81/69 | 23 | 74 | 72 | 71 |
| Virginia | | | | | | | | | | | | |
| Charlottsville | 38°1′ | 78°3′ | 870 | 14 | 18 | 94/74 | 91/74 | 88/73 | 23 | 77 | 76 | 75 |
| Danville AP | 36°3′ | 79°2′ | 590 | 14 | 16 | 94/74 | 92/73 | 90/73 | 21 | 77 | 76 | 75 |
| Fredericksburg | 38°2′ | 77°3′ | 50 | 10 | 14 | 96/76 | 93/75 | 90/74 | 21 | 78 | 77 | 75 |
| Harrisonburg | 38°3′ | 78°5′ | 1,340 | 12 | 16 | 93/72 | 91/72 | 88/71 | 23 | 75 | 74 | 73 |
| Lynchburg AP | 37°2′ | 79°1′ | 947 | 12 | 16 | 93/74 | 90/74 | 88/73 | 21 | 77 | 76 | 75 |
| Norfolk AP | 36°5′ | 76°1′ | 26 | 20 | 22 | 93/77 | 91/76 | 89/76 | 18 | 79 | 78 | 77 |
| Petersburg | 37°1′ | 77°3′ | 194 | 14 | 17 | 95/76 | 92/76 | 90/75 | 20 | 79 | 78 | 77 |
| Richmond AP | 37°3′ | 77°2′ | 162 | 14 | 17 | 95/76 | 92/76 | 90/75 | 21 | 79 | 78 | 77 |
| Roanoke AP | 37°2′ | 80°0′ | 1,174r | 12 | 16 | 93/72 | 91/72 | 88/71 | 23 | 75 | 74 | 73 |
| Staunton | 38°2′ | 78°5′ | 1,480 | 12 | 16 | 93/72 | 91/72 | 88/71 | 23 | 75 | 74 | 73 |
| Winchester | 39°1′ | 78°1′ | 750 | 6 | 10 | 93/75 | 90/74 | 88/74 | 21 | 77 | 76 | 75 |
| Washington | | | | | | | | | | | | |
| Aberdeen | 47°0′ | 123°5′ | 12 | 25 | 28 | 80/65 | 77/62 | 73/61 | 16 | 65 | 63 | 62 |
| Bellingham AP | 48°5′ | 122°3′ | 150 | 10 | 15 | 81/67 | 77/65 | 74/63 | 19 | 68 | 65 | 63 |

| | | | | | | | | | | | | |
|---|---|---|---|---|---|---|---|---|---|---|---|---|
| Bremerton | 47°3′ | 122°4′ | 162 | 21 | 25 | 82/65 | 78/64 | 75/62 | 20 | 66 | 64 | 63 |
| Ellensburg AP | 47°0′ | 120°3′ | 1,729 | 2 | 6 | 94/65 | 91/64 | 87/62 | 34 | 66 | 65 | 63 |
| Everett-Paine AFB | 47°5′ | 122°2′ | 598 | 21 | 25 | 80/65 | 76/64 | 73/62 | 20 | 67 | 64 | 63 |
| Kennewick | 46°0′ | 119°1′ | 392 | 5 | 11 | 99/68 | 96/67 | 92/66 | 30 | 70 | 68 | 67 |
| Longview | 46°1′ | 123°0′ | 12 | 19 | 24 | 88/68 | 85/67 | 81/65 | 30 | 69 | 67 | 66 |
| Moses Lake, Larson AFB | 47°1′ | 119°2′ | 1,183 | 1 | 7 | 97/66 | 94/65 | 90/63 | 32 | 67 | 66 | 64 |
| Olympia AP | 47°0′ | 122°5′ | 190 | 16 | 22 | 87/66 | 83/65 | 79/64 | 32 | 67 | 66 | 64 |
| Port Angeles | 48°1′ | 123°3′ | 99 | 24 | 27 | 72/62 | 69/61 | 67/60 | 18 | 64 | 62 | 61 |
| Seattle-Boeing Fld | 47°3′ | 122°2′ | 14 | 21 | 26 | 84/68 | 81/66 | 77/65 | 24 | 69 | 67 | 65 |
| Seattle CO (S) | 47°4′ | 122°2′ | 14 | 22 | 27 | 85/68 | 82/66 | 78/65 | 19 | 69 | 67 | 65 |
| Seattle-Tacoma AP (S) | 47°3′ | 122°2′ | 386 | 21 | 26 | 84/65 | 80/64 | 76/62 | 22 | 66 | 64 | 63 |
| Spokane AP (S) | 47°4′ | 117°3′ | 2,357 | −6 | 2 | 93/64 | 90/63 | 87/62 | 28 | 65 | 64 | 62 |
| Tacoma-McChord AFB | 47°1′ | 122°3′ | 350 | 19 | 24 | 86/66 | 82/65 | 79/63 | 22 | 68 | 66 | 64 |
| Walla Walla AP | 46°1′ | 118°2′ | 1,185 | 0 | 7 | 97/67 | 94/66 | 90/65 | 27 | 69 | 67 | 66 |
| Wenatchee | 47°2′ | 120°2′ | 634 | 7 | 11 | 99/67 | 96/66 | 92/64 | 32 | 68 | 67 | 65 |
| Yakima AP | 46°3′ | 120°3′ | 1,061 | −2 | 5 | 96/65 | 93/65 | 89/63 | 36 | 68 | 66 | 65 |
| West Virginia | | | | | | | | | | | | |
| Beckley | 37°5′ | 81°1′ | 2,330 | −2 | 4 | 83/71 | 81/69 | 79/69 | 22 | 73 | 71 | 70 |
| Bluefield AP | 37°2′ | 81°2′ | 2,850 | −2 | 4 | 83/71 | 81/69 | 79/69 | 22 | 73 | 71 | 70 |
| Charleston AP | 38°2′ | 81°4′ | 939 | 7 | 11 | 92/74 | 90/73 | 87/72 | 20 | 76 | 75 | 74 |
| Clarksburg | 39°2′ | 80°2′ | 977 | 6 | 10 | 92/74 | 90/73 | 87/72 | 21 | 76 | 75 | 74 |
| Elkins AP | 38°5′ | 79°5′ | 1,970 | 1 | 6 | 86/72 | 84/70 | 82/70 | 22 | 74 | 72 | 71 |
| Huntington CO | 38°2′ | 82°3′ | 565r | 5 | 10 | 94/76 | 91/74 | 89/73 | 22 | 78 | 77 | 75 |
| Martinsburg AP | 39°2′ | 78°0′ | 537 | 6 | 10 | 93/75 | 90/74 | 88/74 | 21 | 77 | 76 | 75 |
| Morgantown AP | 39°4′ | 80°0′ | 1,245 | 4 | 8 | 90/74 | 87/73 | 85/73 | 22 | 76 | 75 | 74 |
| Parkersburg CO | 39°2′ | 81°3′ | 615r | 7 | 11 | 93/75 | 90/74 | 88/73 | 21 | 77 | 76 | 75 |
| Wheeling | 40°1′ | 80°4′ | 659 | 1 | 5 | 89/72 | 86/71 | 84/70 | 21 | 74 | 73 | 72 |
| Wisconsin | | | | | | | | | | | | |
| Appleton | 44°2′ | 88°2′ | 742 | −14 | −9 | 89/74 | 86/72 | 83/71 | 23 | 76 | 74 | 72 |
| Ashland | 46°3′ | 90°5′ | 650 | −21 | −16 | 85/70 | 82/68 | 79/66 | 23 | 72 | 70 | 68 |
| Beloit | 42°3′ | 89°0′ | 780 | −7 | −3 | 92/75 | 90/75 | 88/74 | 24 | 78 | 77 | 75 |
| Eau Claire AP | 44°5′ | 91°3′ | 888 | −15 | −11 | 92/75 | 89/73 | 86/71 | 23 | 77 | 75 | 73 |
| Fond du Lac | 43°5′ | 88°3′ | 760 | −12 | −8 | 89/74 | 86/72 | 84/71 | 23 | 76 | 74 | 72 |
| Green Bay AP | 44°3′ | 88°1′ | 683 | −13 | −9 | 88/74 | 85/72 | 83/71 | 23 | 76 | 74 | 72 |
| La Crosse AP | 43°5′ | 91°2′ | 652 | −13 | −9 | 91/75 | 88/73 | 85/72 | 22 | 77 | 75 | 74 |
| Madison AP (S) | 43°1′ | 89°2′ | 858 | −11 | −7 | 91/74 | 88/73 | 85/71 | 22 | 77 | 75 | 73 |
| Manitowoc | 44°1′ | 87°4′ | 660 | −11 | −7 | 89/74 | 86/72 | 83/71 | 21 | 76 | 74 | 72 |
| Marinette | 45°0′ | | 605 | −15 | −11 | 87/73 | 84/71 | 82/70 | 20 | 75 | 73 | 71 |
| Milwaukee AP | 43°0′ | 87°5′ | 672 | −8 | −4 | 90/74 | 87/73 | 84/71 | 21 | 76 | 74 | 73 |
| Racine | 42°4′ | 87°4′ | 640 | −6 | −2 | 91/75 | 88/73 | 85/72 | 21 | 77 | 75 | 74 |
| Sheboygan | 43°4′ | 87°4′ | 648 | −10 | −6 | 89/75 | 86/73 | 83/72 | 20 | 77 | 75 | 74 |
| Stevens Point | 43°0′ | 89°3′ | 1,079 | −15 | −11 | 92/75 | 89/73 | 86/71 | 23 | 77 | 75 | 73 |
| Waukesha | 43°0′ | 88°1′ | 860 | −9 | −5 | 90/74 | 87/73 | 84/71 | 22 | 76 | 74 | 73 |
| Wausau AP | 44°6′ | 89°4′ | 1,196 | −16 | −12 | 91/74 | 88/72 | 85/70 | 23 | 76 | 74 | 72 |
| Wyoming | | | | | | | | | | | | |
| Casper AP | 42°5′ | 106°3′ | 5,319 | −11 | −5 | 92/58 | 90/57 | 87/57 | 31 | 63 | 61 | 60 |
| Cheyene AP | 41°1′ | 104°5′ | 6,126 | −9 | −1 | 89/58 | 86/58 | 84/57 | 30 | 63 | 62 | 60 |

**TABLE A10 Continued**

| Col. 1 | Col. 2 | Col. 3 | Col. 4 | Col. 5 (Winter[b]) | | Col. 6 (Summer[c]) | | | Col. 7 (Summer[c]) | Col. 8 (Summer[c]) | | |
|---|---|---|---|---|---|---|---|---|---|---|---|---|
| | | | | Design dry-bulb | | Design dry-bulb and mean coincident wet-bulb | | | Mean daily range | Design wet-bulb | | |
| State and station | Latitude[d] | Longitude[d] | Elevation[e] Ft | 99% | 97.5% | 1% | 2.5% | 5% | | 1% | 2.5% | 5% |
| Cody AP | 44°3′ | 109°0′ | 5,090 | −19 | −13 | 89/60 | 86/60 | 83/59 | 32 | 64 | 63 | 61 |
| Evanston | 41°2′ | 111°0′ | 6,860 | −9 | −3 | 86/55 | 84/55 | 82/54 | 32 | 59 | 58 | 57 |
| Lander AP (S) | 42°5′ | 108°4′ | 5,563 | −16 | −11 | 91/61 | 88/61 | 85/60 | 32 | 64 | 63 | 61 |
| Laramie AP (S) | 41°2′ | 105°3′ | 7,266 | −14 | −6 | 84/56 | 81/56 | 79/55 | 28 | 61 | 60 | 59 |
| Newcastle | 43°5′ | 104°1′ | 4,480 | −17 | −12 | 91/64 | 87/63 | 84/63 | 30 | 69 | 68 | 66 |
| Rawlins | 41°5′ | 107°1′ | 6,736 | −12 | −4 | 86/57 | 83/57 | 81/56 | 40 | 62 | 61 | 60 |
| Rock Springs AP | 41°4′ | 109°0′ | 6,741 | −9 | −3 | 86/55 | 84/55 | 82/54 | 32 | 59 | 58 | 57 |
| Sheridan AP | 44°5′ | 107°0′ | 3,942 | −14 | −8 | 94/62 | 91/62 | 88/61 | 32 | 66 | 65 | 63 |
| Torrington | 44°0′ | 104°1′ | 4,098 | −14 | −8 | 94/62 | 91/62 | 88/61 | 30 | 66 | 65 | 63 |

[a]Table prepared by ASHRAE Technical Committee 4.2, Weather Data, from data compiled from official weather stations, where hourly weather observations are made by trained observers.

[b]Percentage of winter design data shows the percent of the three-month period, December through February.

[c]Percentage of summer design data shows the percent of the four-month period, June through September.

[d]Latitude, for use in calculating solar loads, and longitude are given to the nearest 10 minutes.

[e]Elevations are ground elevations for each station. Temperature readings are generally made at an elevation of 5 ft above ground, except for locations marked r, indicating roof exposure of thermometer.

**TABLE A11** Average Monthly and Yearly Degree Days for Cities in the United States (Base 65°F)[a–c]

| State and station | | Avg. winter temp[d] | July | Aug. | Sept. | Oct. | Nov. | Dec. | Jan. | Feb. | Mar. | Apr. | May | June | Yearly total |
|---|---|---|---|---|---|---|---|---|---|---|---|---|---|---|---|
| Alabama | | | | | | | | | | | | | | | |
| Birmingham | A | 54.2 | 0 | 0 | 6 | 93 | 363 | 555 | 592 | 462 | 363 | 108 | 9 | 0 | 2551 |
| Huntsville | A | 51.3 | 0 | 0 | 12 | 127 | 426 | 663 | 694 | 557 | 434 | 138 | 19 | 0 | 3070 |
| Mobile | A | 59.9 | 0 | 0 | 0 | 22 | 213 | 357 | 415 | 300 | 211 | 42 | 0 | 0 | 1560 |
| Montgomery | A | 55.4 | 0 | 0 | 0 | 68 | 330 | 527 | 543 | 417 | 316 | 90 | 0 | 0 | 2291 |
| Alaska | | | | | | | | | | | | | | | |
| Anchorage | A | 23.0 | 245 | 291 | 516 | 930 | 1284 | 1572 | 1631 | 1316 | 1293 | 879 | 592 | 315 | 10864 |
| Fairbanks | A | 6.7 | 171 | 332 | 642 | 1203 | 1833 | 2254 | 2359 | 1901 | 1739 | 1068 | 555 | 222 | 14279 |
| Juneau | A | 32.1 | 301 | 338 | 483 | 725 | 921 | 1135 | 1237 | 1070 | 1073 | 810 | 601 | 381 | 9075 |
| Nome | A | 13.1 | 481 | 496 | 693 | 1094 | 1455 | 1820 | 1879 | 1666 | 1770 | 1314 | 930 | 573 | 14171 |
| Arizona | | | | | | | | | | | | | | | |
| Flagstaff | A | 35.6 | 46 | 68 | 201 | 558 | 867 | 1073 | 1169 | 991 | 911 | 651 | 437 | 180 | 7152 |
| Phoenix | A | 58.5 | 0 | 0 | 0 | 22 | 234 | 415 | 474 | 328 | 217 | 75 | 0 | 0 | 1765 |
| Tucson | A | 58.1 | 0 | 0 | 0 | 25 | 231 | 406 | 471 | 344 | 242 | 75 | 6 | 0 | 1800 |
| Winslow | A | 43.0 | 0 | 0 | 6 | 245 | 711 | 1008 | 1054 | 770 | 601 | 291 | 96 | 0 | 4782 |
| Yuma | A | 64.2 | 0 | 0 | 0 | 0 | 108 | 264 | 307 | 190 | 90 | 15 | 0 | 0 | 974 |
| Arkansas | | | | | | | | | | | | | | | |
| Fort Smith | A | 50.3 | 0 | 0 | 12 | 127 | 450 | 704 | 781 | 596 | 456 | 144 | 22 | 0 | 3292 |
| Little Rock | A | 50.5 | 0 | 0 | 9 | 127 | 465 | 716 | 756 | 577 | 434 | 126 | 9 | 0 | 3219 |
| Texarkana | A | 54.2 | 0 | 0 | 0 | 78 | 345 | 561 | 626 | 468 | 350 | 105 | 0 | 0 | 2533 |
| California | | | | | | | | | | | | | | | |
| Bakersfield | A | 55.4 | 0 | 0 | 0 | 37 | 282 | 502 | 546 | 364 | 267 | 105 | 19 | 0 | 2122 |
| Bishop | A | 46.0 | 0 | 0 | 48 | 260 | 576 | 797 | 874 | 680 | 555 | 306 | 143 | 36 | 4275 |
| Blue Canyon | A | 42.2 | 28 | 37 | 108 | 347 | 594 | 781 | 896 | 795 | 806 | 597 | 412 | 195 | 5596 |
| Burbank | A | 58.6 | 0 | 0 | 6 | 43 | 177 | 301 | 366 | 277 | 239 | 138 | 81 | 18 | 1646 |
| Eureka | C | 49.9 | 270 | 257 | 258 | 329 | 414 | 499 | 546 | 470 | 505 | 438 | 372 | 285 | 4643 |
| Fresno | A | 53.3 | 0 | 0 | 0 | 84 | 354 | 577 | 605 | 426 | 335 | 162 | 62 | 6 | 2611 |
| Long Beach | A | 57.8 | 0 | 0 | 9 | 47 | 171 | 316 | 397 | 311 | 264 | 171 | 93 | 24 | 1803 |
| Los Angeles | A | 57.4 | 28 | 28 | 42 | 78 | 180 | 291 | 372 | 302 | 288 | 219 | 158 | 81 | 2061 |
| Los Angeles | C | 60.3 | 0 | 0 | 6 | 31 | 132 | 229 | 310 | 230 | 202 | 123 | 68 | 18 | 1349 |
| Mt. Shasta | C | 41.2 | 25 | 34 | 123 | 406 | 696 | 902 | 983 | 784 | 738 | 525 | 347 | 159 | 5722 |
| Oakland | A | 53.5 | 53 | 50 | 45 | 127 | 309 | 481 | 527 | 400 | 353 | 255 | 180 | 90 | 2870 |
| Red Bluff | A | 53.8 | 0 | 0 | 0 | 53 | 318 | 555 | 605 | 428 | 341 | 168 | 47 | 0 | 2515 |
| Sacramento | A | 53.9 | 0 | 0 | 0 | 56 | 321 | 546 | 583 | 414 | 332 | 178 | 72 | 0 | 2502 |
| Sacramento | C | 54.4 | 0 | 0 | 0 | 62 | 312 | 533 | 561 | 392 | 310 | 173 | 76 | 0 | 2419 |
| Sandberg | C | 46.8 | 0 | 0 | 30 | 202 | 480 | 691 | 778 | 661 | 620 | 426 | 264 | 57 | 4209 |
| San Diego | A | 59.5 | 9 | 0 | 21 | 43 | 135 | 236 | 298 | 235 | 214 | 135 | 90 | 42 | 1458 |
| San Francisco | A | 53.4 | 81 | 78 | 60 | 143 | 306 | 462 | 508 | 395 | 363 | 279 | 214 | 126 | 3015 |
| San Francisco | C | 55.1 | 192 | 174 | 102 | 118 | 231 | 388 | 443 | 336 | 319 | 279 | 239 | 180 | 3001 |
| Santa Maria | A | 54.3 | 99 | 93 | 96 | 146 | 270 | 391 | 459 | 370 | 363 | 282 | 233 | 165 | 2967 |

**TABLE A11 Continued**

| State and station | | Avg. winter temp[d] | July | Aug. | Sept. | Oct. | Nov. | Dec. | Jan. | Feb. | Mar. | Apr. | May | June | Yearly total |
|---|---|---|---|---|---|---|---|---|---|---|---|---|---|---|---|
| Colorado | | | | | | | | | | | | | | | |
| Alamosa | A | 29.7 | 65 | 99 | 279 | 639 | 1065 | 1420 | 1476 | 1162 | 1020 | 696 | 440 | 168 | 8529 |
| Colorado Springs | A | 37.3 | 9 | 25 | 132 | 456 | 825 | 1032 | 1128 | 938 | 893 | 582 | 319 | 84 | 6423 |
| Denver | A | 37.6 | 6 | 9 | 117 | 428 | 819 | 1035 | 1132 | 938 | 887 | 558 | 288 | 66 | 6283 |
| Denver | C | 40.8 | 0 | 0 | 90 | 366 | 714 | 905 | 1004 | 851 | 800 | 492 | 254 | 48 | 5524 |
| Grand Junction | A | 39.3 | 0 | 0 | 30 | 313 | 786 | 1113 | 1209 | 907 | 729 | 387 | 146 | 21 | 5641 |
| Pueblo | A | 40.4 | 0 | 0 | 54 | 326 | 750 | 986 | 1085 | 871 | 772 | 429 | 174 | 15 | 5462 |
| Connecticut | | | | | | | | | | | | | | | |
| Bridgeport | A | 39.9 | 0 | 0 | 66 | 307 | 615 | 986 | 1079 | 966 | 853 | 510 | 208 | 27 | 5617 |
| Hartford | A | 37.3 | 0 | 12 | 117 | 394 | 714 | 1101 | 1190 | 1042 | 908 | 519 | 205 | 33 | 6235 |
| New Haven | A | 39.0 | 0 | 12 | 87 | 347 | 648 | 1011 | 1097 | 991 | 871 | 543 | 245 | 45 | 5897 |
| Delaware | | | | | | | | | | | | | | | |
| Wilmington | A | 42.5 | 0 | 0 | 51 | 270 | 588 | 927 | 980 | 874 | 735 | 387 | 112 | 6 | 4930 |
| District of Columbia | | | | | | | | | | | | | | | |
| Washington | A | 45.7 | 0 | 0 | 33 | 217 | 519 | 834 | 871 | 762 | 626 | 288 | 74 | 0 | 4224 |
| Florida | | | | | | | | | | | | | | | |
| Apalachicola | C | 61.2 | 0 | 0 | 0 | 16 | 153 | 319 | 347 | 260 | 180 | 33 | 0 | 0 | 1308 |
| Daytona Beach | A | 64.5 | 0 | 0 | 0 | 0 | 75 | 211 | 248 | 190 | 140 | 15 | 0 | 0 | 879 |
| Fort Myers | A | 68.6 | 0 | 0 | 0 | 0 | 24 | 109 | 146 | 101 | 62 | 0 | 0 | 0 | 442 |
| Jacksonville | A | 61.9 | 0 | 0 | 0 | 12 | 144 | 310 | 332 | 246 | 174 | 21 | 0 | 0 | 1239 |
| Key West | A | 73.1 | 0 | 0 | 0 | 0 | 0 | 28 | 40 | 31 | 9 | 0 | 0 | 0 | 108 |
| Lakeland | C | 66.7 | 0 | 0 | 0 | 0 | 57 | 164 | 195 | 146 | 99 | 0 | 0 | 0 | 661 |
| Miami | A | 71.1 | 0 | 0 | 0 | 0 | 0 | 65 | 74 | 56 | 19 | 0 | 0 | 0 | 214 |
| Miami Beach | C | 72.5 | 0 | 0 | 0 | 0 | 0 | 40 | 56 | 36 | 9 | 0 | 0 | 0 | 141 |
| Orlando | A | 65.7 | 0 | 0 | 0 | 0 | 72 | 198 | 220 | 165 | 105 | 6 | 0 | 0 | 766 |
| Pensacola | A | 60.4 | 0 | 0 | 0 | 19 | 195 | 353 | 400 | 277 | 183 | 36 | 0 | 0 | 1463 |
| Tallahassee | A | 60.1 | 0 | 0 | 0 | 28 | 198 | 360 | 375 | 286 | 202 | 36 | 0 | 0 | 1485 |
| Tampa | A | 66.4 | 0 | 0 | 0 | 0 | 60 | 171 | 202 | 148 | 102 | 0 | 0 | 0 | 683 |
| West Palm Beach | A | 68.4 | 0 | 0 | 0 | 0 | 6 | 65 | 87 | 64 | 31 | 0 | 0 | 0 | 253 |
| Georgia | | | | | | | | | | | | | | | |
| Athens | A | 51.8 | 0 | 0 | 12 | 115 | 405 | 632 | 642 | 529 | 431 | 141 | 22 | 0 | 2929 |
| Atlanta | A | 51.7 | 0 | 0 | 18 | 124 | 417 | 648 | 636 | 518 | 428 | 147 | 25 | 0 | 2961 |
| Augusta | A | 54.5 | 0 | 0 | 0 | 78 | 333 | 552 | 549 | 445 | 350 | 90 | 0 | 0 | 2397 |
| Columbus | A | 54.8 | 0 | 0 | 0 | 87 | 333 | 543 | 552 | 434 | 338 | 96 | 0 | 0 | 2383 |
| Macon | A | 56.2 | 0 | 0 | 0 | 71 | 297 | 502 | 505 | 403 | 295 | 63 | 0 | 0 | 2136 |
| Rome | A | 49.9 | 0 | 0 | 24 | 161 | 474 | 701 | 710 | 577 | 468 | 177 | 34 | 0 | 3326 |
| Savannah | A | 57.8 | 0 | 0 | 0 | 47 | 246 | 437 | 437 | 353 | 254 | 45 | 0 | 0 | 1819 |
| Thomasville | C | 60.0 | 0 | 0 | 0 | 25 | 198 | 366 | 394 | 305 | 208 | 33 | 0 | 0 | 1529 |
| Hawaii | | | | | | | | | | | | | | | |
| Lihue | A | 72.7 | 0 | 0 | 0 | 0 | 0 | 0 | 0 | 0 | 0 | 0 | 0 | 0 | 0 |
| Honolulu | A | 74.2 | 0 | 0 | 0 | 0 | 0 | 0 | 0 | 0 | 0 | 0 | 0 | 0 | 0 |

| | | | | | | | | | | | | | | | |
|---|---|---|---|---|---|---|---|---|---|---|---|---|---|---|---|
| Hilo | A | 71.9 | 0 | 0 | 0 | 0 | 0 | 0 | 0 | 0 | 0 | 0 | 0 | 0 | 0 |
| Idaho | | | | | | | | | | | | | | | |
| Boise | A | 39.7 | 0 | 0 | 132 | 415 | 792 | 1017 | 1113 | 854 | 722 | 438 | 245 | 81 | 5809 |
| Lewiston | A | 41.0 | 0 | 0 | 123 | 403 | 756 | 933 | 1063 | 815 | 694 | 426 | 239 | 90 | 5542 |
| Pocatello | A | 34.8 | 0 | 0 | 172 | 493 | 900 | 1166 | 1324 | 1058 | 905 | 555 | 319 | 141 | 7033 |
| Illinois | | | | | | | | | | | | | | | |
| Cairo | C | 47.9 | 0 | 0 | 36 | 164 | 513 | 791 | 856 | 680 | 539 | 195 | 47 | 0 | 3821 |
| Chicago (O'Hare) | A | 35.8 | 0 | 12 | 117 | 381 | 807 | 1166 | 1265 | 1086 | 939 | 534 | 260 | 72 | 6639 |
| Chicago (Midway) | A | 37.5 | 0 | 0 | 81 | 326 | 753 | 1113 | 1209 | 1044 | 890 | 480 | 211 | 48 | 6155 |
| Chicago | C | 38.9 | 0 | 0 | 66 | 279 | 705 | 1051 | 1150 | 1000 | 868 | 489 | 226 | 48 | 5882 |
| Moline | A | 36.4 | 0 | 9 | 99 | 335 | 774 | 1181 | 1314 | 1100 | 918 | 450 | 189 | 39 | 6408 |
| Peoria | A | 38.1 | 0 | 6 | 87 | 326 | 759 | 1113 | 1218 | 1025 | 849 | 426 | 183 | 33 | 6025 |
| Rockford | A | 34.8 | 6 | 9 | 114 | 400 | 837 | 1221 | 1333 | 1137 | 961 | 516 | 236 | 60 | 6830 |
| Springfield | A | 40.6 | 0 | 0 | 72 | 291 | 696 | 1023 | 1135 | 935 | 769 | 354 | 136 | 18 | 5429 |
| Indiana | | | | | | | | | | | | | | | |
| Evansville | A | 45.0 | 0 | 0 | 66 | 220 | 606 | 896 | 955 | 767 | 620 | 237 | 68 | 0 | 4435 |
| Fort Wayne | A | 37.3 | 0 | 9 | 105 | 378 | 783 | 1135 | 1178 | 1028 | 890 | 471 | 189 | 39 | 6205 |
| Indianapolis | A | 39.6 | 0 | 0 | 90 | 316 | 723 | 1051 | 1113 | 949 | 809 | 432 | 177 | 39 | 5699 |
| South Bend | A | 36.6 | 0 | 6 | 111 | 372 | 777 | 1125 | 1221 | 1070 | 933 | 525 | 239 | 60 | 6439 |
| Iowa | | | | | | | | | | | | | | | |
| Burlington | A | 37.6 | 0 | 0 | 93 | 322 | 768 | 1135 | 1259 | 1042 | 859 | 426 | 177 | 33 | 6114 |
| Des Moines | A | 35.5 | 0 | 6 | 96 | 363 | 828 | 1225 | 1370 | 1137 | 915 | 438 | 180 | 30 | 6588 |
| Dubuque | A | 32.7 | 12 | 31 | 156 | 450 | 906 | 1287 | 1420 | 1204 | 1026 | 546 | 260 | 78 | 7376 |
| Sioux City | A | 34.0 | 0 | 9 | 108 | 369 | 867 | 1240 | 1435 | 1198 | 989 | 483 | 214 | 39 | 6951 |
| Waterloo | A | 32.6 | 12 | 19 | 138 | 428 | 909 | 1296 | 1460 | 1221 | 1023 | 531 | 229 | 54 | 7320 |
| Kansas | | | | | | | | | | | | | | | |
| Concordia | A | 40.4 | 0 | 0 | 57 | 276 | 705 | 1023 | 1163 | 935 | 781 | 372 | 149 | 18 | 5479 |
| Dodge City | A | 42.5 | 0 | 0 | 33 | 251 | 666 | 939 | 1051 | 840 | 719 | 354 | 124 | 9 | 4986 |
| Goodland | A | 37.8 | 0 | 6 | 81 | 381 | 810 | 1073 | 1166 | 955 | 884 | 507 | 236 | 42 | 6141 |
| Topeka | A | 41.7 | 0 | 0 | 57 | 270 | 672 | 980 | 1122 | 893 | 722 | 330 | 124 | 12 | 5182 |
| Wichita | A | 44.2 | 0 | 0 | 33 | 229 | 618 | 905 | 1023 | 804 | 645 | 270 | 87 | 6 | 4620 |
| Kentucky | | | | | | | | | | | | | | | |
| Covington | A | 41.4 | 0 | 0 | 75 | 291 | 669 | 983 | 1035 | 893 | 756 | 390 | 149 | 24 | 5265 |
| Lexington | A | 43.8 | 0 | 0 | 54 | 239 | 609 | 902 | 946 | 818 | 685 | 325 | 105 | 0 | 4683 |
| Louisville | A | 44.0 | 0 | 0 | 54 | 248 | 609 | 890 | 930 | 818 | 682 | 315 | 105 | 9 | 4660 |
| Louisiana | | | | | | | | | | | | | | | |
| Alexandria | A | 57.5 | 0 | 0 | 0 | 56 | 273 | 431 | 471 | 361 | 260 | 69 | 0 | 0 | 1921 |
| Baton Rouge | A | 59.8 | 0 | 0 | 0 | 31 | 216 | 369 | 409 | 294 | 208 | 33 | 0 | 0 | 1560 |
| Lake Charles | A | 60.5 | 0 | 0 | 0 | 19 | 210 | 341 | 381 | 274 | 195 | 39 | 0 | 0 | 1459 |
| New Orleans | A | 61.0 | 0 | 0 | 0 | 19 | 192 | 322 | 363 | 258 | 192 | 39 | 0 | 0 | 1385 |
| New Orleans | C | 61.8 | 0 | 0 | 0 | 12 | 165 | 291 | 344 | 241 | 177 | 24 | 0 | 0 | 1254 |
| Shreveport | A | 56.2 | 0 | 0 | 0 | 47 | 297 | 477 | 552 | 426 | 304 | 81 | 0 | 0 | 2184 |
| Maine | | | | | | | | | | | | | | | |
| Caribou | A | 24.4 | 78 | 115 | 336 | 682 | 1044 | 1535 | 1690 | 1470 | 1308 | 858 | 468 | 183 | 9767 |
| Portland | A | 33.0 | 12 | 53 | 195 | 508 | 807 | 1215 | 1339 | 1182 | 1042 | 675 | 372 | 111 | 7511 |
| Maryland | | | | | | | | | | | | | | | |
| Baltimore | A | 43.7 | 0 | 0 | 48 | 264 | 585 | 905 | 936 | 820 | 679 | 327 | 90 | 0 | 4654 |

**TABLE A11 Continued**

| State and station | | Avg. winter temp[d] | July | Aug. | Sept. | Oct. | Nov. | Dec. | Jan. | Feb. | Mar. | Apr. | May | June | Yearly total |
|---|---|---|---|---|---|---|---|---|---|---|---|---|---|---|---|
| Baltimore | C | 46.2 | 0 | 0 | 27 | 189 | 486 | 806 | 859 | 762 | 629 | 288 | 65 | 0 | 4111 |
| Frederich | A | 42.0 | 0 | 0 | 66 | 307 | 624 | 955 | 995 | 876 | 741 | 384 | 127 | 12 | 5087 |
| Massachusetts | | | | | | | | | | | | | | | |
| Boston | A | 40.0 | 0 | 9 | 60 | 316 | 603 | 983 | 1088 | 972 | 846 | 513 | 208 | 36 | 5634 |
| Nantucket | A | 40.2 | 12 | 22 | 93 | 332 | 573 | 896 | 992 | 941 | 896 | 621 | 384 | 129 | 5891 |
| Pittsfield | A | 32.6 | 25 | 59 | 219 | 524 | 831 | 1231 | 1339 | 1196 | 1063 | 660 | 326 | 105 | 7878 |
| Worcester | A | 34.7 | 6 | 34 | 147 | 450 | 774 | 1172 | 1271 | 1123 | 998 | 612 | 304 | 78 | 6969 |
| Michigan | | | | | | | | | | | | | | | |
| Alpena | A | 29.7 | 68 | 105 | 273 | 580 | 912 | 1268 | 1404 | 1299 | 1218 | 777 | 446 | 156 | 8506 |
| Detroit (City) | A | 37.2 | 0 | 0 | 87 | 360 | 738 | 1088 | 1181 | 1058 | 936 | 522 | 220 | 42 | 6232 |
| Detroit (Wayne) | A | 37.1 | 0 | 0 | 96 | 353 | 738 | 1088 | 1194 | 1061 | 933 | 534 | 239 | 57 | 6293 |
| Detroit (Willow Run) | A | 37.2 | 0 | 0 | 90 | 357 | 750 | 1104 | 1190 | 1053 | 921 | 519 | 229 | 45 | 6528 |
| Escanaba | C | 29.6 | 59 | 87 | 243 | 539 | 924 | 1293 | 1445 | 1296 | 1203 | 777 | 456 | 159 | 8481 |
| Flint | A | 33.1 | 16 | 40 | 159 | 465 | 843 | 1212 | 1330 | 1198 | 1066 | 639 | 319 | 90 | 7377 |
| Grand Rapids | A | 34.9 | 9 | 28 | 135 | 434 | 804 | 1147 | 1259 | 1134 | 1011 | 579 | 279 | 75 | 6894 |
| Lansing | A | 34.8 | 6 | 22 | 138 | 431 | 813 | 1163 | 1262 | 1142 | 1011 | 579 | 273 | 69 | 6909 |
| Marquette | C | 30.2 | 59 | 81 | 240 | 527 | 936 | 1268 | 1411 | 1268 | 1187 | 771 | 468 | 177 | 8393 |
| Muskegon | A | 36.0 | 12 | 28 | 120 | 400 | 762 | 1088 | 1209 | 1100 | 995 | 594 | 310 | 78 | 6696 |
| Sault Ste. Marie | A | 27.7 | 96 | 105 | 279 | 580 | 951 | 1367 | 1525 | 1380 | 1277 | 810 | 477 | 201 | 9048 |
| Minnesota | | | | | | | | | | | | | | | |
| Duluth | A | 23.4 | 71 | 109 | 330 | 632 | 1131 | 1581 | 1745 | 1518 | 1355 | 840 | 490 | 198 | 10000 |
| Minneapolis | A | 28.3 | 22 | 31 | 189 | 505 | 1014 | 1454 | 1631 | 1380 | 1166 | 621 | 288 | 81 | 8382 |
| Rochester | A | 28.8 | 25 | 34 | 186 | 474 | 1005 | 1438 | 1593 | 1366 | 1150 | 630 | 301 | 93 | 8295 |
| Mississippi | | | | | | | | | | | | | | | |
| Jackson | A | 55.7 | 0 | 0 | 0 | 65 | 315 | 502 | 546 | 414 | 310 | 87 | 0 | 0 | 2239 |
| Meridian | A | 55.4 | 0 | 0 | 0 | 81 | 339 | 518 | 543 | 417 | 310 | 81 | 0 | 0 | 2289 |
| Vicksburg | C | 56.9 | 0 | 0 | 0 | 53 | 279 | 462 | 512 | 384 | 282 | 69 | 0 | 0 | 2041 |
| Missouri | | | | | | | | | | | | | | | |
| Columbia | A | 42.3 | 0 | 0 | 54 | 251 | 651 | 967 | 1076 | 874 | 716 | 324 | 121 | 12 | 5046 |
| Kansas City | A | 43.9 | 0 | 0 | 39 | 220 | 612 | 905 | 1032 | 818 | 682 | 294 | 109 | 0 | 4711 |
| St. Joseph | A | 40.3 | 0 | 6 | 60 | 285 | 708 | 1039 | 1172 | 949 | 769 | 348 | 133 | 15 | 5484 |
| St. Louis | A | 43.1 | 0 | 0 | 60 | 251 | 627 | 936 | 1026 | 848 | 704 | 312 | 121 | 15 | 4900 |
| St. Louis | C | 44.8 | 0 | 0 | 36 | 202 | 576 | 884 | 977 | 801 | 651 | 270 | 87 | 0 | 4484 |
| Springfield | A | 44.5 | 0 | 0 | 45 | 223 | 600 | 877 | 973 | 781 | 660 | 291 | 105 | 6 | 4900 |
| Montana | | | | | | | | | | | | | | | |
| Billings | A | 34.5 | 6 | 15 | 186 | 487 | 897 | 1135 | 1296 | 1100 | 970 | 570 | 285 | 102 | 7049 |
| Glasgow | A | 26.4 | 31 | 47 | 270 | 608 | 1104 | 1466 | 1711 | 1439 | 1187 | 648 | 335 | 150 | 8996 |
| Great Falls | A | 32.8 | 28 | 53 | 258 | 543 | 921 | 1169 | 1349 | 1154 | 1063 | 642 | 384 | 186 | 7750 |
| Havre | A | 28.1 | 28 | 53 | 306 | 595 | 1065 | 1367 | 1584 | 1364 | 1181 | 657 | 338 | 162 | 8700 |
| Havre | C | 29.8 | 19 | 37 | 252 | 539 | 1014 | 1321 | 1528 | 1305 | 1116 | 612 | 304 | 135 | 8182 |
| Helena | A | 31.1 | 31 | 59 | 294 | 601 | 1002 | 1265 | 1438 | 1170 | 1042 | 651 | 381 | 195 | 8129 |

| | | | | | | | | | | | | | | |
|---|---|---|---|---|---|---|---|---|---|---|---|---|---|---|
| Kalispell | A | 31.4 | 50 | 99 | 321 | 654 | 1020 | 1240 | 1401 | 1134 | 1029 | 639 | 397 | 207 | 8191 |
| Miles City | A | 31.2 | 6 | 6 | 174 | 502 | 972 | 1296 | 1504 | 1252 | 1057 | 579 | 276 | 99 | 7723 |
| Missoula | A | 31.5 | 34 | 74 | 303 | 651 | 1035 | 1287 | 1420 | 1120 | 970 | 621 | 391 | 219 | 8125 |
| Nebraska | | | | | | | | | | | | | | | |
| Grand Island | A | 36.0 | 0 | 6 | 108 | 381 | 834 | 1172 | 1314 | 1089 | 908 | 462 | 211 | 45 | 6530 |
| Lincoln | C | 38.8 | 0 | 6 | 75 | 301 | 726 | 1066 | 1237 | 1016 | 834 | 402 | 171 | 30 | 5864 |
| Norfolk | A | 34.0 | 9 | 0 | 111 | 397 | 873 | 1234 | 1414 | 1179 | 983 | 498 | 233 | 48 | 6979 |
| North Platte | A | 35.5 | 0 | 6 | 123 | 440 | 885 | 1166 | 1271 | 1039 | 930 | 519 | 248 | 57 | 6684 |
| Omaha | A | 35.6 | 0 | 12 | 105 | 357 | 828 | 1175 | 1355 | 1126 | 939 | 465 | 208 | 42 | 6612 |
| Scottsbluff | A | 35.9 | 0 | 0 | 138 | 459 | 876 | 1128 | 1231 | 1008 | 921 | 552 | 285 | 75 | 6673 |
| Valentine | A | 32.6 | 9 | 12 | 165 | 493 | 942 | 1237 | 1395 | 1176 | 1045 | 579 | 288 | 84 | 7425 |
| Nevada | | | | | | | | | | | | | | | |
| Elko | A | 34.0 | 9 | 34 | 225 | 561 | 924 | 1197 | 1314 | 1036 | 911 | 621 | 409 | 192 | 7433 |
| Ely | A | 33.1 | 28 | 43 | 234 | 592 | 939 | 1184 | 1308 | 1075 | 977 | 672 | 456 | 225 | 7753 |
| Las Vegas | A | 53.5 | 0 | 0 | 0 | 78 | 387 | 617 | 688 | 487 | 335 | 111 | 6 | 0 | 2709 |
| Reno | A | 39.3 | 43 | 87 | 204 | 490 | 801 | 1026 | 1073 | 823 | 729 | 510 | 357 | 189 | 6332 |
| Winnemucca | A | 36.7 | 0 | 34 | 210 | 536 | 876 | 1091 | 1172 | 916 | 837 | 573 | 363 | 153 | 6761 |
| New Hampshire | | | | | | | | | | | | | | | |
| Concord | A | 33.0 | 6 | 50 | 177 | 505 | 822 | 1240 | 1358 | 1184 | 1032 | 636 | 298 | 75 | 7383 |
| Mt. Washington Observatory | | 15.2 | 493 | 536 | 720 | 1057 | 1341 | 1742 | 1820 | 1663 | 1652 | 1260 | 930 | 603 | 13817 |
| New Jersey | | | | | | | | | | | | | | | |
| Atlantic City | A | 43.2 | 0 | 0 | 39 | 251 | 549 | 880 | 936 | 848 | 741 | 420 | 133 | 15 | 4812 |
| Newark | A | 42.8 | 0 | 0 | 30 | 248 | 573 | 921 | 983 | 876 | 729 | 381 | 118 | 0 | 4589 |
| Trenton | C | 42.4 | 0 | 0 | 57 | 264 | 576 | 924 | 989 | 885 | 753 | 399 | 121 | 12 | 4980 |
| New Mexico | | | | | | | | | | | | | | | |
| Albuquerque | A | 45.0 | 0 | 0 | 12 | 229 | 642 | 868 | 930 | 703 | 595 | 288 | 81 | 0 | 4348 |
| Clayton | A | 42.0 | 0 | 6 | 66 | 310 | 699 | 899 | 986 | 812 | 747 | 429 | 183 | 21 | 5158 |
| Raton | A | 38.1 | 9 | 28 | 126 | 431 | 825 | 1048 | 1116 | 904 | 834 | 543 | 301 | 63 | 6228 |
| Roswell | A | 47.5 | 0 | 0 | 18 | 202 | 573 | 806 | 840 | 641 | 481 | 201 | 31 | 0 | 3793 |
| Silver City | A | 48.0 | 0 | 0 | 6 | 183 | 525 | 729 | 791 | 605 | 518 | 261 | 87 | 0 | 3705 |
| New York | | | | | | | | | | | | | | | |
| Albany | A | 34.6 | 0 | 19 | 138 | 440 | 777 | 1194 | 1311 | 1156 | 992 | 564 | 239 | 45 | 6875 |
| Albany | C | 37.2 | 0 | 9 | 102 | 375 | 699 | 1104 | 1218 | 1072 | 908 | 498 | 186 | 30 | 6201 |
| Binghamton | A | 33.9 | 22 | 65 | 201 | 471 | 810 | 1184 | 1277 | 1154 | 1045 | 645 | 313 | 99 | 7286 |
| Binghamton | C | 36.6 | 0 | 28 | 141 | 406 | 732 | 1107 | 1190 | 1081 | 949 | 543 | 229 | 45 | 6451 |
| Buffalo | A | 34.5 | 19 | 37 | 141 | 440 | 777 | 1156 | 1256 | 1145 | 1039 | 645 | 329 | 78 | 7062 |
| New York (Central Park) | C | 42.8 | 0 | 0 | 30 | 233 | 540 | 902 | 986 | 885 | 760 | 408 | 118 | 9 | 4871 |
| New York (La Guardia) | A | 43.1 | 0 | 0 | 27 | 223 | 528 | 887 | 973 | 879 | 750 | 414 | 124 | 6 | 4811 |
| New York (Kennedy) | A | 41.4 | 0 | 0 | 36 | 248 | 564 | 933 | 1029 | 935 | 815 | 480 | 167 | 12 | 5219 |
| Rochester | A | 35.4 | 9 | 31 | 126 | 415 | 747 | 1125 | 1234 | 1123 | 1014 | 597 | 279 | 48 | 6748 |
| Schenectady | C | 35.4 | 0 | 22 | 123 | 422 | 756 | 1159 | 1283 | 1131 | 970 | 543 | 211 | 30 | 6650 |
| Syracuse | A | 35.2 | 6 | 28 | 132 | 415 | 744 | 1153 | 1271 | 1140 | 1004 | 570 | 248 | 45 | 6756 |
| North Carolina | | | | | | | | | | | | | | | |
| Asheville | C | 46.7 | 0 | 0 | 48 | 245 | 555 | 775 | 784 | 683 | 592 | 273 | 87 | 0 | 4042 |
| Cape Hatteras | | 53.3 | 0 | 0 | 0 | 78 | 273 | 521 | 580 | 518 | 440 | 177 | 25 | 0 | 2612 |
| Charlotte | A | 50.4 | 0 | 0 | 6 | 124 | 438 | 691 | 691 | 582 | 481 | 156 | 22 | 0 | 3191 |
| Greensboro | A | 47.5 | 0 | 0 | 33 | 192 | 513 | 778 | 784 | 672 | 552 | 234 | 47 | 0 | 3805 |

**TABLE A11 Continued**

| State and station | | Avg. winter temp[d] | July | Aug. | Sept. | Oct. | Nov. | Dec. | Jan. | Feb. | Mar. | Apr. | May | June | Yearly total |
|---|---|---|---|---|---|---|---|---|---|---|---|---|---|---|---|
| Raleigh | A | 49.4 | 0 | 0 | 21 | 164 | 450 | 716 | 725 | 616 | 487 | 180 | 34 | 0 | 3393 |
| Wilmington | A | 54.6 | 0 | 0 | 0 | 74 | 291 | 521 | 546 | 462 | 357 | 96 | 0 | 0 | 2347 |
| Winston-Salem | A | 48.4 | 0 | 0 | 21 | 171 | 483 | 747 | 753 | 652 | 524 | 207 | 37 | 0 | 3595 |
| North Dakota | | | | | | | | | | | | | | | |
| Bismarck | A | 26.6 | 34 | 28 | 222 | 577 | 1083 | 1463 | 1708 | 1442 | 1203 | 645 | 329 | 117 | 8851 |
| Devils Lake | C | 22.4 | 40 | 53 | 273 | 642 | 1191 | 1634 | 1872 | 1579 | 1345 | 753 | 381 | 138 | 9901 |
| Fargo | A | 24.8 | 28 | 37 | 219 | 574 | 1107 | 1569 | 1789 | 1520 | 1262 | 690 | 332 | 99 | 9226 |
| Williston | A | 25.2 | 31 | 43 | 261 | 601 | 1122 | 1513 | 1758 | 1473 | 1262 | 681 | 357 | 141 | 9243 |
| Ohio | | | | | | | | | | | | | | | |
| Akron-Canton | A | 38.1 | 0 | 9 | 96 | 381 | 726 | 1070 | 1138 | 1016 | 871 | 489 | 202 | 39 | 6037 |
| Cincinnati | C | 45.1 | 0 | 0 | 39 | 208 | 558 | 862 | 915 | 790 | 642 | 294 | 96 | 6 | 4410 |
| Cleveland | A | 37.2 | 9 | 25 | 105 | 384 | 738 | 1088 | 1159 | 1047 | 918 | 552 | 260 | 66 | 6351 |
| Columbus | A | 39.7 | 0 | 6 | 84 | 347 | 714 | 1039 | 1088 | 949 | 809 | 426 | 171 | 27 | 5660 |
| Columbus | C | 41.5 | 0 | 0 | 57 | 285 | 651 | 977 | 1032 | 902 | 760 | 396 | 136 | 15 | 5211 |
| Dayton | A | 39.8 | 0 | 6 | 78 | 310 | 696 | 1045 | 1097 | 955 | 809 | 429 | 167 | 30 | 5622 |
| Mansfield | A | 36.9 | 0 | 22 | 114 | 397 | 768 | 1110 | 1169 | 1042 | 924 | 543 | 245 | 60 | 6403 |
| Sandusky | C | 39.1 | 0 | 6 | 66 | 313 | 684 | 1032 | 1107 | 991 | 868 | 495 | 198 | 36 | 5796 |
| Toledo | A | 36.4 | 0 | 16 | 117 | 406 | 792 | 1138 | 1200 | 1056 | 924 | 543 | 242 | 60 | 6494 |
| Youngstown | A | 36.8 | 6 | 19 | 120 | 412 | 771 | 1104 | 1169 | 1047 | 921 | 540 | 248 | 60 | 6417 |
| Oklahoma | | | | | | | | | | | | | | | |
| Oklahoma City | A | 48.3 | 0 | 0 | 15 | 164 | 498 | 766 | 868 | 664 | 527 | 189 | 34 | 0 | 3725 |
| Tulsa | A | 47.7 | 0 | 0 | 18 | 158 | 522 | 787 | 893 | 683 | 539 | 213 | 47 | 0 | 3860 |
| Oregon | | | | | | | | | | | | | | | |
| Astoria | A | 45.6 | 146 | 130 | 210 | 375 | 561 | 679 | 753 | 622 | 636 | 480 | 363 | 231 | 5186 |
| Burns | C | 35.9 | 12 | 37 | 210 | 515 | 867 | 1113 | 1246 | 988 | 856 | 570 | 366 | 177 | 6957 |
| Eugene | A | 45.6 | 34 | 34 | 129 | 366 | 585 | 719 | 803 | 627 | 589 | 426 | 279 | 135 | 4726 |
| Meacham | A | 34.2 | 84 | 124 | 288 | 580 | 918 | 1091 | 1209 | 1005 | 983 | 726 | 527 | 339 | 7874 |
| Medford | A | 43.2 | 0 | 0 | 78 | 372 | 678 | 871 | 918 | 697 | 642 | 432 | 242 | 78 | 5004 |
| Pendleton | A | 42.6 | 0 | 0 | 111 | 350 | 711 | 884 | 1017 | 773 | 617 | 396 | 205 | 63 | 5127 |
| Portland | A | 45.6 | 25 | 28 | 114 | 335 | 597 | 735 | 825 | 644 | 586 | 396 | 245 | 105 | 4635 |
| Portland | C | 47.4 | 12 | 16 | 75 | 267 | 534 | 679 | 769 | 594 | 536 | 351 | 198 | 78 | 4109 |
| Roseburg | A | 46.3 | 22 | 16 | 105 | 329 | 567 | 713 | 766 | 608 | 570 | 405 | 267 | 123 | 4491 |
| Salem | A | 45.4 | 37 | 31 | 111 | 338 | 594 | 729 | 822 | 647 | 611 | 417 | 273 | 144 | 4754 |
| Pennsylvania | | | | | | | | | | | | | | | |
| Allentown | A | 38.9 | 0 | 0 | 90 | 353 | 693 | 1045 | 1116 | 1002 | 849 | 471 | 167 | 24 | 5810 |
| Erie | A | 36.8 | 0 | 25 | 102 | 391 | 714 | 1063 | 1169 | 1081 | 973 | 585 | 288 | 60 | 6451 |
| Harrisburg | A | 41.2 | 0 | 0 | 63 | 298 | 648 | 992 | 1045 | 907 | 766 | 396 | 124 | 12 | 5251 |
| Philadelphia | A | 41.8 | 0 | 0 | 60 | 297 | 620 | 965 | 1016 | 889 | 747 | 392 | 118 | 40 | 5144 |
| Philadelphia | C | 44.5 | 0 | 0 | 30 | 205 | 513 | 856 | 924 | 823 | 691 | 351 | 93 | 0 | 4486 |
| Pittsburgh | A | 38.4 | 0 | 9 | 105 | 375 | 726 | 1063 | 1119 | 1002 | 874 | 480 | 195 | 39 | 5987 |
| Pittsburgh | C | 42.2 | 0 | 0 | 60 | 291 | 615 | 930 | 983 | 885 | 763 | 390 | 124 | 12 | 5053 |

| | | | | | | | | | | | | | | | |
|---|---|---|---|---|---|---|---|---|---|---|---|---|---|---|---|
| Reading | C | 42.4 | 0 | 0 | 54 | 257 | 597 | 939 | 1001 | 885 | 735 | 372 | 105 | 0 | 4945 |
| Scranton | A | 37.2 | 0 | 19 | 132 | 434 | 762 | 1104 | 1156 | 1028 | 893 | 498 | 195 | 33 | 6254 |
| Williamsport | A | 38.5 | 0 | 9 | 111 | 375 | 717 | 1073 | 1122 | 1002 | 856 | 468 | 177 | 24 | 5934 |
| Rhode Island | | | | | | | | | | | | | | | |
| Block Island | A | 40.1 | 0 | 16 | 78 | 307 | 594 | 902 | 1020 | 955 | 877 | 612 | 344 | 99 | 5804 |
| Providence | A | 38.8 | 0 | 16 | 96 | 372 | 660 | 1023 | 1110 | 988 | 868 | 534 | 236 | 51 | 5954 |
| South Carolina | | | | | | | | | | | | | | | |
| Charleston | A | 56.4 | 0 | 0 | 0 | 59 | 282 | 471 | 487 | 389 | 291 | 54 | 0 | 0 | 2033 |
| Charleston | C | 57.9 | 0 | 0 | 0 | 34 | 210 | 425 | 443 | 367 | 273 | 42 | 0 | 0 | 1794 |
| Columbia | A | 54.0 | 0 | 0 | 0 | 84 | 345 | 577 | 570 | 470 | 357 | 81 | 0 | 0 | 2484 |
| Florence | A | 54.5 | 0 | 0 | 0 | 78 | 315 | 552 | 552 | 459 | 347 | 84 | 0 | 0 | 2387 |
| Greenville-Spartenburg | A | 51.6 | 0 | 0 | 6 | 121 | 399 | 651 | 660 | 546 | 446 | 132 | 19 | 0 | 2980 |
| South Dakota | | | | | | | | | | | | | | | |
| Huron | A | 28.8 | 9 | 12 | 165 | 508 | 1014 | 1432 | 1628 | 1355 | 1125 | 600 | 288 | 87 | 8223 |
| Rapid City | A | 33.4 | 22 | 12 | 165 | 481 | 897 | 1172 | 1333 | 1145 | 1051 | 615 | 326 | 126 | 7345 |
| Sioux Falls | A | 30.6 | 19 | 25 | 168 | 462 | 972 | 1361 | 1544 | 1285 | 1082 | 573 | 270 | 78 | 7839 |
| Tennessee | | | | | | | | | | | | | | | |
| Bristol | A | 46.2 | 0 | 0 | 51 | 236 | 573 | 828 | 828 | 700 | 598 | 261 | 68 | 0 | 4143 |
| Chattanooga | A | 50.3 | 0 | 0 | 18 | 143 | 468 | 698 | 722 | 577 | 453 | 150 | 25 | 0 | 3254 |
| Knoxville | A | 49.2 | 0 | 0 | 30 | 171 | 489 | 725 | 732 | 613 | 493 | 198 | 43 | 0 | 3494 |
| Memphis | A | 50.5 | 0 | 0 | 18 | 130 | 447 | 698 | 729 | 585 | 456 | 147 | 22 | 0 | 3232 |
| Memphis | C | 51.6 | 0 | 0 | 12 | 102 | 396 | 648 | 710 | 568 | 434 | 129 | 16 | 0 | 3015 |
| Nashville | A | 48.9 | 0 | 0 | 30 | 158 | 495 | 732 | 778 | 644 | 512 | 189 | 40 | 0 | 3578 |
| Oak Ridge | C | 47.7 | 0 | 0 | 39 | 192 | 531 | 772 | 778 | 669 | 552 | 228 | 56 | 0 | 3817 |
| Texas | | | | | | | | | | | | | | | |
| Abilene | A | 53.9 | 0 | 0 | 0 | 99 | 366 | 586 | 642 | 470 | 347 | 114 | 0 | 0 | 2624 |
| Amarillo | A | 47.0 | 0 | 0 | 18 | 205 | 570 | 797 | 877 | 664 | 546 | 252 | 56 | 0 | 3985 |
| Austin | A | 59.1 | 0 | 0 | 0 | 31 | 225 | 388 | 468 | 325 | 223 | 51 | 0 | 0 | 1711 |
| Brownsville | A | 67.7 | 0 | 0 | 0 | 0 | 66 | 149 | 205 | 106 | 74 | 0 | 0 | 0 | 600 |
| Corpus Christi | A | 64.6 | 0 | 0 | 0 | 0 | 120 | 220 | 291 | 174 | 109 | 0 | 0 | 0 | 914 |
| Dallas | A | 55.3 | 0 | 0 | 0 | 62 | 321 | 524 | 601 | 440 | 319 | 90 | 6 | 0 | 2363 |
| El Paso | A | 52.9 | 0 | 0 | 0 | 84 | 414 | 648 | 685 | 445 | 319 | 105 | 0 | 0 | 2700 |
| Fort Worth | A | 55.1 | 0 | 0 | 0 | 65 | 324 | 536 | 614 | 448 | 319 | 99 | 0 | 0 | 2405 |
| Galveston | A | 62.2 | 0 | 0 | 0 | 6 | 147 | 276 | 360 | 263 | 189 | 33 | 0 | 0 | 1274 |
| Galveston | C | 62.0 | 0 | 0 | 0 | 0 | 138 | 270 | 350 | 258 | 189 | 30 | 0 | 0 | 1235 |
| Houston | A | 61.0 | 0 | 0 | 0 | 6 | 183 | 307 | 384 | 288 | 192 | 36 | 0 | 0 | 1396 |
| Houston | C | 62.0 | 0 | 0 | 0 | 0 | 165 | 288 | 363 | 258 | 174 | 30 | 0 | 0 | 1278 |
| Laredo | A | 66.0 | 0 | 0 | 0 | 0 | 105 | 217 | 267 | 134 | 74 | 0 | 0 | 0 | 797 |
| Lubbock | A | 48.8 | 0 | 0 | 18 | 174 | 513 | 744 | 800 | 613 | 484 | 201 | 31 | 0 | 3578 |
| Midland | A | 53.8 | 0 | 0 | 0 | 87 | 381 | 592 | 651 | 468 | 322 | 90 | 0 | 0 | 2591 |
| Port Arthur | A | 60.5 | 0 | 0 | 0 | 22 | 207 | 329 | 384 | 274 | 192 | 39 | 0 | 0 | 1447 |
| San Angelo | A | 56.0 | 0 | 0 | 0 | 68 | 318 | 536 | 567 | 412 | 288 | 66 | 0 | 0 | 2255 |
| San Antonio | A | 60.1 | 0 | 0 | 0 | 31 | 204 | 363 | 428 | 286 | 195 | 39 | 0 | 0 | 1546 |
| Victoria | A | 62.7 | 0 | 0 | 0 | 6 | 150 | 270 | 344 | 230 | 152 | 21 | 0 | 0 | 1173 |
| Waco | A | 57.2 | 0 | 0 | 0 | 43 | 270 | 456 | 536 | 389 | 270 | 66 | 0 | 0 | 2030 |
| Wichita Falls | A | 53.0 | 0 | 0 | 0 | 99 | 381 | 632 | 698 | 518 | 378 | 120 | 6 | 0 | 2832 |

**TABLE A11 Continued**

| State and station | | Avg. winter temp[d] | July | Aug. | Sept. | Oct. | Nov. | Dec. | Jan. | Feb. | Mar. | Apr. | May | June | Yearly total |
|---|---|---|---|---|---|---|---|---|---|---|---|---|---|---|---|
| Utah | | | | | | | | | | | | | | | |
| Milford | A | 36.5 | 0 | 0 | 99 | 443 | 867 | 1141 | 1252 | 988 | 822 | 519 | 279 | 87 | 6497 |
| Salt Lake City | A | 38.4 | 0 | 0 | 81 | 419 | 849 | 1082 | 1172 | 910 | 763 | 459 | 233 | 84 | 6052 |
| Wendover | A | 39.1 | 0 | 0 | 48 | 372 | 822 | 1091 | 1178 | 902 | 729 | 408 | 177 | 51 | 5778 |
| Vermont | | | | | | | | | | | | | | | |
| Burlington | A | 29.4 | 28 | 65 | 207 | 539 | 891 | 1349 | 1513 | 1333 | 1187 | 714 | 353 | 90 | 8269 |
| Virginia | | | | | | | | | | | | | | | |
| Cape Henry | C | 50.0 | 0 | 0 | 0 | 112 | 360 | 645 | 694 | 633 | 536 | 246 | 53 | 0 | 3279 |
| Lynchburg | A | 46.0 | 0 | 0 | 51 | 223 | 540 | 822 | 849 | 731 | 605 | 267 | 78 | 0 | 4166 |
| Norfolk | A | 49.2 | 0 | 0 | 0 | 136 | 408 | 698 | 738 | 655 | 533 | 216 | 37 | 0 | 3421 |
| Richmond | A | 47.3 | 0 | 0 | 36 | 214 | 495 | 784 | 815 | 703 | 546 | 219 | 53 | 0 | 3865 |
| Roanoke | A | 46.1 | 0 | 0 | 31 | 229 | 549 | 825 | 834 | 722 | 614 | 261 | 65 | 0 | 4150 |
| Washington | | | | | | | | | | | | | | | |
| Olympia | A | 44.2 | 68 | 71 | 198 | 422 | 636 | 753 | 834 | 675 | 645 | 450 | 307 | 177 | 5236 |
| Seattle-Tacoma | A | 44.2 | 56 | 62 | 162 | 391 | 633 | 750 | 828 | 678 | 657 | 474 | 295 | 159 | 5145 |
| Seattle | C | 46.9 | 50 | 47 | 129 | 329 | 543 | 657 | 738 | 599 | 577 | 396 | 242 | 117 | 4424 |
| Spokane | A | 36.5 | 9 | 25 | 168 | 493 | 879 | 1082 | 1231 | 980 | 834 | 531 | 288 | 135 | 6655 |
| Walla Walla | C | 43.8 | 0 | 0 | 87 | 310 | 681 | 843 | 986 | 745 | 589 | 342 | 177 | 47 | 4805 |
| Yakima | A | 39.1 | 0 | 12 | 144 | 450 | 828 | 1039 | 1163 | 868 | 713 | 435 | 220 | 69 | 5941 |
| West Virginia | | | | | | | | | | | | | | | |
| Charleston | A | 44.8 | 0 | 0 | 63 | 254 | 591 | 865 | 880 | 770 | 648 | 300 | 96 | 9 | 4476 |
| Elkins | A | 40.1 | 9 | 25 | 135 | 400 | 729 | 992 | 1008 | 896 | 791 | 444 | 198 | 48 | 5675 |
| Huntington | A | 45.0 | 0 | 0 | 63 | 257 | 585 | 856 | 880 | 764 | 636 | 294 | 99 | 12 | 4446 |
| Parkersburg | C | 43.5 | 0 | 0 | 60 | 264 | 606 | 905 | 942 | 826 | 691 | 339 | 115 | 6 | 4754 |
| Wisconsin | | | | | | | | | | | | | | | |
| Green Bay | A | 30.3 | 28 | 50 | 174 | 484 | 924 | 1333 | 1494 | 1313 | 1141 | 654 | 335 | 99 | 8029 |
| La Crosse | A | 31.5 | 12 | 19 | 153 | 437 | 924 | 1339 | 1504 | 1277 | 1070 | 540 | 245 | 69 | 7589 |
| Madison | A | 30.9 | 25 | 40 | 174 | 474 | 930 | 1330 | 1473 | 1274 | 1113 | 618 | 310 | 102 | 7863 |
| Milwaukee | A | 32.6 | 43 | 47 | 174 | 471 | 876 | 1252 | 1376 | 1193 | 1054 | 642 | 372 | 135 | 7635 |
| Wyoming | | | | | | | | | | | | | | | |
| Casper | A | 33.4 | 6 | 16 | 192 | 524 | 942 | 1169 | 1290 | 1084 | 1020 | 657 | 381 | 129 | 7410 |
| Cheyenne | A | 34.2 | 28 | 37 | 219 | 543 | 909 | 1085 | 1212 | 1042 | 1026 | 702 | 428 | 150 | 7381 |
| Lander | A | 31.4 | 6 | 19 | 204 | 555 | 1020 | 1299 | 1417 | 1145 | 1017 | 654 | 381 | 153 | 7870 |
| Sheridan | A | 32.5 | 25 | 31 | 219 | 539 | 948 | 1200 | 1355 | 1154 | 1051 | 642 | 366 | 150 | 7680 |

[a]Data for United States cities from a publication of the United States Weather Bureau, *Monthly Normals of Temperature, Precipitation and Heating Degree Days*, 1962, are for the period 1931 to 1960 inclusive. These data also include information from the 1963 revisions to this publication, where available.

[b]Data for airport stations, A, and city stations, C, are both given where available.

[c]Data for Canadian cities were computed by the Climatology Division, Department of Transport from normal monthly mean temperatures, and the monthly values of heating degree days data were obtained using the National Research Council computer and a method devised by H. C. S. Thom of the United States Weather Bureau. The heating degree days are based on the period from 1931 to 1960.

[d]For period October to April, inclusive.

# Index